Cardiovascular Genetics and Genomics in Clinical Practice

Cardiovascular Genetics and Genomics in Clinical Practice

Editors

Sanjiv J. Shah, MD
Associate Professor of Medicine
Division of Cardiology, Department of Medicine
Northwestern University Feinberg School of Medicine
Chicago, Illinois

Donna K. Arnett, PhD
Chair and Professor of Epidemiology
School of Public Health
University of Alabama at Birmingham
Birmingham, Alabama

Visit our website at www.demosmedical.com

ISBN: 9781620700143
e-book ISBN: 9781617051784

Acquisitions Editor: Rich Winters
Compositor: Exeter Premedia Services Private Ltd.

Medicine is an ever-changing science. Research and clinical experience are continually expanding our knowledge, in particular our understanding of proper treatment and drug therapy. The authors, editors, and publisher have made every effort to ensure that all information in this book is in accordance with the state of knowledge at the time of production of the book. Nevertheless, the authors, editors, and publisher are not responsible for errors or omissions or for any consequences from application of the information in this book and make no warranty, expressed or implied, with respect to the contents of the publication. Every reader should examine carefully the package inserts accompanying each drug and should carefully check whether the dosage schedules mentioned therein or the contraindications stated by the manufacturer differ from the statements made in this book. Such examination is particularly important with drugs that are either rarely used or have been newly released on the market.

Library of Congress Cataloging-in-Publication Data
Cardiovascular genetics and genomics in clinical practice / editors, Sanjiv J. Shah, Donna K. Arnett.
    p. ; cm.
  Includes bibliographical references.
  ISBN 978-1-62070-014-3—ISBN 978-1-61705-178-4 (e-book)
  I. Shah, Sanjiv, editor.   II. Arnett, Donna K. editor.
  [DNLM: 1. Heart Diseases—genetics.   2. Cardiovascular Diseases—genetics.   3. Genomics—methods.   WG 210]
  RC682
  616.1'2042—dc23

                                                                    2014023501

Printed in the United States of America by Bradford & Bigelow.
14 15 16 17 / 5 4 3 2 1

*To Neera for her love and support, and to Anika, Mila, and Zaan,
who bring me joy and laughter every day.*

Sanjiv J. Shah, MD

*To Steve for his dedication to my success, and to my many trainees
who continue to inspire me.*

Donna K. Arnett, PhD

Contents

Contributors   ix
Preface   xiii
Video Captions   xv

PART I: GENETICS AND GENOMICS: THE BASICS

1. Mendelian Genetics   3
   Sadeep Shrestha

2. Genetics of Complex Traits   12
   Ryan Irvin

3. Genome-Wide Association Studies   23
   Daniel H. Katz and Laura J. Rasmussen-Torvik

4. Bioinformatics   34
   Kiran Musunuru

5. Epigenetics   39
   Lifang Hou

6. MicroRNAs   56
   R. Kannan Mutharasan

7. Gene Expression   67
   Chiang-Ching Huang and Samantha L. Gadd

8. Whole-Exome and Whole-Genome Sequencing   77
   Sanjiv J. Shah and Donna K. Arnett

9. Gene–Environment Interactions   86
   Stella Aslibekyan

10. Genetic Counseling   96
    Kelly M. Bontempo

PART II: GENETICS OF CARDIOVASCULAR DISORDERS/TRAITS

11. Blood Pressure Genomics   111
    Changwei Li and Tanika N. Kelly

12. Genetics of Electrocardiographic Traits   128
    Andrew J. Sauer and Sanjiv J. Shah

13. Atrial Fibrillation Genetics and Genomics   140
    Amir Y. Shaikh, Steven A. Lubitz, Honghuang Lin, Emelia J. Benjamin, and David D. McManus

14. Inherited Ventricular Arrhythmias   153
    Brittany Weber and Rajat Deo

15. Genetics of Cardiac Structure and Function   168
    Sanjiv J. Shah, Sadiya S. Khan, and Donna K. Arnett

16. Genetics of Heart Failure   186
    Wolfgang Lieb and Ramachandran S. Vasan

17. Inherited Cardiomyopathies   195
    Michael A. Burke

18. Genetics of Pulmonary Hypertension   219
    Ankit A. Desai, Julio D. Duarte, Sunit Singla, and Roberto F. Machado

19. Genetics of Blood Lipids, Lipoproteins, and Related Phenotypes   239
    Steven A. Claas and Donna K. Arnett

20. Genetic Applications in Coronary Artery Disease   253
    Stephen Pan, Juyong Brian Kim, Nehal N. Mehta, and Joshua W. Knowles

21. Genetics of Valvular Heart Disease   265
    Andreas C. Mauer, David S. Owens, Christopher J. O'Donnell, Wendy S. Post, and George Thanassoulis

22. Genetics of Congenital Heart Disease   287
    Elizabeth M. Bonachea, Courtney E. Vaughn, and Vidu Garg

Index   299

Contributors

Donna K. Arnett, PhD
Chair and Professor of Epidemiology
School of Public Health
University of Alabama at Birmingham
Birmingham, Alabama

Stella Aslibekyan, PhD
Assistant Professor
Department of Epidemiology
University of Alabama at Birmingham
Birmingham, Alabama

Emelia J. Benjamin, MD, ScM
Professor of Medicine
Boston University School of Medicine;
Professor of Epidemiology
Boston University School of Public Health
Boston, Massachusetts

Elizabeth M. Bonachea, MD
Assistant Professor
Division of Neonatology
Department of Pediatrics
Ohio State University College of Medicine
Nationwide Children's Hospital
Columbus, Ohio

Kelly M. Bontempo, MS, CGC
Certified Genetic Counselor
Advocate Medical Group
Division of Genetics
Park Ridge, Illinois

Michael A. Burke, MD
Cardiovascular Division
Department of Medicine
Brigham and Women's Hospital
Harvard Medical School
VA Boston Healthcare System
Boston, Massachusetts

Steven A. Claas, MS
Program Manager
Department of Epidemiology
School of Public Health
University of Alabama at Birmingham
Birmingham, Alabama

Rajat Deo, MD, MTR
Assistant Professor of Medicine
Section of Electrophysiology
Division of Cardiovascular Medicine
University of Pennsylvania
Philadelphia, Pennsylvania

Ankit A. Desai, MD
Assistant Professor of Medicine
Division of Cardiology
University of Arizona
Sarver Heart Center
Tucson, Arizona

Julio D. Duarte, PharmD, PhD
Assistant Professor
Department of Pharmacy Practice
University of Illinois at Chicago
Chicago, Illinois

Samantha L. Gadd, PhD
Research Assistant Professor
Department of Pathology
Northwestern University Feinberg School of Medicine
Chicago, Illinois

Vidu Garg, MD
Associate Professor
Department of Pediatrics
Division of Cardiology
Department of Molecular Genetics
Ohio State University College of Medicine;
Director of Translational Research
Center for Cardiovascular and Pulmonary Research
The Heart Center
Nationwide Children's Hospital
Columbus, Ohio

Lifang Hou, MD, MS, PhD
Chief
Division of Cancer Epidemiology and Prevention
Associate Professor
Department of Preventive Medicine
Northwestern University Feinberg School of Medicine
Chicago, Illinois

Chiang-Ching Huang, PhD
Associate Professor
Division of Biostatistics
Joseph J. Zilber School of Public Health
University of Wisconsin
Milwaukee, Wisconsin

Ryan Irvin, PhD
Assistant Professor
Department of Biostatistics
University of Alabama at Birmingham School of Public Health
Birmingham, Alabama

Daniel H. Katz, MD
Department of Medicine
Massachusetts General Hospital
Harvard Medical School
Boston, Massachusetts

Tanika N. Kelly, PhD, MPH
Assistant Professor
Department of Epidemiology
Tulane University School of Public Health and Tropical
    Medicine
New Orleans, Louisiana

Sadiya S. Khan, MD
Cardiology Fellow
Division of Cardiology
Department of Medicine
Northwestern University Feinberg School of Medicine
Chicago, Illinois

Juyong Brian Kim, MD, MPH
Fellow
Division of Cardiovascular Medicine
Stanford University School of Medicine
Stanford, California

Joshua W. Knowles, MD, PhD
Instructor
Division of Cardiovascular Medicine
Stanford University School of Medicine
Stanford, California;
Chief Medical Officer
The FH Foundation
South Pasadena, California

Changwei Li, MD, MPH
Doctoral Research Fellow
Department of Epidemiology
Tulane University School of Public Health and Tropical
    Medicine
New Orleans, Louisiana

Wolfgang Lieb, MD, MSc
Professor of Epidemiology
Institute of Epidemiology
Christian-Albrechts-University Kiel
Kiel, Germany

Honghuang Lin, PhD
Assistant Professor
Department of Medicine
Boston University School of Medicine
Boston, Massachusetts

Steven A. Lubitz, MD, MPH
Cardiovascular Research Center and Cardiac Arrhythmia
    Service
Massachusetts General Hospital;
Assistant Professor
Harvard Medical School
Boston, Massachusetts

Roberto F. Machado, MD
Associate Professor of Medicine
Section of Pulmonary, Critical Care, and Allergy
University of Illinois at Chicago
Chicago, Illinois

Andreas C. Mauer, MD, MS
Clinical and Research Fellow
Cardiology Division
Massachusetts General Hospital
Harvard Medical School
Boston, Massachusetts

David D. McManus, MD, ScM
Assistant Professor of Medicine
Cardiology Division
University of Massachusetts Medical School
Worcester, Massachusetts

Nehal N. Mehta, MD, MSCE, FAHA
Chief
Section of Inflammation and Cardiometabolic Diseases
National Heart, Lung and Blood Institute
Bethesda, Maryland

Kiran Musunuru, MD, PhD, MPH
Assistant Professor
Department of Stem Cell and Regenerative Biology
Harvard University
Cambridge, Massachusetts;
Associate Physician
Division of Cardiovascular Medicine
Brigham and Women's Hospital
Boston, Massachusetts

R. Kannan Mutharasan, MD
Assistant Professor of Medicine
Division of Cardiology, Department of Medicine
Feinberg Cardiovascular Research Institute
Northwestern University Feinberg School of Medicine
Chicago, Illinois

Christopher J. O'Donnell, MD, MPH
Chief, Cardiovascular Epidemiology and Human Genomics Branch
Division of Intramural Research
National Heart, Lung and Blood Institute (NHLBI)
Bethesda, Maryland;
Associate Director
NHLBI's Framingham Heart Study
Cardiology Division
Massachusetts General Hospital
Harvard Medical School
Boston, Massachusetts

David S. Owens, MD, MS
Assistant Professor
Division of Cardiology
University of Washington
Seattle, Washington

Stephen Pan, MD, MS
Fellow
Center for Advanced Cardiac Care
Columbia Presbyterian Medical Center
New York, New York

Wendy S. Post, MD, MS
Professor of Medicine and Epidemiology
Johns Hopkins School of Medicine
Baltimore, Maryland

Laura J. Rasmussen-Torvik, PhD, MPH
Assistant Professor
Department of Preventive Medicine
Northwestern University Feinberg School of Medicine
Chicago, Illinois

Andrew J. Sauer, MD
Assistant Professor of Medicine
Division of Cardiology, Department of Medicine
Northwestern University School of Medicine
Chicago, Illinois

Sanjiv J. Shah, MD
Associate Professor of Medicine
Division of Cardiology, Department of Medicine
Northwestern University Feinberg School of Medicine
Chicago, Illinois

Amir Y. Shaikh, MD
Assistant Professor
Department of Medicine
University of Massachusetts Medical School
Worcester, Massachusetts

Sadeep Shrestha, PhD, MHS, MS
Associate Professor
Department of Epidemiology
University of Alabama at Birmingham
Birmingham, Alabama

Sunit Singla, MD
Clinical Instructor
Section of Pulmonary, Critical Care, and Allergy
University of Illinois at Chicago
Chicago, Illinois

George Thanassoulis, MD, MSc
Director of Preventive and Genomic Cardiology
McGill University Health Center;
Assistant Professor of Medicine
Department of Medicine
McGill University
Montreal, Quebec, Canada

Ramachandran S. Vasan, MD
Professor of Medicine
Chief
Section of Preventive Medicine and Epidemiology
Boston University School of Medicine
Boston, Massachusetts

Courtney E. Vaughn, BS
Research Assistant
Department of Molecular Genetics
Ohio State University College of Medicine
Columbus, Ohio

Brittany Weber, MD, PhD
Resident
Department of Medicine
Brigham and Women's Hospital
Boston, Massachusetts

Preface

Although significant progress has been made in recent decades toward preventing and treating cardiovascular disease (CVD), statistics from the World Health Organization indicate that every year more people die because of CVD than any other single disease. Approximately 17.3 million people died from CVD in 2008, amounting to about one-third of all deaths globally (1). In the United States, more than 2,000 people die from CVD each day, and although the prevalence of certain risk factors (eg, smoking) has declined, the prevalence of other risk factors remains disturbingly high: 32% of children and 68% of adults are overweight or obese, 43% of adults have elevated cholesterol, and 33% of adults have high blood pressure (2). It has long been recognized that certain lifestyle factors (such as diet and physical activity) can significantly influence the effects of these and other CVD risk factors, and clinical and public health efforts will continue to rely on modifying these known environmental exposures to further reduce CVD risk in individuals and populations. However, it has also long been known that most types of CVD and their risk factors tend to aggregate in families, and a growing body of evidence not only strengthens the general hypothesis that genetic factors influence CVD risk, but also supports specific claims of particular genetic markers associating with particular risk factors and CVD outcomes. This book concerns itself with our evolving understanding of these heritable factors and the ways this new knowledge is reshaping our understanding of CVD and potentially opening new avenues of CVD risk stratification, diagnosis, and treatment.

We envision the audience of this book to comprise a broad spectrum of those involved in clinical cardiovascular care and/or CVD research. The breakneck pace of advances in molecular genetics over the past two decades has resulted in a large population of practicing clinicians whose formative education may not have included a significant genetic or genomic component. These individuals will find this book presenting a familiar clinical milieu in a radically different and fascinating context. Individuals who are just beginning their clinical training or practice will find these chapters foundational in careers that will doubtless bear witness to advances in medical genomics beyond the imagination of the editors and authors of this volume. For those just beginning research careers in medical genomics and genetic epidemiology, this book can provide both the back-story and a cutting-edge briefing on an extremely active and dynamic field of research. Finally, practicing molecular and medical geneticists and genetic epidemiologists who work in domains outside cardiovascular medicine or who are new to particular CVD phenotypes will find in these chapters outstanding reviews of important clinical and public health metrics.

The first part of the book presents a primer on genetics and genomics. We begin with what will likely be a review for many—a discussion of Mendelian genetics. Additional chapters generally fit into categories that have biological or structural relevance (eg, complex traits, epigenetics, microRNAs, and gene expression), or categories that align with evolving genetic and genomic methods, such as genome-wide association studies and whole exome and genome sequencing. An entire chapter (Chapter 9) is devoted to the critical concept of gene–environment interaction. Part I of the book ends with a review of genetic counseling, a field that remains incredibly important in the genomic era. Although Part I is meant as a general overview (ie, "toolbox") on genetics and genomics, each chapter in this part deliberately gives examples that relate to CVD to put the concepts and lessons into context for the reader. Part II of the book offers chapters on the genetics and genomics of major CVDs and their risk factors, including topics such as inherited cardiomyopathies, coronary artery disease, blood pressure, valvular heart disease, and blood lipids. Each chapter in Part II traces the unfolding story of genetic and genomic research in each of these domains. In additional to useful lists of take home points, every chapter in this section contains a case study that models how current genomic knowledge can be put to use in the clinical setting.

Assembling a book such as this one has proven to be both a daunting and an energizing task. On the one hand, the scope and breadth of CVD-related genomic knowledge that has been generated in the past decades are massive, and finding a sensible way to present it to a diverse audience has been challenging. On the other hand, the translation of research findings into clinically useful tools after the first maps of the human genome were completed has been slower than many anticipated. Despite these hurdles, distilling the large volume of information on cardiovascular genetics and genomics, from basic concepts to clinical application, has reinforced our belief in the need for a book such as ours, and also the promise the field holds for future discovery. We

hope these chapters demonstrate that the recent deluge of genomic knowledge need not overwhelm CVD practitioners and researchers and that this knowledge is already finding a place in the clinic, a practice that will surely only expand in the coming years.

We would like to sincerely thank the chapter authors for taking time from their busy clinical, research, and teaching schedules to create what we believe is a truly engaging and useful book. Their expertise and insights have been a gift to us and, we hope, will be prized by all interested readers.

Sanjiv J. Shah, MD
Donna K. Arnett, PhD

REFERENCES

1. World Health Organization. *Global Status Report on Noncommunicable Diseases 2010.* Geneva, Switzerland: World Health Organization; 2011.
2. Go AS, Mozaffarian D, Roger VL, et al; American Heart Association Statistics Committee and Stroke Statistics Subcommittee. Heart disease and stroke statistics—2014 update: a report from the American Heart Association. *Circulation.* 2014;129(3):e28–e292.

Video Captions

The videos can be viewed on the Demos website by accessing the links provided below. The video links are located below their respective captions.

Chapter 16: Genetics of Heart Failure

Video 16.1 (*referred to on page 187*)
Echocardiographic apical 4-chamber view demonstrating increased left ventricular wall thickness, abnormal myocardial wall texture consistent with infiltrative cardiomyopathy, and severely reduced left ventricular systolic function (ejection fraction = 20%).

To view the video, please visit the following link: http://www.demosmedical.com/media/videos/Shah_Video_16_1.mp4

Chapter 17: Inherited Cardiomyopathies

Video 17.1 (*referred to on page 195*)
Echocardiographic parasternal long-axis view of the patient described in the case study. Note the systolic anterior motion of the mitral valve and the severe asymmetric septal hypertrophy.

To view the video, please visit the following link: http://www.demosmedical.com/media/videos/Shah_Video_17_1.mp4

Video 17.2 (*referred to on page 195*)
Echocardiographic apical five-chamber view of the patient described in the case study. Note the severe asymmetric septal hypertrophy.

To view the video, please visit the following link: http://www.demosmedical.com/media/videos/Shah_Video_17_2.mp4

Chapter 18: Genetics of Pulmonary Hypertension

Video 18.1A (*referred to on page 220*)
Echocardiographic parasternal long axis view demonstrating severe right ventricular dilation, a small left ventricle, and interventricular septal flattening, consistent with right ventricular overload.

To view the video, please visit the following link: http://www.demosmedical.com/media/videos/Shah_Video_18_1A.mp4

Chapter 21: Genetics of Valvular Heart Disease

Video 21.1 (*referred to on page 266*)
Transesophageal echocardiography 2D view of mitral valve at 75 degrees. The 2D image demonstrates myxomatous degeneration of mitral valve with a flail P3 segment of the posterior mitral valve leaflet.

To view the video, please visit the following link: http://www.demosmedical.com/media/videos/Shah_Video_21_1.mp4

Video 21.2 (*referred to on page 266*)
Transesophageal echocardiography 3D view of mitral valve ("surgeon's view" from the left atrium). The 3D image demonstrates the flail P3 segment of the mitral valve with an associated ruptured chord (bottom right side corner of the valve).

To view the video, please visit the following link: http://www.demosmedical.com/media/videos/Shah_Video_21_2.mp4

Chapter 18: Genetics of Pulmonary Hypertension

Video 18.1B (*referred to on page 220*)
Echocardiographic parasternal short axis view demonstrating severe right ventricular dilation and interventricular septal flattening, consistent with right ventricular overload and significant pulmonary hypertension.

To view the video, please visit the following link: http://www.demosmedical.com/media/videos/Shah_Video_18_1B.mp4

Cardiovascular Genetics and Genomics in Clinical Practice

GENETICS AND GENOMICS: THE BASICS

Mendelian Genetics

Sadeep Shrestha

TAKE HOME POINTS

1. Cardiovascular disease (CVD) is often considered a complex disease, but several have a Mendelian inheritance.
2. Mendelian CVDs still pose unique challenges to clinicians, as most have a wide range of phenotypic presentations due to varying degrees of penetrance and expressivity.
3. Interpretation of genetic tests for Mendelian CVD requires careful examination of family pedigrees.

In the era of modern genetics, cardiovascular diseases (CVDs) or traits are often considered multifactorial disorders or complex diseases that result from the interactions among inherited gene variants and environmental factors such as chemical, physical, biological, infectious, behavioral, and nutritional factors. However, there are several CVDs that follow Mendelian patterns of inheritances; for example, cardiomyopathy, inherited arrhythmias, aortic aneurysm, and congenital heart disease (1–4), although their prevalence typically accounts for only a small proportion of the total CVD observed in the population (Table 1.1). There are some forms of complex cardiovascular phenotypes that also follow the Mendelian mode of inheritance (eg, *MEF2A* gene in coronary artery disease) (5). The first Mendelian CVD was reported in 1985 where a homozygous deletion of a 5 kb sequence region that contained several exons of low-density lipoprotein receptor (LDLR)

was found in a patient with familial hypercholesterolemia (6). Since then, progress in several other CVD Mendelian diseases has been made in parallel with rapid advancements in technologies that have provided high-throughput methods such as genome-wide association studies, exome sequencing, and whole-genome sequencing. In the foreseeable future, it is likely that advancements in genomic research in Mendelian CVDs will provide for early and effective diagnosis that can lead to the timely treatment of patients and their family members.

Historically, inheritance as a physical unit had been observed and experimented with the hybridization of animals and plants, but was not well understood. Gregory Mendel, an Augustinian clergyman, was the first to elucidate the gene as the fundamental unit that transmits traits from parents to offspring and formed the foundation of inheritance. He performed several simple series of experiments and examined the inheritance patterns in garden peas. Mendel's 8 years of cross-breeding experiments with the garden pea plants is described in a single research paper *Versuche uber Pflanzen-Hybriden* in 1866 (7). Mendel performed his experiments in plants, which produce large quantities of offspring needed to robustly derive statistical outcomes. Mendel carefully noted and tracked the simple phenotypes of all offspring that resulted from specific matings. He applied mathematical principles to show that the ratios of phenotypes in the offspring followed certain biological patterns. These laws, referred to as Mendel's Laws, can be

TABLE 1.1 Mendelian Cardiovascular Diseases and Syndromes

Disease	Prevalence	Inheritance	Genes
Hypertrophic cardiomyopathy (HCM)	1 in 500	AD	MYH7, MYBPC3, TNNT2, TNNI3, TPM1, MYL2, ACTC, MYL3, CSRP3, MYH6, PRKAG2, TNNC1, TTN, VCL
Dilated cardiomyopathy (DCM)	40–50 in 100,000	AD, AR, XR	ABCC9, ACTC, ACTN2, CSRP3, DES, DSG2, DSP, EYA4, FCMD, LAMP2, LDB3, LMNA, MYBPC3, TTN, LMNA, MYH6, MYH7, NEXN, PLN, PSEN1, PSEN2, RBM20, SCN5A, SGCD, TCAP, TMPO, TNN, TNNC1, TNNI3, TPM1, VCL, TNNT2
Dilated cardiomyopathy with ataxia (DCMA)	Unknown	AR	DNAJC19
Arrhythmogenic right ventricular cardiomyopathy (ARVC)	1–5 in 10,000	AD, AR	ACTN2, DSC2, DSG2, DSP, JUP, TMEM43, LDB3, PKP2, RYR2, TGFB3,
Left ventricular noncompaction (LVNC)		AD, XR, Mt	G4.5, Alpha-dystrobrevin, RYR2, DTNA, TAZ
Restrictive cardiomyopathy (RCM)	< 1 in 1,000,000		MYH7, TNNI3
Noncompaction cardiomyopathy (NCC)	0.12 in 100,000		LDB3, MYH7, MYBPC3
Familial hypercholesterolemia	> 1 in 1,000	AD, AR	LDLR, APOB, ABCG5, ABCG8, ARH, PCSK9
Long QT syndrome (LQTS)	1 in 7,000	AD, AR	KCNQ1, KCNH2, SCN5A, ANK2, KCNE1, KCNE2, KCNJ2, CACNA1C, CAV3, SCN4B, AKAP9, SNTA1
Catecholaminergic polymorphic ventricular tachycardia (CPVT)	1–5 in 10,000	AD, AR	RYR2, CASQ2
Brugada syndrome	1–5 in 10,000	AD	SCN5A, GPD1L, CACNA1C, CACNB2, SCN1B, KCNE3, SCN3B
Marfan syndrome	1–5 in 10,000	AD	FBN1
Loeys–Dietz syndrome	< 1 in 1,000,000	AD	TGFBR1, TGFBR2
Holt–Oram syndrome	1 in 100,000	AD	TBX5
Alagille syndrome	1 in 70,000	AD	JAG1, NOTCH2
Char syndrome	< 1 in 1,000,000	AD	TFAP2B
CHARGE syndrome	Unknown	AD	CHD7, SEMA3E
Costello syndrome	1 in 300,000–1.25 million	AD	HRAS
Ellis–van Creveld syndrome (EVS)	1 in 60,000–200,000	AR	EVC, EVC2

TABLE 1.1 Mendelian Cardiovascular Diseases and Syndromes (*continued*)

Disease	Prevalence	Inheritance	Genes
LEOPARD syndrome	< 1 in 1,000,000	AD	PTPN11, RAF1
Noonan syndrome	Unknown	AD	PTPN11, SOS1, RAF1, KRAS, NRAS, BRAF
Heterotaxia	1–9 in 100,000	AD, AR, XR	ZIC3, CFC1, CRELD1, ACVR2B, LEFTYB, NODAL, INVERSINE
Rubinstein–Taybi syndrome	1–9 in 100,000	AD	CBP, EP300
Smith–Lemli–Opitz syndrome	Unknown	AR	DHCR7
DiGeorge or velocardiofacial syndrome	Unknown	ND	22q11.2 deletion (includes TBX1)
Williams–Beuren syndrome	Unknown	ND	7q11.23 deletion (includes ELN)
Atrial septal defect–atrioventricular conduction delay	< 1 in 1,000,000	AD	NKX2.5
Ventricular septal defect (VSD)		ND	GATA4, TBX5, CITED2
Bicuspid aortic valve (BAV)	> 1/1,000	AD	NOTCH1
Loeys–Dietz syndrome	< 1 in 1,000,000	AD	TGFBR1, TGFBR2
Vascular Ehlers–Danlos syndrome (EDS)	1–9 in 100,000	AD	COL3A1
Familial thoracic aortic aneurysm and aortic dissection (TAAD)	1 in 500	ND	MYH11, ACTA2, TGFBR1, TGFBR2, MYLK, FBN1, FAA1, TAAD1, SMAD3
Familial hypobetalipoproteinemia	> 1/1,000	CD	APOB, PCSK9, ANGPTL3, MTTP
Mendelian low and high blood pressure			SLC12A3, SLC12A1, KCNJ1, CLCNKB, NR3C2, SCNN1A, SCNN1B, SCNN1G, CYP11B2, CYP11B1, HSD11B2, NR3C2, SCNN1B, SCNN1G, WNK1, WNK4, KLHL3, CUL3

Abbreviations: AD, autosomal dominant; AR, autosomal recessive; CD, codominant; Mt, mitochondrial; ND, not determined; XR, X-linked recessive.

extrapolated to humans even though humans yield a much smaller number of offspring compared to plants. Mendel's Laws form the foundation in human genetics and provide the fundamental understanding of inheritance patterns of monogenic diseases that are caused by mutations in the DNA sequence within a single gene.

MENDEL'S LAWS

Mendel's first law is referred to as *the principle of segregation*, which states that one allele (ie, an alternate form of the gene or sequence at a particular locus in the chromosome) of each parent is randomly and independently selected for transmission to the offspring and the two transmitted alleles from each parent unite randomly to form the offspring's genotype. Today, we understand Mendel's Law in the context of meiosis, where the segregation of alleles occurs during the process of sex cell formation. The union of the sperm (male) and ova (female) produces the fertilized egg cells, called zygotes.

Mendel's second law is referred to as *the principle of independent assortment*, which states that alleles underlying two or more different traits are transmitted independently of one another in the formation of gametes. As a result, new combinations of genes can be present in the offspring that are otherwise not possible in either of the parents. These two principles of inheritance and the concept of dominance of one allele over the other established the foundation of our modern science of genetics. Nonetheless, there is an exception to Mendel's Law that is particularly relevant for modern genetic epidemiology: loci in the same chromosomes tend to transmit together rather than independently. This phenomenon is known as linkage disequilibrium.

MODES OF INHERITANCE

Before Mendel's experiments, traits in offspring were believed to result from a hybrid of traits from each parent. However, when Mendel cross-pollinated one type of purebred plant with another, these crosses yielded different offspring: some offspring looked like one parent, others looked like the other parent, and some offspring looked like neither parent. These observations formed the six categories of the modes of inheritance of Mendelian diseases (Figure 1.1). The categories are determined by where the gene is located (autosomal, sex-linked, or mitochondrial) and how many copies of the mutant alleles are required for the trait to be observed (dominant or recessive). These six categories are described below:

1. *Autosomal recessive inheritance.* The locus is on an autosomal chromosome and both alleles, inherited from the mother and father, are mutant alleles. In pedigrees of families with multiple affected generations, autosomal recessive single-gene diseases often show a clear pattern in which the disease "skips" one or more generations (Figure 1.1A). For example, Ellis–van Creveld (EVC) syndrome is an autosomal recessive disorder caused by mutations in *EVC* gene.

2. *Autosomal dominant inheritance.* The locus is on an autosomal chromosome and only one mutant allele, inherited either from the father or the mother, is sufficient to cause the disease. For autosomal dominant diseases, all affected individuals should have an affected parent and both sexes are equally likely to be affected (Figure 1.1B). Most Mendelian CVDs have this mode of inheritance (Table 1.1).

3. *X-linked recessive inheritance.* Since females have two copies of X chromosomes whereas males only have one copy, the inheritance pattern with X-linked recessive inheritance differs from autosomal chromosomes. A female with the X-linked recessive disease inherits one mutant copy from the affected father and a second mutant copy from a mother who is either a carrier (heterozygous) or has the disease (homozygous). A male with the X-linked recessive disease always has a copy of the single mutant allele inherited from the mother (with or without the disease), but the father will not have the disease (Figure 1.1C). Left ventricular noncompaction (LVNC) and heterotaxia are transmitted in this mode of inheritance in some families.

4. *X-linked dominant inheritance.* A female with the X-linked dominant disease inherits one mutant copy from either the affected father (with single copy) or the affected mother who is either heterozygous or homozygous for the mutation. A male with the X-linked disease always has a copy of the single mutant allele inherited from the mother who always has the disease (Figure 1.1D). This mode of inheritance results in a higher disease prevalence in females than males. To date, no CVD is known to be inherited in this mode.

5. *Y-linked dominant inheritance.* This type of inheritance is also very rare. Only males have a single copy of the Y chromosome; thus, the mutant copy is always inherited from the father who also has

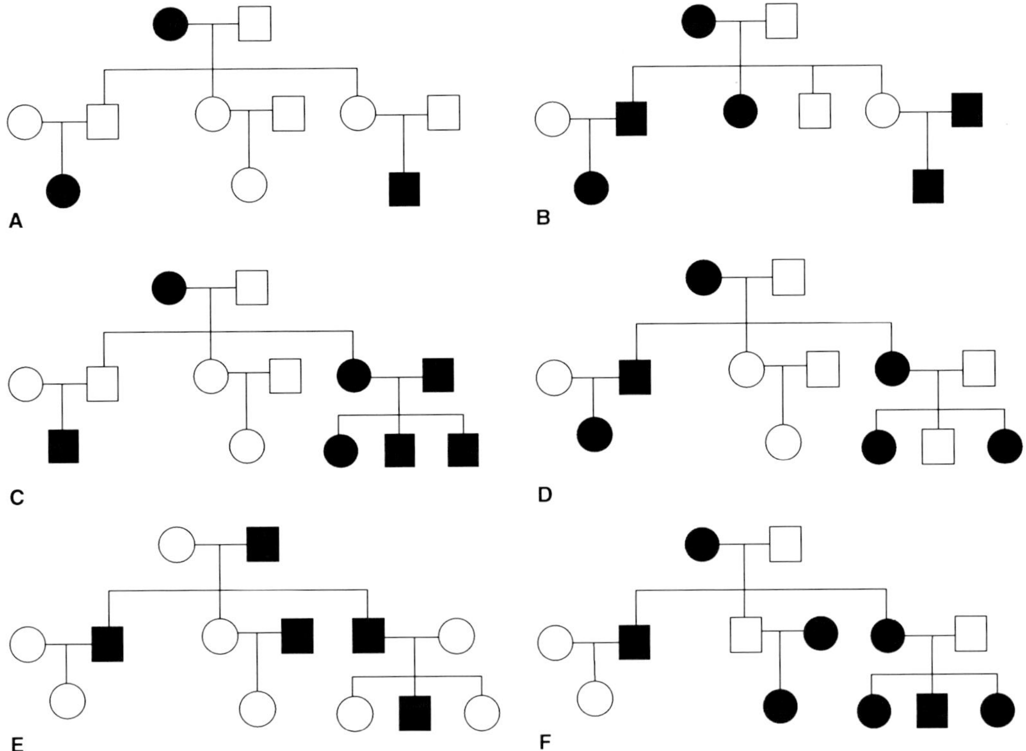

FIGURE 1.1 Basic Mendelian inheritance patterns; (A) autosomal recessive, (B) autosomal dominant, (C) X-linked recessive, (D) X-linked dominant, (E) Y-linked dominant, and (F) mitochondrial. Males are represented by squares and females by circles; shaded represent diseased.

the disease (Figure 1.1E). There is no difference in dominant or recessive mode of inheritance. To date, no CVD is known to be inherited in this mode.

6. *Mitochondrial inheritance.* The mitochondrial DNA encodes about 39 genes and is susceptible to mutations at a higher rate than nuclear DNA. Mitochondrial (Mt) DNA is always inherited from the mother; thus all mothers with the disease transmit the disease to their offspring (Figure 1.1F). LVNC shows a mitochondrial inheritance pattern in some families.

In recent years, the distinction between complex and Mendelian or monogenic diseases has been less obvious. Almost all Mendelian diseases, particularly those adult-onset conditions common in CVD, suggest they result from more than a single genetic defect (Table 1.1). In fact, several CVDs show multiple modes of Mendelian inheritance in different families (eg, arrhythmogenic right ventricular cardiomyopathy [ARVC], dilated cardiomyopathy [DCM], LVNC, familial hypercholesterolemia, long QT syndrome [LQTS], catecholaminergic polymorphic ventricular tachycardia [CPVT], and heterotaxia). Mendelian diseases are often known to have a straightforward genotype–phenotype relationship,

that is, the disease is caused by single-gene mutations. However, sometimes identical genes can produce different expression patterns. Thus, there can be three different relationships between a single gene and a Mendelian phenotype.

Penetrance

Even among carriers of Mendelian mutations, some may exhibit the phenotype or syndromes of the disease, whereas others may not. The proportion of genotypes that actually express the expected phenotype of the disease is called *penetrance*. Several Mendelian CVDs show incomplete penetrance. For example, penetrance of mutations leading to thoracic aortic aneurysm and aortic dissection (TAAD) is about 50%, especially in women.

Expressivity

Quantitative and qualitative variations in phenotypes of Mendelian inheritance observed within families are defined as expressivity. Expressivity can occur with differing degrees of disease severity, heterogeneity in clinical syndromes, or differences in the age of onset. For example, carriers of the same mutations in the fibrillin 1 (*FBN1*) gene that causes Marfan syndrome can display variable clinical manifestations.

Pleiotropy

Mutations in one gene can cause multiple phenotypes. For example, mutations in the *SCN5A* gene can cause LQTS, DCM, and Brugada syndrome (Table 1.1).

METHODS USED TO DETERMINE MODES OF INHERITANCE

Traditionally, the analyses of large family pedigrees with many affected individuals are used to determine a disease-associated gene and its inheritance pattern. Pedigrees are required for segregation and linkage analyses (8).

Segregation Analysis

Segregation analysis is a test for Mendel's first law and determines whether the transmission pattern of the disease in families is consistent with expected Mendel's inheritance (eg, dominant or recessive). The method does not utilize genetic markers. Rather, the proportions of affected and unaffected offspring are estimated based on the phenotypes of the parents, and statistical models test if these proportions segregate in the families according to that expected by the various Mendelian models. With the emergence of genomic markers, segregation analysis is rarely used since recruitment and ascertainment of all family members is difficult. Segregation analysis requires multigeneration family trees preferably with more than one affected member.

Linkage Analysis

Genetic linkage tests for cosegregation of alleles at loci physically adjacent on the same chromosome. Two loci that are unlinked follow Mendel's second law and assort independently. Suppose, in an individual locus 1 has C/T and locus 2 has A/G alleles in a linear order on a chromosome (Figure 1.2). If the two loci are unlinked, then based on the allele combinations, four possible gametes (C-A, C-G, T-A, T-G) can be produced with equal frequencies (0.25). However, two linked loci on the same chromosome will deviate from this distribution and produce predominance of linked allele pairs (eg, C-A, T-G) following nonindependent assortment. Two loci physically close have a small chance of separating by crossover event (recombination) during meiosis. The recombination fraction [θ] is a measure of the genetic distance between two loci. The probability that any two alleles at two randomly selected loci with be inherited together is 0.5. If two loci are closely linked, that is, physically close, then the chances of a crossover or recombination event occurring are less than 0.5.

Genetic linkage analysis tests whether the marker segregates with the disease in pedigrees with multiple affected individuals, according to a Mendelian mode of inheritance, without actually knowing the mutation. However, the marker tested must cosegregate with the gene related to the phenotype. Linkage analyses are performed under the assumption that (a) alleles at two nearby loci on the genome are transmitted together from parent to offspring, and (b) there are no recombination events at meiosis. If a genetic marker transmitted through family occurs more commonly in the affected individuals, then the disease locus that is physically close segregates with it as well.

Two types of linkage analysis are performed: (a) parametric and (b) nonparametric analyses. Parametric linkage analysis tests whether the inheritance pattern fits a specific model. Newton E. Morton developed the statistical test, the logarithm (base 10) of odds (LOD) score that compares the likelihood of observing the segregation pattern of the marker alleles at a given recombination fraction θ (linked) to the likelihood of the same segregation pattern in the absence of linkage (by chance), that is, LOD = L(θ)/L (θ = 0.5) (9). The objective of parametric linkage analysis is to estimate the recombination fraction (θ) and to test whether θ is less than 0.5, which is the case when two loci are genetically linked. Linkage analysis requires several tedious steps: (a) the establishment of pedigree (with history of the phenotype), (b) the estimation of recombination frequency, and (c) the calculation of the LOD score. A positive LOD score supports the presence of linkage whereas a negative LOD score indicates the absence of linkage. Traditionally, a cutoff of LOD more than +3 is used as evidence for linkage. Loci with multiple alleles, such as short tandem repeats (STRs), are more informative for linkage studies than diallelic single nucleotide polymorphisms (SNPs).

Nonparametric linkage (NPL) analysis methods do not make any assumption about the mode of disease inheritance, but rather, evaluate the statistical significance of excess allele sharing for specific markers among affected siblings. With this approach, the inheritance pattern is measured in terms of identical by descent (IBD), where the same allele is inherited from a common ancestor and identical by state (IBS), where the allele is the same but not necessarily inherited from the same ancestor. Thus, these methods are based on the fact that affected relatives have a higher probability of sharing genes IBD at or near a locus of the susceptibility allele/gene to a disease than sharing an unlinked locus.

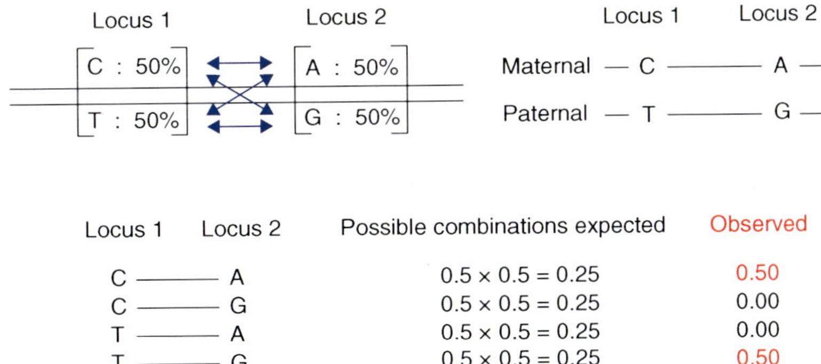

FIGURE 1.2 Genetic linkage at two loci (locus 1 and locus 2). Expected four gametes under independent assortment and observed 2 gametes due to linkage.

The genes contributing to the phenotypic variation have been successfully localized by linkage (cosegregation) analysis for Mendelian diseases that have a strong genetic effect and are relatively rare (eg, *TGFBR1* and *TGFBR2* in Loeys–Dietz syndrome; *FBN1* in Marfan syndrome) (3).

Linkage was first demonstrated in humans by Julia Bell and J. B. S. Haldane who showed that hemophilia and color blindness tend to be inherited together in some families (10). In 1989, the first linkage analysis of CVD was performed with hypertrophic cardiomyopathy (HCM) in a large French Canadian family that identified the chromosomal position of the causal gene in 14q1 and subsequently mutations in the gene in this region that encoded the beta cardiac myosin heavy chain were identified to be associated with the disease (11). Even Mendelian forms of complex diseases such as coronary artery disease and hypertension have been defined using genetic analysis.

Association Analysis

Historically, variants within genes causing Mendelian disease were discovered using parametric linkage analyses of large extended families with multiple affected family members. Association studies have now replaced linkage studies, even for Mendelian diseases, since it is not easy to assemble families and new and cost-effective methods are available. While case–control-based association studies, where one would test if genetic variants are more frequent in cases than controls, are often the preferable design, family-based association studies can also be powerful, particularly for Mendelian diseases. The most widely used family-based association study is the transmission disequilibrium test (TDT), which is based on the comparison of parental marker alleles that are transmitted and those not transmitted to the child with the disease (12). In regard to Mendelian diseases, these variations are often rare and directly differ with the function of the gene such as altered protein structures. The candidate genes can be selected a priori based on the biology or regions mapped by linkage studies. Recently, application of exome sequencing has been successful in finding mutation of Mendelian diseases.

CLINICAL UTILITY OF GENETIC INFORMATION

Inherited Mendelian conditions have a wide range of phenotype presentation, making their management, and the management of the family, a challenge for health care providers. Genetic counseling is inextricably linked to Mendelian diseases and important for patients and their families since advice and recommendation can be made based on the mode of inheritance. Since any first-degree relative of an individual with an autosomal dominant CVD has a 50% chance of also being affected by this trait, screening, diagnosis, and treatment regimens can be targeted based on the expected inheritance pattern in the family. Pregnancy offers another opportunity for screening for Mendelian disease if the mother or fetus is at particular risk of a genetic condition. Approximately 1% to 3% of pregnancies are complicated by a cardiac condition in the mother. As one example, a mother with Marfan syndrome, an autosomal dominant condition, can face sudden death during pregnancy from cardiovascular complications, such as dissection of the aorta. Therefore, any mother who is diagnosed or suspected of Marfan syndrome needs to consult a cardiologist.

Genetic testing is a routine procedure for several Mendelian diseases such as phenylketonuria (PKU), but it is not yet widely recommended for most CVDs. For example, several genetic testing panels consisting of sequencing of 12 to 18 genes are commercially available for HCM; however, none of them have been approved by the U.S. Food and Drug

Administration (FDA). Due to the complexity from varying degrees of penetrance for different mutations, consideration for testing of second- or third-degree relatives is made based on certain conditions and detailed pedigree analysis, as recommended by the consulting medical geneticist. With the availability of next-generation sequencing, all rare and common variants can be cataloged for various Mendelian CVDs, but there are challenges with accurate predictions. Eventually it is hoped that screening for a variety of these diseases in newborns or genetic counseling of new parents will be feasible and that such screenings will have major benefit for management of CVDs. In theory, the availability of results from genetic testing could improve treatment adherence, particularly for those diseases for which there is a standard treatment (eg, familial hypercholesterolemia). However, three major parameters must be evaluated before these genetic tests are transitioned from research into clinical practice:

1. Analytical validity: how good is the test in predicting underlying genotype?
2. Clinical validity: how good is the test in diagnosing the disease?
3. Clinical utility: what are the benefits of and risks from the test?

For instance, heterotaxia is considered a Mendelian disease, but has locus heterogeneity resulting from mutation in several genes. A further complication is that the disease has at least three known inheritance patterns, that is, autosomal dominant, autosomal recessive, and X-linked recessive (Table 1.1). For these reasons, a genetic risk score in one family may not perform as well in another family if they do not represent the actual causal variants. Likewise, policy makers and the FDA have to carefully assess how genetic information can help the management of patients in real life, that is, develop an algorithm to weigh if there is any evidence of benefit for intervention or treatment prescribed according to the genetic profile. All clinical evaluations have to be considered in addition to the genetic profile, since most CVDs do not follow a typical Mendelian pattern.

GLOSSARY

Allele: Alternate form of a gene or DNA sequence. At a sequence level, the nucleotide base pair at a particular position can have four *alleles* A, T, C, or G and often they are diallelic.

Allelic heterogeneity: Multiple alleles at a locus can be associated with the same disease. This phenomenon is known as *allelic heterogeneity* and can be observed with multiallelic locus. This may explain why in some studies one allele is associated with the disease and in other studies it is another allele.

Genotype: The exact combination of alleles found in each locus. In a combination of two alleles, T and C, there are three possible *genotypes*: *TT, TC,* and *CC.*

Heterozygous: *Heterozygous* is the term used to describe the mixed allele combinations, that is, one chromosome has one allele and the second chromosome has the alternate allele (eg, TC).

Homozygous: *Homozygous* is the term that describes where the same allele is present in both chromosomes (eg, TT or CC).

Linkage: The phenomenon of cosegregation where two genetic loci are transmitted together from parent to offspring more often than expected under independent inheritance.

Locus: A point in the genome, identified by a marker, which can be mapped by some means.

Locus heterogeneity: A phenomenon where multiple genes influence the disease independently, possibly through different biological pathways.

Pedigree: A diagram of two or more generations of a family's lineage showing genetic and phenotype information. Pedigrees are illustrated with standard symbols.

Phenotype: An observable trait, for example, familial hypercholesterolemia.

Pure line: Mendel created what is known as a *pure line,* a population of plants in which all progenies from self-fertilization resemble the same trait as the parent with no genetic variation. Mendel cross-fertilized two different *pure lines* and progenies from subsequent generations in his experiments.

REFERENCES

1. Frazier L, Johnson RL, Sparks E. Genomics and cardiovascular disease. *J Nurs Scholarsh.* 2005;37:315–21.
2. Kathiresan S, Srivastava D. Genetics of human cardiovascular disease. *Cell.* 2012;148:1242–1257.
3. McBride KL, Garg V. Impact of mendelian inheritance in cardiovascular disease. *Ann NY Acad Sci.* 2010;1214:122–137.
4. Ware SM, Jefferies JL. New genetic insights into congenital heart disease. *J Clin Exp Cardiol.* 2012;S8:003.

5. Wang L, Fan C, Topol SE, et al. Mutation of mef2a in an inherited disorder with features of coronary artery disease. *Science.* 2003;302:1578–1581.
6. Lehrman MA, Schneider WJ, Sudhof TC, et al. Mutation in LDL receptor: ALU-ALU recombination deletes exons encoding transmembrane and cytoplasmic domains. *Science.* 1985;227:140–146.
7. Mendel G. Versuche über pflanzen-hybriden. *Verh Naturforsch Ver Brünn.* 1866;4:3–47 *(in english in 1901, J R Hortic Soc. 26:1–32).*
8. Khoury MJ, Beaty TH, Cohen BH. *Fundamental of Genetic Epidemiology.* Oxford, UK: Oxford University Press; 1993.
9. Morton NE. Sequential tests for the detection of linkage. *Am J Hum Genet.* 1955;7:277–318.
10. Bell J, Haldane JBS. The linkage between the genes for colour-blindness and haemophilia in man. *Proc R Soc (Lond) B.* 1937;123:119–150.
11. Jarcho JA, McKenna W, Pare JA, et al. Mapping a gene for familial hypertrophic cardiomyopathy to chromosome 14q1. *N Engl J Med.* 1989;321:1372–1378.
12. Spielman RS, McGinnis RE, Ewens WJ. Transmission test for linkage disequilibrium: the insulin gene region and insulin-dependent diabetes mellitus (IDDM). *Am J Hum Genet.* 1993;52:506–516.

2

C H A P T E R

Genetics of Complex Traits

Ryan Irvin

TAKE HOME POINTS

1. Extensive methodology for measuring the familial component of common diseases has established that cardiovascular traits are transmitted within families in a complex manner through multiple genes the function of which may be modified by the environment.
2. Much less is known about the specific combination of genes and environments dictating the established genetic component.
3. The "missing heritability" problem may be due to inflated estimates of heritability stemming from ascertainment bias, nonadditive genetic terms, or phenotypic misclassification. Alternatively, improvements in genotyping technologies such as next-generation sequencing (NGS) are making more comprehensive genomic studies possible, which can uncover additional variants that will explain a larger portion of the estimated heritability.

Substantial progress has been made in human genetics over the past 10 years following the publication of the first draft of the human genome sequence as part of the Human Genome Project (1,2,3). Along with this massive effort, the Human Genome Organization (HUGO) Gene Nomenclature Committee assigned unique gene symbols and names to over 37,000 human loci, of which approximately 19,000 are protein coding. Given the enormous diversity observed, it is not surprising that the majority of genetic diseases have a complex rather than Mendelian mode of inheritance. For instance, congenital birth defects, myocardial infarction, cancer, mental illness, diabetes, and Alzheimer disease cause morbidity and premature mortality in nearly two of every three individuals during their lifetimes whereas Mendelian disorders are much less common (affecting less than 1 in 200 births) (4). Many of these common diseases recur in the relatives of affected individuals more frequently than in the general population although the mode of inheritance does not conform to a traceable pattern from a single gene. In contrast, complex traits and diseases result from interactions among a number of genes and environmental factors that comprise a multifactorial inheritance pattern. For example, blood lipid levels are known to be influenced by multiple genes involved in lipid transport (Figure 2.1), though much more remains to be discovered to fully characterize interindividual variance in blood lipids (5–7). Overall, the existence of a genetic component for most common cardiovascular diseases (CVDs) has been established, but the specific genetic and environmental contributors are not fully defined. In this chapter, we introduce basic concepts in complex trait genetics and describe how studies of familial aggregation, twin studies, and estimates of heritability are used to establish and quantify the genetic contribution to cardiovascular traits and diseases. We also discuss limitations in these measures, potential causes of these limitations, and point to ways advancing technology will improve future efforts.

QUANTITATIVE TRAITS

Quantitative traits are measurable as physical or biochemical quantities (eg, blood pressure, cholesterol, and carotid intima-media thickness). Variation

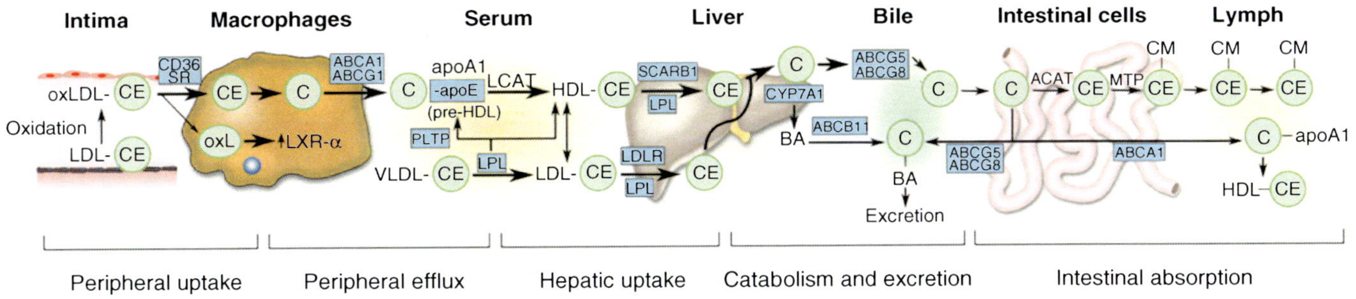

FIGURE 2.1 Genes (boxed) regulating lipid transport. *Abbreviations*: C, cholesterol; CE, cholesterol esters.

among such traits is usually due to differences in genetic as well as nongenetic (ie, environmental) factors. A graph of the number of individuals in the population (*y*-axis) having a particular quantitative value (*x*-axis) often yields the familiar bell-shaped curve known as the normal (or Gaussian) distribution (Figure 2.2). The peak and shape of the curve are governed by the mean (μ) and the variance (σ^2), respectively (Figure 2.2A). The mean is the arithmetic average of the values, and because most people have trait values near the average, the peak is observed at the mean value. The standard deviation (σ, square root of the variance) measures the spread of trait values to either side of the mean and, therefore, determines the span of the curve. Basic statistical theory shows that only 5% of the population will have measurements more than standard deviations above or below the population mean when a trait is normally distributed. The concept of the normal range of a physical quantity is fundamental to CVD identification and management. For example, hypertension, hypercholesterolemia, and obesity (**ie, a discreet or qualitative trait**) are defined when a quantitative trait value exceeds some set threshold (Figure 2.2B). Particular measured physiological traits are described as "normal" or "abnormal" depending on how far the trait falls above or below the mean value. Therefore, the normal distribution provides guidelines for setting the limits of the normal range. Any physiological quantity that can be measured fits the description of a quantitative trait and has a mean and a variance, though not all quantitative traits are normally distributed. The variance of a measured quantity in the population is called the **total phenotypic variance**. It has been argued that research should be focused not on endpoints (ie, disease states), but rather on the intermediate or quantitative phenotypes that determine the endpoint: as it is hypothesized that these phenotypes are "less removed" from the genes that influence them. The major challenge is to determine the extent to which genes contribute to observed variability. Figure 2.2C

demonstrates a very simplistic example of how a cardiovascular quantitative trait may vary by genotype.

FAMILIAL AGGREGATION

To various degrees, almost all diseases tend to aggregate within families (8). Family history has long been used as a risk factor for diseases, in medicine and in medical research (9). As previously noted, the family history can be attributed to genetic similarity (ie, the existence of a genetic etiology), environmental similarity (ie, risk factors common to family members), or a combination of both. The ultimate goal of complex genetic research is to distinguish between genetic susceptibility, environmental impact, and gene–environment interaction, and to quantify their relative importance as risk factors that influence CVD morbidity or mortality in the population (10).

MEASURING FAMILIAL AGGREGATION

Approaches for measuring aggregation depend on the nature of the phenotype (ie, quantitative or qualitative) but the unifying theme in existing approaches is that there is no specific Mendelian genetic model in mind (ie, the inheritance pattern is assumed to be complex and polygenic). The basic design involves sampling families. Usually an index family member (or proband) is identified, often in the clinic, and additional family participants are recruited with the help of the proband, and disease assessment is conducted in all family members.

For quantitative traits like blood pressure, familial aggregation can be assessed using a covariance-based measure. For example, the intrafamily correlation (or the *intraclass correlation coefficient* [ICC]) is a descriptive statistic that can be used when quantitative measurements are made on units that are organized

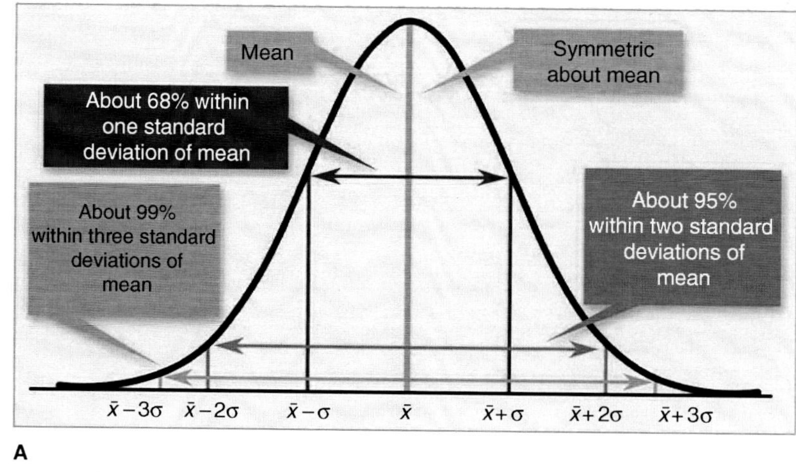

$\bar{x}-3\sigma$ $\bar{x}-2\sigma$ $\bar{x}-\sigma$ $\bar{x}$ $\bar{x}+\sigma$ $\bar{x}+2\sigma$ $\bar{x}+3\sigma$

A

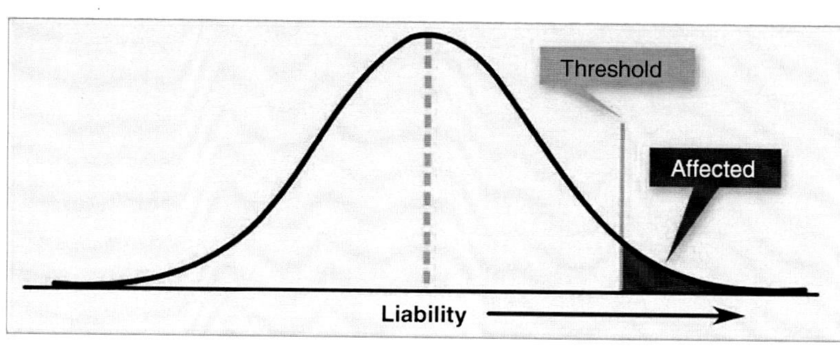

B

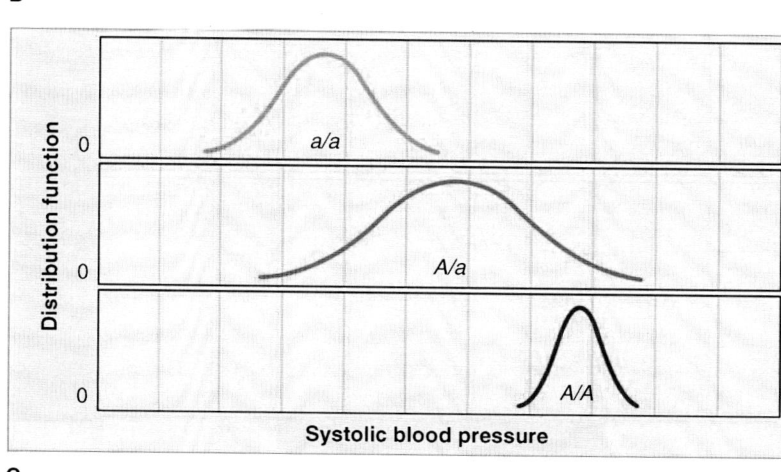

C

FIGURE 2.2   Quantitative traits and genetic contribution to phenotypic variance.

into groups (families). The ICC can be interpreted as the proportion of the total variability in a phenotype that can reasonably be attributed to real variability between families ($\sigma_B/[\sigma_B + \sigma_W]$). When the within-family variability is small, which would occur with a trait that is predominantly genetic, then the between-family variability increases and ICC estimate increases. Techniques such as linear regression and multilevel modeling analysis of variance are useful to derive estimates. Nonrandom ascertainment can seriously bias an ICC and result in inflated estimates of familial aggregation since the sampling is not accounting for the full range of trait values across families in the total population; therefore, heritability is ideally estimated from randomly ascertained families.

For qualitative traits, the familial aggregation of a disease can be measured by comparing the frequency of the disease in the relatives of an affected proband with the frequency of the qualitative trait (prevalence) in the general population. The relative risk (RR) ratio λ_r is defined as:

$$\lambda_r = \text{Prevalence of the disease in relatives of an affected person/Prevalence of the disease in the general population}$$

TABLE 2.1 RR in Relatives of AF Patients, by Degree of Relationship

Degree of Relationship	λ_r (95% CI)	*P*-value
1	1.77 (1.67, 1.88)	<.001
2	1.36 (1.27, 1.44)	<.001
3	1.18 (1.14, 1.23)	<.001
4	1.10 (1.06, 1.13)	<.001
5	1.05 (1.02, 1.07)	<.001

Source: From Ref. (11). Arnar DO, Thorvaldsson S, Manolio TA, et al. Familial aggregation of atrial fibrillation in Iceland. *Eur Heart J*. 2006;27:708–712.

The subscript r for λ is used here to refer to relatives. Alternatively, λ can be measured for a particular class of relatives (eg, r = s for siblings or r = p for parents). Larger values of λ_r correspond to greater familial aggregation and λ_r increases with increasing genetic contribution and decreasing population prevalence. A value of λ_r = 1 indicates that a relative is no more likely to develop the disease than is any individual in the population (8). Examples of λ_r values in relatives of atrial fibrillation patients by degree of relationship are shown in Table 2.1 (11). As detailed in Table 2.1, a first-degree relative of a proband with atrial fibrillation is 77% more likely to have atrial fibrillation than the rates observed in the general population.

CONCORDANCE VERSUS DISCORDANCE

Measures of familial aggregation are derived from measures of concordance and discordance. Specifically, when two individuals in a family have the same disease, they are considered concordant for the disease. Conversely, when only one member of a pair of relatives is affected, they are considered discordant. Concordance within family members suggests a common mechanism that may involve genes, environment, or their interaction (8). However, there are several considerations, which may limit familial aggregation study designs. In complex CVDs, concordance may occur even when two affected relatives have different predisposing genotypes. This situation is defined as a "genocopy"—a phenotype that appears identical to another phenotype but is caused by a different genetic mechanism. Concordance may also result from "phenocopy"—a phenotype that is due to an environmental exposure cause in one family member and a gene in the other. Another factor affecting concordance and discordance between family members is penetrance. Penetrance is defined as the proportion of individuals with a genetic variant who have the corresponding disease. For highly penetrant alleles, the phenotype is generally expressed, but for incompletely penetrant alleles, some members of the family will not express the trait even though they carry the risk allele. Lack of penetrance and frequent genocopies and phenocopies obscure the inheritance pattern in complex cardiovascular genetic diseases. Therefore, measures of familial aggregation alone are insufficient to demonstrate a genetic basis of a trait or disease. When trying to decipher the importance of genetic versus environmental factors, twin studies and heritability studies can provide more information.

ALLELE SHARING

When genes contribute to a disease, the frequency of concordance increases as the degree of relatedness increases. For example, monozygotic (MZ) twins have identical genetic content excluding any de novo mutations that may occur within an individual twin. The next most closely related individuals in a family are first-degree relatives, such as parent/offspring and sibling pairs (including dizygotic [DZ] twins). In a parent–offspring pair, the child has one allele in common with each parent at every locus corresponding to the allele inherited from that parent. Each sibling pair shares on average of one allele in common 50% of the time, and no alleles or two alleles at a specific locus 25% of the time. Inherited alleles shared in common are said to be identical by descent (IBD). Therefore, at any locus, the average number of alleles one sibling is expected to share with another is one (ie, 25% × 2 alleles + 50% × 1 allele + 25% × 0 alleles = 1 allele). For example, if genes predispose to a disease, one would expect λ_r to be greatest for MZ twins, then to decrease for first-degree relatives such as siblings or parent–offspring pairs, and to continue to decrease as allele sharing decreases among the more distant relatives, as demonstrated in Table 2.1 for atrial fibrillation.

TABLE 2.2 Concordance Rates for Cardiovascular-Related Diseases in Monozygotic (MZ) and Dizygotic (DZ) Twins

Disorder	MZ	DZ	Population	Reference
Abdominal aortic aneurysm	24%	5%	Swedish Twin Registry	Wahlgren et al. (14)
Venus thromboembolism	22%	8%	Danish Twin Registry	Larsen et al. (15)
Stroke death or hospitalization	11%	7%	Danish Twin Registry	Bak et al. (16)
Type 2 diabetes	72%	41%	Japan Diabetes Society	Matsuda et al. (17)
LDL-C subclass phenotype	80%	73%	Kaiser Permanente Women Twins Study	Austin et al. (18)

TWIN STUDIES

Studies of twins have played a significant role in helping to assess the relative contributions of genes and environment to disease causation. MZ twins are the result of a single fertilized egg splitting within the first 14 days after fertilization. As a result, MZ twins share 100% of their genes and are always the same sex. Their frequency is approximately 0.3% of all births, without significant differences across racial groups (12). DZ twins are the result of two separate eggs fertilized by two separate sperms. Therefore, DZ twins are siblings who share a womb and, on average, 50% of the alleles at all loci. Like any sibling pair, DZ twins may or may not be of the same sex. DZ twins occur more frequently and with more variability by race. For instance, DZ twins represent 0.2% of births among Asians to more than 1% of births in parts of Africa and among African Americans (13). DZ twins reared together allow geneticists to measure disease concordance in relatives who grow up in a shared environment starting from the earliest developmental stages, but do not share all their genes, whereas MZ twins provide an opportunity to compare siblings who also have identical genotypes and may or may not share the same environment. MZ twins reared apart can help distinguish the genetic versus environmental contributions to CVD.

CONCORDANCE OF MZ VERSUS DZ TWINS

The essence of a classical twin study is comparison of the similarities and differences of MZ and DZ twins under several assumptions. First, MZ twins are genetically identical; second, DZ twins are similar genetically to other siblings, sharing on average 50% of genes; third, both types of twins are sampled from the same gene pool; and finally, the environmental similarities and differences of the two types of twins are equal. MZ and same-sex DZ twins share a common intrauterine environment and are usually reared together in the same household by the same parents.

Thus, a comparison of disease concordance between MZ and DZ twins (preferably same sex) shows how frequently disease occurs when relatives who experience the same environment have all versus half of their genes in common. Greater concordance in MZ versus DZ twins provides evidence of a genetic component to the disease (Table 2.2), although information on the mode of inheritance cannot be obtained. This conclusion is strongest for conditions with early onset, although is still useful for conditions with adult onset.

TWINS REARED APART

Studies of twins reared apart provide a fascinating experiment in nature and a simple approach to disentangle the influence of environmental and genetic factors behind human traits (19). MZ twins separated at birth and raised apart provide the opportunity to observe disease concordance in individuals with identical genotypes reared in different environments. Such studies have been used primarily in psychiatric and behavioral research but are rare in studies of CVD. Studies to date suggest that the effect of being reared in the same home is negligible relative to the profound effect of genetic factors (ie, trait concordance is similar for twins reared apart versus together) (19). One study evaluated the environmental augmentation of genetic risk (Figure 2.3) in 198 MZ Japanese twin pairs reared apart and together. The intra-MZ pair concordance for blood pressure increased from 51% among twins with age of separation 0 to 5 years (ie, reared apart) to 78% with age of separation 26 years (ie, reared together) (20).

LIMITATIONS OF TWIN STUDIES

Ascertainment Bias

Ascertainment bias is a concern in twin studies for CVDs since (with the exception of congenital heart

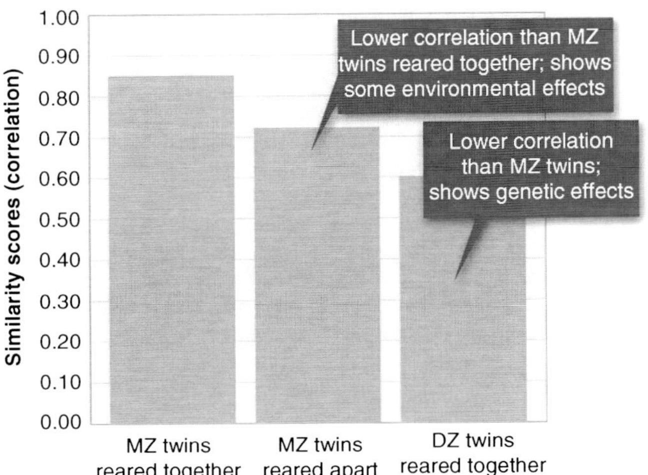

FIGURE 2.3 Example of how twin studies can suggest environmental augmentation of genetic risk.

disease) most CVDs occur later in life. Unless the trait of interest can be diagnosed at birth, great care must be taken to avoid ascertainment bias (21). An approach to minimize this potential bias is to register twins at birth and observe new onset diseases prospectively or to collect all twins in a defined population. When twin pairs are recruited in other ways, the first twin identified becomes the index case and the prevalence of disease in the cotwin becomes the proband concordance rate. Excess of concordant pairs can suggest ascertainment bias in addition to a genetic cause of the disease.

MZ Twin Genes Are Not Actually 100% Identical

It has been assumed that MZ twins are genetically identical, but considerable data suggest this is not actually the case and that genetic differences may explain MZ twin discordance (22). De novo mutations, retrotranspositions, indels, duplications, and chromosomal rearrangements can occur after the embryo has split (23,24). The rate for de novo base substitutions is about 10^{-8} per base pair per generation, making some genetic differences between adult twins likely (25). Copy number variants (CNVs), which account for a major portion of the genome, are strongly polymorphic and relatively unstable, with mutation rates 100 to 10,000 times higher than those for single base substitutions (26). Phenotypic discordance in MZ twins may in part be caused by de novo mutations of CNVs and CNV mosaicism (27). The regulation of gene expression plays a pivotal role in complex phenotypes, and epigenetic mechanisms such as DNA methylation and histone acetylation are essential to this process (22). For example, random

X inactivation (an epigenetic process) after cleavage into two female MZs produces significant differences in the expression of alleles of X-linked genes in different tissues (28). Epigenetic research among MZ twins shows they are very epigenetically similar early in life, though considerable variations accumulate as MZ twins age (29). Until recently, epigenetics has been largely missing from our understanding of how different phenotypes can originate from the same genotype. Recent advances in genomic technology have enabled epigenome-wide studies that will undoubtedly inform future twin research.

Environmental Variability

Finally, environmental exposures may not be the same for twins, especially once the twins reach adulthood and leave their childhood home. Even the intrauterine environment may differ between twins. For example, MZ twins frequently share a placenta, and there may be a disparity between the twins in blood supply, intrauterine development, and birth weight. Despite these limitations, twin studies offer an unusual opportunity to study disease occurrence when genetic influences are held constant (measuring disease concordance in MZ twins reared together or apart) or when genetic differences are present but environmental influences are similar (comparing disease concordance in MZ versus DZ twins).

HERITABILITY

Heritability analyzes the relative contributions of differences in genetic and nongenetic factors to the total phenotypic variance of quantitative traits in a population. High heritability implies a strong genetic component. In this context, heritability sounds like a relatively simple concept, but in fact it is quite complicated as described in the following.

Quantifying Heritability

Heritability of a trait is estimated from the correlation between measurements of that trait among relatives of known degrees of relatedness, such as parents and children, siblings, or, as described next, MZ and DZ twins. Any phenotype can be modeled as the sum of the environmental and genetic effects (Phenotype [P] = Genotype [G] + Environment [E]). To that end, total phenotypic variance is $V_P = V_G + V_E + 2COV_{G,E}$, where $COV_{G,E}$ is the covariance between genes and the environment. The combined effect of all loci, including possible allelic interactions within loci (dominance) and between loci (epistasis), is the genotypic variance (V_G). Heritability is formally defined as the proportion of total phenotypic variation (V_P) that is due to variation in genetic values (V_G). Broad-sense heritability, defined as $H^2 = V_G/V_P$, captures the proportion of phenotypic variation due to genetic values that may include effects due to dominance and epistasis. On the other hand, narrow-sense heritability, $h^2 = V_A/V_P$, captures only that proportion of genetic variation that is due to additive genetic values (V_A). Given its definition as a ratio of variance components, the value of heritability always lies between 0 and 1 (30).

Estimation

Heritability is estimated in populations by partitioning the observed variation into components that reflect unobserved genetic and environmental factors from data on the observed and expected concordance between relatives. The expected concordance between relatives depends on assumptions regarding a trait's underlying environmental and genetic causes. For example, if a trait has a strong genetic component, the average of the parent's quantitative trait value should strongly correlate with the offspring value. In Figure 2.4, examples are given of a scatter plot of progeny phenotypes (*y*-axis) versus the average of two parental phenotypes (mean parent phenotypic value, *x*-axis), for traits with high (0.9) and low (0.1) heritability. The straight line is the best-fit linear relationship between *y* and *x* obtained from a linear regression. The slope of the regression line is an estimate of heritability. This estimate, referred to as narrow-sense heritability, is limited in the fact that parent and child also share the same environment that may confound the heritability estimate.

Twin analysis is an approach to calculating heritability, which largely avoids the confounding of genotype with the shared environment. Comparing the phenotypic concordance of MZ versus DZ twins allows estimation of the effect of sharing an extra half of a genome on phenotypic concordance while fixing the environmental sharing to the greatest extent possible. For example, if trait concordance is measured as Pearson's correlation coefficient (r) and that value is 0.25 between MZ twins and 0.5 between DZ twins, then sharing an extra half genome with your twin explains an additional 20% of the trait, which extends to 40% (ie, 2 × 20%) when sharing a full genome.

This is called Falconer's formula:

$$\text{Heritability} = 2(r_{MZ} - r_{DZ})$$

Traditionally, heritability has been estimated from simple and often balanced designs, including simple functions of the regression of offspring on parental phenotypes, sibling correlations, and Falconer's formula (31). When phenotypic measures are available on individuals across multiple generations or when the design is unbalanced (eg, there are unequal numbers of observations per family), estimates of additive genetic variance and environmental components are most efficiently calculated with statistical methods that use all data simultaneously and take account of the exact properties of the data. Maximum likelihood (ML) methods and software have been developed to estimate genetic (co)variances in pedigrees for both univariate and multivariate models (32–35). What all these methods have in common is that they estimate genetic parameters from observed trait variation between and within families, assuming an underlying model (eg, additive, recessive, or dominant mode of inheritance) for causative components of variance (31,36). More recently, new methods that exploit the use of genetic marker data have been proposed and applied to estimate heritability essentially free of such assumptions (36). These methods utilize the efficient high-density array technology that genotypes millions of single nucleotide polymorphisms (SNPs). Improvements in technology will continue to offer new avenues to estimate heritability and, likely, new analysis methods to shed additional information on the genetic background of CVD (30).

Common Misconceptions Regarding Heritability

Heritability is not the proportion of a phenotype that is genetic, but rather the proportion of phenotypic variance that is due to genetic (ie, familial) factors (30). Therefore, it is important to understand heritability is not constant within and between populations as it depends on population-specific factors. The presence or absence of an environmental modifier of a genetic effect can result in substantially different heritability estimates across two populations. Additionally, heritability can change over time within a population due to changes in allele frequencies (attributed to selection, inbreeding, migration, and mutation), environmental factors, interactions between genes and environment, and even aging of the population. Heritability does not provide insights about the actions and interactions of individual genes. Identification of the specific genes and environmental factors underlying complex CVD are left to further genetic studies of specific genomic markers and groups of markers, much of which is the subject of the remainder of this book.

Heritability of Complex CVD

Many published studies point to a genetic background of complex CVD. Estimates of heritability vary for the

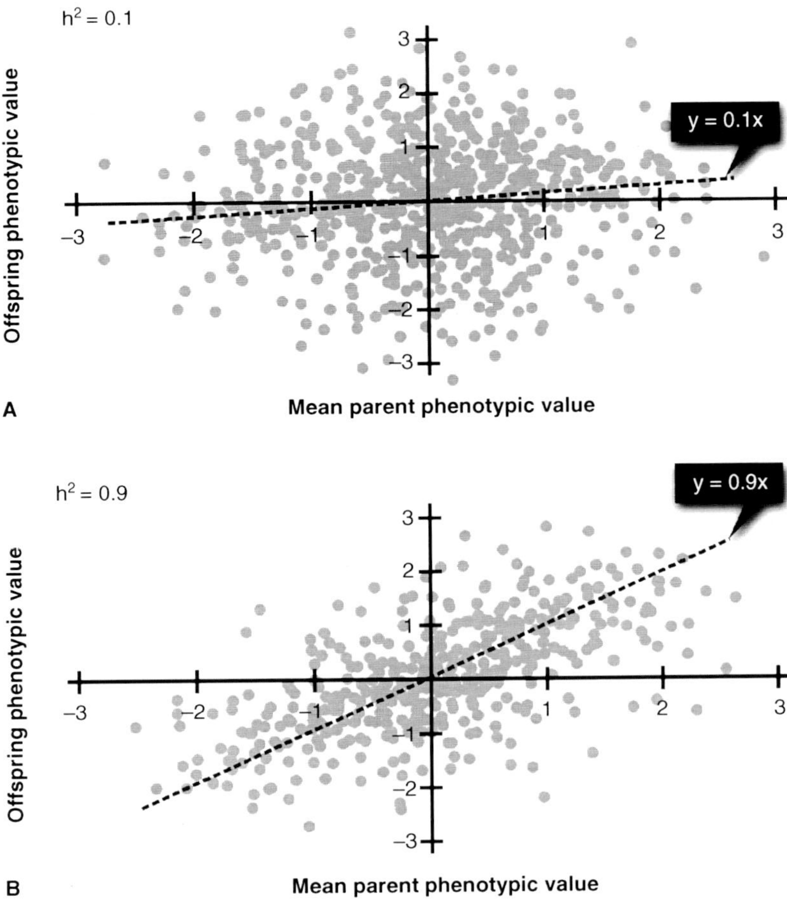

FIGURE 2.4 Heritability estimation from the regression of offspring phenotype on the mean phenotype of the parents. The slope of the regression line is an estimate of the narrow-sense heritability for traits with a heritability of 0.1 (A) and 0.9 (B).

same trait due to differences in the study population as already discussed, trait measurement variability, and/ or differences in study design and estimation methods. Therefore, there is no single true heritability estimate for any cardiovascular trait. For instance, for carotid intima-media thickness, a subclinical measure of atherosclerosis, studies show heritabilities ranging from 12% to 40% (37–42). Other reports have focused on other correlates of atherosclerosis. Flow-mediated dilation (FMD), a measure of endothelial dysfunction, is also a subclinical indicator of atherosclerosis. Heritability of FMD has been estimated between 12% and 14% (43). Heritabilities for cardiovascular risk factors such as body mass index (BMI), blood pressure, and lipid levels have been extensively studied (44–47). A study set within four different twin samples from three countries suggested an important role for genes in determining the variance of apolipoprotein and lipid levels (44). Heritabilities for these lipoprotein and lipid phenotypes ranged from 48% to 87% and most estimates exceeded 60%. A heritability estimate of 50% was reported for total cholesterol levels among American Indians participating in the Strong Family Heart Study (45). Heritability estimates for systolic and diastolic blood pressure range from 30% to 60% (45,46,48,49) and for BMI range from 25% to 80% (46,47,50). These results largely suggest that the familial aggregation of cardiovascular risk factors has a genetic component, supporting expansion of research to find cardiovascular risk and disease predisposing genes.

Missing Heritability in CVD

Despite considerable advances in genomic research, understanding the genetic component of complex CVDs remains poorly understood since a considerable portion of the heritable component remains unexplained. In CVD, as in most complex traits, the fraction of total phenotypic variance explained by individual common SNPs (minor allele frequency [MAF] more than 5%) usually ranges from 1% to 2% and heritability explained collectively by common

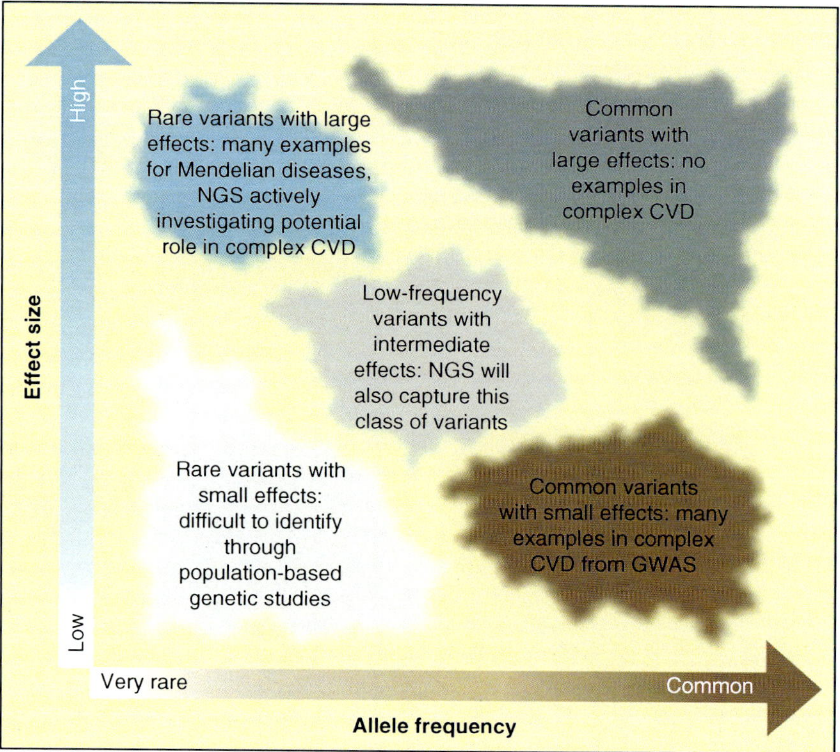

FIGURE 2.5 Potential to identify genetic variants for common CVDs by risk allele frequency and strength of genetic effect.

risk SNPs often remains below 10% to 15%, even for traits for which large numbers of loci have been detected (51). For example, only around one third of the heritable component (ie, 20%–25% of the genetic variance and approximately 12% of the total phenotypic variance) of low-density lipoprotein cholesterol (LDL-C) and high-density lipoprotein cholesterol (HDL-C) levels is explained by upwards of 90 common SNPs measured by genome-wide association studies (GWAS) even with sample sizes in the tens of thousands of people (52). There are several potential contributors to elements of missing heritability in CVD, many of which we will discuss here and will be expanded upon in the remainder of this text.

The accuracy of heritability estimates is potentially one important contributor to missing heritability. Narrow-sense heritability estimates in humans can be inflated if family resemblance is influenced by non-additive genetic factors (dominance and gene–gene interactions), shared environments, and/or by gene–environment interactions (30,31). Heritability estimates of complex traits may continue to be improved with the increasing availability of genome-wide markers that can provide more accurate estimates of allele sharing due to common ancestry between pairs of relatives (36,53).

Phenotypic measurement issues, such as measurement error or imprecision in defining the phenotype, may also explain some of the missing heritability. Many CVD-related traits are likely to result from many different contributing genetic pathways (eg, inflammatory, central nervous system, and salt-handling components of hypertension). These phenotypes might be aided by refining the phenotype into subphenotypes (eg, diastolic and systolic heart failure, or ischemic versus thrombotic stroke). The topic of phenotypic measurement in genetic studies of complex disease is comprehensively discussed by van der Sluis et al (54). The authors show that suboptimal classification of the phenotypic model can greatly dilute the genetic signal and that optimized modeling of the phenotypic component of genotype–phenotype data improves the power to detect genetic variants. Therefore, phenotype reclassification given expert input from clinicians and genetic researchers alike may improve upon the missing heritability problem.

The missing heritability problem may be addressed by rapidly advancing improvement in genomic technology including increasing availability of assays targeting rare variants, microRNAs and epigenetic factors. Chapters 3, 5, and 8 on GWAS, epigenetics, microRNAs, exome (the protein-coding region of the genome), and whole-genome sequencing (often referred to as NGS studies), respectively, will discuss how new technologies can and will facilitate future

CVD research. In an elegant review of the problem of missing heritability, Manolio et al discuss strategies to bridge gaps in genomic research of complex diseases. A large focus of the review is on rare (MAF less than 0.5%) and low-frequency genetic variation (0.5% less than MAF less than 5%) that was not comprehensively captured by prior GWAS. The primary technology for the detection of rare variants is sequencing, which may target regions of interest (eg, candidate genes, the exome), or may examine the whole genome. Importantly, low-frequency variants could have substantial effect sizes compared to more common SNPs assayed by GWAS without demonstrating clear Mendelian segregation, and could contribute substantially to missing heritability in CVD (Figure 2.5).In sum, the "missing heritability" is likely an overestimated enigma with multiple contributing constituents. Well-designed, large population-based studies that continue to capitalize on new technologies will continue to shed light on the genetic architecture of complex CVDs.

CONCLUSIONS

The study of the genetic basis of human disease is a relatively new scientific discipline. Much progress has been made in understanding the complexity of the human genome, and rapidly advancing technology allows us to dig ever deeper into its multi-layered components. Continued understanding of genomic complexity reflects the difficulties dissecting the genetic component of complex CVDs. Future efforts will undoubtedly benefit from integration of information from multiple genomic discovery approaches (ie, integration of data from GWAS, sequencing, microRNA, and epigenetic discovery). Discovery must be complemented by functional studies in vitro and in vivo that demonstrate gene expression and protein changes linked to discovered variants. In particular, functionally relevant loci will be essential to elucidate the molecular mechanisms of genomic changes that contribute to disease pathogenesis. Functionally validated variants will be necessary to inform the development of new therapeutic and diagnostic agents. Much published research suggests that cardiovascular traits are transmitted within families in a complex manner through multiple genes whose function may be modified by the environment. This chapter has focused on basic concepts in complex disease genetics, preliminary studies necessary to justify gene discovery efforts, and their limitations. The remainder of this text will expand upon gene discovery efforts largely targeted for complex traits.

REFERENCES

1. Lander ES, Linton LM, Birren B, et al. Initial sequencing and analysis of the human genome. *Nature*. 2001;409:860–921.
2. Venter JC, Adams MD, Myers EW, et al. The sequence of the human genome. *Science*. 2001;291:1304–1351.
3. Naidoo N, Pawitan Y, Soong R, et al. Human genetics and genomics a decade after the release of the draft sequence of the human genome. *Hum Genomics*. 2011;5:577–622.
4. Nussbaum R, McInnes R, Willard H. *Thompson & Thompson: Genetics in Medicine*. Philadelphia, PA: Saunders Elsevier; 2007.
5. Garcia-Rios A, Perez-Martinez P, Delgado-Lista J, et al. Nutrigenetics of the lipoprotein metabolism. *Mol Nutr Food Res*. 2012;56:171–183.
6. Corella D, Ordovas JM. Single nucleotide polymorphisms that influence lipid metabolism: interaction with dietary factors. *Annu Rev Nutr*. 2005;25:341–390.
7. Lovegrove JA, Gitau R. Nutrigenetics and CVD: what does the future hold? *Proc Nutr Soc*. 2008;67:206–213.
8. Guo SW. Familial aggregation of environmental risk factors and familial aggregation of disease. *Am J Epidemiol*. 2000;151:1121–1131.
9. Lilienfeld AM. Formal discussion of: genetic factors in the etiology of cancer: an epidemiologic view. *Cancer Res*. 1965;25:1330–1335.
10. Harper AE. Potential contributions of genetic epidemiology. *Genet Epidemiol*. 1988;5:203–206.
11. Arnar DO, Thorvaldsson S, Manolio TA, et al. Familial aggregation of atrial fibrillation in iceland. *Eur Heart J*. 2006;27:708–712.
12. MacGillivray I, Nylander PPS, Corney G. *Human Multiple Reproduction*. London, UK: WB Sanders; 1975.
13. Knox G, Morley D. Twinning in Yoruba women. *J Obstet Gynaecol Br Emp*. 1960;67:981–984.
14. Wahlgren CM, Larsson E, Magnusson PK, et al. Genetic and environmental contributions to abdominal aortic aneurysm development in a twin population. *J Vasc Surg*. 2010;51:3–7; discussion 7.
15. Larsen TB, Sørensen HT, Skytthe A, et al. Major genetic susceptibility for venous thromboembolism in men: a study of Danish twins. *Epidemiology*. 2003;14:328–332.
16. Bak S, Gaist D, Sindrup SH, et al. Genetic liability in stroke: a long-term follow-up study of Danish twins. *Stroke*. 2002;33:769–774.
17. Matsuda A, Kuzuya T. Relationship between obesity and concordance rate for type 2 (non-insulin-dependent) diabetes mellitus among twins. *Diabetes Res Clin Pract*. 1994;26:137–143.
18. Austin MA, Newman B, Selby JV, et al. Genetics of LDL subclass phenotypes in women twins. Concordance, heritability, and commingling analysis. *Arterioscler Thromb*. 1993;13:687–695.
19. Bouchard TJ, Lykken DT, McGue M, et al. Sources of human psychological differences: the Minnesota study of twins reared apart. *Science*. 1990;250:223–228.
20. Hayakawa K, Shimizu T. Blood pressure discordance and lifestyle: Japanese identical twins reared apart and together. *Acta Genet Med Gemellol (Roma)*. 1987;36:485–491.
21. Bundey S. Uses and limitations of twin studies. *J Neurol*. 1991;238:360–364.

22. Czyz W, Morahan JM, Ebers GC, Ramagopalan SV. Genetic, environmental and stochastic factors in monozygotic twin discordance with a focus on epigenetic differences. *BMC Med.* 2012;10:93.

23. Erickson RP. Somatic gene mutation and human disease other than cancer: an update. *Mutat Res.* 2010; 705:96–106.

24. Hirschhorn R. In vivo reversion to normal of inherited mutations in humans. *J Med Genet.* 2003;40:721–728.

25. Abecasis GR, Altshuler D, Auton A, et al. A map of human genome variation from population-scale sequencing. *Nature.* 2010;467:1061–1073.

26. Itsara A, Wu H, Smith JD, et al. De novo rates and selection of large copy number variation. *Genome Res.* 2010;20:1469–1481.

27. Bruder CE, Piotrowski A, Gijsbers AA, et al. Phenotypically concordant and discordant monozygotic twins display different DNA copy-number-variation profiles. *Am J Hum Genet.* 2008;82:763–771.

28. Youssoufian H, Pyeritz RE. Mechanisms and consequences of somatic mosaicism in humans. *Nat Rev Genet.* 2002;3:748–758.

29. Fraga MF, Ballestar E, Paz MF, et al. Epigenetic differences arise during the lifetime of monozygotic twins. *Proc Natl Acad Sci USA.* 2005;102:10604–10609.

30. Visscher PM, Hill WG, Wray NR. Heritability in the genomics era—concepts and misconceptions. *Nat Rev Genet.* 2008;9:255–266.

31. Falconer DS, Mackay TFC. *Introduction to Quantitative Genetics.* Longman, Harlow; 1996.

32. Patterson HD, Thompson R. Recovery of inter-block information when block sizes are unequal. *Biometrika.* 1971;58:545–554.

33. Lange K, Westlake J, Spence MA. Extensions to pedigree analysis. III. Variance components by the scoring method. *Ann Hum Genet.* 1976;39:485–491.

34. Gilmour AR, Thompson R, Cullis BR. Average information reml: an efficient algorithm for variance parameter estimation in linear mixed models. *Biometrics.* 1995;51:1440–1450.

35. Bochud M. Estimating heritability from nuclear family and pedigree data. *Methods Mol Biol.* 2012;850:171–186.

36. Visscher PM, Medland SE, Ferreira MA, et al. Assumption-free estimation of heritability from genome-wide identity-by-descent sharing between full siblings. *PLoS Genet.* 2006;2:e41.

37. Xiang AH, Azen SP, Buchanan TA, et al. Heritability of subclinical atherosclerosis in Latino families ascertained through a hypertensive parent. *Arterioscler Thromb Vasc Biol.* 2002;22:843–848.

38. Hunt KJ, Duggirala R, Göring HH, et al. Genetic basis of variation in carotid artery plaque in the San Antonio Family Heart Study. *Stroke.* 2002;33:2775–2780.

39. Swan L, Birnie DH, Inglis G, et al. The determination of carotid intima medial thickness in adults—a population-based twin study. *Atherosclerosis.* 2003; 166:137–141.

40. Fox CS, Polak JF, Chazaro I, et al. Genetic and environmental contributions to atherosclerosis phenotypes in men and women: heritability of carotid intima-media thickness in the Framingham Heart Study. *Stroke.* 2003;34:397–401.

41. Juo SH, Lin HF, Rundek T, et al. Genetic and environmental contributions to carotid intima-media thickness and obesity phenotypes in the Northern Manhattan Family Study. *Stroke.* 2004;35:2243–2247.

42. Moskau S, Golla A, Grothe C, Boes M, Pohl C, Klockgether T. Heritability of carotid artery atherosclerotic lesions: an ultrasound study in 154 families. *Stroke.* 2005;36:5–8.

43. Benjamin EJ, Larson MG, Keyes MJ, et al. Clinical correlates and heritability of flow-mediated dilation in the community: the Framingham Heart Study. *Circulation.* 2004;109:613–619.

44. Beekman M, Heijmans BT, Martin NG, et al. Heritabilities of apolipoprotein and lipid levels in three countries. *Twin Res.* 2002;5:87–97.

45. North KE, Howard BV, Welty TK, et al. Genetic and environmental contributions to CVD risk in American Indians: the strong heart family study. *Am J Epidemiol.* 2003;157:303–314.

46. Knuiman MW, Divitini ML, Welborn TA, Bartholomew HC. Familial correlations, cohabitation effects, and heritability for cardiovascular risk factors. *Ann Epidemiol.* 1996;6:188–194.

47. Schousboe K, Willemsen G, Kyvik KO, et al. Sex differences in heritability of BMI: a comparative study of results from twin studies in eight countries. *Twin Res.* 2003;6:409–421.

48. Levy D, DeStefano AL, Larson MG, et al. Evidence for a gene influencing blood pressure on chromosome 17. Genome scan linkage results for longitudinal blood pressure phenotypes in subjects from the Framingham Heart Study. *Hypertension.* 2000;36:477–483.

49. Williams RR, Hunt SC, Hasstedt SJ, et al. Genetics of hypertension: what we know and don't know. *Clin Exp Hypertens A.* 1990;12:865–876.

50. Elks CE, den Hoed M, Zhao JH, et al. Variability in the heritability of body mass index: a systematic review and meta-regression. *Front Endocrinol (Lausanne).* 2012;3:29.

51. Park JH, Gail MH, Weinberg CR, et al. Distribution of allele frequencies and effect sizes and their interrelationships for common genetic susceptibility variants. *Proc Natl Acad Sci USA.* 2011;108:18026–18031.

52. Teslovich TM, Musunuru K, Smith AV, et al. Biological, clinical and population relevance of 95 loci for blood lipids. *Nature.* 2010;466:707–713.

53. Meuwissen TH, Hayes BJ, Goddard ME. Prediction of total genetic value using genome-wide dense marker maps. *Genetics.* 2001;157:1819–1829.

54. van der Sluis S, Verhage M, Posthuma D, Dolan CV. Phenotypic complexity, measurement bias, and poor phenotypic resolution contribute to the missing heritability problem in genetic association studies. *PLoS One.* 2010;5:e13929.

Genome-Wide Association Studies

Daniel H. Katz and Laura J. Rasmussen-Torvik

TAKE HOME POINTS

1. Single nucleotide polymorphisms (SNPs) are frequent across the genome and allow for finer resolution for gene mapping of disease.
2. Technological developments in high-throughput SNP assay methods have provided the ability to conduct large-scale genome-wide association studies (GWAS) for identifying genomic regions of interest for a variety of diseases.
3. GWAS are useful for the detection of common variants that contribute to common diseases, but require large sample sizes for locus identification. Because of the large number of SNPs evaluated, GWAS must balance between the power to identify true positives and the risk of advancing false negatives.

As covered in Chapter 2, most phenotypes, including common diseases, are not the result of a single-gene variation. Rather, they are the result of a complex interplay among multiple genes, their multiple gene products, and the environment in which those products exist. Recent advances in human genomics now allow fast and reliable assessment of variation across the entire genome. This allows for high-throughput analyses of associations between common genetic variants and complex traits. Such analyses are termed **genome-wide association studies (GWAS)**. In this chapter, we discuss the genetics and technology underlying these studies, the methods they utilize, and the keys to interpreting and assessing GWAS in the literature. Throughout this chapter, references are made to an open-access study, "Genome-wide association studies of the PR interval in African

Americans" by Smith et al (1). This chapter helps to elucidate some of the fundamental elements of a modern GWAS. It is recommended that you read the full article after reading this chapter to assess your understanding.

WHY DO WE NEED GWAS?

Common cardiovascular diseases such as hypertension, heart failure, and coronary artery disease (CAD) are the end result of a complex interplay among various organ systems, tissues, cellular products, and human behavior. They also result from complex genetics: multiple variants in multiple genes. Thus, it is no surprise that the methods used to identify gene defects for single-gene disorders have been unable to determine genetic components for many common diseases. Nonetheless, heritability studies suggest that these diseases have a genetic component, and understanding the complex genetics of common disease is essential to modern cardiovascular medicine (1–3).

Previous Techniques

Before the advent of recent large-scale genotyping technologies, two different techniques were often used to look for genetic variations associated with common diseases. The first, **candidate gene studies**, requires a hypothesis that a particular gene is associated with the trait or disease of interest. This hypothesis might be generated from biochemical data about protein–protein interactions, gene manipulation in animal studies, or other types of studies (4,5). Many known variants in the candidate gene are then genotyped in all study subjects, and traditional statistical tests of association with the disease or phenotype of interest are performed. When this design is employed in family studies, a transmission disequilibrium test (TDT) is sometimes employed to determine the significance of associations (6,7). Unfortunately, candidate gene studies are limited by our current understanding of cellular and molecular pathways, preventing the identification of unanticipated disease genes. Furthermore, the low thresholds of significance utilized in most candidate gene studies historically translated to numerous false positive results, which failed to be replicated in subsequent studies (5).

Another technique that has been used to determine genetic variations associated with common disease is the **family linkage study**. These studies do not presuppose a candidate gene for association with a disease. Instead they use related individuals and track the cosegregation between disease and genetic markers called microsatellites, which vary between individuals and are spaced widely throughout the genome (8). The coinheritance of certain microsatellites and disease within families suggests that a gene variation of interest might be nearby (9). While these studies can "scan" the genome without a hypothesis about a specific gene, they are still limited. The need for related individuals makes recruitment complicated and can result in small sample sizes, reducing study power. The low power only permits the identification of genes with large effects; genes conferring small effects can be easily missed (10). Family linkage also has limited use in late-onset disease, since some designs require that parents and children be enrolled. Perhaps most importantly, the widely spaced markers used in family linkage studies identify regions including hundreds of genes, and so the identification of causative variants requires considerable additional research (11,12).

GENETIC BASIS OF GWAS

While candidate gene and family linkage studies have led to the identification of some novel disease genes, GWAS expand scientific capabilities with new technology based on a deeper understanding of human genetic variation. Understanding GWAS requires a basic understanding of the following underlying genetic principles.

Single Nucleotide Polymorphisms

While the vast majority of genetic material (99.9%) is identical among humans, the enormity of the genome leaves room for substantial variation (13). Genetic variation comes in a lot of forms including insertions, inversions, short tandem repeats, and copy number variants. However, the variations most commonly studied in GWAS are **single nucleotide polymorphisms** (SNPs). SNPs have arisen in the genome over millions of years of evolution, and are simply a single nucleotide site that differs between chromosomes. For example, whereas at some point in history all human chromosomes may have had thymine (T) at a particular locus on a chromosome, a mutation arose such that over generations a small percentage of chromosomes carried a cytosine (C) at that location. In this example, C is considered to be the **minor allele** since it is rarer than T, which is the major allele (Figure 3.1A). Since all humans carry two copies of all autosomes, an individual may be homozygous for the major allele for a SNP (ie, TT), heterozygous (TC), or homozygous for the minor allele (CC). It is estimated that there are 10 million such sites in human genomes that have a minor allele frequency (MAF) of greater than 1% and that these sites account for 90% of the genetic variation in the population (14). Compared to other genetic markers, such as microsatellites, SNPs are far more frequent across the genome and allow far finer resolution for genetic mapping of disease. The Human Genome Project, with the complete sequencing of multiple genomes, set the groundwork for GWAS by identifying the location of a large number of human SNPs (14).

Linkage Disequilibrium

Adjacent SNPs on a chromosome are often in **linkage disequilibrium** (LD). LD occurs because during meiosis, genetic recombination does not happen with equal frequency everywhere along a chromosome. The alleles of SNPs in high LD do not commonly experience recombination between them, and therefore generally get passed on to the next generation together. In general, the closer the two SNPs are on a chromosome, the more likely they are to be transmitted together to the next generation. If a new SNP arises in the population, it is likely that an allele of that SNP will be carried through generations with alleles from other nearby SNPs on that chromosome. The level of LD between two SNPs can be calculated by comparing the frequency with which two alleles appear together in a given population (15). These calculations can be presented graphically across a range of SNPs in a single chromosomal region (Figure 3.2). The LD between two SNPs generally decreases as the distance between the two SNPs increases (16).

LD is also population specific. For example, individuals of African descent have lower LD in their genome than Caucasians (17). The International HapMap Project has sought to document LD patterns across the genome in several different world populations (14).

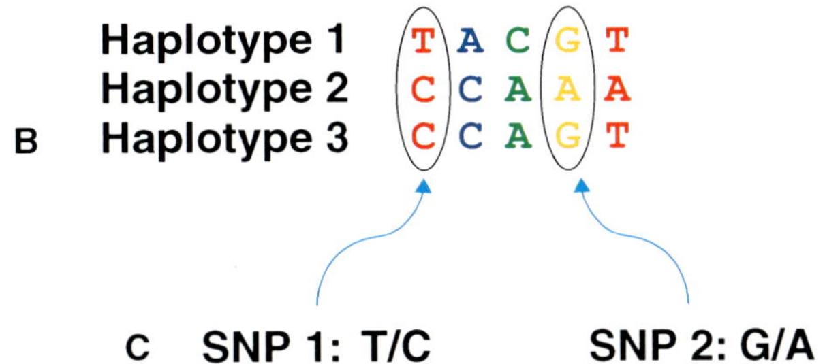

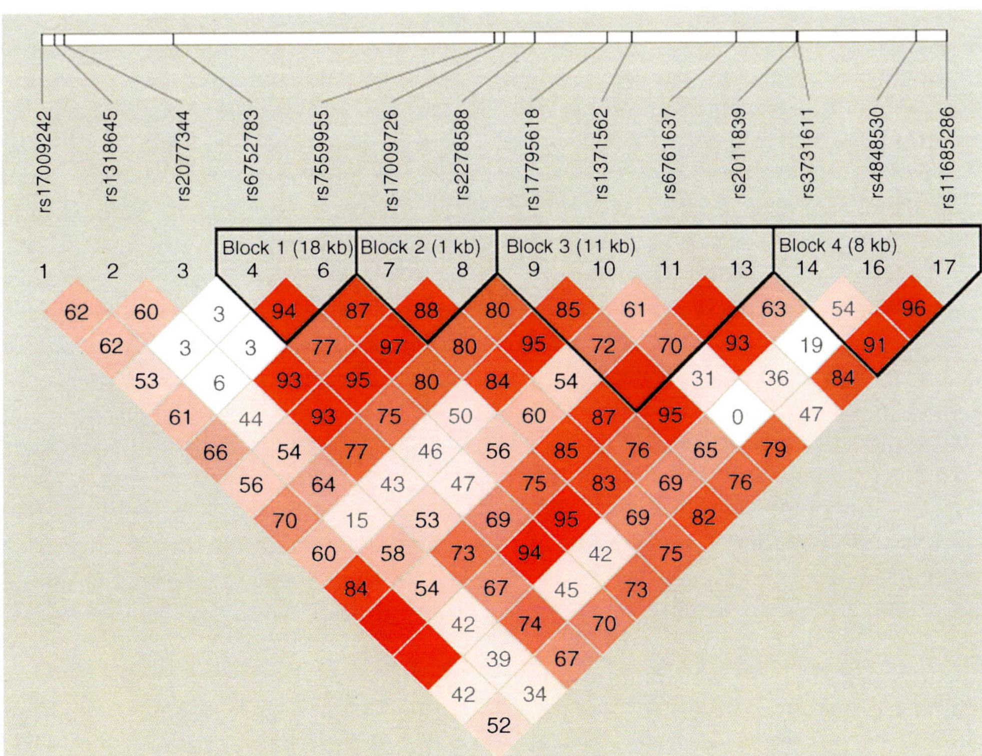

FIGURE 3.1 Single nucleotide polymorphisms and haplotypes. (A) Shown are stretches of single-stranded DNA from one copy of one chromosome in six individuals. While most of the chromosome is identical among individuals, the colored bases identify single nucleotide polymorphisms (SNPs)—places where the nucleotide varies among chromosomes in the population. (B) A block of SNPs that are consistently found together in this population is called a haplotype. Haplotypes occur because there is infrequent recombination between SNPs that are close together on the chromosome. As a result, adjacent SNPs are generally inherited together from generation to generation in these haplotype blocks. (C) Since there are a limited number of observed combinations, just the two circled SNPs need to be genotyped to correctly identify the other SNPs in the haplotype.

FIGURE 3.2 Linkage disequilibrium mapping. Shown is a region of chromosome 2 with each SNP labeled with its unique identifier. A single nucleotide polymorphism's (SNP's) relative location in the stretch of the chromosome is indicated on the segment above. Below, the D′ estimates (a measure of linkage disequilibrium [LD]) between SNPs are shown. The darker red shading indicates stronger LD, and the D′ value is written within the square. The clusters of dark red indicate regions of higher LD.

Source: Ma MJ, Wang HB, Li H, et al. Genetic variants in MARCO are associated with the susceptibility to pulmonary tuberculosis in Chinese Han population. *PLoS One*. 2011;6:e24069, under the Creative Commons Attribution License (CCAL).

They have created maps of lengthy segments of the genome frequently inherited together as blocks. These blocks are referred to as **haplotypes** (Figure 3.1B). *The understanding of LD patterns in different populations has greatly benefited GWAS*; when two SNPs are in high LD, only one SNP needs to be genotyped because it serves as a proxy for the other—one SNP "tags" the other. Therefore, the variation in millions of SNPs can be well approximated by just 500,000 (Figure 3.1C). In any GWAS, the genotyped SNPs that associate most strongly with a disease might also be the cause of phenotypic variation, or they may simply be in LD with (tagging) a nearby causal variant (14).

SIMPLE GWAS

In general, a GWAS can be conducted like any other observational study; numerous designs can be used and GWAS are subject to similar biases as other observational studies. The primary difference is that GWAS collect significantly more exposure measurements (in the form of millions of genotypes) than most other observational studies. This volume of data means that analyzing GWAS requires special attention to data management and statistical analysis.

Subjects and Design

Before conducting a GWAS, it is important to determine the experimental design that suits the proposed question, as GWAS can take on numerous designs. Case–control studies are common in the GWAS literature (18–20). They are useful when the disease of interest is rare. Critically, cases and controls should be drawn from the same source population for the design to be valid.

Many GWAS are also performed in cohort studies; large groups of individuals are recruited and data on dozens of phenotypes are collected at the time of enrollment. Individuals are then followed for a number of years and more phenotypic data are gathered, typically regarding disease incidence. Using studies of this type, GWAS can examine cross-sectional associations between SNPs and prevalent disease or traits, or they can examine SNP associations with incident disease information gathered from longitudinal follow-up. The paper from Smith et al presents cross-sectional data gathered from multiple observational studies, including cohort studies, on the PR interval. The Methods section features a description of each study included in the analysis (1).

A specialized design sometimes employed for GWAS is the **trio design**, which generally features two parents and their offspring (21). Trio designs are far less common owing to the challenges of enrolling families.

Regardless of the study type, a table describing the characteristics of the study participants is very important to properly assess a paper. Table 1 in the paper from Smith et al includes baseline characteristics by cohort (1).

Phenotypes: The Outcome Variable

Again, in GWAS the outcome of interest is no different than it is in any other study. Outcomes can be continuous (eg, the PR interval) or binary (eg, presence of disease). Using clinical cutoffs of continuous variables to create binary variables is not usually recommended as continuous outcome variables can increase the power to detect an association. It is critical that the methods for determining any phenotype of interest be clearly stated and consistent across studies if more than one is included in the GWAS. Smith et al explain in the Methods section how electrocardiograms (ECGs) were recorded in each cohort (1).

Genotypes: The Predictor Variable

A key element of executing a GWAS is genotyping thousands or millions of SNPs in each subject. The biochemical platforms for genotyping thousands of SNPs simultaneously are outside the scope of this chapter, but there are numerous techniques, and they vary by manufacturer (22). GWAS platforms continue to decrease in cost, making these studies more and more affordable. Smith et al feature a description of their genotyping process in their paper. They used the Affymetrix Genome-Wide Human SNP Array, which genotypes 906,600 SNPs (1).

While the genotyped SNPs on this array will span the genome, they are not necessarily spaced evenly throughout the genome. LD can vary by 100-fold across the genome, as well as among ethnic groups (14). This affects the **coverage** of any given array. Coverage is measured as the percentage of SNPs that are either directly genotyped or have a proxy SNP (in strong LD) that is directly genotyped. Coverage will vary depending on the number of SNPs on the array, their location in the genome, and the LD patterns in the population being studied. As mentioned previously, African Americans typically have shorter stretches of LD than Caucasians, and thus arrays typically cover a lower percentage of the genome in African Americans (17).

To increase statistical power, GWAS now commonly combine multiple studies together in meta-analyses, as is seen in the Smith et al paper (1). However, these meta-analyses are only feasible if there is a considerable overlap among the SNPs that are genotyped across studies. If each study were originally genotyped on a separate platform, the overlap in genotyped SNPs may be insufficient. Imputation is a process that uses the available SNP genotypes to predict additional SNPs not directly genotyped (23). Using known LD patterns as a guide, specialized softwares can impute the value of missing SNPs from the genotypes of nearby SNPs (24). Imputation greatly increases the volume of available SNP data, and permits meta-analyses across studies genotyped on different platforms.

Data Analysis

GWAS are different from many other observational studies because of the sheer volume of data involved. Smith et al worked with data on approximately one million SNPs in 6,247 individuals (1). Such volumes require specialized softwares for efficient analysis, and many such programs are available including SNPTEST, PLINK, and ProbABEL (25–27).

Despite the fact that softwares are required to efficiently manage analysis of the data, the statistical analysis at the level of a single SNP locus is relatively straightforward. While existing softwares can model recessive or dominant SNPs, most studies assume an additive model. For analytic purposes, in this model, homozygosity for one allele can be scored a 0, heterozygosity can be scored a 1, and homozygosity for the other allele can be scored a 2. This value is coded for each individual, and then the trait of interest regressed on the coded SNP value. GWAS softwares can handle a number of different types of regression analyses that may be indicated by different traits. Continuous traits can be analyzed through linear regression, dichotomous traits though logistic regression, and disease incidence data (with follow-up time) can be analyzed with Cox regression. Smith et al use linear regression with an additive model for analysis of the PR interval (1). Covariates can be added for adjustment, just as they would in any observational study. Most GWAS are adjusted for age, sex, study, and principal components of ancestry (28). Principal components of ancestry control for a unique kind of confounding seen in genetic studies and are covered in another section. The regression process is repeated (by the computer software) for each genotyped SNP. The result is thousands (or millions) of regression analyses.

To conveniently display the results of the regression for each SNP, GWAS use a special graph called a Manhattan plot (Figure 3.3). The chromosomal location is plotted along the x-axis, and the adjusted −log(10) P-value for the association between a SNP and the phenotype of interest is plotted along the y-axis. This is repeated for every genotyped SNP, with markings to separate chromosomes. Higher

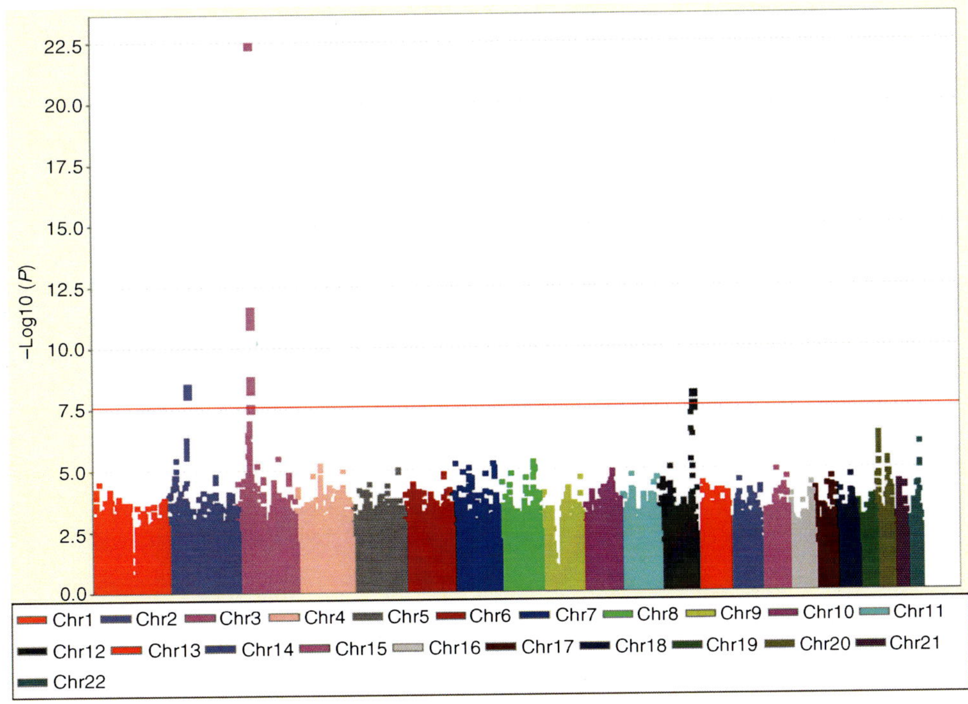

FIGURE 3.3 Manhattan plot. A Manhattan plot displays the statistical significance (P-value) for the association of all SNPs typed in a GWAS with the outcome of interest. Each SNP is represented by one point, and is plotted according to its genomic location across the x-axis, with chromosomes separated by color coding. The P-value is displayed on the y-axis by plotting −log (P-value). Thus, higher points represent more significant SNP associations. The red line indicates the threshold for genome-wide significance as set by the authors.

Source: From Ref. (1). Smith JG, Magnani JW, Palmer C, et al. Genome-wide association studies of the PR interval in African Americans. *PLoS Genet.* 2011;7:e1001304, under the Creative Commons Attribution License (CCAL).

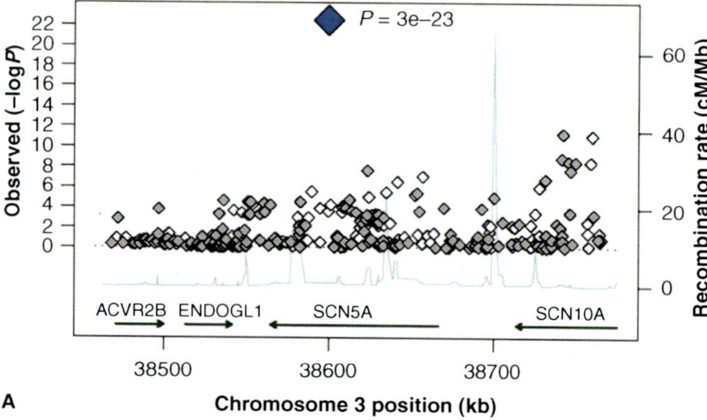

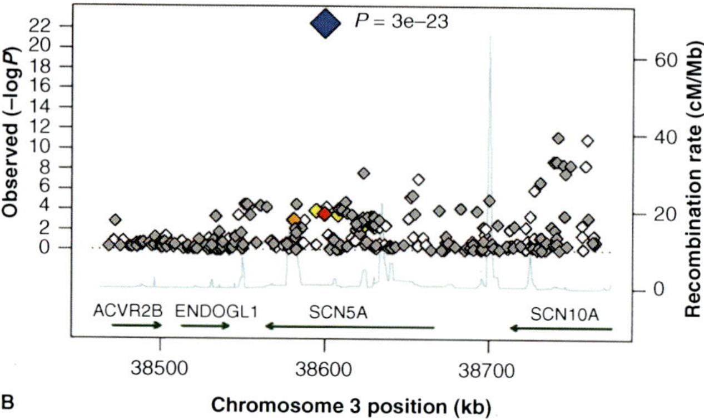

FIGURE 3.4 Local Manhattan plot. This local Manhattan plot shows a very small segment of the genome relative to the whole genome Manhattan plot in Figure 3.3. The most significant hit is plotted with a large blue diamond. Additionally, recombination rate is plotted with a blue line and measured on the right *y*-axis. Known genes in the genomic segment are displayed at the bottom.

Source: From Ref. (1). Smith JG, Magnani JW, Palmer C, et al. Genome-wide association studies of the PR interval in African Americans. *PLoS Genet.* 2011;7:e1001304, under the Creative Commons Attribution License (CCAL).

points indicate greater statistical significance (smaller *P*-values). Local Manhattan plots can also be used to show SNPs in a particular stretch of chromosome with greater resolution (Figure 3.4). Looking at the whole genome Manhattan plot from Smith et al, it is apparent that regions of very significant association with the PR interval lie on chromosomes 2, 3, and 12 (1).

Reaching Significance

In many observational studies, the accepted level of significance is *P* is less than .05. However, it is acknowledged that this is likely too liberal a threshold when multiple exposures are examined in a single study (29). This problem of "multiple comparisons" is especially acute in GWAS where over 1 million

SNPs (exposures) are examined. To correct for this, GWAS simply use a more stringent threshold for significance, often referred to as **genome-wide significance**. An approach to determine this threshold is to use a **Bonferroni correction**, where the commonly accepted value of .05 is divided by the number of comparisons performed (12). An accepted level of genome-wide significance is *P* is less than 5×10^{-8} (30). Some have argued that this correction is overly conservative because it assumes 1 million independent SNP tests, which, given the existence of LD, may not be accurate in many genome-wide scans (31). Nonetheless, it has stood up over several years as a method for identifying results likely to be replicated. In any case, it is critical that the authors provide their *P*-value threshold for genome-wide significance and their justification in their methods. Smith et al report their prespecified level of genome-wide significance to be 2.5×10^{-8} in their Methods section (1). This is indicated on their Manhattan plot by a horizontal line (Figure 3.3).

BEYOND GWAS BASICS: CHALLENGES AND LIMITATIONS

While GWAS are theoretically straightforward, in practice these studies have several challenges that their authors must be aware of and must address. Like any epidemiological study, one should attempt to eliminate measurement errors and biases. For GWAS in particular, it is important to maximize power in order to detect true positive results. Efforts to achieve these aims are undertaken both during data gathering and statistical analysis. This section focuses on some of these efforts.

Genotyping Quality Control

In any observational study it is critical that measurements are precise and accurate. This is no different in GWAS. Before any GWAS can proceed to statistical analysis, the quality of the genotyping must be assessed by standard methods. Fortunately, many computer software platforms, including PLINK, PLATO, and R have methods for carrying out this process (27,32,33).

Genotyping errors can occur for multiple reasons. Sample errors occur when samples are mislabeled or duplicated (34). Smith et al control for misidentified samples by cross-checking the identity of 24 SNPs against genotyping on a different platform (1). They eliminate duplicates using identity by descent methods. In addition, some DNA samples may simply be of poor quality and have many SNPs that are not genotyped at all. Smith et al filtered DNA samples with less than 95% genotyping success (1).

Issues can also arise with marker quality—the genotyping of specific SNPs across samples. One such concern is **call rate**. Across multiple samples, it is possible for the genotyping of a particular SNP to perform poorly, suggesting that genotyping this SNP is potentially difficult on the platform. Genotypes generated for these markers should be considered unreliable. Smith et al filtered SNPs with less than 90% genotyping success (1). Many GWAS also filter SNPs with low MAFs. SNPs with a low MAF have low power to detect true positives and may be more prone to genotyping error (33). Smith et al filter SNPs with MAF less than 1% (1).

Some studies will filter SNPs that are out of the **Hardy–Weinberg equilibrium (HWE)**. The HWE is based on the Hardy–Weinberg law that in a large, stable population, allele and genotype frequencies are constant across generations and there is a simple relationship between allele frequencies and genotype frequencies. Using this law, one can test if genotype frequencies are as expected (in HWE), given allele frequencies (35). Some have suggested that rather than filtering using HWE thresholds, those SNPs that violate HWE could be flagged for later analysis (33). Smith et al do not filter using HWE because their sample contains admixture, which violates an assumption required for HWE (1).

Imputed SNPs also need quality control. When imputation is used to analyze two cohorts originally genotyped on different platforms, tests for strand orientation and estimates of imputation performance should be made (23). Smith et al note that 50,000 imputed genotypes were genotyped on a separate platform and each sample had approximately 95% concordance rate (1). In conclusion, any GWAS should assess the quality of its genotyping in a systematic way and state with some detail these steps in the Methods section.

Power and Significance

As already mentioned, one critical issue in any GWAS study is where to set the threshold for significance. A Bonferroni correction has been effective at identifying true positives in GWAS, but it has been noted as conservative (31). While this stringency is helpful at weeding out false positives, it adversely affects power—the study is more likely to miss true positives. Many of the common variants that affect common disease have a weak effect (17,36). Given this, and stringent thresholds for significance, very large sample sizes are often required to detect significant associations in GWAS. When stringent thresholds are not used, many of the results initially found to be significant often are not replicated.

One approach to address power issues is to use a multistage study design. These studies use less stringent significance thresholds in the early stage in the hope of having the power to catch weak but interesting associations. The SNPs of interest are then genotyped in a larger sample population and analyzed with higher significance thresholds. Multistage studies at one time had the benefit of reduced cost, as not all individuals needed GWAS genotyping (6). With the reduction of costs in GWAS genotyping, multistage studies are being employed less frequently.

More recently GWAS meta-analyses are frequently used to improve power. By taking advantage of multiple studies and combining their results meta-analytically, power can be increased and generalizability can be improved (31). This is the technique utilized by Smith et al, as the PR interval is analyzed across four observational studies (1).

Genomic Inflation

The essential goal of any GWAS is to find genetic variants significantly associated with disease or phenotypes. Hopefully, the statistically significant associations in the Manhattan plot are true positives, particularly if appropriate significance thresholds are used. However, there are a number of reasons that test statistics can be inflated, resulting in false-positive associations. To identify if there has been a systematic inflation of test statistics in the GWAS analysis, investigators should examine whether the total distribution of test statistics departs dramatically from what would be expected if no associations existed. This is often described as testing for genomic inflation.

One visual approach to assess genomic inflation is a quantile–quantile plot, or a **Q–q plot** (37). For a Q–q plot, the *observed* (−log 10) P-values of all tested SNPs are plotted against the *expected* (−log 10) P-values under the null hypothesis (no association between SNPs and disease). Ideally, for GWAS, Q–q plots should only deviate from the null distribution (represented by dashed line on the plot) at the smallest P-values (the right of the graph), which suggests a few highly significant SNP associations. If the Q–q plot deviates from the identity line for higher (less significant) P-values, this indicates that something may be systematically inflating the test statistics for the regressions in the GWAS analysis. It can be seen in Figure 3.5 that P-values recorded by Smith et al only deviate from the expected distribution for very small P-values, suggesting that there is no systematic inflation of test statistics in this analysis (1,12,37). To control for any systematic inflation, most large GWAS (including Smith et al) use **genomic control** (1,38,39). The details of genomic control are outside the scope of this chapter, but the method essentially estimates

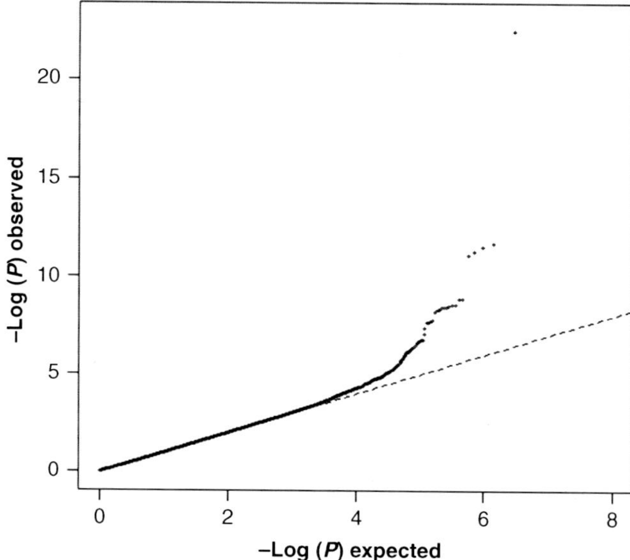

FIGURE 3.5 Q–q plot. The observed (–log 10) *P*-values from the association analysis of all SNPs with the trait of interest are plotted against the (–log 10) *P*-value distribution expected under the null hypothesis. Upward deviations from the dashed line indicate more *P*-values of a given size than expected. In this Q–q plot, deviation occurs only at the smallest *P*-values, suggesting the deviation is due to true positive results. Earlier deviations from the line could suggest genomic inflation.

Source: From Ref. (1). Smith JG, Magnani JW, Palmer C, et al. Genome-wide association studies of the PR interval in African Americans. *PLoS Genet.* 2011;7:e1001304, under the Creative Commons Attribution License (CCAL).

the amount of genomic inflation, usually denoted by λ_{GC}, and then uses that value to adjust test statistics from GWAS analysis. It can help address the systematic inflation that arises for many reasons, notably population stratification.

Population Stratification

Population stratification is one potential cause of genomic inflation in GWAS that deserves special discussion. GWAS results (and all genetic association results) can be confounded when genetically distinct populations with different allele frequencies and different trait distributions are combined in a single sample. When this happens, the alleles that vary in frequency between the distinct populations will spuriously correlate with the trait of interest (40). For example, an association discovered between an allele in the D_2 dopamine receptor and alcoholism was found to be caused by differences in the frequency of the allele and the frequency of alcoholism in the included ethnic populations, and did not hold up to more rigorous analyses (41).

As mentioned previously, population stratification can be controlled for with the use of principal components of ancestry. Since patterns of genetic variation identify different ethnic populations, statistical software (like EIGENSTRAT) can recognize these patterns in the genotype data, and determine the likely population structure (42). The structure identified by these patterns can then be used to estimate the principal components of ancestry for a given individual and these numeric estimates can be used to adjust regression analyses. Both genomic control and adjustment for principal components of ancestry can reduce power, and they do not negate the need for rigorous study design (43).

LIMITATIONS OF GWAS

GWAS are a critical tool in modern genetic analysis, but there are inherent limitations to the information they can provide. It is worth mentioning again the delicate trade-off between true and false positives (power and significance) in GWAS. The National Cancer Institute-National Human Genome Research Institute (NCI-NHGRI) Working Group on Replication in Association Studies made clear in their report in 2007 that most GWAS at that time were underpowered for common alleles with weak effect sizes (31). Since then, using large-scale meta-analyses of multiple studies has served to alleviate this problem somewhat by providing greater power for detecting weaker alleles. Alleles with small effect size may not seem important, but if common enough, the population-attributable risk can be very large, making them a target for investigation (31). Smith et al acknowledge this limitation in their Discussion (1).

Additionally, GWAS are extremely underpowered for rare mutations and not useful in detecting associations between the many different very rare variants that can occur in a given gene (sometimes called "private mutations") and a single disease (44,45). This is well documented in the case of *BRCA1* and *BRCA2* (46,47).

As GWAS uses known LD patterns from the HapMap project (14) and other sequencing projects to identify proxy SNPs and for SNP imputation, GWAS can be limited in the populations that are not well characterized. The HapMap project currently includes people from 11 populations, and most large sequencing projects to date have been about Caucasians or African Americans (14). Extending known LD patterns to other ethnicities may not be valid (48).

Finally, GWAS, like other genetic association studies, do not assess function or mechanism. GWAS are hypothesis-generating studies, designed to identify genomic regions of interest. The SNPs that are found to associate with the phenotype of interest identify a region that may be in LD with a causal variant. Determination of the actual causal variant must be done in other types of studies.

USING GWAS RESULTS

As mentioned, a great benefit of GWAS is that they identify narrow regions likely to contain causal genetic variants. These regions are far narrower than those identified by linkage studies. Once the region is identified, fine mapping can be performed in these regions. In this process, many or all of the SNPs in the identified region are examined (48).

Often GWAS also report all the known genes in LD with the index SNP. If experimental data support one of these genes as a likely contributor to the phenotype of interest, it can be prioritized for further examination including testing in animal models or deep sequencing to capture all of the genetic information in that gene. If all genes in LD with the index SNP are poorly understood, they may all be the target of later studies. Smith et al identify *SCN5A*, *SCN10A*, *MEIS1*, and *TBX5* as the genes associated with their SNP results and show their locations in local Manhattan plots (see Figure 3.4) (1).

GWAS SUCCESS IN CARDIOVASCULAR DISEASE

The works cited in this section represent some of the important GWAS in cardiovascular disease to date and are recommended for further reading. Not surprisingly, multiple GWAS have examined genetic risk factors for CAD. Some of the earliest work on this subject was published in 2007, with three groups commonly identifying the 9p21 region as associated with CAD (49–51). This region is near the genes *CDKN2A* and *CDKN2B*. Since then, numerous studies have been published examining this region (52). In 2011, the region was determined to be involved in altering the activation of *CDK2NA/B* in response to inflammation in a vascular cell, providing a functional link for disease susceptibility (53). CAD GWAS have identified dozens of other SNPs, suggesting a complex genetic contribution to disease (54–59).

A large GWAS for lipid levels was published in 2010 (60). This meta-analysis identified 95 SNP loci associated with lipid levels, with several of these SNPs located near genes known to be associated with lipids or that are targets of existing lipid therapy, including *HMGCR* (which is targeted by statins (61)) and *NPC1L1* (which is targeted by ezetimibe (62)). SNPs also tagged *GALNT2*, *PPP1R3B*, and *TTC39B*, which the authors linked to lipid levels in mouse models, making these potential future targets for investigation or therapy. GWAS have also been used to identify regions of interest in heart failure (63), blood pressure (64,65), and metabolic syndrome (66,67).

Another area of research using GWAS involves assessing medication response phenotypes. The recent Randomized Evaluation of Long-Term Anticoagulation Therapy (RE-LY) trial demonstrated the efficacy of dabigatran for stroke prevention in non-valvular atrial fibrillation (68). GWAS from that trial identified a minor SNP allele in *CES1* associated with a lower risk of bleeding while using dabigatran. Studies like this may soon allow for tailored pharmacological treatment.

SUMMARY

Recent technological developments including high-throughput SNP genotyping, mapping of SNP LD patterns in various ethnic populations, and large-scale observational studies have allowed for the advent of GWAS as a method for identifying genomic regions of interest for a variety of diseases. While GWAS must balance the power to identify true positives and the risk of advancing false negatives, it is nonetheless an important tool for guiding future genetic investigation and has already begun to transform our understanding of cardiovascular disease.

GLOSSARY

Bonferroni correction: A method of correcting for multiple comparisons; divides a *P*-value of .05 by the number of comparisons.

Call rate: The percentage of samples in which a particular SNP has been genotyped by the selected platform.

Candidate gene study: A type of study that examines the association of genetic variants dispersed throughout a particular gene with a phenotype or disease.

Coverage (array): The percentage of known SNPs that are either directly genotyped by a SNP array or are well proxied by a SNP on the array.

Family linkage study: A type of study that uses genetic variants dispersed widely throughout the genome, usually in the form of microsatellites, as markers to identify regions that cosegregate with phenotypes or diseases. Coinheritance of markers and disease within families identifies a large region that might contain a causal variant.

Genome-wide association study (GWAS): A type of study that examines the association of genetic variants dispersed throughout the genome, usually SNPs, with phenotypes or diseases.

Genome-wide significance: A threshold for significance set in GWAS to account for multiple comparisons. Commonly 5×10^{-8} is used.

Genomic control: A technique to control for the inflation of test statistics in a GWAS.

Haplotypes: Blocks of DNA that are frequently inherited together. A pattern of genetic variantions can define a particular haplotype for a segment of DNA (see Figure 3.1).

Hardy–Weinberg equilibrium: In a large, stable population, it is assumed that generally allele and genotype frequencies are constant across generations and there is a simple expected relationship between allele frequencies and genotype frequencies.

Linkage disequilibrium (LD): The nonrandom association of alleles at two or more loci. SNPs near each other on a chromosome tend to be in LD because of infrequent recombination between them.

Minor allele frequency: The frequency of the less common SNP allele in a sample.

Minor allele: The base for a SNP that is rarer in a given population.

Quantile–quantile plot (Q–q plot): Plots observed ($-\log 10$) P-values against those expected under the null hypothesis.

Single nucleotide polymorphism (SNP): A single nucleotide site whose DNA base varies between the chromosomes in a population.

Trio study: A special type of genetic study that generally enrolls two parents and one affected child.

REFERENCES

1. Smith JG, Magnani JW, Palmer C, et al. Genome-wide association studies of the PR interval in African Americans. *PLoS Genet.* 2011;7:e1001304.
2. Bella JN, Goring HH. Genetic epidemiology of left ventricular hypertrophy. *Am J Cardiovasc Dis.* 2012;2: 267–278.
3. Joe B, Shapiro JI. Molecular mechanisms of experimental salt-sensitive hypertension. *J Am Heart Assoc.* 2012;1:e002121.
4. Lindpaintner K, Pfeffer MA, Kreutz R, et al. A prospective evaluation of an angiotensin-converting-enzyme gene polymorphism and the risk of ischemic heart disease. *N Engl J Med.* 1995;332:706–711.
5. Tabor HK, Risch NJ, Myers RM. Candidate-gene approaches for studying complex genetic traits: practical considerations. *Nat Rev Genet.* 2002;3:391–397.
6. Hirschhorn JN, Daly MJ. Genome-wide association studies for common diseases and complex traits. *Nat Rev Genet.* 2005;6:95–108.
7. Spielman RS, McGinnis RE, Ewens WJ. Transmission test for linkage disequilibrium: the insulin gene region and insulin-dependent diabetes mellitus (IDDM). *Am J Hum Genet.* 1993;52:506–516.
8. Evans DM, Cardon LR. Guidelines for genotyping in genomewide linkage studies: single-nucleotide-polymorphism maps versus microsatellite maps. *Am J Hum Genet.* 2004;75:687–692.
9. Gyapay G, Morissette J, Vignal A, et al. The 1993–94 Genethon human genetic linkage map. *Nat Genet.* 1994;7:246–339.
10. Risch N, Merikangas K. The future of genetic studies of complex human diseases. *Science.* 1996;273:1516–1517.
11. Altmuller J, Palmer LJ, Fischer G, et al. Genomewide scans of complex human diseases: true linkage is hard to find. *Am J Hum Genet.* 2001;69:936–950.
12. Pearson TA, Manolio TA. How to interpret a genome-wide association study. *JAMA.* 2008;299:1335–1344.
13. Kruglyak L, Nickerson DA. Variation is the spice of life. *Nat Genet.* 2001;27:234–236.
14. International HapMap Consortium. The International HapMap Project. *Nature.* 2003;426:789–796.
15. Xiong M, Guo SW. Fine-scale genetic mapping based on linkage disequilibrium: theory and applications. *Am J Hum Genet.* 1997;60:1513–1531.
16. Jorde LB. Linkage disequilibrium and the search for complex disease genes. *Genome Res.* 2000;10:1435–1444.
17. Wang WY, Barratt BJ, Clayton DG, Todd JA. Genome-wide association studies: theoretical and practical concerns. *Nat Rev Genet.* 2005;6:109–118.
18. Link E, Parish S, Armitage J, et al. SLCO1B1 variants and statin-induced myopathy—a genomewide study. *N Engl J Med.* 2008;359:789–799.
19. Lyons PA, Rayner TF, Trivedi S, et al. Genetically distinct subsets within ANCA-associated vasculitis. *N Engl J Med.* 2012;367:214–223.
20. arcOGEN Consortium, arcOGEN Collaborators. Identification of new susceptibility loci for osteoarthritis (arcOGEN): a genome-wide association study. *Lancet.* 2012;380:815–823.
21. Smyth DJ, Cooper JD, Bailey R, et al. A genome-wide association study of nonsynonymous SNPs identifies a type 1 diabetes locus in the interferon-induced helicase (IFIH1) region. *Nat Genet.* 2006;38:617–619.
22. Syvanen AC. Accessing genetic variation: genotyping single nucleotide polymorphisms. *Nat Rev Genet.* 2001;2:930–942.
23. de Bakker PI, Ferreira MA, Jia X, et al. Practical aspects of imputation-driven meta-analysis of genome-wide association studies. *Hum Mol Genet.* 2008;17:R122–R128.
24. Li Y, Willer C, Sanna S, Abecasis G. Genotype imputation. *Annu Rev Genomics Hum Genet.* 2009;10: 387–406.
25. Aulchenko YS, Struchalin MV, van Duijn CM. ProbABEL package for genome-wide association analysis of imputed data. *BMC Bioinformatics.* 2010;11:134.
26. Marchini J, Howie B, Myers S, et al. A new multipoint method for genome-wide association studies by imputation of genotypes. *Nat Genet.* 2007;39:906–913.
27. Purcell S, Neale B, Todd-Brown K, et al. PLINK: a tool set for whole-genome association and population-based linkage analyses. *Am J Hum Genet.* 2007;81: 559–575.

28. Cordell HJ, Clayton DG. Genetic association studies. *Lancet.* 2005;366:1121–1131.
29. Hunter DJ, Kraft P. Drinking from the fire hose—statistical issues in genomewide association studies. *N Engl J Med.* 2007;357:436–439.
30. Frayling TM. Genome-wide association studies provide new insights into type 2 diabetes aetiology. *Nat Rev Genet.* 2007;8:657–662.
31. Chanock SJ, Manolio T, Boehnke M, et al. Replicating genotype-phenotype associations. *Nature.* 2007;447:655–660.
32. Gogarten SM, Bhangale T, Conomos MP, et al. GWASTools: an R/Bioconductor package for quality control and analysis of genome-wide association studies. *Bioinformatics.* 2012;28:3329–3331.
33. Turner S, Armstrong LL, Bradford Y, et al. Quality control procedures for genome-wide association studies. *Curr Protoc Hum Genet.* 2011;Chapter 1, Unit 1.19.
34. Weale ME. Quality control for genome-wide association studies. *Methods Mol Biol.* 2010;628:341–372.
35. Falconer DS, Mackay TFC. *Quantitative Genetics.* Harlow, UK: Pearson; 1996:5.
36. Reich DE, Lander ES. On the allelic spectrum of human disease. *Trends Genet.* 2001;17:502–510.
37. Wellcome Trust Case Control Consortium. Genome-wide association study of 14,000 cases of seven common diseases and 3,000 shared controls. *Nature.* 2007;447:661–678.
38. Devlin B, Roeder K. Genomic control for association studies. *Biometrics.* 1999;55:997–1004.
39. Zheng G, Freidlin B, Gastwirth JL. Robust genomic control for association studies. *Am J Hum Genet.* 2006;78:350–356.
40. Cardon L, Palmer L. Population stratification and spurious allelic association. *Lancet.* 2003;361:598–604.
41. Gelernter J, Goldman D, Risch N. The A1 allele at the D2 dopamine receptor gene and alcoholism. A reappraisal. *JAMA.* 1993;269:1673–1677.
42. Patterson N, Price AL, Reich D. Population structure and eigenanalysis. *PLoS Genet.* 2006;2:e190.
43. Price AL, Patterson NJ, Plenge RM, Weinblatt ME, Shadick NA, Reich D. Principal components analysis corrects for stratification in genome-wide association studies. *Nat Genet.* 2006;38:904–909.
44. Manolio TA, Collins FS, Cox NJ, et al. Finding the missing heritability of complex diseases. *Nature.* 2009;461:747–753.
45. Carvajal-Carmona LG. Challenges in the identification and use of rare disease-associated predisposition variants. *Curr Opin Genet Dev.* 2010;20:277–281.
46. Shuen AY, Foulkes WD. Inherited mutations in breast cancer genes—risk and response. *J Mammary Gland Biol Neoplasia.* 2011;16:3–15.
47. Risch HA, McLaughlin JR, Cole DE, et al. Population BRCA1 and BRCA2 mutation frequencies and cancer penetrances: a kin-cohort study in Ontario, Canada. *J Natl Cancer Inst.* 2006;98:1694–1706.
48. Musunuru K, Kathiresan S. HapMap and mapping genes for cardiovascular disease. *Circ Cardiovasc Genet.* 2008;1:66–71.
49. Helgadottir A, Thorleifsson G, Manolescu A, et al. A common variant on chromosome 9p21 affects the risk of myocardial infarction. *Science.* 2007;316:1491–1493.
50. McPherson R, Pertsemlidis A, Kavaslar N, et al. A common allele on chromosome 9 associated with coronary heart disease. *Science.* 2007;316:1488–1491.
51. Samani NJ, Erdmann J, Hall AS, et al. Genomewide association analysis of coronary artery disease. *N Engl J Med.* 2007;357:443–453.
52. Roberts R, Stewart AF. 9p21 and the genetic revolution for coronary artery disease. *Clin Chem.* 2012;58:104–112.
53. Harismendy O, Notani D, Song X, et al. 9p21 DNA variants associated with coronary artery disease impair interferon-gamma signalling response. *Nature.* 2011;470:264–268.
54. Deloukas P, Kanoni S, Willenborg C, et al. Large-scale association analysis identifies new risk loci for coronary artery disease. *Nat Genet.* 2013;45:25–33.
55. Coronary Artery Disease Genetics C. A genome-wide association study in Europeans and South Asians identifies five new loci for coronary artery disease. *Nat Genet.* 2011;43:339–344.
56. Davies RW, Wells GA, Stewart AF, et al. A genome-wide association study for coronary artery disease identifies a novel susceptibility locus in the major histocompatibility complex. *Circ Cardiovasc Genet.* 2012;5:217–225.
57. Lettre G, Palmer CD, Young T, et al. Genome-wide association study of coronary heart disease and its risk factors in 8,090 African Americans: the NHLBI CARe Project. *PLoS Genet.* 2011;7:e1001300.
58. Mehta NN. A genome-wide association study in Europeans and South Asians identifies 5 new loci for coronary artery disease. *Circ Cardiovasc Genet.* 2011;4:465–466.
59. Schunkert H, Konig IR, Kathiresan S, et al. Large-scale association analysis identifies 13 new susceptibility loci for coronary artery disease. *Nat Genet.* 2011;43:333–338.
60. Teslovich TM, Musunuru K, Smith AV, et al. Biological, clinical and population relevance of 95 loci for blood lipids. *Nature.* 2010;466:707–713.
61. Istvan ES. Structural mechanism for statin inhibition of 3-hydroxy-3-methylglutaryl coenzyme A reductase. *Am Heart J.* 2002;144:S27–S32.
62. von Bergmann K, Sudhop T, Lutjohann D. Cholesterol and plant sterol absorption: recent insights. *Am J Cardiol.* 2005;96:10D–14D.
63. Smith NL, Felix JF, Morrison AC, et al. Association of genome-wide variation with the risk of incident heart failure in adults of European and African ancestry: a prospective meta-analysis from the cohorts for heart and aging research in genomic epidemiology (CHARGE) consortium. *Circ Cardiovasc Genet.* 2010;3:256–266.
64. Ehret GB, Munroe PB, Rice KM, et al. Genetic variants in novel pathways influence blood pressure and cardiovascular disease risk. *Nature.* 2011;478:103–109.
65. Wain LV, Verwoert GC, O'Reilly PF, et al. Genome-wide association study identifies six new loci influencing pulse pressure and mean arterial pressure. *Nat Genet.* 2011;43:1005–1011.
66. Fox CS, White CC, Lohman K, et al. Genome-wide association of pericardial fat identifies a unique locus for ectopic fat. *PLoS Genet.* 2012;8:e1002705.
67. Kim YJ, Go MJ, Hu C, et al. Large-scale genome-wide association studies in East Asians identify new genetic loci influencing metabolic traits. *Nat Genet.* 2011;43:990–995.
68. Pare G, Eriksson N, Lehr T, et al. Genetic determinants of dabigatran plasma levels and their relation to bleeding. *Circulation.* 2013;127:1404–1412.

Bioinformatics

Kiran Musunuru

TAKE HOME POINTS

1. Bioinformatic tools use mapping information to identify the gene affected by the causal DNA variant, that is, the causal gene involved in disease pathogenesis (or protection).
2. A straightforward bioinformatics tool, the Genome Browser is from the University of California, Santa Cruz (UCSC) Genome Informatics website (genome.ucsc.edu). This tool allows users to access genome data from a variety of species, including humans (genome.ucsc.edu/cgi-bin/hgGateway) simply by searching a specific DNA variant or a genomic region.
3. To understand the relevance of a genetic variant in disease causation, there are in silico functional prediction algorithms to help. Among the most commonly used are SIFT (sift.jcvi.org), PolyPhen-2 (genetics.bwh.harvard.edu/pph2), and MutationTaster (www.mutationtaster.org). Because these algorithms are distinct, the predictions from these various websites are not necessarily concordant, and it is best to query multiple algorithms when assessing the functional consequences of a novel DNA variant.

Genome-wide association studies (GWAS), as described in Chapter 3, are designed to localize a DNA variant that contributes to a disease to a particular region on a chromosome. They rarely identify the exact location of the causal DNA variant. Typically, a linkage analysis will narrow down to a region no smaller than a few megabases (million DNA bases), and a GWAS to a region no smaller than tens to hundreds of kilobases (thousand DNA bases). The next step is to use this mapping information to further pinpoint the causal DNA variant and, with luck, identify the gene affected by the causal DNA variant, that is, the causal gene involved in disease pathogenesis (or protection). There are a variety of bioinformatic tool(s) that can help in this process.

This chapter outlines several publicly available, web-based bioinformatic tool(s) that are commonly used by researchers to follow up on the results of GWAS and family-based studies. To demonstrate the utility of these tools, I will discuss a real-life example taken from genetic studies of blood lipid levels. (The reader is advised to have an Internet-connected computer handy while reading this chapter so as to follow along as various tools are described.) While this discussion is by no means comprehensive, it should provide a good starting point for those wishing to dig more deeply into the results of modern human genetic studies.

FOLLOWING UP ON LINKAGE ANALYSES

Let us begin by considering a rare blood lipid disorder that has been termed familial combined hypolipidemia (FCH). Individuals with FCH have extremely low blood levels of low-density lipoprotein cholesterol (LDL-C), high-density lipoprotein cholesterol (HDL-C), and triglycerides (TGs). A family in which several members had this condition was first described in 1998 (1). Thirty-eight family members were recruited into a research study, DNA samples were taken, and genotyping of a panel of 378 simple sequence repeat (SSR) markers across the genome was performed. The marker genotypes were used to perform linkage analyses, that is, calculations were performed to assess which chromosomal markers best correlated ("segregated") with affected family members with FCH (low lipid levels) versus unaffected members without FCH.

A linkage analysis using LDL-C levels as a quantitative trait identified a region on chromosome 1 with a suggestive linkage "peak" (2). The best marker, GATA61A06, had a logarithm (base 10) of odds (LOD) score of 3.9 (LOD scores greater than 3.0 are generally considered to be statistically significant). The next best marker, not surprisingly, was a neighboring (proximal) marker, GATA165C03, with an LOD score of 3.1. The neighboring marker in the opposite (distal) direction, GATA109, had an LOD score of only 0.5.

Armed with this information, we can localize the precise interval on chromosome 1 corresponding to the linkage peak. Perhaps the most straightforward way to do this is to use the UCSC Genome Informatics website (genome.ucsc.edu). In this case, the relevant tool is the Genome Browser, which allows users to access genome data from a variety of species, including humans (genome.ucsc.edu /cgi-bin/hgGateway). Pinpointing the exact location of a DNA marker is as simple as entering the name into the box labeled **search term**, although the nucleotide number will vary depending on which "build" of the human genome is used—for our purposes, we should use the most recent build from February 2009, also known as "hg19," by setting **genome** to *human* and **assembly** to *Feb. 2009 (GRCH37/hg19)* using the pull-down menus.

In this case, the linkage peak is defined by the best marker (GATA61A06) and the two flanking markers (GATA165C03 and GATA109). Thus, we wish to define the range bounded by GATA165C03 and GATA109. Submitting "GATA165C03" in the Genome Browser will pull up a small interval around the marker, *chr1:60,571,863-60,772,183*. Doing the same with "GATA109" will pull up *chr1:81,898,598-82,098,757*. We can therefore specify our linkage peak as ranging from nucleotides 60,571,863 to 82,098,757 on chromosome 1. The causal DNA variant(s) and, presumably, the causal gene responsible for FCH must lie in this interval.

We can now enter this range in the Genome Browser by typing "chr1:60,571,863-82,098,757" in the search box. Depending on which default options are selected, various types of information will be displayed, the most important for our current purpose being the genes in the interval. There are many dozens of genes in the region defined by the linkage peak. In principle, any of these genes could be the causal gene. One can use the **move, zoom in,** and **zoom out** buttons near the top of the page to focus in on particular genes.

Prior to 2009, the next step would have entailed selecting a subset of these genes for resequencing in the affected family members with FCH—the number of genes being determined by the available budget. The hope would be that the resequencing would turn up a "smoking gun" mutation or mutations (such as a nonsense or frameshift mutation) that would clearly affect the gene's function and thus nominate that gene as the likely causal gene. A mutation with a less obvious effect on gene function (such as a missense mutation) would make a weaker case for the gene being causal, but certainly worth further consideration. In this present example, few researchers would have the budget to resequence all of the genes in the chromosomal region, and so one would typically have to choose a few that seem of highest interest. This could be done by selecting genes based on previously published literature that might suggest the genes being involved in the phenotype of interest (in this case, cholesterol levels); of course, this would not be helpful if the causal gene is a novel gene for which there is no published information.

After 2009, the problem has become simpler with the availability of exome sequencing, which allows for resequencing of all of the approximately 20,000 genes in the human genome for an affordable cost. Upon obtaining exome-sequencing data, one could then focus on the genes in the linkage intervals, nominate candidate genes based on the presence of mutations, and carry out further studies to assess which of the genes is causal for the phenotype.

FOLLOWING UP ON GENOME-WIDE ASSOCIATION STUDIES

In contrast to family-based studies, GWAS use large numbers of individuals—in some cases, as many as hundreds of thousands from a population. The goal of a GWAS is to detect associations between DNA variants and phenotypes. In a standard GWAS design, millions of single nucleotide polymorphisms (SNPs) across the genome are tested. Each SNP typically has two different alleles (occasionally three, and rarely four). The allele that appears more frequently in a population is called the major allele, and the less frequent allele is the minor allele. The minor allele frequency (MAF) of an SNP often varies widely among different ethnic groups.

A typical GWAS analyzes whether, for each of the interrogated SNPs, the MAF is statistically significantly different between a group of individuals with a phenotype (cases) and a group of control individuals, with a successful GWAS identifying one or several SNPs meeting this criterion. Each SNP defines a genomic interval within which a causal DNA variant is present; rarely is the GWAS SNP (also known as a "tag" SNP) also a causal SNP. The genomic interval identified by GWAS is generally much smaller than the interval identified by a linkage study, due to (a) the much larger study population and (b) the phenomenon of linkage disequilibrium (LD).

LD occurs when two SNPs (or other types of DNA variants) are very close together on a chromosome, so that their alleles are linked as they are transmitted through many generations of parents to offspring. It is accentuated by the existence of "hot spots" where recombination between paired chromosomes during meiosis almost always occurs. Two SNPs that lie within an interval bounded by two consecutive recombination hot spots will be in a high degree of LD and remain linked together, whereas SNPs separated by one or more hot spots will be in a low degree of LD or may not be linked at all.

The degree of LD is expressed by the metric r^2, which ranges from 0, which indicates no linkage, to 1, which indicates perfect linkage. Two SNPs with r^2 is greater than 0, particularly if r^2 is greater than 0.5, are likely to be in an interval between two recombination hot spots or, put another way, in a single LD "block." An LD block is typically of the order of tens to hundreds of kilobases in size—contrasting with linkage peaks/intervals (as described in the last section), which are megabases in size. Of note, the genomic locations of recombination hot spots vary across ethnic groups, and so too do the sizes and boundaries of LD blocks. For reasons of evolutionary history, individuals of African descent tend to have smaller LD blocks than individuals of European descent.

For the most part, any two SNPs in the same LD block will have some degree of LD, and one of the SNPs will therefore "tag" the other SNP. GWAS take advantage of this phenomenon by selectively interrogating one or a few tag SNPs for each LD block throughout the genome, limiting the number of SNPs that need to be directly genotyped to a few hundred thousand in total, rather than every single SNP in the genome. A corollary is that if a GWAS finds a tag SNP to be convincingly associated with a phenotype, the SNP must be in LD with the causal DNA variant, which itself must be located somewhere with the LD block.

Let us illustrate these principles by returning to the example of blood lipid levels. A GWAS reported in 2010 was performed in a total of approximately 100,000 individuals of European descent (3). A total of 95 tag SNPs were identified as having MAFs that were associated with LDL-C, HDL-C, TG, and/or total cholesterol levels. If one were to ask whether any of the tag SNPs lies in the linkage interval defined by the study of the FCH family discussed previously (flanked by the markers GATA165C03 and GATA109), one would identify the SNP rs2131925, which is associated with both LDL-C and TG.

One could type rs2131925 into the UCSC Genome Browser and then zoom out from the location of the SNP to see which genes lie nearby. But a more useful tool for defining an LD block around a tag SNP is provided by the Broad Institute's SNP Annotation and Proxy Search (SNAP) server (www.broadinstitute.org/mpg/snap). Clicking the **Plots** option on the home page brings up a page in which one can enter an SNP of interest and generate a graphical plot of all of the surrounding SNPs (within a user-specified distance) with their r^2 values relative to the original SNP, as well as boundaries defined by an r^2 threshold. The plot thereby shows the boundaries of the LD block, along with the genes contained within the LD block.

If we do this with rs2131925, setting the distance at 500 kilobases and using a threshold of r^2 is greater than 0.5 in Europeans (CEU), we obtain a graph in which the LD block surrounding rs2131925 is defined as ranging from nucleotides 62,673,831 to 62,975,809 on chromosome 1. There are a very large number of SNPs that have a high degree of LD with rs2131925; in principle, any of these SNPs could be the causal DNA variant. The LD block itself harbors three genes—*USP1*, *DOCK7*, and *ANGPTL3*—making it likely that one of these three genes is the causal gene affected by the causal DNA variant (although it cannot be ruled out that the causal DNA variant in the LD block is acting on a gene at a large distance away, for example, by affecting a transcriptional enhancer whose influence extends hundreds of kilobases). The list of genes is certainly more manageable than the large number of genes marked out by the linkage interval defined by the study of the FCH family (flanked by the markers GATA165C03 and GATA109), illustrating the much higher resolution afforded by GWAS compared to linkage analyses.

The Broad Institute's SNAP server offers other tools that can be useful in following up on the results of a GWAS. For example, the **Pairwise LD** option on the home page brings up a page through which one can generate a list of all of the SNPs that are in LD with a tag SNP, along with the locations and r^2 values (ie, a textual equivalent of the graphical plot generated by the **Plots** page).

OBTAINING INFORMATION ON CANDIDATE GENES

Quite often the end goal of family-based studies and GWAS is to identify the causal gene responsible for the phenotype. While these studies can generate lists of candidate genes by virtue of pinpointing specific chromosomal regions, some type of follow-up genetic or functional study is typically needed to validate a gene as causal. If the number of candidate genes precludes further experimentation on all of the genes—whether resequencing or experiments in animal models—then some method is needed to pare down the list to a manageable number of genes.

One clue as to whether a gene might be causal is if that gene has previously been reported to be connected in some way with the phenotype in question. In the FCH example described earlier, one could look through the large list of genes in the vicinity of markers GATA165C03 and GATA109 on chromosome 1 to identify genes previously implicated in cholesterol metabolism. One could do the same for the smaller list of genes—three in total—for the LD block around rs2131925.

A useful tool for assessing whether a gene has previously been implicated in disease is the Online Mendelian Inheritance in Man (OMIM) catalog (omim.org). At this website, one simply needs to enter in the name of a gene of interest, and the server will return articles that pertain to the gene in any way. For example, in looking through the FCH linkage interval on chromosome 1, one finds the gene *LEPR* (chr1:65,886,131-66,103,176), which encodes the leptin receptor. Entering this gene into OMIM brings up several hits, the first of which focuses specifically on the leptin receptor. In reading the provided information, one learns that the gene is involved in the regulation of body weight, and that mutations in the gene result in obesity in humans. While dysregulated body weight could plausibly lead to disturbances in blood lipid levels, no specific mention of altered lipid levels is made.

Another nearby gene that seems more promising is *PCSK9* (chr1:55,505,149-55,530,526), which encodes proprotein convertase subtilisin/kexin type 9. Upon entering this gene into OMIM, it quickly becomes apparent that mutations in the gene cause familial hypercholesterolemia, in which patients have extremely high LDL-C levels—the opposite of FCH. Thus, it seems quite plausible that other mutations in *PCSK9* could be responsible for FCH. Yet another gene, *ANGPTL3*, is present in both the FCH interval as well as the LD block around rs2131925. In OMIM, the page on *ANGPTL3* describes the characterization of a mouse with abnormally low plasma lipid levels (hypolipidemia) in which the causal mutation was mapped to the *ANGPTL3* gene. So *ANGPTL3* is a strong candidate for both the gene responsible for FCH as well as the GWAS finding of rs2131925 being associated with LDL-C and TG.

Another means by which to find information on candidate genes is to search websites such as WikiGenes (www.wikigenes.org) and GeneCards (www.genecards.org), which seek to aggregate information from a variety of different sources and provide useful links to obtain more data on gene function, as well as inviting researchers to curate or otherwise contribute to the information on the webpages.

CHARACTERIZING DNA VARIANTS FOUND WITHIN GENES

Even if one is able to narrow down a list of candidate genes from a linkage analysis or GWAS to a few genes or even one gene, the causal DNA variant(s) remains to be identified. With the emergence of exome sequencing (and, soon, whole-genome sequencing) as a standard technique in human genetics, it has become extremely straightforward to search through candidate genes for coding DNA variants. Indeed, the challenge of identifying a causal DNA variant has been entirely reversed—whereas prior to the Human Genome Project it was difficult to hunt down any DNA variants due to the prohibitive nature of gene resequencing, nowadays next-generation sequencing studies of cohorts are uncovering a plethora of DNA variants of unclear significance. What does it mean to discover novel DNA variants in a group of patients with a disease, or in a group of control individuals? How can one be sure that the DNA variants are causal for the disease or, conversely, protect against disease?

A critical step is to assess the functional consequences of a given DNA variant on gene function. This is much more straightforward when dealing with a coding variant in a gene, rather than a noncoding variant, but even for coding variants there is not yet a 100% reliable in silico method to predict their effects. Whereas nonsense and frameshift mutations can reliably be expected to alter a gene's function, missense mutations are much harder to interpret. The definitive means to assess function is to perform experimentation in model systems, though screening a large number of DNA variants in this fashion remains prohibitive.

There exist in silico methods that, while far from perfect, can make rational guesses as to the functional consequences of DNA variants. These methods rely on a variety of principles. For example, a missense mutation may alter an amino acid that lies in a predicted secondary structure element such as an alpha helix. If the substituted amino acid is not one that is well accommodated in an alpha helix, it might be expected to disrupt the helix and cause that portion of the protein to be misfolded, altering the protein's function. Another example is phylogenetic conservation. A missense mutation may alter an amino acid that is identical in all versions of the protein found across all species, arguing that the amino acid has some important functional role that will not tolerate a different amino acid in the same location. The missense mutation could then be inferred to disrupt the protein's function. More generally, a conservative missense mutation—for example, the small hydrophobic leucine being changed to the small hydrophobic isoleucine—is less likely to affect protein function than a nonconservative mutation—for example, the large positively charged arginine being changed to the smaller negatively charged aspartic acid.

A number of functional prediction algorithms have been developed using some of these principles and instantiated as publicly accessible websites. Among the most commonly used are SIFT (sift.jcvi.org), PolyPhen-2 (genetics.bwh.harvard.edu/pph2), and MutationTaster (www.mutationtaster.org). Because the underlying algorithms are distinct, the predictions from these various websites are not necessarily concordant, and if expedient it is best to query multiple algorithms when assessing the functional consequences of a novel DNA variant.

As an example of the use of functional prediction algorithms, let us return to the FCH family described earlier in this chapter. There are two approaches by which we could have tried to identify the causal gene. First, a linkage analysis of the family pointed to a region on chromosome 1, and searching through functional annotations of genes in the region identified two promising candidates, *PCSK9* and *ANGPTL3*. We might have performed targeted resequencing of the coding regions of those two genes in the family members. Second, we might have performed exome sequencing of a few of the most severely affected family members and sought out mutations shared by all of these individuals. Regardless of which approach we had taken, we would have discovered two novel mutations in the *ANGPTL3* gene in affected family members—S17X (serine in position 17 altered to a stop codon) and E129X (glutamic acid in position 129 altered to a stop codon) (2). In both cases, these are nonsense mutations that would cause severe truncation of the protein (which has a total of 460 amino acids) and presumably a complete loss of its function.

Having discovered the putative causal gene in this FCH family, a natural question is whether other unrelated FCH individuals have mutations in *ANGPTL3*. An actual search of this kind in an Italian population discovered three different missense mutations—F295L, G56V, and R332Q—along with nonsense and frameshift mutations (4). To assess each of the missense mutations, we can use the PolyPhen-2 server (genetics.bwh.harvard.edu/pph2). On the home page, we can enter the protein identifier (ANGPTL3), the position of the altered amino acid (295), and mark the amino acid substitution (AA1 = F, AA2 = L). After clicking the **Submit Query** button and a short wait, the results are reported: **Probably Damaging**. Testing each of the other two mutations returns the same result. Further inspection of data provided on the results pages shows that in each case, the original amino acid is highly conserved across species, arguing that a nonconservative amino acid substitution is likely to disrupt protein function. Upon entering these same mutations into another functional prediction algorithm server (eg, MutationTaster), one receives similar results (in the case of MutationTaster, **disease causing**). The discovery of different, likely function-disrupting mutations in the *ANGPTL3* gene in unrelated FCH individuals argues strongly that *ANGPTL3* is a causal gene for FCH. The case is further bolstered when considering that an independently conducted GWAS on LDL-C and TG, as described earlier in this chapter, identified an SNP in close proximity to *ANGPTL3*. Hence, we have confirmed a new disease gene!

REFERENCES

1. Pulai JI, Neuman RJ, Groenewegen AW, et al. Genetic heterogeneity in familial hypobetalipoproteinemia: linkage and non-linkage to the apoB gene in Caucasian families. *Am J Med Genet.* 1998;76(1):79–86.
2. Musunuru K, Pirruccello JP, Do R, et al. Exome sequencing, ANGPTL3 mutations, and familial combined hypolipidemia. *N Engl J Med.* 2010;363(23):2220–2227.
3. Teslovich TM, Musunuru K, Smith AV, et al. Biological, clinical and population relevance of 95 loci for blood lipids. *Nature.* 2010;466(7307):707–713.
4. Noto D, Cefalù AB, Valenti V, et al. Prevalence of ANGPTL3 and APOB gene mutations in subjects with combined hypolipidemia. *Arterioscler Thromb Vasc Biol.* 2012;32(3):805–809.

5

C H A P T E R

Epigenetics

Lifang Hou

TAKE HOME POINTS

1. Rapidly growing evidence has linked cardiovascular disease (CVD) with epigenetic variations, including changes in DNA methylation, microRNAs, and histone modifications. In humans, most studies have been conducted on systemic epigenetic markers using white blood cells (WBCs). DNA methylation is the most studied epigenetic marker. More than 900 miRNAs have been reported in humans. The aberrant expression of miRNAs has been linked to various CVDs. Post-translational histone modifications have emerged in recent years as key regulators for gene expression. Study of histone modification in relation to CVD is largely limited.

2. Epigenetic markers are related by an intricate series of interactions that generate a self-reinforcing cycle of epigenetic events that control gene expression. Future studies that include comprehensive investigations of multiple epigenetic mechanisms might help elucidate the timing and participation of DNA methylation, histone modifications, and miRNAs to determine environmental effects on disease development.

3. The epigenotype is dynamic and varies over time as a function of environmental exposure, aging, and diseases and other factors. Quantifying WBC epigenetic changes in longitudinally collected biological samples may offer a convenient "molecular archive" of CVD risk burden as well as mechanistic insights into the translation of these risk factors to subclinical and clinical CVD.

4. Epigenetic changes are reversible. This feature offers novel insights to develop new preventive and therapeutic strategies.

Cardiovascular disease (CVD) represents a substantial disease burden, accounting for 32% of the total mortality in the United States (1). CVD is a complex trait that can be affected by genetic and nongenetic factors. Previous studies have reported the role of genetic variants, mainly single nucleotide polymorphisms (SNPs), in determining individuals' susceptibility to CVD (2,3). However, genetic factors are thought to explain only a small portion of CVD risk. In recent years, emerging evidence indicates that epigenetic factors may play an integral role in the development of chronic diseases. Although epigenetics has been mainly evaluated in cancer (4), a growing number of studies suggest the involvement of epigenetic-mechanism-regulated gene expression changes in CVD development and progression (5–7). Dynamic chromatin remodeling is required for the initial steps in gene transcription, which can be achieved by altering the accessibility of gene promoters and regulatory regions (8). Epigenetic factors, including DNA methylation, microRNAs (miRNAs), and histone modifications (Figure 5.1) participate in these regulatory processes, thus controlling gene expression (9,10).

Epigenetic factors, unlike genetic factors, are thought to be tissue specific. However, it is often impractical to study diseased tissues (eg, heart or vessels) for CVD epigenetic studies. Surrogate tissues, such as white blood cells (WBCs), have been used and represent a biologically relevant and easily accessible tissue for identifying epigenetic biomarkers for CVD. Another advantage of using surrogate tissues is that these biomarkers can be measured repeatedly during the disease development and progression (11). Currently,

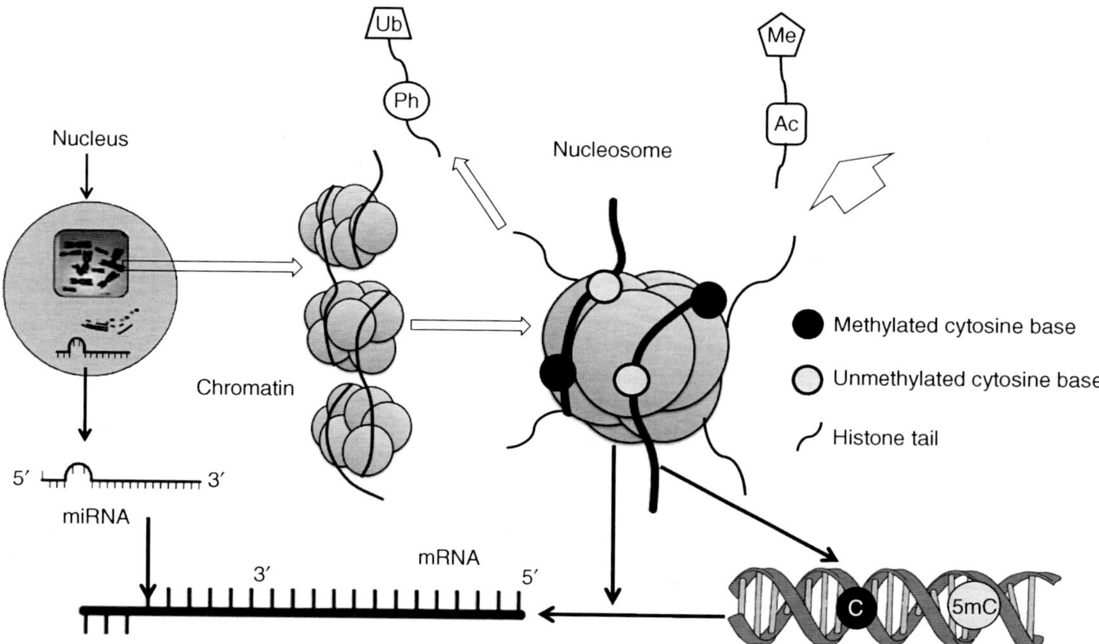

FIGURE 5.1 Transcriptional regulation at the epigenetic level. Epigenetic mechanisms, including DNA methylation, histone modifications and miRNAs, regulate chromatin compaction and gene expression. DNA methylation at CpG sites usually suppresses gene expression. Histones are globular proteins that undergo posttranslational modifications, such as Ac methylation and phosphorylation, thus influencing chromatin structure and gene expression. Active genes are usually characterized by low DNA methylation and highly acetylated chromatin configuration that allow access to transcription factors. MiRNAs are a set of small, nonprotein-coding RNAs that negatively regulate expression of target genes at the posttranscriptional level by binding to 30-untranslated regions of target mRNAs.

most CVD epigenetic studies have been conducted using WBC DNA. Cardiovascular risk factors cause persistent, low-grade systemic inflammation and subsequent inflammatory responses, including increased WBC count and the release of inflammatory mediators (12,13). WBCs are involved in atherosclerosis by participating in the formation of atherosclerotic plaques, the production of inflammation, oxidative and proteolytic damage to endothelial cells, and hypercoagulability (13). Epidemiology studies have consistently shown that the total and subtype of WBC counts independently associate with the future cardiovascular events in those with and without pre-existing CVD (13–21). WBC-derived circulating cytokines can affect the biological function of vascular endothelium as well as adipocytes and hepatocytes, leading to the generation of a spectrum of proatherogenic changes that include insulin resistance (IR), dyslipidemia, prothrombotic effects, pro-oxidative stress, and endothelial dysfunction (22–26). Thus, WBCs play a pivotal role in CVD and represents an important and accessible surrogate tissue for studying epigenetic biomarkers of CVD.

DNA METHYLATION

DNA methylation, including global methylation and gene-specific methylation at promoter CpG islands, is a well-defined reversible epigenetic mechanism that plays a key role in a wide variety of fundamental biological processes including gene expression and maintenance of genomic stability (27). In mammals, methylation involves addition of methyl groups to cytosine to form 5-methyl cytosine (5mC), and whole genomic DNA methylation derives from the overall level of 5mC in the genome. DNA methylation patterns are largely established in utero or during early life and stably maintained during later development, but can be changed in response to endogenous and exogenous exposure factors (28,29).

Evidence for DNA methylation in surrogate tissues and CVD has begun to accumulate. To date, a number of human studies have examined DNA methylation, including global and gene-specific DNA methylation in relation to CVD and CVD-related factors summarized in Table 5.1.

Global DNA Methylation and CVD

Global methylation is the most studied DNA methylation marker in relation to CVD. However, the results are inconsistent with most studies reporting global hypomethylation in CVD patients or individuals with risk factors for CVD, and some studies reporting global hypermethylation in CVD patients or individuals with risk factors for CVD.

TABLE 5.1 DNA Methylation, CVD, and CVD-Related Factors

CVD and CVD Markers	Study Design	Tissue Type	DNA Methylation Measurement	Association
Global methylation				
Vascular disease (30)	Case–control	WBC	SAH, SAM/SAH	Negative
CHD (31)	Case–control (mothers)	WBC	WBC	Negative
CHD (32)	Case–control (mothers)	WBC	LINE-1	Negative
CHD (33)	Case–control (mothers)	WBC	SAH, SAM/SAH	Negative
CHD (34)	Case–control (children)	WBC	WBC	Negative
IHD and stroke (35)	Cohort study	WBC	LINE-1	Negative
VCAM-1 (36)	Cohort study	WBC	LINE-1	Negative
Fibrinogen, IL-6 (37)	Cohort study	WBC	5-methylcytosine	Negative
LDL and HDL (38)	Cohort study	WBC	LINE-1	Negative
Obesity (39)	Cohort study	PBMC	LINE-1	Negative
Birth weight (40)	Cohort study	Cord blood	LINE-1	Negative
Hypertension (41)	Case–control	WBC	TLC analysis for 5mC level	Negative
Inflammation markers (77)	Case–control	WBC	*Hpa*II/*Msp*I ratio in LUMA	Positive
CAD (78)	Case–control	WBC	*Hpa*II/*Msp*I	Positive
Insulin resistance (IR) (79)	Twin study	WBC	Alu	Positive
CVD and risk factors (80)	Prospective study	WBC	ALU and AS	Positive
Gene specific				
MI (91)	Case–control	WBC	Mass spectrometry	*GNASAS* ↑, *INSIGF* ↑
Major depression (92)	Case–control	WBC	PCR	*ACE* ↑

Abbreviations: AdoHcy/SAH, *S*-adenosylhomocysteine; AdoMet/SAM, *S*-adenosylmethionine; AS, 2 satellite 2 repetitive element; CHD, congenital heart defect; CVD, cardiovascular disease; ICAM-1, intercellular adhesion molecule-1; IHD, ischemic heart disease; LUMA, luminometric methylation assay; PBMC, peripheral blood mononuclear cells; tHcy, total homocysteine; TLC, thin layer chromatography; VCAM-1, vascular cell adhesion molecule-1; WBC, white blood cells.

Global DNA Hypomethylation and CVD

Global hypomethylation in WBC has been associated with CVD (30–35) and CVD-related factors, including inflammation markers (36,37), obesity (38–40), and hypertension (41).

Atherosclerosis. S-adenosylmethionine (SAM), the methyl group donor, is converted to S-adenosylhomocysteine (SAH), and the ratio of these two measures (SAM/SAH) has been applied as an indicator of cellular methylation capacity. In an early study with 17 patients with atherosclerotic vascular disease and 15 controls, Castro et al measured plasma total homocysteine (tHcy), SAH, and SAM, and assessed global DNA methylation by a cytosine extension assay (30). It was found that atherosclerosis patients had significantly higher tHcy and SAH concentrations, lower SAM/SAH ratios, and lower global DNA methylation, indicating that the pathogenic role of hyperhomocysteinemia in vascular disease might be caused by SAM accumulation and global DNA hypomethylation. In the Castro et al study, DNA methylation is correlated with tHcy and SAH, but not with SAM or SAM/SAH, which is consistent with previous studies in humans (42) and animals (43), suggesting that SAH is a more reliable biomarker of methylation status.

Congenital Heart Defects. Congenital heart defects (CHDs) account for more than 1 million affected newborns worldwide; however, the etiology remains largely unknown (44). To date, four studies have observed the association of global hypomethylation with CHD risks in mothers whose pregnancies were affected by CHD (31,32), mothers of CHD children (33), and CHD children (34).

In a case–control study, Hobbs et al observed that mothers whose pregnancies were affected by CHD had higher concentration of homocysteine and SAH, and lower concentration of SAM and methionine compared with controls (31). It is explained that homocysteine may have an embryotoxic effect possible through oxidative stress and SAH accumulation, thus leading to inhibition of DNA methyltransferase (DNMT) reactions, DNA hypomethylation, and altered gene expression (45). However, the folic acid concentrations were similar between cases and controls, indicating that cases were either more genetically susceptible to increases in homocysteine and SAH, or were more likely to have reduced activity of relevant enzymes (31). Thus, further studies are needed to understand the association of CHD with genetic variants that may increase homocysteine.

Consistently, Chowdhury et al reported that LINE-1 DNA methylation was significantly lower in mothers of CHD-affected pregnancies than controls (32). It was speculated that the increases in cellular proliferation and carbon metabolism during pregnancy, as well as increased demand for methyl groups during embryo development may contribute together to the deleterious changes resulting in CHDs and other birth defects (32). It has been shown that CHD shared similar pathophysiological pathways as neural tube defects (NTDs), possibly due to their common neural epithelial cell origin (46). An increased risk of NTDs has been found to be associated with a decrease in LINE-1 (47).

Van Driel et al found higher tHcy, SAM, and lower SAM/SAH ratio in mothers of children with CHD than mothers of nonmalformed children. The result indicates that a maternal status of hypomethylation is associated with an increased risk of having a child with CHD and Down syndrome (33). Similarly, in very young children with CHD, Obermann-Borst et al found higher tHcy and lower SAM/SAH ratio compared with controls (34). In particular, the subgroup of children with Down syndrome, the most common chromosomal abnormality associated CHDs, showed a significantly lower SAM/SAH ratio, which is in line with previous findings in mothers (33). Assuming that the embryo was exposed to the maternal global hypomethylation status, it is possible that the global hypomethylation state in children was caused by heritability and intrauterine adaptations with consequences for cardiogenesis (48,49). Additionally, differences in vitamin intake, medication use (50), and physical activity (51) may all contribute to the low methylation status in CHD children. Taken together, global hypomethylation may play a key role in the development of CHD.

Ischemic Heart Disease and Stroke. In a study of 712 elderly individuals from the Boston-Area Normative Aging Study, Baccarelli et al observed that individuals with prevalent ischemic heart disease (IHD) and stroke exhibited lower LINE-1 methylation. In longitudinal analyses, persons with lower LINE-1 methylation were at higher risk for incident IHD and ischemic stroke, and total CVD-related mortality (52). The authors suggested that plaque formation in advanced atherosclerotic lesions would parallel the global hypomethylation, which is characterized by proliferating cells with clonal origin accompanied by systemic inflammation (53,54). As atherosclerotic

lesions often proceed to CVD, blood DNA hypomethylation might be a good marker to identify persons at risk for cardiovascular events. In addition, hypomethylated DNA has been shown to be prone to aberrant gene expression patterns in vascular tissues, leading to vascular fibrocellular lesions (55). Further research is needed to examine whether blood LINE-1 hypomethylation is associated with aberrant methylation in such genes in diseased tissues. In this study, LINE-1 methylation showed very few associations with several established CVD risk factors, and no association with plasma homocysteine that has been inversely related with global or LINE-1 methylation in previous studies (30,56). The older age might have contributed to discrepancies with other studies.

Inflammatory Markers. Growing evidence indicates that atherosclerosis is largely an inflammatory disease and circulating inflammatory factors might serve as predictors of CVD (57). In a study consisting of 593 white males, Baccarelli et al examined whether LINE-1 hypomethylation is associated with three inflammatory markers that have been identified as independent predictors of CVD, including vascular cell adhesion molecule-1 (VCAM-1), intracellular adhesion molecule-1 (ICAM-1), and C-reactive protein (CRP) (36). It was found that VCAM-1 increased progressively in association with LINE-1 hypomethylation. Moreover, the association was significant in individuals without IHD or stroke, but not in those with prevalent disease, further indicating that it may represent an early event in the etiology of CVD. Inflammation involves recruitment of inflammatory cells from blood and migration mediated by adhesion molecules, such as VCAM-1 (58), which is rapidly expressed in pro-atherosclerotic conditions and plays a critical role in atherosclerosis development (59,60). However, no association of LINE-1 with ICAM-1 or CRP was observed in this study (36). It is explained that DNA sequence of the VCAM-1 gene region contains eight LINE-1 repetitive elements, while ICAM-1 does not contain any LINE-1 sequence, and CRP contains only one (36). In addition, it remains unclear whether the LINE-1 hypomethylation in blood correlates with its level in vascular tissues. Further studies are needed in both blood DNA and vascular tissues to better understand the mechanisms linking LINE-1 hypomethylation with various inflammatory markers.

In a study of Psychological, Social, and Biological Determinants of Ill Health Cohort, the association of socioeconomic status and inflammation markers with global DNA methylation was examined (37). Global hypomethylation was associated with socioeconomic deprivation, and two inflammation biomarkers, IL-6 and fibrinogen (37). Chronic inflammation may link to a wide range of pathologies (61), which could be influenced by epigenetic reprogramming in the etiology of chronic inflammatory diseases (62). Another explanation at a mechanistic level is that inflammation has a direct impact on the activity of enzymes involved in the epigenetic status, which was supported by the regulation of IL-6 on human DNMT (63).

Obesity. Previous studies have examined the association of global hypomethylation with obesity and related biomarkers or factors, including low-density lipoprotein (LDL) and high-density lipoprotein (HDL) (38), body mass index (BMI) in women of child-bearing age (39), and birth weight in newborns and infants (40).

American Samoans have a higher prevalence of obesity, which has led to rapid rises in obesity-related diseases, such as CVD (64). Cash et al examined the relationship between LINE-1 methylation and CVD risk factors in blood DNA of 355 Samoan islanders from Samoan Family Study of Overweight and Diabetes, finding that high LDL and low HDL were associated with lower LINE-1 level (38). Increased levels of LDL and decreased levels of HDL are known risk factors for atherosclerosis (65), which may result in various cardiovascular morbidities and related mortalities (66). Altered DNA methylation profiles were observed in peripheral blood mononuclear cells (PBMCs) from mice lacking apolipoprotein E with higher LDL and lower HDL levels, indicating the influence on LINE-1 by lipoproteins (67).

Epigenetic alterations during pregnancy have emerged as important factors for obesity-related diseases in children as well as in adults. Piyathilake et al examined the association of global methylation with body weight in women of child-bearing age, finding women with higher BMI, waist circumference, or percentage of body fat have lower LINE-1 methylation (68). In addition, women with higher plasma folate concentrations were less likely to have lower LINE-1 methylation, suggesting the beneficial effects of folate status on obesity-related health outcomes (68).

Low or higher birth weight has been related to future risk of CVD and CVD markers, such as hypertension and diabetes (69–71). However the mechanisms underlying these associations are poorly understood. Fetal epigenetic programming may reset the growth hormone/insulin-like growth factor axis, allowing adjustment to a range of intrauterine conditions. Adverse conditions beyond this range may result in reprogramming leading to future chronic diseases. Recent studies have indicated that epigenetic mechanisms may contribute to fetal programming (72,73). In the Epigenetic Birth Cohort at Brigham and Women's Hospital at Harvard Medical School, Boston, the largest birth weight and global methylation study conducted to date, it was reported that newborns with low or high birth weight had lower LINE-1 methylation levels in cord blood, further suggesting fetal epigenetic programming may play a role in adulthood CVD risk (40).

Hypertension. Hypertension is a common disease of the cardiovascular system and accounts for 13% of the total mortality in the world (74). Smolarek et al examined the association of hypertension with global DNA methylation assessed by thin layer chromatography analysis, finding that the mean level of 5-methylcytosine was higher in the healthy subjects than in the patients with stage 1 and stage 2 hypertension, and all patients combined (41). Similarly, higher levels of 5-methylcytosine were found in patients with stage 1 compared to stage 2 hypertension (41). Aberrant promoter methylation of multiple genes involved in blood pressure regulation may therefore participate in blood pressure changes (75,76). Although there is no report on the influence of the antihypertensive drugs with 5-methylcytosine levels, the influence of antihypertensive drugs on global methylation level should be considered in future studies.

A challenge common to these global hypomethylation–CVD studies is that they have retrospective designs, wherein samples were collected after the disease diagnosis. Therefore, the results from these studies reporting low global methylation levels could represent reverse causality.

Global Hypermethylation and CVD

Although most studies observed global DNA hypomethylation, global DNA hypermethylation has also been associated with CVD and CVD risk factors in case–control studies (77,78), twin study (79), and one prospective study (80).

Coronary Artery Disease. Methylation alterations with age play a triggering role in age-related diseases, including atherosclerosis (81). DNA global methylation is significantly higher in coronary artery disease (CAD) elder patients compared with controls (78). In these CAD patients with the mean age of 54, age-induced global DNA hypomethylation is offset by promoter-specific hypermethylation, resulting in increased overall DNA methylation in CAD patients, especially in the older age group (78).

Inflammatory Markers. Stenvinkel et al found global DNA hypermethylation is associated with inflammation markers, including CRP, and increased mortality in chronic kidney disease (CKD) stage 5 patients (77). It was hypothesized that chronic inflammation may cause aberrant DNA methylation, which may accelerate atherosclerosis by upregulation of atherosclerosis-susceptible and downregulation of atherosclerosis-protective genes. This hypothesis was supported by the finding of methylation-associated inactivation of the estrogen receptor alpha gene (82) and monocarboxylate transport (*MCT*) gene (83) in human cardiovascular tissue.

Other CVD-Related Factors. In a monozygotic twin study, a higher Alu level is associated with increased risk of IR (79), a characteristic feature of CVD (84). One possible reason is that alterations in DNA methylation as reflected by Alu would promote genomic instability and lead to inflammation, which are involved in IR and its complications (85). Another possible reason is that the association between global methylation and IR may be attributed, at least partly, to telomere shortening, which could be regulated epigenetically (86), and may play a role in genomic instability and IR (87).

In the population-based prospective Singapore Chinese cohort study, Kim et al observed higher methylation levels associated with the prevalence of CVD and CVD risk factors including hypertension, diabetes, and BMI (80). This is the first report on association of CVD prevalence and risk factors with global DNA methylation assessed by Alu in a relatively lean and elderly population. More studies are needed in other Asians, as well as more distinct ethnic groups, including wider ranges of BMI to confirm the association.

Gene-Specific DNA Methylation and CVD

To date, limited studies have examined gene-specific methylation in relation with CVD in heart tissues from animals and humans, showing hypo- or hypermethylation in specific genes that are involved in CVD development (88–90). Only two previous human studies have investigated the association of CVD with gene-specific methylation in WBC DNA samples (91,92).

Talens et al recently investigated the association between DNA methylation and myocardial infarction (MI) in WBC DNA in six genes previously implicated in CVD and reported that DNA methylation level at *GNASAS* and *INS* was higher in cases than controls, suggesting methylation at specific loci in these two genes may be an epigenetic marker for the risk of CAD (91). *INS* and *GNASAS* are regulators of fetal growth, and higher DNA methylation at *INS* and *GNASAS* is associated with lower expression of the insulin gene in pancreatic beta cells and higher expression of Galphas in adipose and pituitary tissue (93–95). These data suggest that WBC methylation in genes involved in CVD may be directly involved in CAD (91).

Major depression (MD) represents an important CVD marker, and CVD also in turn influences MD (96). The angiotensin-converting enzyme (ACE) has been suggested to play a critical role in regulating this bidirectional relationship (97). Zill et al reported *ACE* gene promoter hypermethylation in the blood of MD patients compared with controls, suggesting that *ACE* promoter DNA hypermethylation may be a pathogenic factor for the bidirectional relationship between MD and CVD (92).

MicroRNAs

MicroRNAs (miRNAs) are short single-stranded ribonucleic acids (RNAs) of nearly 20 to 24 nucleotides in length that are transcribed from DNA but not translated into proteins. MiRNAs regulate expression of target genes at the posttranscriptional level by binding to three untranslated regions of target mRNAs (98). Each mature miRNA is partially complementary to multiple target mRNAs and directs the RNA-induced silencing complex (RISC) to identify the target mRNAs for inactivation (99). MiRNAs are initially transcribed as longer primary transcripts (pri-miRNAs) and processed first by the RNase enzyme complex, and then by Dicer, leading to incorporation of a single strand into the RISC. MiRNAs guide RISC to interact with mRNAs and determine posttranscriptional repression. Compared with other mechanisms involved in gene expression, miRNAs act directly before protein synthesis and may be more directly involved in fine-tuning of gene expression or quantitative regulation (100,101). Moreover, miRNAs also play key roles in modifying chromatin structure and participating in the maintenance of genome stability (102). MiRNAs can regulate various physiological and pathological processes, such as cell growth, differentiation, proliferation, apoptosis, and metabolism (98,103). More than 900 miRNAs have been reported in humans by using computational and experimental methods in miRNA-related public databases. The aberrant expression of miRNAs has been linked to various CVDs (104,105). We summarize the epidemiology studies of circulating miRNA and CVD and CVD biomarkers as follows (Table 5.2).

MiRNAs and CVD

Myocardial Infarction. Acute MI is the world's leading cause of morbidity and mortality. A rapidly increasing number of studies have shown that circulating miRNAs are altered in MI, including miR-1, -133a, -133b, -208a, -208b, -328, and -499 (98,106–114).

Ai et al have observed significantly elevated miR-1 in the blood of MI patients compared with controls showing undetectable miR-1 (98). miR-1 is a muscle-specific miRNA, and is primarily expressed in cardiac and skeletal muscles. Therefore, miR-1 in

TABLE 5.2 miRNA, CVD, and CVD Biomarkers

CVD and CVD Markers	Study Design	Tissue Type	Findings
AMI (98)	Case–control	Plasma	miR-1 ↑
AMI (106)	Case–control	Serum	miR-1 ↑
AMI (107)	Case–control	Serum	miR-1, -133 ↑
AMI (108)	Case–control	Plasma/whole blood	miR-133, -128 ↑
AMI (109)	Case–control	Plasma	miR-1, -133a, -499, -208a ↑
AMI (110)	Case–control	Plasma	miR-1, -133a, -133b, -499-5p ↑
AMI (111)	Case–control	Plasma	miR-208b, -499 ↑
AMI (112)	Cohort study	Plasma	MiR-133a, -208b associated with disease
AMI (113)	Case–control	Plasma	miR499 ↑
AMI (114)	Case–control	WBC	miR-30c, -145 ↑
HF (113)	Case–control	Plasma	miR499 not detected
HF (121)	Case–control	Plasma	miR-126 ↓
HF (122)	Case–control	Plasma	mi-499, -122 ↑
HF (113)	Case–control	Plasma	miR499 not detected
CAD (125)	Case–control	Plasma	133a, -208a ↑ miR-126, -17, 92a, -155, -145 ↓
CAD (126)	Case–control	Whole blood	miR-140 ↓ and -182 ↑
CAD (127)	Case–control	PBMC	miR-135a ↑ miR-147 ↓
Stroke (134)	Cohort study	WBC	Multiple miRNAs
Type 2 diabetes (135)	Case–control	Plasma	miR-126, -15a, -29b, -223 ↓ miR-28-3p ↑
Type 2 diabetes (136)	Case–control	Serum	miR-9, -29a, -30d, -34a, -124a, -146a, -375 ↑
Type 1 diabetes (143)	Case–control	Serum	miR-25 ↑
Hypertension (138)	Case–control	Plasma	miR-296-5p, let-7e, -hcmv-miR-UL112 ↑

Abbreviations: AMI, acute myocardial infarction; CAD, coronary artery disease; CVD, cardiovascular disease; HF, heart failure; PBMC, peripheral blood mononuclear cells; WBC, white blood cells.

the bloodstream is hypothesized to be released from necrotic myocytes, which was confirmed in a rat study by Cheng et al showing that cardiac miR-1 is released, and the released amount is associated with the extent of cardiac cell damage (106). Similarly, Cheng et al also observed increased miR-1 levels in serum from patients with MI, and the level of miR-1 may be related to MI size (106). Since miR-1 is the most abundant miRNA in the normal heart, the trace amount released into the circulating blood could serve as an early marker of MI (106).

Kuwabara found that serum levels of miR-1 and miR-133a were increased significantly in patients with MI (107). miR-133, a key regulator of vascular smooth muscle cell phenotypic switch, is expressed in smooth, skeletal, and cardiac muscles, and contributes to the progression of atherosclerosis (115,116). Similarly, markedly increased level of miR-133 in

plasma and whole blood was observed in MI patients compared with controls by Wang et al (108). In this study, miR-328 level is also increased in MI patients, and the level of miR-133 or miR-328 was correlated with cardiac troponin I, the myocardial damage marker, indicating that the miR-133 and miR-328 may also be released from damaged myocardium into circulating blood when heart injury occurs (108).

Wang et al detected significantly increased levels of miR-1, miR-133a, and miR-499 in the plasma from MI patients compared with controls in whom the levels of these miRNAs were present with very low abundance (109). Interestingly, miR-208a was absent in controls, but was readily detectible in 91% of the MI patients (109). However, in a study by D'Alessandra, the level of miR-208a was undetectable in MI patients, although increased levels of miR-1, miR-133a, and miR-499 in acute myocardial infarction (AMI) patients were observed (110). It was explained that the different time points for blood collection, as well as low level of miR-208a in the blood even after MI, may all contribute to the discrepancy of the results (105). Corsten et al reported a markedly increased level of miR-208b, another miR-208 family member, in the plasma of MI patients compared with controls (111). Widera et al confirmed the association between miR-208b with MI in a large study with 444 acute coronary syndrome (ACS) patients, and further indicated that miR-208b levels were predictive of 6-month mortality (112). Taken together, miR-208, a miRNA mainly expressed in the heart that is involved in cardiomyocyte hypertrophy, fibrosis, and regulation of other cardiac muscle gene expressions and functions, might serve as a reliable biomarker for MI diagnosis (117,118). Corsten et al also found increased level of miR-499 (111), which is in line with a previous finding reporting an increased level of miR-499 (113).

Recently, Meder et al have assessed the whole-genome miRNA expression in blood, finding 20 miRNAs as a unique signature that predicts MI, including miR-30c and miR-145 (114). In addition, miR-145 and miR-30c levels significantly correlate with troponin levels, the biomarker for MI diagnosis (114). Furthermore, miRNA profiles of MI patients showed characteristics of miRNA signature when the troponin proteins are still negative, suggesting that miRNA signatures might improve biomarker-based early diagnosis of MI.

Heart Failure. Recently, several studies have shown that miRNAs modulate cardiac functions in heart development and also in heart failure (HF) (119). Although many miRNAs are tissue specific, three studies have demonstrated that certain circulating miRNAs, including miR-423-5p, -126, -122, -499, may be biomarkers in HF patients (120–122).

In an array study, the miR423-5p level has been demonstrated to be upregulated in heart tissues from HF patients (123). Similarly, Tijsen et al have identified six miRNAs that are elevated in patients with HF, among which miR423-5p is most strongly related to the clinical diagnosis of HF, and the miR423-5p level is related to NT-proBNP, a marker of HF (120). The results are less likely to be influenced by major cardiac cell loss because individuals with recent cardiac ischemia or infarction were excluded. Therefore, miR-423-5p may serve as a possible biomarker specific for HF. However, it remains unclear what cell type is responsible for the release of miR-423-5p into plasma, and whether other mechanisms exist to elevate miR-423-5p in blood (120).

MiR-126 level in plasma was found to be decreased with age and downregulated by poor conditions related to HF (121). Since miR-126 is enriched in endothelial cells and reflects vascular endothelial cell conditions, the observed decreased level of miR-126 in HF patients is surprising. The reduced metabolic activity and renewal of endothelial cells caused by perfusion defect and cellular aging, as well as degradation of circulating miRNAs caused by the endothelial activation, may all contribute to the decreased miR-126 level in HF patients. Further studies are needed to determine whether miR-126 is a promising new biomarker for various vascular diseases including HF. In HF patients, increased levels of miR-122 and -499 were observed (111). miR-122 is a liver-specific miR that is known to correlate with hepatic damage (122); the observed association in HF may reflect hepatic venous congestion, a common complication of HF (124).

CAD. Previous studies have also reported a set of circulating miRNAs as biomarkers for CAD, including miR-126, -17, 92a, -155, -145, -140, -182, -135, -147 (125–127).

Fichtlscherer et al investigated the plasma miRNA levels in patients with CAD and demonstrated significantly decreased levels of endothelial-expressed miRNAs (ie, miR-126, -17, 92a), inflammation-associated miRNA-155, and smooth muscle–enriched miRNA-145 in CAD patients compared with controls (125). Because an elevation of the miRNAs level in CAD patients is expected due to the release of microparticles and remnants of apoptotic cells induced by endothelial activation, the observed decreased level is unexpected: plasma concentrations of miR-126 and -155 were negatively correlated with age, and CAD patients were 30 years older than the control subjects. miR-155 levels were also lower in males than females, suggesting that gender and age may all contribute to the observed decreased level of miR-155.

Taurino et al examined the miRNAs level in whole blood in CAD patients by microarray analysis and observed increased level of miR-140 and -182 (126). In a rat model of vascular injury, miR-92 modulated the process of neointima formation (128). In Taurino's study, expression of miR-92 increased after cardiac rehabilitation (126), which is consistent with the study by Fichtlscherer et al that found reduced miR-92 in CAD patients (125).

Hoekstra et al studied the miRNA signature in CAD patients using PBMCs: they found increased miR-135a and decreased miR-147 in CAD patients compared to controls (127). An miRNA/target gene/biological function analysis suggested that these changes might be associated with a change in intracellular cadherin/Wnt signaling (127). This observation is further supported by their recent finding of enhanced expression of Dickkopf-1, an endogenous inhibitor of Wnt signaling in CAD, suggesting a link between Wnt signaling and atherogenesis (129).

miR-135a expression in lymphoma cells has been associated with decreased JAK2 expression, a tyrosine kinase involved in innate and adaptive immunity, and enhanced apoptosis rate, which may result in decreased cell growth (130,131). In macrophages, the inflammatory-associated miR-147 expression could repress pro-inflammatory cytokine secretion, such as IL-6 (132). Therefore, the observed changes in miRNA may reflect the altered immune status in the CAD patients, indicating that the PBMCs in CAD have an altered miRNA repertoire, and possibly shift to a more pro-inflammatory phenotype (127). Interestingly, in this study, it was also observed that patients with unstable angina pectoris could be distinguished from stable patients on the basis of relatively high levels of three circulating miRNAs, including miR-134, -198, and -370 (127). miR-370 expression levels in liver were increased in response to ischemia and correlate to the severity of ischemic injury in a recent mice study, suggesting that the high level of miR-370 in patients with ACSs could be a direct result of systemic ischemia (133). However, the exact function of these observed microRNAs in mononuclear cell remains to be elucidated.

Stroke. Stroke patients have differentially regulated 157 microRNAs (134). Some of these miRNAs have been shown to be involved in endothelial function and angiogenesis (ie, hsa-let-7f, miR-130a, -150, -17, -19a, -19b, -20a, -222, and -378), vascular remodeling (ie, miR-21, -126, -150), regulation of hematopoiesis (miR-223), and immune response (miR-20a, -17-92, -101, -150, -106a, -181a, and -223). Furthermore, the miRNA profile of small artery strokes showed a distinctly different pattern from that of the large artery strokes, indicating miRNA profiling could be used as a potential tool for clinicians to determine the stroke type (134).

MiRNAs and CVD-Related Markers

Although still limited, evidence is beginning to accumulate to link miRNA with CVD risk factors, such as diabetes (135–137) and hypertension (138).

Diabetes. Type 2 diabetes mellitus (T2D), characterized by chronic elevations of blood glucose levels and IR, is one of the major risk factors for CVD (139). Zampetaki et al examined the plasma miRNA signature for T2D in a large population-based cohort and observed reduced levels of miR-126, -15a, -29b, and -223 and a modest increase of miR-28-3p (135). Based on principal component analysis, miR-15a has the highest score. Although the function of miR-15 in T2D or CVD is largely unknown, it has been implicated in cell cycle control and apoptosis in cancer cells (140). Among these miRNAs, miR-126, one of the downregulated miRNAs in atherosclerotic CAD (125), was most consistently associated with T2D and was further validated in a prospective population of 822 individuals in the Bruneck study in Italy. MiR-126 was observed before the onset of overt T2D and was associated with vascular complications (135). MiR-126 is highly enriched in endothelial cells, and has been shown to play a pivotal role in maintaining endothelial homeostasis and vascular integrity by facilitating vascular endothelial growth factor signaling (141,142).

Kong et al analyzed seven diabetes-related miRNAs (miR-9, miR-29a, miR-30d, miR34a, miR-124a, miR146a, and miR375) in serum from newly diagnosed T2D patients, prediabetic and healthy subjects (136). The expression levels of seven miRNAs were all significantly elevated in T2D patients, while prediabetes and healthy subjects featured similar expression levels of these miRNAs, suggesting that expressing patterns of diabetes-related miRNAs had not changed dramatically in the prediabetic stage. Among the seven miRNAs, miR-34a was shown to be the strongest predictor of T2D. In addition, triglyceride also showed significantly higher stepwise trends from control to prediabetes and then T2D patients, suggesting that hypertriglyceridemia, one of the characteristic blood lipid abnormalities in diabetes, may favor the onset of diabetes via microRNA miR-34a (136).

Type 1 diabetes (T1D), an autoimmune disease resulting from an immune-mediated destruction of the insulin producing beta cells, has been associated with increased risk of CVD (137). Recently, Nielsen et al compared the expression level of miRNAs from new onset T1D children and controls, and related the miRNA expression levels to beta-cell function and glycemic control (143). Twelve differentially expressed miRNAs were identified, some of which are involved in the regulation of apoptosis (miR-181a, -24, -25, -210, and -26a) and beta-cell regulatory networks (miR-24, -148a, -200a, and -29a), and some have unknown functions in T1D (miR-152 and -30a-5p). Out of these studied, miR-25 has been associated with better glycemic control and residual beta-cell function, which is in line with a previous animal study showing miR-25 was significantly reduced in kidney from diabetic rats (144).

Hypertension. In a Chinese population, Li et al found 27 miRNAs were differentially expressed in plasma samples from hypertensive patients compared with controls (138). The expressions of selected miRNAs, including miR-296-5p, let-7e, and hcmv-miR-UL112 (a human cytomegalovirus [HCMV]-encoded miRNA), were validated independently, and further determined for an absolute fold change. Li et al also measured the HCMV virus titers, finding higher titers in the hypertension group than in control subjects, which is consistent with previous findings in an animal study (145). It was further demonstrated that the interferon regulatory factor 1 (IRF-1) is a direct target of hcmv-miR-UL112. IRF-1 has been shown to be involved in immunologic, inflammatory, and anti-infection responses (146).

HISTONE MODIFICATIONS

In humans, protection and packaging of the genetic material are largely performed by histone proteins, which also offer a mechanism for regulating DNA transcription, replication, and repair (147). Histones are nuclear globular proteins that can be covalently modified by acetylation, methylation, phosphorylation, glycosylation, sumoylation, ubiquitination, and ADP ribosylation (148,149), thus influencing chromatin structure and gene expression (150,151). Histone acetylation (Ac), with usually only a single acetyl group added to each amino acid residue, increases gene transcriptional activity (152–155), whereas histone methylation (Me), found as mono (Me), dimethyl (Me2), and trimethyl (Me3) group states (156), can inhibit or increase gene expression depending on the amino acid position that is modified (157–159). Histone modifications have emerged in recent years as key regulators for gene expression. The most common histone modifications that have been shown to be associated with various diseases are acetylation and methylation of lysine residues in the amino terminal of histone 3 (H3) and histone 4 (H4) (153,155), especially in cancer (152). However, the study of histone modification in relation to CVD is largely limited. In the following we discussed some experimental and human studies of histone modifications and CVD.

Caveolins are an integral part of lipid rafts, a subcompartment of the plasma membrane. Recent evidence demonstrated the cardioprotection role of caveolins against ischemic injury; however, the mechanism remains unclear (160,161). Following ischemia reperfusion injury in caveolin knockout mice, Das et al found that the increased histone methylation was significantly correlated with an increased level of histone methyltransferase G9a protein and increased level of histone deacetylase (HDAC) activity. Additionally, the decreased translocation of forkhead transcription factor (FOXO3a) to the nucleus and the reduced induction of the expression of SIRT-1 in the preconditioned heart was also observed. Furthermore, several biomarkers, such as reduced ventricular function, increased cardiomyocyte apoptosis, increased expression of janus kinase (JNK) and Bax and decreased expression of phospho-adenosine

monophosphate-activated protein kinase (AMPK), phospho-AKT, and B cell lymphoma-2 (Bcl-2) were also observed to confirm the protection function of Caveolin-1 (162). This study, for the first time, suggests that Caveolin-1 may induce cardioprotection through histone modification mechanism.

Previous studies of histone modifications studied in CVD have been focused on lysine residues in the amino terminal of histone 3 (H3) (163–165). Kaneda et al conducted a genome-wide investigation on the histone methylation profile for HF in both human and mice cardiac myocytes and found that trimethylation of histone H3 on lysine-4 (H3K4) or lysine-9 (H3K9) is markedly affected in cardiomyocytes (163). This finding is biologically plausible because the genes positioned close to the H3K4 or H3K9 marks were highly enriched in those that encode components of signaling pathways related to cardiac function, such as *RYR2*, *CACNA2D1*, and *CACNB2* that are directly implicated in the regulation of intracellular calcium concentration and muscle contraction (166,167). Stein et al also studied the role of H3K4 in a mouse model that could inducibly ablate PAX interacting protein 1 (PTIP), a key component of the H3K4me complex (164). The loss of H3K4 methylation in differentiated cardiomyocytes altered gene expression profiles, such as downregulation of Kv channel-interacting protein 2 (KCNIP2), a regulator of multiple facets of the electrical cardiac phenotype (168). It is still unclear whether these epigenetic changes are associated with disease states or contribute to the disease initiation and progression. Based on the findings in PTIP-null mice, it was hypothesized that the histone modifications can be sufficient to cause disease by altering the transcriptional profile of fully differentiated cardiomyocytes, which may ultimately sensitize the heart to arrhythmias (164). To understand the mechanism underlying the development of cardiac hypertrophy and failure through histone methylation, Zhang et al investigated the role of JMJD2 protein, a lysine trimethyl-specific histone demethylase that catalyzes the demethylation of trimethylated H3K9 and H3K36 in a mouse model (165). It was demonstrated that JMJD2A could mediate cardiac hypertrophy through regulation of its target FHL1, a key component of the mechanotransducer machinery in the heart (169), suggesting that epigenetic regulation of cardiac hypertrophy could be a possible novel mechanism.

CHALLENGES AND OPPORTUNITIES FOR HUMAN CVD EPIGENETIC STUDIES

The evidence on the role of epigenetics in CVD is rapidly accumulated. However, a variety of issues and challenges still remain to be addressed, such as investigation of epigenetic factors in different tissues, study of interaction between genetic and epigenetic factors, longitudinal study designs with multiple biological sample collections for studying the temporal relationships of CVD risk factor → epigenetic changes → disease outcome, ethnicity-specific epigenetic changes (170–172), and identification of novel markers using ever-changing high-throughput technologies.

What Are Suitable Surrogate Tissues for CVD Epigenetic Studies?

Because epigenetic changes are thought to be tissue specific, WBC DNA methylation markers in individuals with CVD may differ from DNA methylation patterns in local lesion tissues. However, obtaining atheroma tissue from asymptomatic community-based individuals is clearly both impractical and unethical. Thus, studies aiming at identifying WBC DNA epigenetic signatures at the systemic level may provide important information for the development of preventive measures of CVD. Additionally, because each subtype of WBC plays its unique role in CVD etiology, it has been proposed that studying epigenetic markers in specific WBC type may provide more specific information. Our group has previously shown that neutrophil and lymphocyte proportions modified the LINE-1 methylation level measured in total WBC DNA (173). Therefore, measurement of DNA methylation after separation of specific WBC subpopulations based on their functions should be considered in future DNA methylation–CVD studies.

However, most of the population studies have used total WBC as surrogate tissue for epigenetic studies for several reasons: (a) enabling study of the immune system in its true context, that is, with all cellular components represented and interacting together, providing a more realistic reflection of in vivo events; (b) eliminating the need for cell separation, a time-, labor-, and cost-intensive process requiring large volumes of blood, particularly when a specific relatively rare cell population, for example, monocyte, is required; (c) cell separation can also result in inadvertent cell activation (174); and (d) in large population-based studies, use of total WBC enables a larger number of samples to be studied in a time- and cost-effective manner.

What Are Suitable Study Designs and Approaches for CVD Epigenomic Studies?

The epigenotype is dynamic and varies over time as a function of environmental exposure, aging, and diseases and other factors (175–178). It has been shown that, in humans, both global and gene-specific methylation patterns change progressively with aging. A recent study demonstrated a longitudinal decline of WBC global DNA methylation levels with increasing age over 8-year follow-up (179). Another recent study examined WBC global and gene-specific DNA methylation using two time points' DNA samples with average 11 to 16 years apart in an Icelandic cohort and in a Utah cohort. Icelandic individuals showed greater than 10% methylation change over time (180). The family-based sample also showed significant intraindividual changes over time in both global and gene-specific methylation patterns (180). In addition, within-pair differences in DNA methylation were greater in older than in younger monozygotic twins (181). These data indicate that, in addition to age, the accumulation of DNA methylation aberrations with increased age may also, at least partially, be due to long-term exposure to risk factors, which mediates long-term cellular memory and determines the cellular repertoire of expressed genes, thus leading to subsequent diseases. However, because of reverse causation or confounding, cross-sectional assessment of DNA methylation alterations in relation to diseases does not permit any inference regarding direction of causation. Prospective investigations with longitudinally collected risk factor data and multiple DNA samples collected before disease phenotype identification are needed to determine the temporal relationships of risk factor exposure, DNA methylation, and chronic disease development. Quantifying WBC epigenetic changes in longitudinally collected biological samples may offer a convenient "molecular archive" of CVD risk burden as well as mechanistic insights into the translation of these risk factors to subclinical and clinical CVD.

The Role of Diet and Genetic Factors in Epigenetic Studies

Dietary intakes of folate and B vitamins modify DNA methylation by determining SAM availability, DNMT activity, and reactive oxygen species (ROS) persistence (30,182). Thus, it is crucial that intakes and plasma levels of folate, vitamin B_{12} and B_6, as well as plasma homocysteine levels, be considered while evaluating potential effects on DNA methylation. Additionally, methyl transfer to DNA is dependent on individual genetic characteristics such as the SNPs identified in the methylenetetrahydrofolate reductase (*MTHFR* C677CT) and cytoplasmic serine hydroxymethyltransferase (*cSHMT* C1420T) genes. MTHFR converts 5,10-methylene-tetrahydrofolate to 5-methyl-tetrahydrofolate, the primary circulatory form of folate (183). MTHFR thereby regulates the remethylation of homocysteine into methionine by 5-methyltetrahydrofolate. cSHMT is a related enzyme that reversibly converts serine and tetrahydrofolate

(THF) to glycine and 5,10-methylenetetrahydrofolate (184). In the Normative Aging Study (NAS) cohort, functional SNPs in *MTHFR* (C677CT) and cSHMT (C1420T) were recently found to interact with each other to determine highly increased risk of CVD (185).

The Potential Interactions Between Different Forms of Epigenetic Modification

Epigenetic markers are related by an intricate series of interactions that may generate a self-reinforcing cycle of epigenetic events directed to control gene expression (186). For example, histone deacetylation and methylation at specific amino acid residues contribute to the establishment of DNA methylation patterns. miRNA expression is controlled by DNA methylation in miRNA-encoding genes, and, in turn, miRNAs have been shown to modify DNA methylation (187). However, currently, most studies of CVD only examine one type of epigenetic mechanism. Future studies, including comprehensive investigations of multiple epigenetic mechanisms, would help elucidate their roles in CVD development.

Discovery of Novel Biomarkers

Although DNA methylation is the best-studied epigenetic mechanism, most current studies on gene promoter methylation have been focused on cancer-related genes, whereas studies of CVD-related genes remains limited to a small set of candidate methylation markers. However, the data has begun to accumulate. For example, with the development of next-generation sequencing (NGS) technologies, several novel miRNAs have been discovered (188,189). It was revealed that mature miRNAs have an extensive degree of variation at the terminal nucleotides, which might influence miRNA half-life, subcellular localization, and miRNA target specificity. NGS also revealed that the most abundant miRNA species do not match the sequence listed in public databases, such as miRBase, indicating that probe or primer design may not be suitable for all miRNAs (190). Therefore, application of NGS to identify other epigenetic factors, that is, methylation alterations and histone modifications in CVD studies, will shed new light on novel target discovery.

Can Epigenomic Markers Be Used for Prevention?

Numerous clinical and preclinical studies showed that most of the epigenetic changes are reversible, which offers novel insights to develop new preventive

and therapeutic strategies that might take advantage of molecules that modify the activities of epigenetic enzymes, such as DNMTs and HDACs, as well as of the growing field of RNAi therapeutics. In addition, future epigenomic research may provide information for developing preventive strategies, including exposure reduction, as well as pharmacological, dietary, or lifestyle interventions.

Further investigations, taking into account the above-mentioned and other factors, are needed to improve our understanding of epigenetic factors as a biomarker for CVD etiology and progression. Large prospective and longitudinal human studies are required to study these factors in surrogate tissues. Ultimately, study in easily accessible surrogate tissues could lead to the development of epigenetic-related measures for CVD prevention and therapy.

REFERENCES

1. Roger VL, Go AS, Lloyd-Jones DM, et al. Heart disease and stroke statistics—2012 update: a report from the American Heart Association. *Circulation.* 2012;125(1):e2–e220.
2. Esparragon FR, Companioni O, Bello MG, er al. Replication of relevant SNPs associated with cardiovascular disease susceptibility obtained from GWAs in a case-control study in a Canarian population. *Dis Markers.* 2012;32(4):231–239.
3. Thanassoulis G, Peloso GM, Pencina MJ, et al. A genetic risk score is associated with incident cardiovascular disease and coronary artery calcium: the Framingham Heart Study. *Circ Cardiovasc Genet.* 2012;5(1):113–121.
4. Jones PA, Laird PW. Cancer epigenetics comes of age. *Nat Genet.* 1999;21(2):163–167.
5. Baccarelli A, Rienstra M, Benjamin EJ. Cardiovascular epigenetics: basic concepts and results from animal and human studies. *Circ Cardiovasc Genet.* 2010; 3(6):567–573.
6. Handy DE, Castro R, Loscalzo J. Epigenetic modifications: basic mechanisms and role in cardiovascular disease. *Circulation.* 2011;123(19):2145–2156.
7. Turunen MP, Aavik E, Yla-Herttuala S. Epigenetics and atherosclerosis. *Biochim Biophys Acta.* 2009; 1790(9):886–891.
8. Vaissiere T, Sawan C, Herceg Z. Epigenetic interplay between histone modifications and DNA methylation in gene silencing. *Mutat Res.* 2008;659(1–2): 40–48.
9. Grewal SI, Moazed D. Heterochromatin and epigenetic control of gene expression. *Science.* 2003;301(5634): 798–802.
10. Reik W, Dean W, Walter J. Epigenetic reprogramming in mammalian development. *Science.* 2001; 293(5532):1089–1093.
11. Terry MB, Delgado-Cruzata L, Vin-Raviv N, et al. DNA methylation in white blood cells: association with risk factors in epidemiologic studies. *Epigenetics.* 2011;6(7):828–837.

12. McInnes IB. Rheumatoid arthritis. From bench to bedside. *Rheum Dis Clin North Am.* 2001;27(2):373–387.

13. Madjid M, Awan I, Willerson JT, Casscells SW. Leukocyte count and coronary heart disease: implications for risk assessment. *J Am Coll Cardiol.* 2004;44(10):1945–1956.

14. Giugliano G, Brevetti G, Lanero S, et al. Leukocyte count in peripheral arterial disease: a simple, reliable, inexpensive approach to cardiovascular risk prediction. *Atherosclerosis.* 2010;210(1):288–293.

15. Li C, Engstrom G, Hedblad B. Leukocyte count is associated with incidence of coronary events, but not with stroke: a prospective cohort study. *Atherosclerosis.* 2010;209(2):545–550.

16. Elkind MS, Sciacca RR, Boden-Albala B, et al. Relative elevation in baseline leukocyte count predicts first cerebral infarction. *Neurology.* 2005;64(12):2121–2125.

17. Margolis KL, Manson JE, Greenland P, et al. Leukocyte count as a predictor of cardiovascular events and mortality in postmenopausal women: the Women's Health Initiative Observational Study. *Arch Intern Med.* 2005;165(5):500–508.

18. Sweetnam PM, Thomas HF, Yarnell JW, et al. Total and differential leukocyte counts as predictors of ischemic heart disease: the Caerphilly and Speedwell studies. *Am J Epidemiol.* 1997;145(5):416–421.

19. Huang ZS, Chien KL, Yang CY, et al. Peripheral differential leukocyte counts and subsequent mortality from all diseases, cancers, and cardiovascular diseases in Taiwanese. *J Formos Med Assoc.* 2003;102(11):775–781.

20. Kawaguchi H, Mori T, Kawano T, et al. Band neutrophil count and the presence and severity of coronary atherosclerosis. *Am Heart J.* 1996;132(1, pt 1):9–12.

21. Olivares R, Ducimetiere P, Claude JR. Monocyte count: a risk factor for coronary heart disease? *Am J Epidemiol.* 1993;137(1):49–53.

22. Sattar N, McCarey DW, Capell H, McInnes IB. Explaining how "high-grade" systemic inflammation accelerates vascular risk in rheumatoid arthritis. *Circulation.* 2003;108(24):2957–2963.

23. Zhou X. CD4+ T cells in atherosclerosis. *Biomed Pharmacother.* 2003;57(7):287–291.

24. Laurat E, Poirier B, Tupin E, et al. In vivo downregulation of T helper cell 1 immune responses reduces atherogenesis in apolipoprotein E-knockout mice. *Circulation.* 2001;104(2):197–202.

25. Davenport P, Tipping PG. The role of interleukin-4 and interleukin-12 in the progression of atherosclerosis in apolipoprotein E-deficient mice. *Am J Pathol.* 2003;163(3):1117–1125.

26. Daugherty A, Rateri DL. T lymphocytes in atherosclerosis: the Yin-Yang of Th1 and Th2 influence on lesion formation. *Circ Res.* 2002;90(10):1039–1040.

27. Ohgane J, Yagi S, Shiota K. Epigenetics: the DNA methylation profile of tissue-dependent and differentially methylated regions in cells. *Placenta.* 2008;29(Suppl A):S29–S35.

28. Gluckman PD, Hanson MA, Cooper C, Thornburg KL. Effect of in utero and early-life conditions on adult health and disease. *N Engl J Med.* 2008;359(1):61–73.

29. Hsieh CL. Dynamics of DNA methylation pattern. *Curr Opin Genet Dev.* 2000;10(2):224–228.

30. Castro R, Rivera I, Struys EA, et al. Increased homocysteine and S-adenosylhomocysteine concentrations and DNA hypomethylation in vascular disease. *Clin Chem.* 2003;49(8):1292–1296.

31. Hobbs CA, Cleves MA, Melnyk S, et al. Congenital heart defects and abnormal maternal biomarkers of methionine and homocysteine metabolism. *Am J Clin Nutr.* 2005;81(1):147–153.

32. Chowdhury S, Cleves MA, MacLeod SL, et al. Maternal DNA hypomethylation and congenital heart defects. *Birth Defects Res A Clin Mol Teratol.* 2011;91(2):69–76.

33. van Driel LM, de Jonge R, Helbing WA, et al. Maternal global methylation status and risk of congenital heart diseases. *Obstet Gynecol.* 2008;112(2, Pt 1):277–283.

34. Obermann-Borst SA, van Driel LM, Helbing WA, et al. Congenital heart defects and biomarkers of methylation in children: a case-control study. *Eur J Clin Invest.* 2011;41(2):143–150.

35. Baccarelli A, Wright R, Bollati V, et al. Ischemic heart disease and stroke in relation to blood DNA methylation. *Epidemiology.* 2010;21(6):819–828.

36. Baccarelli A, Tarantini L, Wright RO, et al. Repetitive element DNA methylation and circulating endothelial and inflammation markers in the VA normative aging study. *Epigenetics.* 2010;5(3):222–228.

37. McGuinness D, McGlynn LM, Johnson PC, et al. Socio-economic status is associated with epigenetic differences in the pSoBid cohort. *Int J Epidemiol.* 2012;41(1):151–160.

38. Cash HL, McGarvey ST, Houseman EA, et al. Cardiovascular disease risk factors and DNA methylation at the LINE-1 repeat region in peripheral blood from Samoan Islanders. *Epigenetics.* 2011;6(10):1257–1264.

39. Piyathilake C, Badiga S, Johanning G, et al. Predictors and health consequences of epigenetic changes associated with excess body weight in women of child-bearing age. *Cancer Epidemiol Biomark.* 2011;20(4):719.

40. Michels KB, Harris HR, Barault L. Birthweight, maternal weight trajectories and global DNA methylation of LINE-1 repetitive elements. *PLoS One.* 2011;6(9):e25254.

41. Smolarek I, Wyszko E, Barciszewska AM, et al. Global DNA methylation changes in blood of patients with essential hypertension. *Med Sci Monit.* 2010;16(3):CR149–CR155.

42. Yi P, Melnyk S, Pogribna M, et al. Increase in plasma homocysteine associated with parallel increases in plasma S-adenosylhomocysteine and lymphocyte DNA hypomethylation. *J Biol Chem.* 2000;275(38):29318–29323.

43. Caudill MA, Wang JC, Melnyk S, et al. Intracellular S-adenosylhomocysteine concentrations predict global DNA hypomethylation in tissues of methyl-deficient cystathionine beta-synthase heterozygous mice. *J Nutr.* 2001;131(11):2811–2818.

44. *March of Dimes Birth Defects Foundation: The Hidden Toll of Dying and Disabled Children, in Global Report on Birth Defects. The Hidden Toll of Dying and Disabled Children.* White Plains, NY: March of Dimes Birth Defects Foundation; 2006:28.

45. Perna AF, Ingrosso D, Lombardi C, et al. Possible mechanisms of homocysteine toxicity. *Kidney Int Suppl.* 2003;63:S137–S140.

46. Rosenquist TH, Finnell RH. Genes, folate and homocysteine in embryonic development. *Proc Nutr Soc.* 2001;60(1):53–61.

47. Wang L, Wang F, Guan J, et al. Relation between hypomethylation of long interspersed nucleotide elements and risk of neural tube defects. *Am J Clin Nutr.* 2010;91(5):1359–1367.

48. Gluckman PD, Hanson MA, Cooper C, Thornburg KL. Effect of in utero and early-life conditions on adult health and disease. *N Engl J Med.* 2008. 359(1):61–73.

49. Waterland RA, Garza C. Potential mechanisms of metabolic imprinting that lead to chronic disease. *Am J Clin Nutr.* 1999;69(2):179–197.

50. Minet JC, Bissé E, Aebischer CP, et al. Assessment of vitamin B-12, folate, and vitamin B-6 status and relation to sulfur amino acid metabolism in neonates. *Am J Clin Nutr.* 2000;72(3):751–757.

51. Konig D, Bissé E, Deibert P, et al. Influence of training volume and acute physical exercise on the homocysteine levels in endurance-trained men: interactions with plasma folate and vitamin B12. *Ann Nutr Metab.* 2003;47(3–4):114–118.

52. Baccarelli A, Wright R, Bollati V, et al. Ischemic heart disease and stroke in relation to blood DNA methylation. *Epidemiology.* 2010;21(6):819–828.

53. Zaina S, Lindholm MW, Lund G. Nutrition and aberrant DNA methylation patterns in atherosclerosis: more than just hyperhomocysteinemia? *J Nutr.* 2005; 135(1):5–8.

54. Turunen MP, Aavik E, Yla-Herttuala S. Epigenetics and atherosclerosis. *Biochim Biophys Acta.* 2009; 1790(9):886–891.

55. Newman PE. Can reduced folic acid and vitamin B12 levels cause deficient DNA methylation producing mutations which initiate atherosclerosis? *Med Hypotheses.* 1999;53(5):421–424.

56. Jiang YD, Zhang JZ, Huang Y, et al. Homocysteine-mediated expression of SAHH, DNMTs, MBD2, and DNA hypomethylation potential pathogenic mechanism in VSMCs. *DNA Cell Biol.* 2007;26(8):603–611.

57. Ross R. Mechanisms of disease—atherosclerosis—an inflammatory disease. *N Engl J Med.* 1999;340(2): 115–126.

58. Blankenberg S, Barbaux S, Tiret L. Adhesion molecules and atherosclerosis. *Atherosclerosis.* 2003; 170(2):191–203.

59. Cybulsky MI, Iiyama K, Li H, et al. A major role for VCAM-1, but not ICAM-1, in early atherosclerosis. *J Clin Invest.* 2001;107(10):1255–1262.

60. Malik I, Danesh J, Whincup P, et al. Soluble adhesion molecules and prediction of coronary heart disease: a prospective study and meta-analysis. *Lancet.* 2001;358(9286):971–975.

61. Deans KA, Bezlyak V, Ford I, et al. Differences in atherosclerosis according to area level socioeconomic deprivation: cross sectional, population based study. *BMJ.* 2009;339:b4170.

62. Backdahl L, Bushell A, Beck S. Inflammatory signalling as mediator of epigenetic modulation in tissue-specific chronic inflammation. *Int J Biochem Cell Biol.* 2009;41(1):176–184.

63. Hodge DR, Cho E, Copeland TD, et al. IL-6 enhances the nuclear translocation of DNA cytosine-5-methyltransferase 1 (DNMT1) via phosphorylation of the nuclear localization sequence by the AKT kinase. *Cancer Genom Proteom.* 2007;4(6):387–398.

64. DiBello JR, McGarvey ST, Kraft P, et al. Dietary patterns are associated with metabolic syndrome in adult Samoans. *J Nutr.* 2009;139(10):1933–1943.

65. Krauss RM. Lipoprotein subfractions and cardiovascular disease risk. *Curr Opin Lipidol.* 2010;21(4): 305–311.

66. Lusis AJ. Atherosclerosis. *Nature.* 2000;407(6801): 233–241.

67. Lund G, Andersson L, Lauria M, et al. DNA methylation polymorphisms precede any histological sign of atherosclerosis in mice lacking apolipoprotein E. *J Biol Chem.* 2004;279(28):29147–29154.

68. Piyathilake C, Badiga S, Johanning G, et al. Predictors and health consequences of epigenetic changes associated with excess body weight in women of child-bearing age, in 35th Annual Meeting of the American Society of Preventive Oncology, Las Vegas, NV. *Cancer Epidemiol Biomarkers Prev.* 2011:719.

69. Barker DJ, Osmond C. Low birth weight and hypertension. *BMJ.* 1988;297(6641):134–135.

70. Barker DJ, Winter PD, Osmond C, et al. Weight in infancy and death from ischaemic heart disease. *Lancet.* 1989;2(8663):577–580.

71. Rich-Edwards JW, Colditz GA, Stampfer MJ, et al. Birthweight and the risk for type 2 diabetes mellitus in adult women. *Ann Intern Med.* 1999;130(4, Pt 1): 278–284.

72. Waterland RA, Michels KB. Epigenetic epidemiology of the developmental origins hypothesis. *Annu Rev Nutr.* 2007;27:363–388.

73. Hanson M, Godfrey KM, Lillycrop KA, et al. Developmental plasticity and developmental origins of non-communicable disease: theoretical considerations and epigenetic mechanisms. *Prog Biophys Mol Biol.* 2011;106(1):272–280.

74. Rodgers A. Chapter 4: Quantifying selected major risks to health. *World Health Rep.* 2002:47–97.

75. Attwood JT, Yung RL, Richardson BC. DNA methylation and the regulation of gene transcription. *Cell Mol Life Sci.* 2002;59(2):241–257.

76. Harrap SB, Davidson HR, Connor JM, et al. The angiotensin I converting enzyme gene and predisposition to high blood pressure. *Hypertension.* 1993;21(4): 455–460.

77. Stenvinkel P, Karimi M, Johansson S, et al. Impact of inflammation on epigenetic DNA methylation—a novel risk factor for cardiovascular disease? *J Intern Med.* 2007;261(5):488–499.

78. Sharma P, Kumar J, Garg G, et al. Detection of altered global DNA methylation in coronary artery disease patients. *DNA Cell Biol.* 2008;27(7):357–365.

79. Zhao J, Goldberg J, Bremner JD, Vaccarino V. Global DNA methylation is associated with insulin resistance: a monozygotic twin study. *Diabetes.* 2011;61(2):542–546.

80. Kim M, Long TI, Arakawa K, et al. DNA methylation as a biomarker for cardiovascular disease risk. *PLoS One.* 2010;5(3):e9692.

81. Issa JP. Age-related epigenetic changes and the immune system. *Clin Immunol.* 2003;109(1):103–108.

82. Post WS, Goldschmidt-Clermont PJ, Wilhide CC, et al. Methylation of the estrogen receptor gene is associated with aging and atherosclerosis in the cardiovascular system. *Cardiovasc Res.* 1999;43(4): 985–991.

83. Zhu S, Goldschmidt-Clermont PJ, Dong C. Inactivation of monocarboxylate transporter MCT3 by DNA methylation in atherosclerosis. *Circulation.* 2005;112(9):1353–1361.

84. Ginsberg HN. Insulin resistance and cardiovascular disease. *J Clin Invest.* 2000;106(4):453–458.

85. Shoelson SE, Lee J, Goldfine AB. Inflammation and insulin resistance. *J Clin Invest.* 2006;116(7):1793–1801.

86. Blasco MA. The epigenetic regulation of mammalian telomeres. *Nat Rev Genet.* 2007;8(4):299–309.

87. Sampson MJ, Winterbone MS, Hughes JC, et al. Monocyte telomere shortening and oxidative DNA damage in type 2 diabetes. *Diabetes Care.* 2006;29(2):283–289.

88. Laukkanen MO, Mannermaa S, Hiltunen MO, et al. Local hypomethylation in atherosclerosis found in rabbit EC-SOD gene. *Arterioscler Thromb Vasc Biol.* 1999;19(9):2171–2178.

89. Losordo DW, Kearney M, Kim EA, et al. Variable expression of the estrogen receptor in normal and atherosclerotic coronary arteries of premenopausal women. *Circulation.* 1994;89(4):1501–1510.

90. Movassagh M, Choy MK, Goddard M, et al. Differential DNA methylation correlates with differential expression of angiogenic factors in human heart failure. *PLoS One.* 2010;5(1):e8564.

91. Talens RP, Jukema JW, Trompet S, et al. Hypermethylation at loci sensitive to the prenatal environment is associated with increased incidence of myocardial infarction. *Int J Epidemiol.* 2012;41(1):106–115.

92. Zill P, Baghai TC, Schüle C, et al. DNA methylation analysis of the angiotensin converting enzyme (ACE) gene in major depression. *PLoS One.* 2012;7(7):e40479.

93. Kuroda A, Rauch TA, Todorov I, et al. Insulin gene expression is regulated by DNA methylation. *PLoS One.* 2009;4(9):e6953.

94. Williamson CM, Turner MD, Ball ST, et al. Identification of an imprinting control region affecting the expression of all transcripts in the Gnas cluster. *Nat Genet.* 2006;38(3):350–355.

95. Weinstein LS, Xie T, Zhang QH, Chen M. Studies of the regulation and function of the Gs alpha gene Gnas using gene targeting technology. *Pharmacol Ther.* 2007;115(2):271–291.

96. Grippo AJ, Johnson AK. Biological mechanisms in the relationship between depression and heart disease. *Neurosci Biobehav Rev.* 2002;26(8):941–962.

97. Phillips MI, de Olivcira EM. Brain renin angiotensin in disease. *J Mol Med (Berl).* 2008;86(6):715–722.

98. Ai J, Zhang R, Li Y, et al. Circulating microRNA-1 as a potential novel biomarker for acute myocardial infarction. *Biochem Biophys Res Commun.* 2010;391(1):73–77.

99. Matkovich SJ, Van Booven DJ, Eschenbacher WH, Dorn GW. RISC RNA sequencing for context-specific identification of in vivo microRNA targets. *Circ Res.* 2011;108(1):18–26.

100. Ying SY, Chang DC, Lin SL. The microRNA (miRNA): overview of the RNA genes that modulate gene function. *Mol Biotechnol.* 2008;38(3):257–268.

101. Bartel DP, Chen CZ. Micromanagers of gene expression: the potentially widespread influence of metazoan microRNAs. *Nat Rev Genet.* 2004;5(5):396–400.

102. Guil S, Esteller M. DNA methylomes, histone codes and miRNAs: tying it all together. *Int J Biochem Cell Biol.* 2009;41(1):87–95.

103. Backes C, Meese E, Lenhof HP, Keller A. A dictionary on microRNAs and their putative target pathways. *Nucleic Acids Res.* 2010;38(13):4476–4486.

104. Papageorgiou N, Tousoulis D, Androulakis E, et al. The role of microRNAs in cardiovascular disease. *Curr Med Chem.* 2012;19(16):2605–2610.

105. Creemers EE, Tijsen AJ, Pinto YM. Circulating microRNAs: novel biomarkers and extracellular communicators in cardiovascular disease? *Circ Res.* 2012;110(3):483–495.

106. Cheng Y, Tan N, Yang J, et al. A translational study of circulating cell-free microRNA-1 in acute myocardial infarction. *Clin Sci (Lond).* 2010;119(2):87–95.

107. Kuwabara Y, Ono K, Horie T, et al. Increased microRNA-1 and microRNA-133a levels in serum of patients with cardiovascular disease indicate myocardial damage. *Circ Cardiovasc Genet.* 2011;4(4):446–454.

108. Wang R, Li N, Zhang Y, et al. Circulating microRNAs are promising novel biomarkers of acute myocardial infarction. *Intern Med.* 2011;50(17):1789–1795.

109. Wang GK, Zhu JQ, Zhang JT, et al. Circulating microRNA: a novel potential biomarker for early diagnosis of acute myocardial infarction in humans. *Eur Heart J.* 2010;31(6):659–666.

110. D'Alessandra Y, Devanna P, Limana F, et al. Circulating microRNAs are new and sensitive biomarkers of myocardial infarction. *Eur Heart J.* 2010;31(22):2765–2773.

111. Corsten MF, Dennert R, Jochems S, et al. Circulating microRNA-208b and microRNA-499 reflect myocardial damage in cardiovascular disease. *Circ Cardiovasc Genet.* 2010;3(6):499–506.

112. Widera C, Gupta SK, Lorenzen JM, et al. Diagnostic and prognostic impact of six circulating microRNAs in acute coronary syndrome. *J Mol Cell Cardiol.* 2011;51(5):872–875.

113. Adachi T, Nakanishi M, Otsuka Y, et al. Plasma microRNA 499 as a biomarker of acute myocardial infarction. *Clin Chem.* 2010;56(7):1183–1185.

114. Meder B, Keller A, Vogel B, et al. MicroRNA signatures in total peripheral blood as novel biomarkers for acute myocardial infarction. *Basic Res Cardiol.* 2011;106(1):13–23.

115. Han M, Toli J, Abdellatif M. MicroRNAs in the cardiovascular system. *Curr Opin Cardiol.* 2011;26(3):181–189.

116. Torella D, Iaconetti C, Catalucci D, et al. MicroRNA-133 controls vascular smooth muscle cell phenotypic switch in vitro and vascular remodeling in vivo. *Circ Res.* 2011;109(8):880–893.

117. van Rooij E, Sutherland LB, Qi X, et al. Control of stress-dependent cardiac growth and gene expression by a microRNA. *Science.* 2007;316(5824):575–579.

118. Callis TE, Pandya K, Seok HY, et al. MicroRNA-208a is a regulator of cardiac hypertrophy and conduction in mice. *J Clin Invest.* 2009;119(9):2772–2786.

119. Oliveira-Carvalho V, da Silva MM, Guimarães GV, et al. MicroRNAs: new players in heart failure. *Mol Biol Rep.* 2013;40(3):2663–2670.

120. Tijsen AJ, Creemers EE, Moerland PD, et al. MiR423-5p as a circulating biomarker for heart failure. *Circ Res.* 2010;106(6):1035–1039.

121. Fukushima Y, Nakanishi M, Nonogi H, et al. Assessment of plasma miRNAs in congestive heart failure. *Circ J.* 2011;75(2):336–340.

122. Wang K, Zhang S, Marzolf B, et al. Circulating microRNAs, potential biomarkers for drug-induced liver injury. *Proc Natl Acad Sci USA.* 2009;106(11):4402–4407.

123. Thum T, Galuppo P, Wolf C, et al. MicroRNAs in the human heart: a clue to fetal gene reprogramming in heart failure. *Circulation.* 2007;116(3):258–267.

124. Giallourakis CC, Rosenberg PM, Friedman LS. The liver in heart failure. *Clin Liver Dis.* 2002;6(4):947–967, viii–ix.

125. Fichtlscherer S, De Rosa S, Fox H, et al. Circulating microRNAs in patients with coronary artery disease. *Circ Res.* 2010;107(5):677–684.

126. Taurino C, Miller WH, McBride MW, et al. Gene expression profiling in whole blood of patients with coronary artery disease. *Clin Sci (Lond).* 2010;119(8):335–343.

127. Hoekstra M, van der Lans CA, Halvorsen B, et al. The peripheral blood mononuclear cell microRNA signature of coronary artery disease. *Biochem Biophys Res Commun.* 2010;394(3):792–797.

128. Ji R, Cheng Y, Yue J, et al. MicroRNA expression signature and antisense-mediated depletion reveal an essential role of MicroRNA in vascular neointimal lesion formation. *Circ Res.* 2007;100(11):1579–1588.

129. Ueland T, Otterdal K, Lekva T, et al. Dickkopf-1 enhances inflammatory interaction between platelets and endothelial cells and shows increased expression in atherosclerosis. *Arterioscler Thromb Vasc Biol.* 2009;29(8):1228–1234.

130. Navarro A, Diaz T, Martinez A, et al. Regulation of JAK2 by miR-135a: prognostic impact in classic Hodgkin lymphoma. *Blood.* 2009;114(14):2945–2951.

131. Ghoreschi K, Laurence A, O'Shea JJ. Janus kinases in immune cell signaling. *Immunol Rev.* 2009;228:273–287.

132. Liu G, Friggeri A, Yang Y, et al. miR-147, a microRNA that is induced upon Toll-like receptor stimulation, regulates murine macrophage inflammatory responses. *Proc Natl Acad Sci USA.* 2009;106(37):15819–15824.

133. Xu CF, Yu CH, Li YM. Regulation of hepatic microRNA expression in response to ischemic preconditioning following ischemia/reperfusion injury in mice. *OMICS.* 2009;13(6):513–520.

134. Tan KS, Armugam A, Sepramaniam S, et al. Expression profile of microRNAs in young stroke patients. *PLoS One.* 2009;4(11):e7689.

135. Zampetaki A, Kiechl S, Drozdov I, et al. Plasma microRNA profiling reveals loss of endothelial miR-126 and other microRNAs in type 2 diabetes. *Circ Res.* 2010;107(6):810–817.

136. Kong L, Zhu J, Han W, et al. Significance of serum microRNAs in pre-diabetes and newly diagnosed type 2 diabetes: a clinical study. *Acta Diabetol.* 2011;48(1):61–69.

137. Orchard TJ, Costacou T, Kretowski A, Nesto RW. Type 1 diabetes and coronary artery disease. *Diabetes Care.* 2006;29(11):2528–2538.

138. Li S, Zhu J, Zhang W, et al. Signature microRNA expression profile of essential hypertension and its novel link to human cytomegalovirus infection. *Circulation.* 2011;124(2):175–184.

139. Nathan DM. Long-term complications of diabetes mellitus. *N Engl J Med.* 1993;328(23):1676–1685.

140. Bandi N. Zbinden S, Gugger M, et al. miR-15a and miR-16 are implicated in cell cycle regulation in a Rb-dependent manner and are frequently deleted or down-regulated in non-small cell lung cancer. *Cancer Res.* 2009;69(13):5553–5559.

141. Fish JE, Santoro MM, Morton SU, et al. miR-126 regulates angiogenic signaling and vascular integrity. *Dev Cell.* 2008;15(2):272–284.

142. Wang SS, Aurora AB, Johnson BA, et al. The endothelial-specific microRNA miR-126 governs vascular integrity and angiogenesis. *Dev Cell.* 2008;15(2):261–271.

143. Nielsen LB, Wang C, Sørensen K, et al. Circulating levels of microRNA from children with newly diagnosed type 1 diabetes and healthy controls: evidence that miR-25 associates to residual beta-cell function and glycaemic control during disease progression. *Exp Diabetes Res.* 2012;2012:896362.

144. Fu YB, Zhang Y, Wang Z, et al. Regulation of NADPH oxidase activity is associated with miRNA-25-mediated NOX4 expression in experimental diabetic nephropathy. *Am J Nephrol.* 2010;32(6):581–589.

145. Cheng J, Ke Q, Jin Z, et al. Cytomegalovirus infection causes an increase of arterial blood pressure. *PLoS Pathog.* 2009;5(5):e1000427.

146. Miyamoto M, Fujita T, Kimura Y, et al. Regulated expression of a gene encoding a nuclear factor, IRF-1, that specifically binds to IFN-beta gene regulatory elements. *Cell.* 1988;54(6):903–913.

147. Shahbazian MD, Grunstein M. Functions of site-specific histone acetylation and deacetylation. *Annu Rev Biochem.* 2007;76:75–100.

148. Suganuma T, Workman JL. Crosstalk among histone modifications. *Cell.* 2008;135(4):604–607.

149. Zheng YG, Wu J, Chen Z, Goodman M. Chemical regulation of epigenetic modifications: opportunities for new cancer therapy. *Med Res Rev.* 2008;28(5):645–687.

150. Kouzarides T. Chromatin modifications and their function. *Cell.* 2007;128(4):693–705.

151. Luger K, Mäder AW, Richmond RK, et al. Crystal structure of the nucleosome core particle at 2.8 A resolution. *Nature.* 1997;389(6648):251–260.

152. Glozak MA, Seto E. Histone deacetylases and cancer. *Oncogene.* 2007;26(37):5420–5432.

153. Sterner DE, Berger SL. Acetylation of histones and transcription-related factors. *Microbiol Mol Biol Rev.* 2000;64(2):435–459.

154. Cress WD, Seto E. Histone deacetylases, transcriptional control, and cancer. *J Cell Physiol.* 2000;184(1):1–16.

155. Wang Z, Zang C, Rosenfeld JA, et al. Combinatorial patterns of histone acetylations and methylations in the human genome. *Nat Genet.* 2008;40(7):897–903.

156. Klose RJ, Zhang Y. Regulation of histone methylation by demethylimination and demethylation. *Nat Rev Mol Cell Biol.* 2007;8(4):307–318.

157. Martin C, Zhang Y. The diverse functions of histone lysine methylation. *Nat Rev Mol Cell Biol.* 2005;6(11):838–849.

158. Wysocka J, Allis CD, Coonrod S. Histone arginine methylation and its dynamic regulation. *Front Biosci.* 2006;11:344–355.

159. Meissner A, Mikkelsen TS, Gu H, et al. Genome-scale DNA methylation maps of pluripotent and differentiated cells. *Nature.* 2008;454(7205):766–770.

160. Cohen AW, Park DS, Woodman SE, et al. Caveolin-1 null mice develop cardiac hypertrophy with hyperactivation of p42/44 MAP kinase in cardiac fibroblasts. *Am J Physiol Cell Physiol.* 2003;284(2):C457–C74.

161. Woodman SE, Park DS, Cohen AW, et al. Caveolin-3 knock-out mice develop a progressive cardiomyopathy and show hyperactivation of the p42/44 MAPK cascade. *J Biol Chem.* 2002;277(41):38988–38997.

162. Das M, Das S, Lekli I, Das DK. Caveolin induces cardioprotection through epigenetic regulation. *J Cell Mol Med.* 2012;16(4):888–895.

163. Kaneda R, Takada S, Yamashita Y, et al. Genome-wide histone methylation profile for heart failure. *Genes Cells.* 2009;14(1):69–77.

164. Stein AB, Jones TA, Herron TJ, et al. Loss of H3K4 methylation destabilizes gene expression patterns and physiological functions in adult murine cardiomyocytes. *J Clin Invest.* 2011;121(7):2641–2650.

165. Zhang QJ, Chen HZ, Wang L, et al. The histone trimethyllysine demethylase JMJD2A promotes cardiac hypertrophy in response to hypertrophic stimuli in mice. *J Clin Invest.* 2011;121(6):2447–2456.

166. Cataldi M, Secondo A, D'Alessio A, et al., Studies on maitotoxin-induced intracellular Ca(2+) elevation in chinese hamster ovary cells stably transfected with cDNAs encoding for L-type Ca(2+) channel subunits. *J Pharmacol Exp Ther.* 1999;290(2):725–730.

167. Marx SO, Reiken S, Hisamatsu Y, et al. PKA phosphorylation dissociates FKBP12.6 from the calcium release channel (ryanodine receptor): defective regulation in failing hearts. *Cell.* 2000;101(4):365–376.

168. Deschenes I, Armoundas AA, Jones SP, Tomaselli GF. Post-transcriptional gene silencing of KChIP2 and Navbeta1 in neonatal rat cardiac myocytes reveals a functional association between Na and Ito currents. *J Mol Cell Cardiol.* 2008;45(3):336–346.

169. Sheikh F, Raskin A, Chu PH, et al. An FHL1-containing complex within the cardiomyocyte sarcomere mediates hypertrophic biomechanical stress responses in mice. *J Clin Invest.* 2008;118(12):3870–3880.

170. Nielsen DA, Hamon S, Yuferov V, et al. Ethnic diversity of DNA methylation in the OPRM1 promoter region in lymphocytes of heroin addicts. *Hum Genet.* 2010;127(6):639–649.

171. Niv Y, Vilkin A, Brenner B, et al. hMLH1 promoter methylation and JC virus T antigen presence in the tumor tissue of colorectal cancer Israeli patients of different ethnic groups. *Eur J Gastroenterol Hepatol.* 2010;22(8):938–941.

172. Tian C, Gregersen PK, Seldin MF. Accounting for ancestry: population substructure and genome-wide association studies. *Hum Mol Genet.* 2008;17(R2):R143–R150.

173. Zhu ZZ, Hou L, Bollati V, et al. Predictors of global methylation levels in blood DNA of healthy subjects: a combined analysis. *Int J Epidemiol.* 2012;41(1):126–139.

174. Reilly SJ, Odeberg J, Tornvall P. Use of the whole leucocyte population in the study of the NFkappaB pathway. *Scand J Immunol.* 2011;73(4):338–343.

175. Brown SE, Fraga MF, Weaver IC, et al. Variations in DNA methylation patterns during the cell cycle of HeLa cells. *Epigenetics.* 2007;2(1):54–65.

176. Kangaspeska S, Stride B, Métivier R, et al. Transient cyclical methylation of promoter DNA. *Nature.* 2008;452(7183):112–115.

177. Metivier R, Gallais R, Tiffoche C, et al. Cyclical DNA methylation of a transcriptionally active promoter. *Nature.* 2008;452(7183):45–50.

178. Foley DL, Craig JM, Morley R, et al. Prospects for epigenetic epidemiology. *Am J Epidemiol.* 2009;169(4):389–400.

179. Bollati V, Schwartz J, Wright R, et al. Decline in genomic DNA methylation through aging in a cohort of elderly subjects. *Mech Ageing Dev.* 2009;130(4):234–239.

180. Bjornsson HT, Sigurdsson MI, Fallin MD, et al. Intra-individual change over time in DNA methylation with familial clustering. JAMA. 2008;299(24):2877–2883.

181. Fraga MF, Ballestar E, Paz MF, et al. Epigenetic differences arise during the lifetime of monozygotic twins. *Proc Natl Acad Sci USA.* 2005;102(30):10604–10609.

182. Yi P, Melnyk S, Pogribna M, et al. Increase in plasma homocysteine associated with parallel increases in plasma S-adenosylhomocysteine and lymphocyte DNA hypomethylation. *J Biol Chem.* 2000;275(38):29318–29323.

183. Frederiksen J, Juul K, Grande P, et al. Methylenetetrahydrofolate reductase polymorphism (C677T), hyperhomocysteinemia, and risk of ischemic cardiovascular disease and venous thromboembolism: prospective and case-control studies from the Copenhagen City Heart Study. *Blood.* 2004;104(10):3046–3051.

184. Girgis S, Nasrallah IM, Suh JR, et al. Molecular cloning, characterization and alternative splicing of the human cytoplasmic serine hydroxymethyltransferase gene. *Gene.* 1998;210(2):315–324.

185. Lim U, Peng K, Shane B, et al. Polymorphisms in cytoplasmic serine hydroxymethyltransferase and methylenetetrahydrofolate reductase affect the risk of cardiovascular disease in men. *J Nutr.* 2005;135(8):1989–1994.

186. Fuks F. DNA methylation and histone modifications: teaming up to silence genes. *Curr Opin Genet Dev.* 2005;15(5):490–495.

187. Chuang JC, Jones PA. Epigenetics and microRNAs. *Pediatr Res.* 2007;61(5, Pt 2):24R–29R.

188. Chen X, Ba Y, Ma L, et al. Characterization of microRNAs in serum: a novel class of biomarkers for diagnosis of cancer and other diseases. *Cell Res.* 2008;18(10):997–1006.

189. Hu Z, Chen X, Zhao Y, et al. Serum microRNA signatures identified in a genome-wide serum microRNA expression profiling predict survival of non-small-cell lung cancer. *J Clin Oncol.* 2010;28(10):1721–1726.

190. Lee LW, Zhang S, Etheridge A, et al. Complexity of the microRNA repertoire revealed by next-generation sequencing. *RNA.* 2010;16(11):2170–2180.

6

C H A P T E R

MicroRNAs

R. Kannan Mutharasan

TAKE HOME POINTS

1. MicroRNAs, small ribonucleic acid (RNA) transcripts that are approximately 22 nucleotides in length, are important regulators of gene expression; the microRNA mechanism of gene regulation is highly conserved across life forms, and in humans microRNA-mediated gene regulation impacts nearly every major cellular pathway, including those involved in cardiovascular homeostasis.
2. Although there is great enthusiasm for microRNAs as biomarkers and therapeutics, much work remains to be done in the budding field, including dissecting gene networks impacted by individual microRNAs, and validating targets of individual microRNAs; larger studies are needed to further evaluate the role of microRNAs as biomarkers in various cardiovascular diseases (CVDs).
3. In considering microRNA drug development, an important hurdle is drug delivery because microRNA-based therapeutics exhibit little oral bioavailability, and thus must be administered by alternative routes (most commonly intravenously) and may also need to be directed toward a target organ; fortunately, there are several carriers of microRNAs in the blood, including exosomes, microparticles, and lipoproteins that could serve as vehicles for novel therapeutics.

In order for organisms to grow, develop, and respond appropriately to environmental stressors, it is critical that systems exist that allow for control over gene expression. Humans are endowed with a multitude of mechanisms that allow for control of gene expression at both the transcriptional and translational levels. Among these mechanisms of control is expression of microRNA. In this chapter, we discuss how microRNA works, highlight some of its mechanistic roles in mediating response to cardiovascular disease (CVD), describe the work being done on using microRNA as biomarkers, anticipate microRNA therapeutics in the context of CVD, and offer future directions for the field.

MicroRNA STRUCTURE AND FUNCTION

MicroRNAs are RNA transcripts that are approximately 22 nucleotides in length. They were first discovered in the nematode *Caenorhabditis elegans* in 1993 by Ambrose and colleagues (1), and soon were found to be present in other animals including flies and mammals (2,3). Although the machinery is somewhat different, microRNAs are also important regulators of gene expression in plants (4). Thus the microRNA mechanism of gene regulation is highly conserved across life forms, and in humans microRNA-mediated gene regulation impacts nearly every major cellular pathway.

MicroRNA biogenesis refers to the process of generating mature microRNA from precursor forms (Figure 6.1). Like any RNA, microRNAs originate after being transcribed from genes. To date, approximately 1,400 microRNA genes have been identified in humans (5,6). The vast majority of these microRNAs carry numerical designations (eg, microRNA-210), in keeping with the rapid pace of their discovery, discovery that has been driven by cloning, deep sequencing, and computational techniques. MicroRNA genes are found throughout the human genome, on every gene with the exception of chromosome Y. Approximately 50% of known human microRNA genes are clustered with other microRNA genes, and are transcribed simultaneously in one

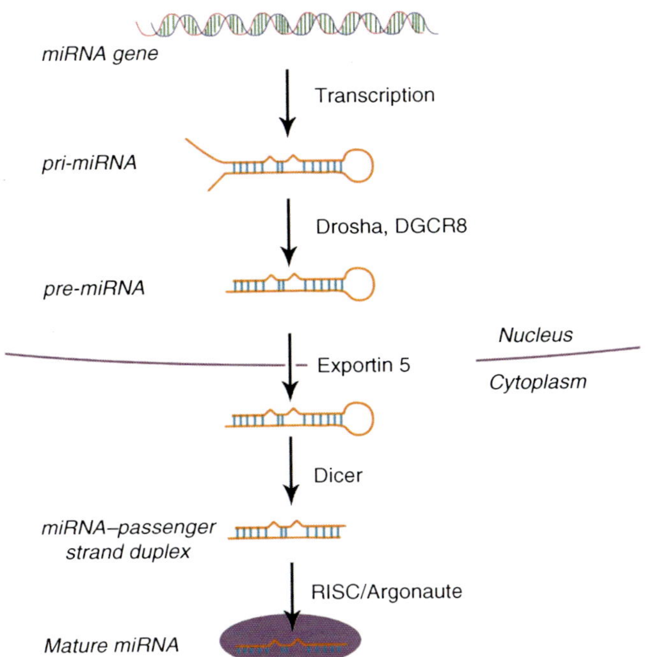

FIGURE 6.1 MicroRNA biogenesis. MicroRNA biogenesis proceeds in a stepwise fashion, beginning with transcription of microRNAs from microRNA genes, leading to primary microRNA, or pri-microRNA. The action of the microprocessor complex, the key protein components of which are Drosha and DGCR8, then processes the transcript further to pre-microRNA. This RNA, which contains a stem-loop structure, is then exported to the cytoplasm by Exportin 5. Dicer further processes the RNA, leaving a microRNA-passenger strand duplex. The correct strand of this duplex is then loaded into the RNA-induced silencing complex (RISC), a major protein component of which is Argonaute.

polycistronic transcript (7). MicroRNAs, like protein-coding genes, can also originate from single transcripts. Very interestingly, a large number are found intronic to protein-coding genes and noncoding transcripts (8). In general, these microRNAs are transcribed in conjunction with their host genes, and further processed in a manner similar to other microRNAs as discussed in the following. However, interesting variations of the canonical processing pathway exist, highlighting the diversity and complexity of the microRNA regulatory mechanism.

Most microRNAs are transcribed by RNA polymerase II, and as such have a 5′ cap and a 3′ poly-A tail (9,10). The initial transcript is termed primary microRNA or pri-microRNA (7). These pri-microRNAs are then further processed by a protein complex termed the microprocessor complex, the two most important protein constituents of which are Drosha and Di George Syndrome critical region 8 (DGCR8) (11). Drosha is a protein with endoribonuclease activity that cleaves the primary miRNA transcript at a location approximately 11 base pairs

from the end of the microRNA and its complementary strand, leaving a two-nucleotide overhang (12). This more mature species is termed pre-microRNA, and is typically around 70 nucleotides in length.

Pre-microRNAs are then recognized by and transported by Exportin 5 out of the nucleus into the cytoplasm (13). There, the Dicer enzyme further processes the microRNA by cleaving the loop off the stem-loop structure, leaving an oligonucleotide approximately 22 nucleotides in length bound to a partially complementary passenger strand of similar length, with two nucleotide overhangs on the 3′ end of either strand of the duplex (14,15). This duplex is then loaded into a protein complex called RNA-induced silencing complex (RISC). RISC contains Argonaute, a protein for which four isoforms exist in humans (16). RISC unloads the passenger strand and retains the correct microRNA strand, generating an RNA–protein complex that is ready to bind messenger RNA and repress translation (17,18). Though the mechanistic details have not been fully elucidated, it is thought that thermodynamic stability guides RISC to retaining the correct microRNA strand (18), though in several cases the passenger strand, often designated, for example, miR-33*, has been shown to exert biological activity (19).

The steps described represent the canonical microRNA-processing pathway. A number of variations on this central theme have been reported, including mirtrons, whose pre-microRNAs are generated by splicing events rather than Drosha-mediated cleavage (20,21), and microRNAs that are cleaved by Argonaute proteins rather than Dicer (22,23). These variations highlight the complexity of microRNA biology.

MicroRNA binding to messenger RNA is an interesting process whose features allow some element of bioinformatic prediction (Figure 6.2). A general feature of microRNA binding to messenger RNA is that binding occurs most typically in the 3′ untranslated region (UTR) of the messenger RNA transcript, though binding of microRNAs to the 5′ region (24) and to the coding regions (25) of the mRNA has also been reported. Another general feature is that binding between base pairs 2 and 8 of the microRNA region—a region called the seed sequence—and the messenger RNA typically occurs with perfect complementarity. This is in contrast to binding between the microRNA and mRNA elsewhere along the length of the interaction, which is typified by imperfect complementarity (26). There appear to be at least two mechanisms by which this interaction represses translation. One is by inhibition of the translational machinery itself, without degradation of the mRNA per se, and the other is mediating destabilization of the miRNA transcript. This destabilization is thought to proceed through removal of the 5′ cap, an event that leaves the transcript susceptible to clearance (27).

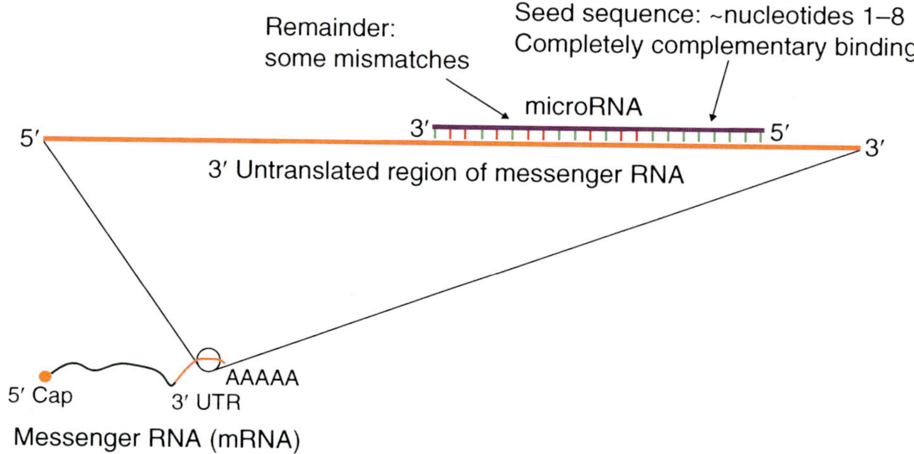

FIGURE 6.2 MicroRNA binding to messenger RNA. Shown in the bottom left is a schematic of a messenger RNA molecule, which comprises a 5' cap (yellow ball), a coding region (green), and a 3' untranslated region (3' UTR), which ends in a polyadenosine tail. MicroRNAs bind to the 3' UTR. Shown in the top half of the diagram is microRNA binding to the 3' UTR. The first eight nucleotides or so of the microRNA, beginning from the 5' end, are considered the seed sequence, which typically binds to the messenger RNA with complete complementarity. The remainder of the microRNA–messenger RNA duplex binds with incomplete complementarity. The perfect base pairing in the seed sequence forms the basis for bioinformatic prediction of microRNA targets.

This perfect complementarity along a stretch of approximately seven base pairs forms a basis for bioinformatic prediction of microRNA–mRNA target pairs. However, simple calculation reveals that there exist only 4^7 unique combinations of 7-mers (16,384), a value that is not sufficient in itself to confer sufficient targeting specificity in a genome comprising approximately 3 billion base pairs. Thus, other factors are also involved in mediating binding between mRNA and microRNA. Among these factors is the notion of UTR context, which implies that the location of the microRNA site within the 3' UTR matters (26). Experimental evidence suggests that locations somewhat away from the stop codon, away from the center of the UTR, and in an area rich in AU base pairs (which enhances site accessibility) are all features that enhance the likelihood that a given stretch of RNA is indeed targeted (26).

Several miRNA target prediction algorithms, such as TargetScan, PicTar, EMBL, and Miranda are freely available. These target prediction algorithms all yield somewhat different results, but the algorithms are largely based on first computing possible targets by checking for seed matches in the 3' UTR of mRNAs, and then comparing these results to cognate stretches in other organisms. This screen for evolutionary conservation increases the specificity of the computed matches (26). Strikingly, each microRNA is typically predicted to target hundreds of protein-coding genes, and conversely approximately 50% of protein-coding genes are predicted to be regulated by microRNA (26).

Because microRNAs have the potential for widespread regulation of cellular pathways, it is not surprising that microRNAs themselves are highly regulated at several levels (28). In addition to regulation at the level of transcription, which occurs for miRNA just as it does for any transcript, microRNAs are regulated at multiple processing steps along their path to maturity. For example, the tumor suppressor p53 associates with the Drosha complex and facilitates Drosha-mediated pri-miRNA processing of several microRNAs that suppress growth (29).

MicroRNA AND CARDIAC FUNCTION

MicroRNAs have been implicated as having a role in nearly every major cell physiology pathway. Cardiovascular pathways are no exception. To help develop a sense of how microRNAs function in pathways and networks, a survey of microRNA roles in two mechanisms of fundamental importance in CVD follows.

Cardiac Hypertrophy

Initial data demonstrating a role for microRNA in cardiac hypertrophy stemmed from a series of microarray studies that demonstrated differential upregulation of specific microRNAs in several experimental models of hypertrophy (30–33). The first such study was reported by van Rooij et al (34). In this study, the differential expressions of microRNAs in two experimental models of hypertrophy were compared. In the first model, mice transgenically overexpressing Calcineurin A were compared to control mice. In the second model, mice that underwent transverse aortic constriction (TAC)—a mouse model of pressure overload hypertrophy—were compared to sham-operated mice. In these comparisons, a number of microRNAs were similarly up- or downregulated in the experimental conditions compared to controls, suggesting that a microRNA "signature" of cardiac hypertrophy might exist. Sayed et al also studied microRNA profiles in mice undergoing TAC, measuring microRNA levels in cohorts of mice at baseline, 7 days after TAC and 14 days after TAC. A progressive evolution of the microRNA signature was seen (31). Thum et al undertook a complementary approach, performing miRNA microarray and transcriptomic analysis on normal, fetal, and failing human hearts (32). This study showed remarkable overlap between the microRNA profile of failing hearts and fetal myocardium, extending the concept that heart failure gene signatures and protein expression patterns represent a recapitulation of the fetal gene program.

These and other studies thus established the groundwork for a role for microRNA in cardiac hypertrophy and heart failure, studying patterns of expression at the whole-genome level. Mechanistic dissection of the role of microRNAs in cardiac hypertrophy has also been undertaken. These studies have demonstrated multiple insights into how microRNAs integrate into signaling cascades already known to play roles in hypertrophy, adding another layer of complexity to the control of cellular response. Among the most elegant examples of this is the gene regulatory network discovered by van Rooij et al involving a set of microRNAs embedded in myosin heavy chain (MYH) genes, genes that encode for the major contractile protein of both cardiac and skeletal muscles (30).

There are several isoforms of MYH genes, and at least three of these isoforms have microRNAs embedded within them. MiR-208a is found intronic to myosin heavy chain 6 (MYH6), also termed MHC-alpha. This isoform predominates in rodent hearts, while myosin heavy chain 7 (MYH7), also known as MHC-beta and which contains miR-208b, is the predominant form in human hearts. The third isoform, miR-499, is found intronic to myosin heavy chain 7B (MYH7B), the major isoform of myosin in slow skeletal muscle. Interestingly, thyroid hormone regulates MYH expression, such that repression of thyroid hormone with propylthiouracil (PTU) decreases MHC-alpha expression and increases MHC-beta expression.

van Rooij et al found that PTU treatment decreased pre-miR-208a in parallel with a decrease in the expression of its host gene, MHC-alpha, showing that the two genes are coregulated (30). Studies in miR-208a knockout mice showed that these mice were less susceptible to fibrosis and hypertrophy, and failed to upregulate MHC-beta in response to TAC in contrast to controls. Linking these two arms of the study, PTU treatment of miR-208a knockout mice failed to significantly upregulate MHC-beta, demonstrating that the thyroid-hormone-induced changes in MHC-beta are largely driven by miR-208a. Thus these studies demonstrated a role for microRNAs in ventricular remodeling induced by thyroid hormone fluctuations.

Another well-studied group of microRNAs with regard to hypertrophic signaling in the heart is miR-1 and miR-133, two different microRNAs that are expressed from a bicistronic transcript and thus are regulated similarly. In the heart, two bicistronic clusters are expressed: one that encodes miR-133a-1/miR-1-2 and one that encodes miR-133a-2/miR-1-1. Downregulation of miR-1 expression appears to be one of the earliest changes that occur in the development of cardiac hypertrophy (31). This has been shown in multiple models of cardiac hypertrophy, including in TAC, overexpression of Akt (a protein kinase that leads to hypertrophic signaling), and in the setting of exercise (35,36). Conversely, adenoviral overexpression of miR-1 has been shown to attenuate cardiac hypertrophy in response to infusion of endothelin-1 (35). Several mRNA targets of miR-1 have been identified. Insulin growth factor-1 appears to be directly regulated by miR-1: overexpression of miR-1 in neonatal rat cardiomyocytes leads to decreased expression of IGF-1 (37). Furthermore, the luciferase assay, an experimental technique to directly assess binding of a microRNA to the 3′ UTR, provides further evidence of a direct effect of miR-1 on IGF-1. Interestingly it appears that IGF-1 also regulates miR-1; thus these two gene products regulate each other in a reciprocal fashion. Patients with acromegaly, a primary disorder of IGF-1 overexpression, were demonstrated to have reduced levels of miR-1 expression on cardiac biopsy.

The role of miR-133, in cardiac hypertrophy, coexpressed with miR-1, has also been studied. Like miR-1, miR-133a is downregulated in cardiac hypertrophy (35). However, the precise physiological role of this protein and whether miR-133a downregulation has a causative role in hypertrophy pathways are less well defined. Care et al demonstrated that overexpression of miR-133a with adenovirus in neonatal rat cardiomyocytes reduces hypertrophy and fetal gene expression. Conversely, knockdown of miR-133a expression with antagomiRs (antisense oligonucleotides perfectly complementary to a microRNA causing partial inactivation) appeared to sensitize myocardium to growth. However, in another study, genetic ablation of one of two copies of miR-133a did not lead to hypertrophy, though it resulted in fibrosis and diminished cardiac function (38). The precise reason for this discrepancy is unclear, but may have to do with differences in technique between genetic and pharmacological approaches to microRNA manipulation. Several other microRNAs and families of microRNAs have been studied in hypertrophy, including miR-21, miR-23a, miR-92, miR-100, miR-199, and miR-92. These microRNAs integrate with a variety of signaling pathways; the reader is referred to excellent recent reviews for further information (38).

Cholesterol Metabolism

Cholesterol homeostasis represents a critical physiological system, as cholesterol is an essential molecule for a number of cellular metabolic and structural functions. In atherosclerosis, cholesterol represents a major driving force for inflammation, vessel wall damage, and ultimately luminal obstruction (39). Given the centrality of cholesterol homeostasis in physiology and pathophysiology, it is not surprising that an important role for microRNAs in cholesterol regulation has emerged.

Steroid response element binding proteins are a key class of transcription factors critical for regulating cholesterol (40). In humans, there are three isoforms encoded by two genes. SREBP1a and SREBP1c are generated from the *SREBP1* gene, and SREBP2 is encoded by the *SREBP2* gene. Isoforms SREBP1a and SREBP2 broadly regulate transcription of genes involved in the biosynthesis of cholesterol, while SREBP1c broadly regulates genes of importance in fatty acid metabolism. Very intriguingly, there is a microRNA intronic to each of the *SREBP* genes: miR-33a in SREBP2 and miR-33b in SREBP1. These microRNAs have been shown to be cotranscribed and thus coregulated with their host genes, and play critical roles in modulating cholesterol and fatty acid metabolism (41–44). Prior to discussing this work, it is important to note a difference between mice and humans: while humans have two *SREBP* and miR-33 genes, mice only have one, meaning that studies referring to mice note only miR-33 rather than miR-33a and miR-33b.

In vitro experiments in mice have shown that miR-33 overexpression decreases expression of ABCA1 (41–44), a major transporter responsible for movement of cholesterol out of cells to high-density lipoproteins (HDLs) in a process termed reverse cholesterol transport. In keeping with this decreased level of ABCA1 protein expression, functional assays of cellular reverse cholesterol transport, wherein the ability of cells to export cholesterol to cholesterol acceptors is assayed, demonstrate that miR-33 overexpression reduces cellular cholesterol efflux. In vivo experiments in mice show that knockdown of miR-33 leads to increased plasma HDL-C levels (41–44). Moreover, miR-33 knockdown promotes atherosclerosis regression in the context of low-density

lipoprotein (LDL) receptor knockout mice, a major model of atherosclerosis (45).

Further studies have extended these initial observations to nonhuman primates, where knockdown of miR-33 has been shown to raise HDL-cholesterol levels significantly (46). These initial observations on the role of miR-33 in cholesterol metabolism have been extended, and a role for this microRNA in fatty acid metabolism has been described (47). A number of genes relevant to beta oxidation of fatty acids are direct targets of either miR-33a or miR-33b, and central metabolic signaling proteins like AMP kinase and Sirtuin 6 are also regulated by miR-33. Furthermore, recent evidence points to a role for the passenger strand, miR-33* in lipid metabolism gene expression as well (19). Thus miR-33 represents a critical metabolic link for organismal cholesterol and fatty acid metabolism.

Cholesterol metabolism is also regulated by miR-122. This microRNA is the most abundantly expressed microRNA in the liver, accounting for approximately 70% of liver microRNA transcripts (48). Though direct targets for miR-122 relevant to cholesterol metabolism have not yet been described, studies have shown that knockdown of miR-122 with antagomiRs led to reduced hepatic cholesterol synthesis and reduced plasma levels of cholesterol (49). Overexpression of miR-122 led to increases in expression of proteins important in cholesterol synthesis, though these proteins are not direct targets of miR-122. It is possible that the increase in expression of these proteins is governed by a central direct target of miR-122 that has yet to be identified. Finally, as an interesting note highlighting the remarkable diversity of microRNA function, miR-122 has been shown to be necessary for propagation of hepatitis C virus (HCV), and mediates this function by binding the 5′ UTR of the HCV genome (50).

Several other microRNAs, including miR-144 (51) and miR-758 (52), have also been demonstrated to have roles in cholesterol metabolism; excellent recent reviews cover this topic in greater detail (53).

MicroRNA AS BIOMARKERS

There is tremendous interest in the concept of using microRNAs as biomarkers for a number of different CVD states, driven by the key observation that despite the presence of RNAses, microRNAs are actually remarkably stable in blood and other body fluids. This is because a variety of structures participate in transport of microRNAs through the blood (Figure 6.3). In this section, we discuss mechanisms by which microRNAs are transported in the blood,

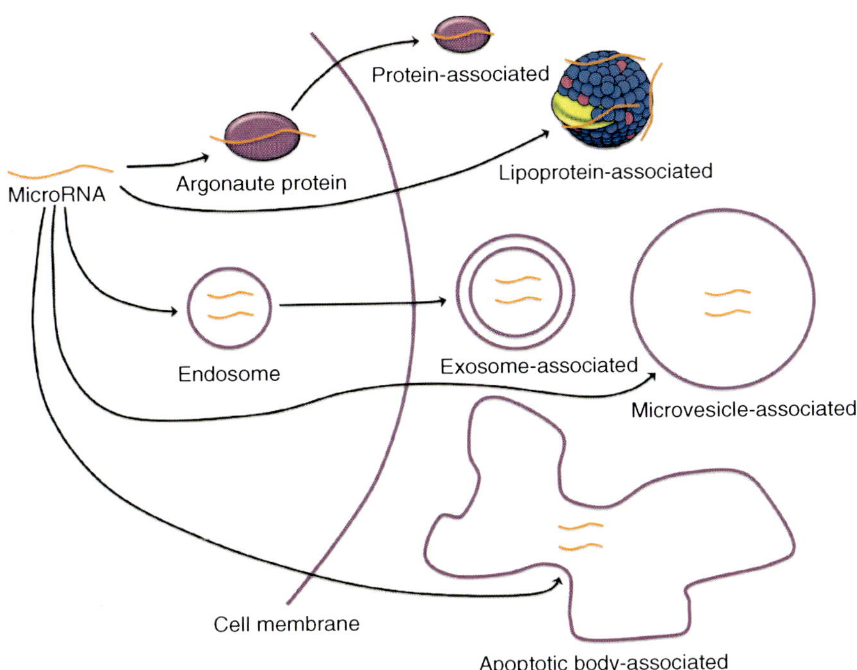

FIGURE 6.3 Structures complexed with circulating microRNA. Circulating microRNAs are stable in the circulation because they are complexed with structures that protect them from RNAse degradation. Structures containing microRNA that have been identified include: vesicular structures, such as exosomes, microparticles, and apoptotic bodies; and nonvesicular structures such as the Argonaute protein and lipoproteins.

potential roles for microRNA in long-range signaling, and finally profile work evaluating cardiac microRNAs as biomarkers in two cardiovascular conditions.

Circulating MicroRNAs

Several mechanisms are responsible for transport of microRNAs in the blood. Research in this area continues to evolve rapidly, but studies thus far indicate that different cell types may use these mechanisms in different ways. Circulating microRNA cargo, how microRNAs are loaded into various structures, and the potential of these structures to mediate long-range intercellular signaling, are areas of intense investigation.

A number of vascular structures have been shown to contain and mediate intracellular transfer of microRNA. Exosomes are small 40 to 100 nm vesicles of endosomal origin characterized by a double plasma membrane and the presence of specific surface proteins (54). In the cardiovascular context, they have been shown to mediate the beneficial effects of endothelial progenitor cells on angiogenesis (55). Valadi et al have shown that mast cell line derived exosomes contain not only messenger RNAs but also microRNAs, and that the levels of microRNAs differed between the exosomes and the host cells, suggesting the potential for selective packaging of certain microRNAs (56). These investigators further demonstrated that these RNAs are transferred to other mast cells, though transfer of microRNAs specifically was not demonstrated.

Microparticles are larger vesicular structures, approximately 100 to 1,000 nm in size and generated by the budding off of a portion of the plasma membrane (57). The bulk of circulating microparticle content is thought to be platelet derived. Several groups have demonstrated that microparticles contain microRNAs (58–60). Zhang et al showed that microRNA-150 is present in microparticles secreted by the THP-1 monocyte cell line in vitro, and that these microparticles can be taken up by the HMEC-1 endothelial cell line and cause more cell migration, a marker of enhanced endothelial cell function (59). Microparticles from embryonic stem cells were shown by Yuan et al to be capable of transferring microRNAs to fibroblasts (58).

Apoptotic bodies, generated as blebs of cells undergoing apoptosis, represent a third class of vesicles shown to mediate microRNA signaling. Zernecke et al showed that delivery of miR-126 by apoptotic bodies induces vascular protection in ApoE knockout mice fed a high-fat diet (61). Thus exosomes, microparticles, and apoptotic bodies have all been shown to contain microRNA, and some classes have been shown to mediate microRNA transfer in vivo.

These modes of miRNA transport were thought to represent the major mechanism through which microRNAs circulated in the bloodstream. However, studies by Turchinovich et al (62) and Arroyo et al (63) demonstrated that a significant amount of microRNA—probably the majority—was found outside of vesicular structures. Furthermore, these extravesicular microRNAs were found to be protease sensitive, suggesting that the microRNAs were complexed with a protein carrier. In immunoprecipitation experiments, one of these protein carriers was identified as Argonaute 2 protein, the same protein described earlier involved in assembly of the RISC. Later work by Turchinovich et al (64) also demonstrated a role for Argonaute 1 protein in binding microRNAs. Interestingly this work showed differential loading of microRNAs between Argonaute 1 and 2, lending further weight to the idea that some degree of specificity dictates the precise composition of circulating microRNAs. Whether other Argonaute family members or RNA-binding proteins outside the Argonaute family also participate in stabilization of microRNAs remain to be determined.

A role for lipoproteins in the transport of microRNA has also been described. Vickers et al used fast protein liquid chromatography gel filtration to size-separate human plasma after density gradient ultracentrifugation (65). This process allowed examination of the RNA content of different lipoprotein fractions. Both LDLs and HDLs were found to carry microRNA, though the microRNA cargo of LDL exhibited significant overlap with the cargo of exosomes. Thus further evaluation of the HDL fraction was undertaken. Several microRNAs were abundantly expressed in HDL; and in serum from patients with familial hypercholesterolemia, differential expression of a panel of microRNAs was found compared to healthy controls. In vitro experiments demonstrated that HDL was capable of transporting microRNAs to cells and caused changes in expression of protein targets. This transfer was shown to be dependent on scavenger receptor B1 protein (SR-B1), a protein known to be involved in bidirectional transport of cholesterol from and to HDL (66). Interestingly, it appears that the ceramide signaling pathway, via the enzyme neutral sphingomyelinase 2 (nSMase 2) reciprocally regulates endosome- and HDL-mediated microRNA export from cells, downregulating the former and upregulating the latter. Overall this work by Vickers et al points to the intriguing possibility that lipoproteins participate in microRNA signaling, and that HDLs in particular may have a unique microRNA signature as compared to exosomes and microparticles.

Thus, several mechanisms exist that allow for microRNAs to exist and potentially be transported in blood. Evidence thus far suggests that circulating microRNAs do indeed participate in intracellular and potentially long-range signaling. However, a more complete understanding of the magnitude and importance of this mechanism awaits further investigation. However, another approach has been to use the potentially rich biological information available from circulating microRNAs by evaluating its potential as a biomarker for disease in various cardiovascular contexts.

Clinical Examples of MicroRNA Biomarkers in CVD

Coronary Artery Disease

Several groups have performed biomarker studies assessing the utility of microRNAs for detecting the presence of coronary artery disease (CAD). These studies have been performed in a variety of blood fractions of potential relevance, including whole blood (67,68), serum/plasma (69,70), the peripheral blood mononuclear fraction (71–73), and the platelet pool (74). In all of these studies, various panels of microRNAs predicted the presence of coronary disease. While some of these studies undertook the approach of identifying a priori a set of microRNAs with biological plausibility (72,73), the majority of studies completed to date have taken an unbiased approach and performed studies in both derivation and validation cohorts. For example, Fichtlscherer et al analyzed the microRNA profile in the serum of eight healthy volunteers as compared to eight patients with stable CAD (69). Several microRNAs were significantly downregulated, and many of these, such as miR-126, are known from prior studies to be expressed highly in the vascular endothelium. The expression levels of the microRNAs were then checked in a validation cohort, with the results comparing favorably.

From available data it is not clear that microRNA biomarker profiling with one or a panel of microRNAs adds prognostic value after adjusting for known clinical risk factors. Thus far, one study has included data on assay performance after multivariable risk adjustment. Takahashi et al measured the expression of miR-146a/b in a coronary disease study primarily designed to assess the effects of angiotensin converting enzyme (ACE) inhibitors and angiotensin II receptor blockers (ARBs) on expression levels of miR-146a/b in peripheral blood monocytes (72). They found that after adjusting for clinical parameters, miR-146a levels remained an independent predictor of cardiovascular events. Thus further studies in large cohorts are needed to determine whether microRNA profiling adds to assessment of CVD risk.

Myocardial Infarction

MicroRNAs have also been studied as biomarkers in the setting of myocardial infarction (MI), as the direct release of myocardial content into the bloodstream upon myocyte necrosis provides a compelling rationale that measuring cardiac-specific microRNAs may be a highly specific strategy. Based on results from preclinical studies in animal models of MI, several groups have studied this problem. Broadly, two approaches have been used: (a) microarray-based discovery of all detectable microRNAs in plasma and serum, with follow-up studies on the performance of specific microRNAs as compared to a gold standard such as troponin; and (b) assessment of the performance of a small panel of candidate microRNAs based on results from preclinical studies indicating biological plausibility.

Studies taking the first approach include those performed by Meder et al (75) and D'Alessandra et al (76). D'Alessandra et al studied 33 patients with acute ST-segment elevation myocardial infarction (STEMI) as compared to healthy controls. They found that five microRNAs, including miR-1, miR-133a, miR-133b, and miR-499-5p were upregulated and two microRNAs were downregulated in the setting of STEMI. Interestingly miR-1, miR-133a, and miR-133b are striated muscle specific, highly expressed in cardiomyocytes and skeletal muscle, while miR-499 is specific to cardiomyocytes. Of the microRNAs that were downregulated, miR-122 was found to remain downregulated in patients even 30 days after MI. Meder et al also compared patients with STEMI to normal controls; however, their analysis was conducted in whole blood (75). They found miR-1291 and miR-663b to have the best performance on receiver-operating curve (ROC) analysis, and showed that among the microRNAs tested, miR-30c and miR-145 correlated best with infarct size. Using a machine-learning statistical approach, a combination of microRNAs was found to have excellent capability of discriminating cases from controls—however this analysis was not repeated in a separate validation cohort. Nonetheless these data suggest that combining information from levels of multiple microRNAs may be superior to data gathered from one microRNA alone.

In studies on the use of microRNAs chosen a priori for detection of MI, the muscle-specific microRNAs miR-1, miR-133a, miR-133b, miR-208, and miR-499 are the best studied (77–81). All of these microRNAs have some degree of diagnostic capability for detecting acute MI. However in the ROC analyses conducted as part of these studies, cardiac troponin has generally been shown to be a superior biomarker to these microRNAs, with areas under the curve (AUCs) for single microRNAs typically in the 0.70 to 0.90 range. Yet in these studies two notable exceptions emerged with regard to microRNA diagnostic performance. Wang et al found that miR-208a had an AUC

of 0.97, and a cut point could be designed, which was 100% specific and 91% sensitive (80). Similarly, Devaux et al found that miR-499, collected within 3 hours of symptom onset, had an AUC of 0.96 (79). These values for miR-208a and miR-499 compare very favorably to that of cardiac troponin. However, because of the already excellent diagnostic power of troponin for acute MI, these microRNAs may not add diagnostic value.

Based on the available data, there are two clinical scenarios in which diagnostic microRNA testing may find utility clinically. The first is in hyperacute diagnostic testing for STEMI. Available data show that some microRNAs can perform as well as troponin at early time points (within 3–4 hours). Whether microRNAs can outperform troponin at even earlier time points remains to be determined. For microRNAs to perform better, the kinetics of microRNA release would have to be more rapid than that of troponin; given that microRNAs (molecular weight approximately 7 kDa) are smaller molecules than either Troponin I (molecular weight approximately 28 kDa) or Troponin T (molecular weight approximately 35 kDa), and that in the setting of ischemia transport of biomarkers into the bloodstream is likely to be diffusion limited, this is biophysically plausible. Another consideration is that ultrarapid microRNA testing in clinical practice would require the development of faster quantitative real-time polymerase chain reaction (PCR) protocols so that laboratory turnaround time is minimized.

The second clinical scenario in which diagnostic microRNA testing for MI may become clinically useful is in the setting of renal dysfunction, a condition that can lead to chronically elevated troponin levels, and thus diagnostic challenges in patients with renal dysfunction who present with chest pain. Devaux et al noted that in the three patients in their study on dialysis, levels of miR-499 did not appear to be elevated (79). This is plausible because microRNA clearance is not likely to be mediated by renal clearance, but rather by RNAse digestion of free microRNAs not bound to a carrier structure such as Argonaute protein, HDL, or vesicular structures (63). Further studies are needed to confirm this preliminary observation. If true, miR-499 testing in patients with chronic renal dysfunction as an alternative to troponin measurement would represent a real translational advance.

Thus, in summary, microRNAs may be useful as biomarkers for CVDs such as CAD and acute MI. Similar biomarker studies have also been conducted in other cardiac and metabolic conditions including heart failure, diabetes, and hypertension. The data for these studies have been recently reviewed (82,83).

MicroRNA THERAPEUTICS

MicroRNAs represent highly attractive targets for gene therapy. Several factors drive this interest. First, microRNAs are each predicted to target dozens to hundreds of genes, making them highly leveraged nodal entry points for modifying whole patterns and networks of gene expression (26). Second, in contrast to development of drugs for protein targets, where time-consuming and expensive small molecule screens must be undertaken before lead compounds are found, design of a new compound to target a specific microRNA, once the overall class of activators or inhibitors is developed, is relatively straightforward. This is because the agonists and antagonists can be straightforwardly chemically synthesized by following canonical Watson–Crick base-pairing rules. Third, a variety of microRNAs, as discussed, have been shown in preclinical studies to be active participants in the development of cardiac pathology when the heart is exposed to pathological stress, and modulation of microRNAs has been shown to significantly impact animal models of cardiac disease.

From a technical standpoint, the field of microRNA inhibition has progressed more rapidly than the field of microRNA mimicry. While robust systems to deliver microRNA to cells remain under development, technologies to knockdown miRNA levels have proceeded more rapidly. These miRNA knockdown technologies rely on chemical modification of RNA to enhance stability. These modifications enhance stability by engendering resistance to RNAse degradation, usually by substitution of some component of the nucleotide backbone; for example, a sulfur atom in place of an oxygen in a phosphorothioate linkage (34). Some of these modifications, like the locked nucleic acid (LNA) modification, also serve to increase the melting temperature of the antagonist-miRNA duplex, a thermodynamic feature that renders the complex more stable and thus causes greater and more irreversible inhibition of the miRNA. To date, no clinical trials in cardiology have used anti-microRNA technology, but given the robust preclinical data for a variety of targets, the first trials should shortly be underway.

In considering microRNA drug development, one important hurdle is drug delivery. MicroRNA-based therapeutics exhibit little oral bioavailability, and thus must be administered by alternative routes, most commonly intravenously. As detailed in the preceding, there are several carriers of microRNAs in the blood, including exosomes, microparticles, and lipoproteins. In addition to the microRNAs themselves, manipulating the contents of these microRNA-carrying structures could offer a complementary way forward for novel therapeutics and drug delivery mechanisms.

CONCLUSIONS

Many exciting possibilities lie ahead in all areas of microRNA research. Much work remains to be done in the field of dissecting gene networks impacted by individual microRNAs, and in validating targets of individual microRNAs. Larger studies as well as studies evaluating combinatorial approaches are needed to further evaluate the role of microRNAs as biomarkers in various CVDs. Long-range signaling by microRNA-carrying structures is a fascinating field, and one that not only has broad implications for fundamental biology, but one that also opens up possibilities for new platforms for drug delivery. Finally, in the next several years the first cardiovascular clinical trials using microRNA therapeutics will launch, marking the completion of one turn of the translational cycle, and yielding new insights to be taken back to the bench, where another cycle of discovery can begin.

REFERENCES

1. Lee RC, Feinbaum RL, Ambros V. The *C. elegans* heterochronic gene lin-4 encodes small RNAs with antisense complementarity to lin-14. *Cell.* 1993;75(5):843–854.
2. Lagos-Quintana M, Rauhut R, Lendeckel W, Tuschl T. Identification of novel genes coding for small expressed RNAs. *Science.* 2001;294(5543):853–858.
3. Lim LP, Glasner ME, Yekta S, et al. Vertebrate microRNA genes. *Science.* 2003;299(5612):1540.
4. Reinhart BJ, Weinstein EG, Rhoades MW, et al. MicroRNAs in plants. *Genes Dev.* 2002;16(13):1616–1626.
5. Kozomara A, Griffiths-Jones S. miRBase: integrating microRNA annotation and deep-sequencing data. *Nucleic Acids Res.* 2011;39(Database issue):D152–D157.
6. Griffiths-Jones S, Saini HK, van Dongen S, Enright AJ. miRBase: tools for microRNA genomics. *Nucleic Acids Res.* 2008;36(Database issue):D154–D158.
7. Lee Y, Jeon K, Lee JT, et al. MicroRNA maturation: stepwise processing and subcellular localization. *EMBO J.* 2002;21(17):4663–4670.
8. Lin SL, Chang D, Wu DY, Ying SY. A novel RNA splicing-mediated gene silencing mechanism potential for genome evolution. *Biochem Biophys Res Commun.* 2003;310(3):754–760.
9. Cai X, Hagedorn CH, Cullen BR. Human microRNAs are processed from capped, polyadenylated transcripts that can also function as mRNAs. *RNA.* 2004;10(12):1957–1966.
10. Rodriguez A, Griffiths-Jones S, Ashurst JL, Bradley A. Identification of mammalian microRNA host genes and transcription units. *Genome Res.* 2004;14(10A):1902–1910.
11. Gregory RI, Yan KP, Amuthan G, et al. The microprocessor complex mediates the genesis of microRNAs. *Nature.* 2004;432(7014):235–240.

12. Lee Y, Ahn C, Han J, et al. The nuclear RNase III Drosha initiates microRNA processing. *Nature.* 2003;425(6956):415–419.

13. Yi R, Qin Y, Macara IG, Cullen BR. Exportin-5 mediates the nuclear export of pre-microRNAs and short hairpin RNAs. *Genes Dev.* 2003;17(24):3011–3016.

14. Hutvagner G, McLachlan J, Pasquinelli AE, et al. A cellular function for the RNA-interference enzyme Dicer in the maturation of the let-7 small temporal RNA. *Science.* 2001;293(5531):834–838.

15. Ketting RF, Fischer SE, Bernstein E, et al. Dicer functions in RNA interference and in synthesis of small RNA involved in developmental timing in *C. elegans. Genes Dev.* 2001;15(20):2654–2659.

16. Ender C, Meister G. Argonaute proteins at a glance. *J Cell Sci.* 2010;123(Pt 11):1819–1823.

17. Schwarz DS, Hutvágner G, Du T, et al. Asymmetry in the assembly of the RNAi enzyme complex. *Cell.* 2003;115(2):199–208.

18. Khvorova A, Reynolds A, Jayasena SD. Functional siRNAs and miRNAs exhibit strand bias. *Cell.* 2003;115(2):209–216.

19. Goedeke L, Vales-Lara FM, Fenstermaker M, et al. A regulatory role for microRNA 33* in controlling lipid metabolism gene expression. *Mol Cell Biol.* 2013;33(11):2339–2352.

20. Okamura K, Hagen JW, Duan H, et al. The mirtron pathway generates microRNA-class regulatory RNAs in Drosophila. *Cell.* 2007;130(1):89–100.

21. Ruby JG, Jan CH, Bartel DP. Intronic microRNA precursors that bypass Drosha processing. *Nature.* 2007;448(7149):83–86.

22. Cifuentes D, Xue H, Taylor DW, et al. A novel miRNA processing pathway independent of Dicer requires Argonaute 2 catalytic activity. *Science.* 2010;328(5986):1694–1698.

23. Cheloufi S, Dos Santos CO, Chong MM, Hannon GJ. A dicer-independent miRNA biogenesis pathway that requires Ago catalysis. *Nature.* 2010;465(7298):584–589.

24. Shimakami T, Yamane D, Jangra RK, et al. Stabilization of hepatitis C virus RNA by an Ago2-miR-122 complex. *Proc Natl Acad Sci USA.* 2012;109(3):941–946.

25. Duursma AM, Kedde M, Schrier M, et al. miR-148 targets human DNMT3b protein coding region. *RNA.* 2008;14(5):872–877.

26. Bartel DP. MicroRNAs: target recognition and regulatory functions. *Cell.* 2009;136(2):215–233.

27. Huntzinger E, Izaurralde E. Gene silencing by microRNAs: contributions of translational repression and mRNA decay. *Nat Rev Genet.* 2011;12(2):99–110.

28. Krol J, Loedige I, Filipowicz W. The widespread regulation of microRNA biogenesis, function and decay. *Nat Rev Genet.* 2010;11(9):597–610.

29. Suzuki HI, Yamagata K, Sugimoto K, et al. Modulation of microRNA processing by p53. *Nature.* 2009;460(7254):529–533.

30. van Rooij E, Sutherland LB, Liu N, et al. A signature pattern of stress-responsive microRNAs that can evoke cardiac hypertrophy and heart failure. *Proc Natl Acad Sci USA.* 2006;103(48):18255–18260.

31. Sayed D, Hong C, Chen IY, et al. MicroRNAs play an essential role in the development of cardiac hypertrophy. *Circ Res.* 2007;100(3):416–424.

32. Thum T, Galuppo P, Wolf C, et al. MicroRNAs in the human heart: a clue to fetal gene reprogramming in heart failure. *Circulation.* 2007;116(3):258–267.

33. Naga Prasad SV, Duan ZH, Gupta MK, et al. Unique microRNA profile in end-stage heart failure indicates alterations in specific cardiovascular signaling networks. *J Biol Chem.* 2009;284(40):27487–27499.

34. van Rooij E, Olson EN. MicroRNA therapeutics for cardiovascular disease: opportunities and obstacles. *Nat Rev Drug Discov.* 2012;11(11):860–872.

35. Care A, Catalucci D, Felicetti F, et al. MicroRNA-133 controls cardiac hypertrophy. *Nat Med.* 2007;13(5):613–618.

36. Soci UP, Fernandes T, Hashimoto NY, et al. MicroRNAs 29 are involved in the improvement of ventricular compliance promoted by aerobic exercise training in rats. *Physiol Genomics.* 2011;43(11):665–673.

37. Elia L, Contu R, Quintavalle M, et al. Reciprocal regulation of microRNA-1 and insulin-like growth factor-1 signal transduction cascade in cardiac and skeletal muscle in physiological and pathological conditions. *Circulation.* 2009;120(23):2377–2385.

38. Liu N, Bezprozvannaya S, Williams AH, et al. MicroRNA-133a regulates cardiomyocyte proliferation and suppresses smooth muscle gene expression in the heart. *Genes Dev.* 2008;22(23):3242–3254.

39. Libby P, Ridker PM, Maseri A. Inflammation and atherosclerosis. *Circulation.* 2002;105(9):1135–1143.

40. Brown MS, Goldstein JL. The SREBP pathway: regulation of cholesterol metabolism by proteolysis of a membrane-bound transcription factor. *Cell.* 1997;89(3):331–340.

41. Rayner KJ, Suárez Y, Dávalos A, et al. MiR-33 contributes to the regulation of cholesterol homeostasis. *Science.* 2010;328(5985):1570–1573.

42. Horie T, Ono K, Horiguchi M, et al. MicroRNA-33 encoded by an intron of sterol regulatory element-binding protein 2 (Srebp2) regulates HDL in vivo. *Proc Natl Acad Sci USA.* 2010;107(40):17321–17326.

43. Marquart TJ, Allen RM, Ory DS, Baldán A. miR-33 links SREBP-2 induction to repression of sterol transporters. *Proc Natl Acad Sci USA.* 2010;107(27):12228–12232.

44. Najafi-Shoushtari SH, Kristo F, Li Y, et al. MicroRNA-33 and the SREBP host genes cooperate to control cholesterol homeostasis. *Science.* 2010;328(5985):1566–1569.

45. Rayner KJ, Esau CC, Hussain FN, et al. Inhibition of miR-33a/b in non-human primates raises plasma HDL and lowers VLDL triglycerides. *Nature.* 2011;478(7369):404–407.

46. Rayner KJ, Sheedy FJ, Esau CC, et al. Antagonism of miR-33 in mice promotes reverse cholesterol transport and regression of atherosclerosis. *J Clin Invest.* 2011;121(7):2921–2931.

47. Davalos A, Goedeke L, Smibert P, et al. miR-33a/b contribute to the regulation of fatty acid metabolism and insulin signaling. *Proc Natl Acad Sci USA.* 2011;108(22):9232–9237.

48. Lagos-Quintana M, Rauhut R, Yalcin A, et al. Identification of tissue-specific microRNAs from mouse. *Curr Biol.* 2002;12(9):735–739.

49. Esau C, Davis S, Murray SF, et al. miR-122 regulation of lipid metabolism revealed by in vivo antisense targeting. *Cell Metab.* 2006;3(2):87–98.

50. Jopling CL, Yi M, Lancaster AM, et al. Modulation of hepatitis C virus RNA abundance by a liver-specific microRNA. *Science.* 2005;309(5740):1577–1581.

51. Ramirez CM, Rotllan N, Vlassov AV, et al. Control of cholesterol metabolism and plasma high-density lipoprotein levels by microRNA-144. *Circ Res.* 2013; 112(12):1592–1601.

52. Ramirez CM, Dávalos A, Goedeke L, et al. MicroRNA-758 regulates cholesterol efflux through posttranscriptional repression of ATP-binding cassette transporter A1. *Arterioscler Thromb Vasc Biol.* 2011; 31(11):2707–2714.

53. Rotllan N, Fernandez-Hernando C. MicroRNA regulation of cholesterol metabolism. *Cholesterol.* 2012; 2012:847–849.

54. Bobrie A, Colombo M, Raposo G, Théry C. Exosome secretion: molecular mechanisms and roles in immune responses. *Traffic.* 2011;12(12):1659–1668.

55. Sahoo S, Klychko E, Thorne T, et al. Exosomes from human CD34(+) stem cells mediate their proangiogenic paracrine activity. *Circ Res.* 2011;109(7):724–728.

56. Valadi H, Ekström K, Bossios A, et al. Exosome-mediated transfer of mRNAs and microRNAs is a novel mechanism of genetic exchange between cells. *Nat Cell Biol.* 2007;9(6):654–659.

57. Mause SF, Weber C. Microparticles: protagonists of a novel communication network for intercellular information exchange. *Circ Res.* 2010;107(9):1047–1057.

58. Yuan A, Farber EL, Rapoport AL, et al. Transfer of microRNAs by embryonic stem cell microvesicles. *PLoS One.* 2009;4(3):e4722.

59. Zhang Y, Liu D, Chen X, et al. Secreted monocytic miR-150 enhances targeted endothelial cell migration. *Mol Cell.* 2010;39(1):133–144.

60. Hunter MP, Ismail N, Zhang X, et al. Detection of microRNA expression in human peripheral blood microvesicles. *PLoS One.* 2008;3(11):e3694.

61. Zernecke A, Bidzhekov K, Noels H, et al. Delivery of microRNA-126 by apoptotic bodies induces CXCL12-dependent vascular protection. *Sci Signal.* 2009;2(100): ra81.

62. Turchinovich A, Weiz L, Langheinz A, Burwinkel B. Characterization of extracellular circulating microRNA. *Nucleic Acids Res.* 2011;39(16):7223–7233.

63. Arroyo JD, Chevillet JR, Kroh EM, et al. Argonaute2 complexes carry a population of circulating microRNAs independent of vesicles in human plasma. *Proc Natl Acad Sci USA.* 2011;108(12):5003–5008.

64. Turchinovich A, Burwinkel B. Distinct AGO1 and AGO2 associated miRNA profiles in human cells and blood plasma. *RNA Biol.* 2012;9(8):1066–1075.

65. Vickers KC, Palmisano BT, Shoucri BM, et al. MicroRNAs are transported in plasma and delivered to recipient cells by high-density lipoproteins. *Nat Cell Biol.* 2011;13(4):423–433.

66. Valacchi G, Sticozzi C, Lim Y, Pecorelli A. Scavenger receptor class B type I: a multifunctional receptor. *Ann NY Acad Sci.* 2011;1229:E1–E7.

67. Taurino C, Miller WH, McBride MW, et al. Gene expression profiling in whole blood of patients with coronary artery disease. *Clin Sci (Lond).* 2010;119(8): 335–343.

68. Weber M, Baker MB, Patel RS, et al. MicroRNA expression profile in CAD patients and the impact of ACEI/ARB. *Cardiol Res Pract.* 2011;2011:532915.

69. Fichtlscherer S, De Rosa S, Fox H, et al. Circulating microRNAs in patients with coronary artery disease. *Circ Res.* 2010;107(5):677–684.

70. Diehl P, Fricke A, Sander L, et al. Microparticles: major transport vehicles for distinct microRNAs in circulation. *Cardiovasc Res.* 2012;93(4):633–644.

71. Hoekstra M, van der Lans CA, Halvorsen B, et al. The peripheral blood mononuclear cell microRNA signature of coronary artery disease. *Biochem Biophys Res Commun.* 2010;394(3):792–797.

72. Takahashi Y, Satoh M, Minami Y, et al. Expression of miR-146a/b is associated with the Toll-like receptor 4 signal in coronary artery disease: effect of renin-angiotensin system blockade and statins on miRNA-146a/b and Toll-like receptor 4 levels. *Clin Sci (Lond).* 2010; 119(9):395–405.

73. Minami Y, Satoh M, Maesawa C, et al. Effect of atorvastatin on microRNA 221/222 expression in endothelial progenitor cells obtained from patients with coronary artery disease. *Eur J Clin Invest.* 2009; 39(5):359–367.

74. Sondermeijer BM, Bakker A, Halliani A, et al. Platelets in patients with premature coronary artery disease exhibit upregulation of miRNA340* and miRNA624*. *PLoS One.* 2011;6(10):e25946.

75. Meder B, Keller A, Vogel B, et al. MicroRNA signatures in total peripheral blood as novel biomarkers for acute myocardial infarction. *Basic Res Cardiol.* 2011; 106(1):13–23.

76. D'Alessandra Y, Devanna P, Limana F, et al. Circulating microRNAs are new and sensitive biomarkers of myocardial infarction. *Eur Heart J.* 2010;31(22):2765–2773.

77. Wang R, Li N, Zhang Y, et al. Circulating microRNAs are promising novel biomarkers of acute myocardial infarction. *Intern Med.* 2011;50(17):1789–1795.

78. Ai J, Zhang R, Li Y, et al. Circulating microRNA-1 as a potential novel biomarker for acute myocardial infarction. *Biochem Biophys Res Commun.* 2010; 391(1):73–77.

79. Devaux Y, Vausort M, Goretti E, et al. Use of circulating microRNAs to diagnose acute myocardial infarction. *Clin Chem.* 2012;58(3):559–567.

80. Wang GK, Zhu JQ, Zhang JT, et al. Circulating microRNA: a novel potential biomarker for early diagnosis of acute myocardial infarction in humans. *Eur Heart J.* 2010;31(6):659–666.

81. Corsten MF, Dennert R, Jochems S, et al. Circulating MicroRNA-208b and MicroRNA-499 reflect myocardial damage in cardiovascular disease. *Circ Cardiovasc Genet.* 2010;3(6):499–506.

82. Tijsen AJ, Pinto YM, Creemers EE. Circulating microRNAs as diagnostic biomarkers for cardiovascular diseases. *Am J Physiol Heart Circ Physiol.* 2012; 303(9):H1085–H1095.

83. Fichtlscherer S, Zeiher AM, Dimmeler S. Circulating microRNAs: biomarkers or mediators of cardiovascular diseases? *Arterioscler Thromb Vasc Biol.* 2011; 31(11):2383–2390.

Gene Expression

Chiang-Ching Huang and Samantha L. Gadd

TAKE HOME POINTS

1. Microarray gene expression studies can provide tremendous insight into the molecular basis of cardiovascular disorders; however, the large volume of data obtained from a single study and the recent increase in the number of cardiovascular gene expression studies necessitate the use of valid statistical and bioinformatic algorithms for normalizing, analyzing, and comparing the data among different studies and different platforms.

2. Once gene expression microarray experiments have been conducted, there are several bioinformatic tools that can be used to further analyze the data, including cluster analysis, principal component analysis (PCA), multidimensional scaling, and ontology/pathway analysis.

3. Reducing bias and chance findings (eg, false positive differential gene expression due to a large number of statistical tests) is essential for producing high-quality, valid gene expression studies.

In the past decade, there has been an explosion of microarray gene expression studies performed in cardiovascular diseases (CVDs). Genomics technology holds great potential for enhancing our understanding of specific molecular functions, global gene regulation, and the cellular response to environmental stimuli or medication treatment. It also represents an unbiased approach for the discovery of novel biomarkers to monitor disease progression, refine risk prediction, and develop therapeutic targets. However, the wealth of data obtained from microarray experiments poses a great challenge to clinicians and investigators—namely, how to digest and retrieve the most essential information related to specific physiological conditions. To address this issue, tremendous efforts have been put toward the development of bioinformatic methods, genomics databases, and analytic software to accelerate translational discovery. In this chapter, we will (a) describe the essential components of microarray techniques for gene expression analysis; and (b) discuss some of the most popular bioinformatic tools and resources for analysis of microarray gene expression data. Because genomics technologies and analytic methods rapidly evolve, readers will need to stay abreast of new knowledge and consult with their local experts.

MICROARRAY TECHNOLOGY

Several different microarray technologies are currently available. The term gene microarray is typically used to refer to a system in which an oligonucleotide corresponding to a complementary region in a gene of interest is attached to a solid support, such as a chip or a glass slide (1). To determine the expression level of the gene in a sample of interest, ribonucleic acid (RNA) that has been prepared from a sample is first converted to cDNA and then labeled with a system-specific dye, usually a fluorescent molecule. The complementary regions between the oligonucleotide on the surface and the labeled sample hybridize, and the label is detected as the signal, with the intensity of the signal corresponding to the degree of expression (2). The number of oligonucleotides on the solid surface varies according to the manufacturer and type of microarray, ranging from less than 100 genes on a glass slide to chips that cover multiple variants for every known gene in the human genome.

Commercial arrays can generally be described as one-channel or two-channel detection systems. In two-channel systems, two different samples for comparison are labeled with two different dyes and are applied to the same surface (3). The intensities for each of the dyes are compared, thus allowing a comparison between samples to be made on the same chip (ie, normal vs atherosclerotic tissue samples from the same individual). Therefore, the two-channel system is cost efficient because two samples are applied to the same chip. Examples of this type of technology include the Agilent Dual-Mode system and the Eppendorf Dualchip system. Conversely, only one sample is applied per chip in the one-channel system and the relative probe intensities for each gene can be compared to the relative probe intensities for samples run on different chips.

The advantages of this system are that a single bad sample would not affect the data derived for the corresponding comparison sample, as is the case with the two-channel system, and the intensities from one-channel systems are more readily compared among different experiments because, in contrast to the two-channel system, they have not been derived as a value relative to the other sample on the chip. Examples of one-channel systems include the Affymetrix GeneChip and the Illumina BeadChip. The Affymetrix and Illumina systems are the two most popular commercial microarray platforms, and most bioinformatic and statistical software has been developed to analyze the gene expression data derived from these two technologies. Therefore, this chapter will emphasize analysis of gene expression data from these two platforms. However, most of the methods described here can be applied to two-channel microarray data. In general, the analysis flow of gene expression profiling data is identical and is depicted in Figure 7.1.

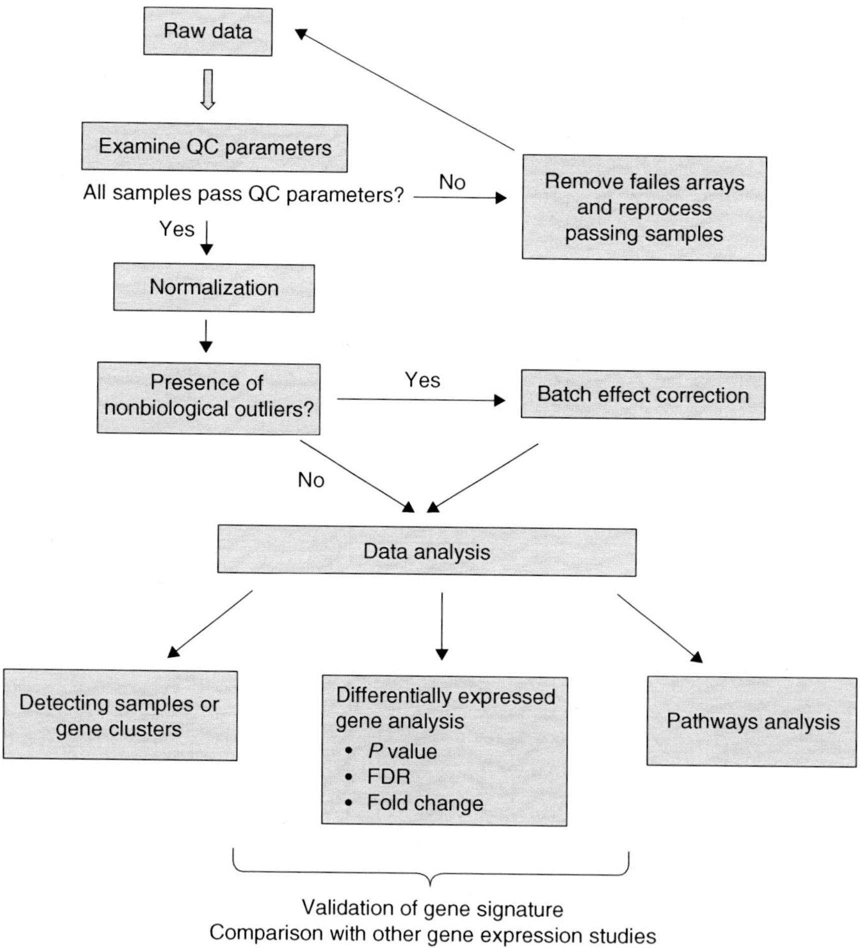

FIGURE 7.1 Schematic of data analysis in microarray gene expression profiling.
Abbreviation: FDR, false discovery rate.

QUALITY CONTROL

Quality control of microarray data is important at the level of the individual laboratory/researcher as well as at the level of the research community in general. Quality control is generally thought of as the parameters that are used to determine whether or not data from a specific sample/chip are acceptable for use, and these parameters may vary based on the manufacturer, the type of sample assessed (ie, tissue samples vs cell lines), and the personal preference of the researcher. Microarray manufacturers typically provide software that allows conversion of the raw data (eg, CEL files in Affymetrix) into actual values per probe and provides output values for quality assessment, usually in text file format. Passing chips should generally have a low level of background noise, adequate probe signal intensity when the background is subtracted, and a high proportion of transcripts that are present and detected reliably. The Affymetrix program assigns each probeset a designation of present ("P"), marginal ("M"), and absent ("A"), with "P" indicating that the transcript was detected, "M" indicating that the transcript was detected at a marginal level, and "A" indicating that the transcript was not detected. The Illumina BeadStudio software provides a detection P value for each probe that indicates the confidence level of the detection of each transcript in the microarray system. The detection P value is, in general, inversely correlated with gene expression. As such, ad hoc approaches have been suggested to remove any probes that have high detection P values (ie, P greater than .05) or absent calls prior to data analysis. This procedure can reduce the possibility of identifying genes that are potentially differentially expressed between pathophysiological conditions, but in fact are not expressed at all. The assessment of quality control parameters allows the researchers to examine the usability of the data for each individual sample.

Quality control measures can also refer to additional details of microarray data, including procedural technique, experimental setup, and data analysis. As the use of microarrays has become more widespread over the past 15 years, data variability has been observed across different microarray platforms, between different laboratories, and even sometimes within the same laboratory. The fact that these differences resulted in data variations for the same cell/tissue type has called the reliability of microarray data into question, and has made the prospect of using microarray data for clinical purposes seem unlikely. In response to these concerns, the Food and Drug Administration (FDA) began the Microarray Quality Control (MAQC) Consortium, which brought together numerous researchers from several institutions to address these issues. Several journals have dedicated entire issues to MAQC reports, including *Nature Biotechnology* (September 2006; August 2010) and *Pharmacogenomics Journal* (August 2010). By using seven different microarray platforms to evaluate data from commercially available cell lines, the MAQC Consortium has released guidelines to help researchers obtain high-quality microarray data, beginning with experimental design and covering the assessment of the quality control outputs provided by each individual manufacturer as well as data analysis.

DATA NORMALIZATION

Data normalization is an essential process for comparison of microarray data among samples that may show varying probe intensities due to differences in procedure, technique, and dye intensities. Further, samples that are processed in different batches at separate periods within the same setting or samples processed in different laboratories are likely to show variations in data that are due to differences in technique or reagents rather than true biological variations, a phenomenon known as the "batch effect." These differences are addressed by the use of normalization techniques. For Affymetrix GeneChip microarrays, DNA-Chip Analyzer (dChip) (4,5) and Robust Multi-array Average (RMA) (6) are two widely used model-based algorithms. dChip estimates gene expression indexes by taking information from both perfect match (PM) and mismatch (MM) probes, while RMA summarizes expression intensities of all PM probes through the median polish procedure. Affymetrix also provides an algorithm, MAS5. These algorithms are available through a Bioconductor package, *affy*, although dChip and MAS5 are available as standalone software. For Illumina BeadChip, expression data can be normalized through the variance stabilization algorithm and quantile normalization procedure that are implemented in the *lumi* bioconductor package (7). In some situations in which the batch effect is pronounced, a statistical technique ComBat (8,9) will be required to correct the nonbiological variation. PCA and multidimensional scaling (MDS) are useful diagnostic tools to visualize the relationship among samples and to identify potential outliers. In the example shown in Figure 7.2, PCA plots are used to illustrate how the batch effect is eliminated after the normalization procedure in conjunction with ComBat. Although statistical techniques may be used to adjust for the batch effect, care should be taken at the level of the experimental design to avoid any confounding factors. For example, in a case–control

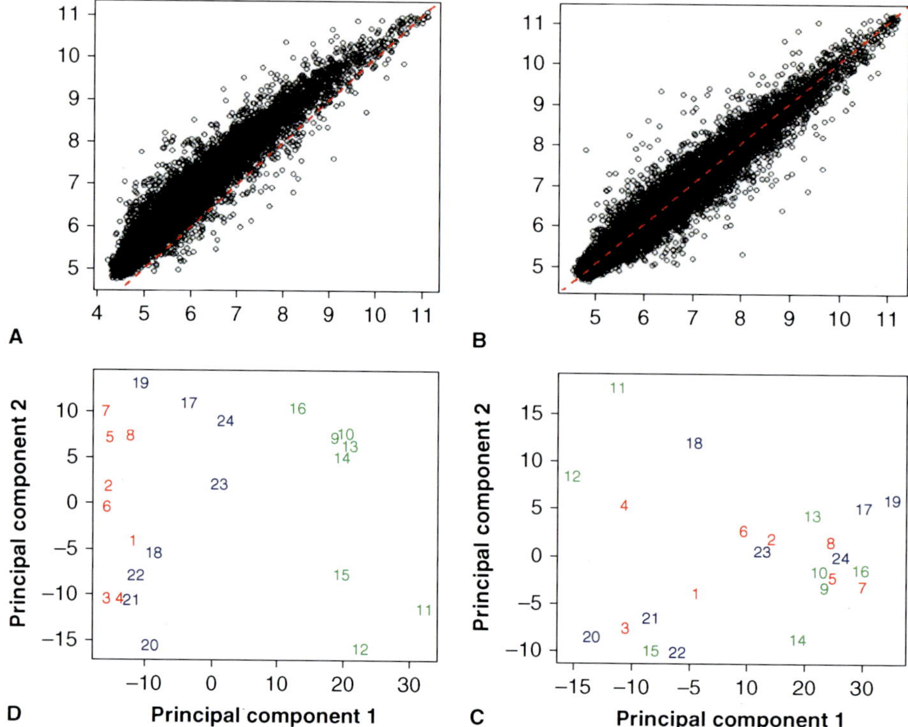

FIGURE 7.2 Diagnostic plots for detecting nonbiological effects in microarray experiments. (A) Scatter plot of gene expression levels from two experimental samples. Each dot represents a gene. The majority of genes have higher expression levels in sample 2 (y-axis) than in sample 1 (x-axis) (eg, above the diagonal line), indicating a systematic measurement bias as a result of instrumental scanning, microarray chips, or other batch effects. (B) After data normalization, most of the nonbiological effects are removed, and the majority of genes are randomly scattered around the diagonal line. (C) Principal component analysis (PCA) plot of gene expression profiling. Each dot represents an experimental sample. Samples from three Illumina chips (ie, batches) are labeled with three colors: red, green, and blue. The separation of samples among the three batches suggests a need for data normalization. (D) After quantile normalization and ComBat batch adjustment, the PCA plot shows no apparent batch effect.

study, case and control samples should be mixed on any single array and run on the same day to avoid any confounding issues.

DATA ANALYSIS

One of the most useful tasks in microarray data analysis is to group genes into different classes based on their expression patterns. This approach can provide much insight into the biological relevance and molecular regulation related to pathophysiological status. Similarly, subsets of patient samples may be grouped based on their expression profiling in order to identify subtypes of heterogeneous disorders that have variable risks or factors of clinical relevance. Several mathematical algorithms and computational tools have been developed, each of which may reveal a different aspect of the data. Therefore, it is important to choose the correct analysis methods to answer a specific question. Some of the most commonly used techniques will be briefly described in the following sections.

Clustering Analysis

Hierarchical clustering algorithms have become the most popular tools for data visualization in gene expression analysis (10). The algorithms cluster both the sample set and the genes into specific clusters based on similar gene expression profiles. Hierarchical clustering can be carried out in either a supervised (analysis to determine ways to accurately split samples/genes or to predict groups of samples or diseases) or an unsupervised (analysis looking for the characterization of the components of a dataset, without a priori input on cases or genes) manner. Unsupervised clustering is usually carried out during the discovery phase in order to find biological subsets within a set of samples and their corresponding gene signatures. In order to obtain the best separation among samples, the samples are often clustered by the genes that satisfy a coefficient of variation (CV) cutoff within the entire set. The algorithm organizes samples or genes into a nested sequence of clusters that can be graphically represented with a tree, called a dendrogram. The similarity among subsets or genes is determined

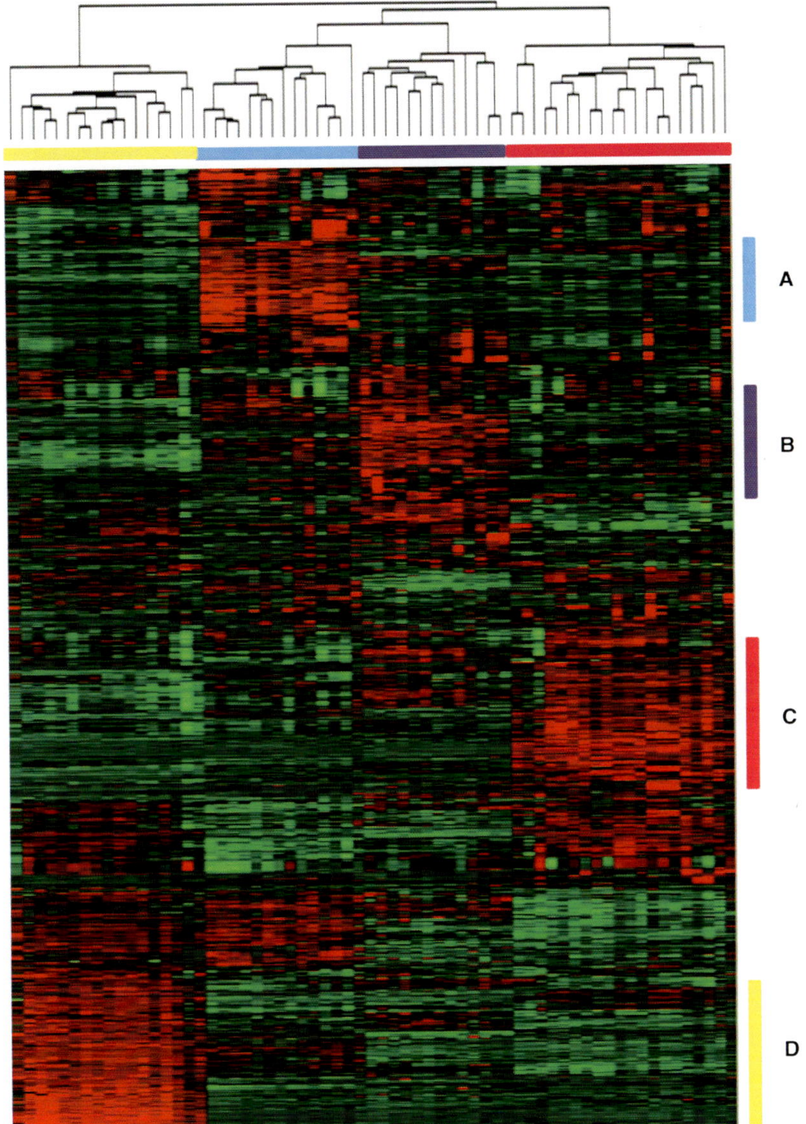

FIGURE 7.3 Cluster analysis and heatmap of microarray gene expression data. Each row in the heatmap represents a gene, and each column represents an experimental sample. Genes and samples are arranged according to their similarity from a hierarchical clustering method. The length of each branch in the cluster dendrogram is a measure of the similarity between genes or samples. Letters A–D corresponding to the color bars on the right side of the heatmap represent gene signatures associated with sample subclusters (color bar beneath the upper dendrogram). The expression values of each gene are standardized with mean 0 and standard deviation 1. In the heatmap, red color represents an increase in gene expression and green color represents a decrease in gene expression.

by the length of the dendrogram bars (see Figure 7.3). Correlation coefficients and Euclidean distance are two similarity metrics that are often used to determine the distance between two branches of a tree. In addition, the hierarchical structure represented in a dendrogram may vary depending on the choice of the agglomeration method (ie, average linkage or complete linkage) and clustering algorithm. Therefore, experienced bioinformaticians are needed to determine the appropriate algorithm among the numerous software programs and to interpret the results (see Table 7.1). For many of the available software programs, the cluster analysis is accompanied by

a heatmap that is a graphical representation of the expression data where individual values contained in the data matrix are represented as colors. The heatmap is a very powerful visualization image that allows researchers to easily identify specific gene signatures pertaining to a subset sample. To obtain greater contrast among the heatmap colors, one may want to center and standardize the expression levels within each gene. This could also avoid potential pitfalls in cluster analysis where the similarity metrics are dominated by genes that are highly expressed but have no biological relevance. In contrast, supervised clustering is performed by clustering the biological samples

TABLE 7.1 Widely Used Bioinformatic Software and Genomics Databases

Provider	Website	Program	Purpose
Broad Institute	genepattern.broadinstitute.org/gp/pages/index.jsf	ExpressionFileCreator ComBat HierarchicalClustering GSEA ARACNE	Data Normalization Batch correction Cluster analysis Ontology and network analysis Network construction
Bioconductor/R	www.bioconductor.org www.r-project.org	SAM/LIMMA/qvalue PCA/MDS ComBat Affy/lumi WGCNA	Differentially expressed gene analysis Data visualization Batch correction Data normalization Network construction
BioDiscovery	www.biodiscovery.com	Nexus Expression	Data normalization, cluster analysis, differentially expressed gene analysis, and pathway analysis
DAVID	david.abcc.ncifcrf.gov	Functional annotation	Ontology and pathway analysis
PANTHER	www.pantherdb.org	Functional annotation	Ontology and pathway analysis
Ingenuity	www.ingenuity.com/products/ipa	IPA	Ontology and pathway analysis
Thomson Reuters	thomsonreuters.com/metacore	MetaCore	Ontology and pathway analysis
Gene Expression Omnibus (GEO)	www.ncbi.nlm.nih.gov/geo		Publicly available gene expression database
Stanford Microarray Database (SMD)	smd.princeton.edu/index.shtml		Publicly available gene expression database
European Bioinformatics Institute (EBI)	www.ebi.ac.uk/arrayexpress	ArrayExpress	Publicly available gene expression database

according to a predefined gene list. Unlike unsupervised clustering, supervised clustering assumes that the biological samples are classified and the objective is to identify clusters that have high probability density with respect to a single class. This method is particularly useful because it can refine a current clinical predefined category and identify a subset of patients who may respond to a treatment or have an elevated risk for a disease. Supervised clustering is an extension of the unsupervised clustering method and, in some ways, it is similar to classification methods such as support vector machine and significance analysis of microarrays (SAM).

Principal Component Analysis and Multidimensional Scaling

PCA is a standard multivariate technique to explore the variability in gene expression data. Each principal component (PC) is a linear combination of genes that represents the whole expression data well. Their ability to represent the original expression data can be measured in terms of the percentage of total variation explained. The first PC explains most of the data variation, and the second PC explains most of the remaining variation, and so on. PCA is a useful dimension reduction technique for gene expression analysis. Normally, this technique can summarize the gene expression variation in the first few PCs. In most cases, the sample grouping in PCA is very similar to the cluster membership in cluster analysis. MDS is another popular visualization method. MDS projects samples from high-dimensional data into a low-dimensional space (normally in two dimensions) in such a way that the distance between any pair of samples will be preserved as much as possible. Therefore, the MDS plot can be used to determine the similarity of gene expression profiles between experimental samples.

Differentially Expressed Gene Analysis

Microarray experiments are often conducted to determine differences in gene expression patterns among different biological classes, such as normal versus atherosclerotic vasculature, early stage versus advanced atherosclerotic lesions, and normal versus failing myocardium. Several statistical methods can be used to identify the most significantly differentially expressed gene for further bioinformatic analysis. This filtering procedure is important for gene prioritization to investigate gene function and biomarker discovery. The T-test, in which the samples are divided into two known subsets and the average value of each probe for the specific subset is compared to the average value of the other subset, is the simplest method for determining differentially expressed genes in known biological sets. Variations of T-tests using empirical Bayes or other shrinkage methods to borrow information across genes have been developed that are particularly useful in small microarray studies (11,12). Nonparametric methods such as the Wilcoxon–Mann–Whitney test are used when gene expression data are not normally distributed. One may want to use fold change (eg, the ratio of average expression levels between two experimental conditions) for further filtering, especially in experimental studies. Those genes that satisfy specific P value and fold change criteria are selected as the significant differentially expressed genes within the subset of interest. An important statistical issue in differentially expressed gene analysis is the inflation of false positives. This is because thousands of statistical comparisons can, by chance alone, lead to the discovery of genes with small P values even though there are, indeed, no differentially expressed genes between the two experimental conditions. Therefore, correction of P values for multiple testing is essential to guard against misleading results. The Bonferroni correction is a traditional method used to counteract the problem of multiple testing. However, it is very stringent and does not conform to the independence assumption in gene expression data because the expression levels of genes are correlated. Thus, false discovery rate (FDR) is normally used instead of the Bonferroni correction to provide a more precise evaluation. The FDR is defined as the expected proportion of false positives among all discoveries (eg, rejected null hypotheses). For example, if 100 genes are considered differentially expressed (ie, P is less than .05), and the maximum FDR level (eg, q-value) for these genes is 0.1, then less than 10 of these genes would be expected to be false positives. Methods to estimate FDR include those developed by Benjamini and Hochberg (13) and John Storey (14,15). SAM (16) is a nonparametric approach to estimate FDR. SAM can also be used in other study designs such as multiple group comparison, quantitative response, time course data, and survival data. This method

has become popular in microarray and any high-throughput data analysis.

ONTOLOGY AND PATHWAY ANALYSIS

Although a list of differentially expressed genes is an important step for the characterization of a specific set of samples, it is often the biological context of differentially expressed genes that will be of greatest use to the researcher. Numerous software tools are now available that detect the biological themes among the genes of interest by determining which pathways are overrepresented in the gene list. For most of these programs, the list of differentially expressed genes is uploaded into the program and various bioinformatic methods are used to determine the overlap between the list of differentially expressed genes and curated gene lists (ie, gene ontology [GO], canonical pathways, gene lists curated from the literature, and bioinformatically derived gene sets). These programs may be freely available or require a purchase, and several of the most frequently used systems are described in the following.

Examples of freely available programs include Protein ANalysis THrough Evolutionary Relationships (PANTHER) (17,18) and Database for Annotation, Visualization and Integrated Discovery (DAVID) (19,20), which compare a list of differentially expressed genes against gene sets that were curated via various mechanisms, including GO terms and canonical pathways. These programs provide a statistical correlation regarding overlap between the gene list and the gene set of interest, allowing biological classification of the particular subset. Commercial software programs such as BioDiscovery Nexus Expression, Ingenuity Pathway Analysis (IPA), and MetaCore also carry out these functions, but have several other features. IPA is the most popular tool on the market. Over the years, substantial features have been embedded into IPA to allow cross-omic data analysis. For example, IPA contains active interface modules that allow the researcher to build potential pathway models within the sample set of interest, to predict which upstream molecules, such as transcription factors or even microRNAs, may be causing the gene expression patterns in the subgroup of interest, and to even predict downstream effects. IPA also provides high-quality representations of pathways and networks for download and publication. MetaCore provides features that are similar to those of IPA, but the license fee is less expensive. Both software programs are PC-platform independent and are run entirely using a web browser. On the other hand, BioDiscovery Nexus Expression software does not provide pathway/network representations for download, but it does accept raw data files (ie, CEL files for Affymetrix chips, iDAT files for Illumina chips, and raw files from a variety of other platforms); therefore, this software enables the user to perform all steps involved in microarray data analysis (raw data normalization, batch correction, hierarchical clustering, and pathway analysis) in sequential steps in the same program. Further, this program can be used to analyze data from platforms for which the available analysis resources are more limited, such as Agilent arrays or even custom arrays.

In contrast to the programs described, which generally measure the overrepresentation of specific pathways in a provided gene list, Gene Set Enrichment Analysis (GSEA) performs biological classification within an entire set of data (21). GSEA comprises various gene sets that are related to curated lists from GO, publications, databases, and bioinformatically derived gene sets, which are referred to as modules. The user first defines a sample phenotype to be used for the comparison, which can consist of two different types of samples (ie, disease vs healthy control) or a continuous phenotype such as a temporal course (ie, the effect of drug dosing over time). In the first step of the analysis, GSEA ranks the expression of each gene within the provided data according to its association with the defined phenotypes, and subsequently ranks the position of the gene within the specific queried gene set. An enrichment score is then calculated for each gene set within the module that is based on the extent to which the genes within that set are enriched with the phenotype of interest. The GSEA output file ranks the enrichment of each gene set within the specific module, providing a normalized enrichment score based on the number of genes within the set and a specified FDR value. Finally, GSEA also provides a gene rank text file that gives a score for each gene within the provided dataset based on its enrichment in the phenotype of interest.

In contrast to the aforementioned enrichment analysis for which prior biological or network knowledge is implemented, several statistical algorithms have been developed for network reconstruction and identification of hub genes between specific genes within the dataset of interest. This approach especially pertains to the discovery of novel biomarkers and therapeutic development due to the fact that many genes are not well annotated in current bioinformatic databases because our knowledge in genome is far from complete. Therefore, this type of "agnostic" analysis could facilitate our understanding of molecular interactions in different physiological states as well as gene prioritization for further experimental investigation. In general, these methods use variations of the correlation analysis (ie, Pearson correlation coefficient), which evaluates the relationship between the expression levels of a gene of interest compared to the expression levels of other genes. This analysis calculates the correlation matrix for each gene in a dataset relative to a gene(s) of interest, resulting in a positive (the expression levels of the two genes show a concordant association) or a negative (the expression levels of the two genes have an opposite association) value in which the absolute magnitude of the value reflects the strength of the association. This type of analysis might allow the researcher to discover novel connections between genes that are significantly associated in the sample set of interest but would not have been associated by using any of the various pathway analysis mechanisms described. This approach can be applied to generate a local network by performing pairwise analysis among a large number of genes. Weighted Gene Co-expression Network Analysis (WGCNA) (22) is such a network analysis program that is gaining in popularity. WGCNA uses topological overlap measures to identify cohesive modules and produces a complex map of interactions called an "adjacency matrix," which indicates the strength of the connection between genes. The software works in the R environment and is available through a Bioconductor package (www.bioconductor.org): WGCNA. The Algorithm for the Reconstruction of Accurate Cellular Networks (ARACNE) (23) is another technique to deal with situations in which the associations between genes are not monotonic or are more subtle. ARACNE uses an information theoretic approach to calculate the interactions among genes relative to a provided "hub" gene(s) in a specific sample type, and like WGCNA generates an adjacency matrix output file. The hub gene and each gene in its corresponding network that has passed a specific cutoff threshold are listed in a row in the output file. A systematic review of these network algorithms and other methods is provided by Allen et al (24).

VALIDATION AND META-ANALYSIS

High-throughput gene expression profiling technologies have gradually become routine research tools for biomarker screening and prioritization. However, the cost of performing such experiments remains high when compared to other traditional assays such as quantitative reverse transcription polymerase chain reaction (RT-PCR). Therefore, most gene expression profiling studies have a limited number of samples and the statistical analysis of such studies could lead to spurious biological findings. Therefore, validation of these results by independent samples is warranted.

To perform such an analysis, one may want to identify publicly available microarray data that were derived from experimental conditions similar to those of the researcher's study. One of the largest curated functional genomics data repositories is Gene Expression Omnibus (GEO; www.ncbi.nlm.nih.gov/geo), which accepts both array- and sequencing-based data. GEO provides tools to help users query and download curated expression profiling data. Once the data are downloaded, the statistical and bioinformatic methods mentioned in this chapter can be used for data analysis. The Stanford Microarray Database (SMD; smd.princeton.edu) (25–27) also provides a large amount of microarray data as well as bioinformatic tools. Recently, the widely used GenePattern bioinformatic suite (28) has been directly incorporated into SMD, making data access and analysis a streamlined process (29). These gene expression repositories also allow researchers to perform meta-analysis of specific genes or pathways. Gene Expression Atlas (www.ebi.ac.uk/gxa), developed by the European Bioinformatics Institute (EBI), provides some useful tools for this type of analysis (30,31).

CONCLUSIONS

In summary, gene expression analysis has the potential to be a powerful tool for evaluating CVD, providing the investigator with the ability to classify patients into distinct subsets, to predict outcome, and even to evaluate the response to treatment. However, the large volume of data obtained from a single study and the recent increase in the number of cardiovascular gene expression studies necessitate the use of valid statistical and bioinformatical algorithms for normalizing, analyzing, and comparing the data among different studies and different platforms. In particular, reducing bias and chance findings (eg, false positive differential gene expression due to a large number of statistical tests) is essential for producing high-quality, valid gene expression studies.

We have introduced the reader to the basic concepts of microarray data analysis, and we have provided brief descriptions of the most commonly used bioinformatic tools. Newer genome-sequencing technologies can generate even more massive datasets by providing whole transcriptome information in coding and noncoding RNAs and splicing variants. This sheer amount of data will be available for researchers to decipher the complexity of transcription regulation for years to come. Hence, a great demand for the development of novel bioinformatic methods, computational infrastructure, and genomic databases is required. Such development will bring forth new knowledge in molecular biology that will eventually open up novel avenues for the prevention and treatment of CVD.

REFERENCES

1. Lockhart DJ, Dong H, Byrne MC, et al. Expression monitoring by hybridization to high-density oligonucleotide arrays. *Nat Biotechnol.* 1996;14(13):1675–1680.
2. Jaluria P, Konstantopoulos K, Betenbaugh M, Shiloach J. A perspective on microarrays: current applications, pitfalls, and potential uses. *Microb Cell Fact.* 2007;6:4.
3. Shalon D, Smith SJ, Brown PO. A DNA microarray system for analyzing complex DNA samples using two-color fluorescent probe hybridization. *Genome Res.* 1996;6(7):639–645.
4. Li C, Hung Wong W. Model-based analysis of oligonucleotide arrays: model validation, design issues and standard error application. *Genome Biol.* 2001;2(8):RESEARCH0032.
5. Li C, Wong WH. Model-based analysis of oligonucleotide arrays: expression index computation and outlier detection. *Proc Natl Acad Sci USA.* 2001;98(1):31–36.
6. Bolstad BM, Irizarry RA, Astrand M, Speed TP. A comparison of normalization methods for high density oligonucleotide array data based on variance and bias. *Bioinformatics.* 2003;19(2):185–193.
7. Du P, Kibbe WA, Lin SM. Lumi: a pipeline for processing illumina microarray. *Bioinformatics.* 2008;24(13):1547–1548.
8. Johnson WE, Li C, Rabinovic A. Adjusting batch effects in microarray expression data using empirical Bayes methods. *Biostatistics.* 2007;8(1):118–127.
9. Chen C, Grennan K, Badner J, et al. Removing batch effects in analysis of expression microarray data: an evaluation of six batch adjustment methods. *PloS one* 2011;6(2):e17238.
10. Eisen MB, Spellman PT, Brown PO, Botstein D. Cluster analysis and display of genome-wide expression patterns. *Proc Natl Acad Sci USA.* 1998;95(25):14863–14868.
11. Smyth GK. Linear models and empirical bayes methods for assessing differential expression in microarray experiments. *Stat Appl Genet Mol Biol.* 2004;3(1):1544–6115.
12. Smyth GK, Michaud J, Scott HS. Use of within-array replicate spots for assessing differential expression in microarray experiments. *Bioinformatics.* 2005;21(9):2067–2075.
13. Benjamini Y, Hochberg Y. Controlling the false discovery rate: a practical and powerful approach to multiple testing *J Roy Statist Soc. Ser B.* 1995;57(1):289–300.
14. Storey J. A direct approach to false discovery rates. *J Roy Statist Soc. Ser B.* 2002;64:479–498.
15. Storey J. The positive false discovery rate: a Bayesian interpretation and the q-value. *Ann Stat.* 2003;31:2013–2035.
16. Tusher VG, Tibshirani R, Chu G. Significance analysis of microarrays applied to the ionizing radiation response. *Proc Natl Acad Sci USA.* 2001;98(9):5116–5121.

17. Mi H, Lazareva-Ulitsky B, Loo R, et al. The PANTHER database of protein families, subfamilies, functions and pathways. *Nucleic Acids Res.* 2005;33(Database issue):D284–D288.

18. Thomas PD, Campbell MJ, Kejariwal A, et al. PANTHER: a library of protein families and subfamilies indexed by function. *Genome Res.* 2003;13(9):2129–2141.

19. Huang da W, Sherman BT, Lempicki RA. Systematic and integrative analysis of large gene lists using DAVID bioinformatics resources. *Nat Protoc.* 2009;4(1):44–57.

20. Huang da W, Sherman BT, Lempicki RA. Bioinformatics enrichment tools: paths toward the comprehensive functional analysis of large gene lists. *Nucleic Acids Res.* 2009;37(1):1–13.

21. Subramanian A, Tamayo P, Mootha VK, et al. Gene set enrichment analysis: a knowledge-based approach for interpreting genome-wide expression profiles. *Proc Natl Acad Sci USA.* 2005;102(43):15545–15550.

22. Zhang B, Horvath S. A general framework for weighted gene co-expression network analysis. *Stat Appl Genet Mol Biol.* 2005;4:Article17.

23. Margolin AA, Nemenman I, Basso K, et al. ARACNE: an algorithm for the reconstruction of gene regulatory networks in a mammalian cellular context. *BMC Bioinformatics.* 2006;7(Suppl 1):S7.

24. Allen JD, Xie Y, Chen M, et al. Comparing statistical methods for constructing large scale gene networks. *PLoS One.* 2012;7(1):e29348.

25. Demeter J, Beauheim C, Gollub J, et al. The Stanford Microarray Database: implementation of new analysis tools and open source release of software. *Nucleic Acids Res.* 2007;35(Database issue):D766–D770.

26. Gollub J, Ball CA, Sherlock G. The Stanford Microarray Database: a user's guide. *Methods Mol Biol.* 2006;338:191–208.

27. Sherlock G, Hernandez-Boussard T, Kasarskis A, et al. The Stanford Microarray Database. *Nucleic Acids Res.* 2001;29(1):152–155.

28. Kuehn H, Liberzon A, Reich M, Mesirov JP. Using GenePattern for gene expression analysis. *Curr Protoc Bioinformatics.* 2008;Chapter 7:Unit 7 12.

29. Hubble J, Demeter J, Jin H, et al. Implementation of GenePattern within the Stanford Microarray Database. *Nucleic Acids Res.* 2009;37(Database issue):D898–D901.

30. Kapushesky M, Adamusiak T, Burdett T, et al. Gene Expression Atlas update—a value-added database of microarray and sequencing-based functional genomics experiments. *Nucleic Acids Res.* 2012;40(Database issue):D1077–D1081.

31. Kapushesky M, Emam I, Holloway E, et al. Gene expression atlas at the European bioinformatics institute. *Nucleic Acids Res.* 2010;38(Database issue):D690–D698.

Whole-Exome and Whole-Genome Sequencing

Sanjiv J. Shah and Donna K. Arnett

TAKE HOME POINTS

1. Whole-exome-sequencing (WES) studies are revolutionizing the approach to identifying genes that cause Mendelian disorders; exome sequencing is also enhancing our ability to identify rare variants with large effects in studies of complex traits.
2. It is critical to match the optimal exome-sequencing study design (including the addition of nonexome-sequencing genetic analytic techniques and bioinformatics) to the available samples, inheritance pattern, and specific disease or trait.
3. Next-generation sequencing technologies such as WES and whole-genome sequencing (WGS) may be useful for clinical diagnostic use; however, several questions surrounding the ethics, financial costs, accuracy, and clinical implications of these tests remain unanswered.

When the Human Genome Project was completed in 2000, there was great hope that the "common disease, common variant" hypothesis would be borne out (1,2). With advances in genotyping of single nucleotide polymorphisms (SNPs), genome-wide association studies (GWAS) subsequently proliferated (3,4), and although several important discoveries were realized, for most complex traits the identified loci using GWAS only explained a fraction of the heritability of these traits (5). For these reasons, investigators began to wonder whether the alternate "common disease, rare variant" hypothesis was in fact more accurate (6). Meanwhile, because the vast majority of SNPs covered by GWAS platforms were noncoding variants, these technologies were of lower utility for Mendelian disorders, the majority of which are due to rare mutations that result in major changes in protein structure.

As the cost of genotyping and sequencing steadily dropped through the GWAS era, and massively parallel sequencing technologies emerged, the prospect that WES and WGS would be available for use in routine studies of human genetics soon became a reality (7–10). In 2009, the first study to use exome sequencing for a Mendelian disease was published (11). In this study, Ng et al demonstrated the feasibility of WES by identifying causal variants for Freeman–Sheldon syndrome, which is associated with congenital joint contractures and myopathy (11). The number of WES studies, especially for the study of Mendelian disorders, began to rise rapidly thereafter. In 2010, WES was first reported to be used for clinical diagnoses, especially Mendelian disorders, followed by a report by Worthey et al in 2011, which demonstrated the utility of WES for both clinical diagnosis and targeted treatment based on the identified causal variant (12,13). In this study, a 15-month-old child with a Crohn-disease-like illness was diagnosed with X-linked inhibitor of apoptosis deficiency based on the clinical presentation, WES genetics, and functional testing, leading to treatment with an allogeneic hematopoietic progenitor cell transplant for the prevention of hemophagocytic lymphohistiocytosis, a life-threatening condition associated with X-linked

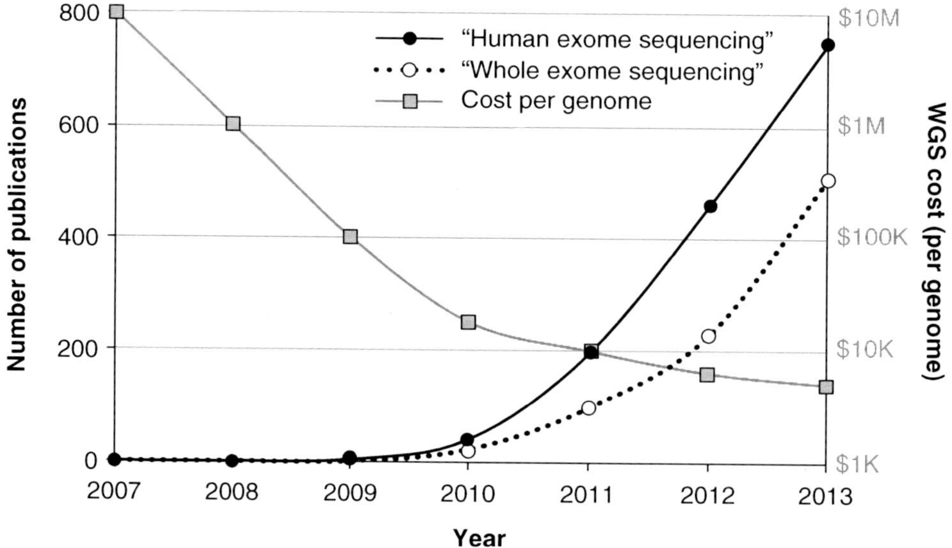

FIGURE 8.1 Rise in the number of exome-sequencing publications with declining sequencing costs. Publications were identified using the following search terms in PubMed: "human exome sequencing" and "whole exome sequencing." *Abbreviation*: WGS, whole-genome sequencing.

inhibitor of apoptosis deficiency (13). This dramatic example illustrates the power of next-generation sequencing technologies such as WES for personalized medicine.

Although examples such as the study by Worthey et al are not yet commonplace, the prospect of pinpointing the genetic causes of disease (and subsequently matching the mechanism of disease with targeted treatment) using technologies such as WES and WGS is promising. Since the initial publication by Ng et al in 2009, the number of exome-sequencing publications has increased exponentially in step with the rapidly dropping costs of sequencing (Figure 8.1). Given the emergence of WES and WGS, it is vital for cardiovascular clinicians and researchers to understand the strengths, limitations, challenges, and clinical utility of these important next-generation sequencing technologies. Here we review the following aspects of WES and WGS: (a) the technological aspects of next-generation sequencing technologies; (b) available strategies for the conduct of next-generation sequencing studies; (c) the application of next-generation sequencing to cardiovascular disease; (d) next-generation sequencing for clinical diagnostics; and (e) future directions.

OVERVIEW OF EXOME CHIP, WES, AND WGS TECHNOLOGIES

The advent of GWAS after the Human Genome Project was made possible due to the explosion of DNA microarray technologies. However, these arrays require a priori knowledge of the genetic variation (eg, SNPs) to be studied. Traditional Sanger sequencing allows for the determination of known and unknown genetic variation but is quite slow and expensive because each nucleotide must be analyzed sequentially. Next-generation sequencing, an umbrella term that encompasses WES and WGS, constituted a major advance in the field of genomics because it allowed for much more rapid sequencing since millions of DNA fragments are analyzed simultaneously (ie, massively parallel sequencing) (10). The remainder of this chapter focuses predominantly on WES, but many of the principles discussed are also applicable to WGS. In WES, as shown in Figure 8.2, genomic DNA is first randomly sheared into DNA fragments (in vitro shotgun library). This library then undergoes enrichment for sequences corresponding to exons using a technique termed hybridization capture. The hybridized fragments are then recovered, after which massively parallel sequencing occurs. Sequences then undergo mapping, alignment, and variant calling to identify the genetic variation in the exome (7).

Notably, exome chip microarray technology is different than WES. Using data from greater than 12,000 exomes and whole-genome sequences from individuals of multiple race/ethnicities, exome chips that contain rare exonic variants that are thought to be functional (ie, result in a change in protein structure) have been constructed (14). The exome chip therefore spans the gap between GWAS SNP arrays and WES. The exome chips have the advantages of being (a) a cost-effective and quick method for genotyping

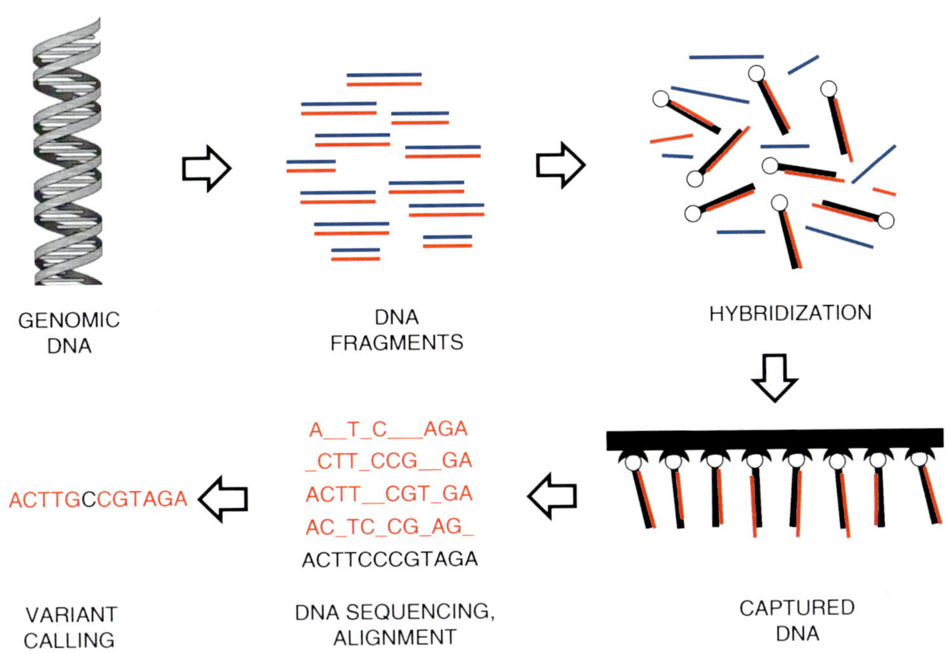

FIGURE 8.2 Whole-exome-sequencing process. An in vitro shotgun library is first created from randomly sheared genomic DNA. The library is then enriched for sequences corresponding to exons by hybridization capture. Pull-down is used to recover hybridized fragments. These fragments then undergo massively parallel sequencing, mapping, alignment, and variant calling.

exonic variants, and (b) feasible for large studies. However, the major disadvantages of the exome chip are the lack of identifying very rare variants and de novo variation in the human genome.

Given the rapidly decreasing costs of next-generation sequencing (Figure 8.1), it is likely that WES will render the exome chip obsolete in the near future. Thus, a paradigm shift has occurred, and researchers now view next-generation sequencing technologies as an essential component of the study of both rare and common diseases and traits (10). WES, in particular, may be the "sweet spot" between GWAS and WGS, given its current cost and focus on protein-coding variants (15). However, certain limitations and pitfalls must be considered when designing and interpreting research studies and clinical diagnostics that use WES.

First, although WES is appealing because of its ability to sequence the majority of the coding variants in the genome, studies from the ENCODE group demonstrate that up to 80% of the human genome (the majority of which is noncoding) has functional significance (16). Variation in noncoding regions, including highly conserved regulatory regions, may be critical for disease pathogenesis (16). For these cases, WGS will be required to fully understand the genetic architecture of some diseases and traits.

Second, when designing experiments that benefit from next-generation sequencing technologies, it is assumed that mutations in a single gene cause a single disease. However, there are many examples of allelic heterogeneity (a single disease or trait caused by mutations in different genes) and phenotypic heterogeneity (different mutations in the same gene causing different phenotypes). There are several known cardiovascular examples of these phenomena; hypertrophic cardiomyopathy is an example of marked allelic heterogeneity (17,18), and mutations in the *SCN5A* gene, which cause a variety of electrophysiological disorders, including the long QT syndrome and Brugada syndrome, is an example of phenotypic heterogeneity (19).

Third, although WES continues to improve and mature as a next-generation sequencing technology, there are still technical problems. Initial WES platforms required relatively large amounts of high-quality DNA, which made it difficult to use previously collected samples. In addition, a significant portion of the exome (5%–10%) is not well covered by exome-sequencing technology, and coverage of insertions and deletions (indels) and assessment of copy number variation are problematic with WES.

Fourth, next-generation sequencing studies are computationally intensive, and narrowing down the large number of identified variants in order to pinpoint the causal mutation remains challenging. Fortunately, continual advances in bioinformatics, and computational and statistical approaches, will hopefully bridge the gap between rapid advances in sequencing technologies and the accurate identification of causal variants (20).

AVAILABLE STRATEGIES FOR THE CONDUCT OF WES STUDIES

Mendelian Disorders

Monogenic disorders and traits that follow a Mendelian pattern of inheritance have benefitted the most from WES. Studies of Mendelian disorders conducted prior to the WES era were limited because traditional gene-discovery techniques did not work well in the setting of any of the following conditions: reduced penetrance, locus heterogeneity, and small numbers of cases or families available to study a particular disorder (8,9). With the advent of technologies such as WES, the challenge of identifying the causal variant for Mendelian disorders resides in the determination of which variant is pathogenic. Countless variants will no doubt be present in every individual's genome—therefore, strategies to prioritize and hone in on the causal variant are crucial. Thus far, for WES studies of Mendelian disorders, the filtering of variants has been based on several assumptions about the causative mutation: (a) it is unique to the study participants or very rare in the population; (b) it has a major effect on the disease or trait of interest; and (c) it resides in an exon and directly alters the structure or function of a protein.

Once causal variants or variants are identified, further evidence of pathogenicity involves evaluating for conservation of the locus, which is likely to be beneficial given that most pathogenic missense variants affect highly conserved nucleotides (8,9).

Complex Traits

Compared to Mendelian traits, the use of next-generation sequencing tools for the identification of causal variants for common, complex diseases and traits is still in its infancy. The use of WES for complex traits is hampered by the genetic heterogeneity of these traits; studies involving large numbers of individuals will most likely be necessary in order to have enough statistical power to detect rare variants (21). Thus, cost considerations are a primary obstacle for WES studies of complex traits. At the present time, studying extremes of the distribution for complex traits or diseases, in order to decrease the number of individuals who undergo WES, is the most commonly used strategy.

Overall Study Design Considerations

For next-generation sequencing studies, selection of samples will be critically important both for keeping costs down and to optimize the ability to identify causal variants (21). Choosing the correct study strategy or strategies, outlined below, is equally important for optimal exome-sequencing study design. Finally, in contrast to GWAS, where accounting for population structure and ancestry is now routine and well-developed, WES studies (particularly those that include unrelated individuals or the study of complex traits) require meticulous matching of ancestry among cases and controls, given the high probability of false positive signals due to small differences in population structure (21).

Specific Study Design and Analytic Strategies for Exome-Sequencing Studies

Several strategies, which have been reviewed in detail by Gilissen et al and others (7–9), exist for exome-sequencing studies. The choice of strategy or strategies depends on several factors, including the type of disease/trait (Mendelian vs complex), the available samples, and the inheritance pattern.

In a *linkage strategy*, investigators must have access to a family with a heritable disorder that appears to be monogenic. Affected family members undergo WES and shared variation among these individuals is identified, thereby honing in on the location of the causal variant (and reducing the number of private variants). If cost is a factor in deciding how many affected family members can undergo exome sequencing, the most distantly related affected family members should be selected. The closer the relatedness between sequenced family members, the more difficult it will be to find the causal variants (for example, siblings share 50% of their DNA sequence, making differentiation between causal and private variants [ie, benign variation] quite difficult).

In the *overlap strategy*, multiple unrelated affected individuals are studied using WES, and shared variation among cases is used as a way to determine the causal variant. The overlap strategy is helpful for disorders or traits that follow an autosomal dominant inheritance pattern (ie, heterozygous causal variant) because most private, benign mutations are heterozygous, so finding the causal variant is more difficult than for homozygous mutations (ie, recessive disorders). It is important to note that the overlap strategy will not work if there is genetic heterogeneity (unless large samples of affected patients are recruited and studied using WES).

In the de novo *strategy*, the working hypothesis is that the genetic mutation causing a rare disorder is not present in either parent of the case patient (ie, the causal variant is a de novo mutation within the affected patient). The de novo strategy also assumes that the disease of interest is monogenic and caused by a coding variant. Once the case patient and parents undergo WES, filtering out all inherited variants in the case patient will typically result in only 0 to 3 de novo mutations, thereby helping to narrow down considerably the list of potential pathogenic variants. The de novo strategy is particularly useful for disorders such as congenital heart disease where the genetic mutation is associated with reduced reproductive fitness.

The *homozygosity-based strategy* is most useful for rare, autosomal recessive disorders in families where consanguinity is known or suspected. This approach works on the assumption that the disease or trait is due to a homozygous variant inherited from both parents, and that the causal variant resides within a large homozygous region. By prioritizing homozygous variants situated in homozygous regions—typically using a combination of both SNP microarrays and WES, though WES alone may be sufficient (22)—investigators can narrow down the causative locus/variant. The homozygosity-based approach allows the use of WES in a single affected individual from a consanguine family for initial identification of the causal variant.

The *double-hit strategy* can also result in identification of the causal variant through WES in a single patient. This strategy works for disorders with an autosomal recessive inheritance pattern and no known or suspected consanguinity in the family. By looking for nonsynonymous homozygous or compound heterozygous variants (ie, two heterogeneous recessive alleles at a particular locus that can be pathogenic in the heterozygous state), which are typically rare in outbred individuals, investigators can narrow down the list of potential causal variants.

In situations where only a single patient is available for study and the genetic disorder is thought to follow an autosomal dominant pattern, the *candidate-based strategy* can be used. In this approach, exome sequencing is performed on the case patient, and variants that are expected to result in a major impact on protein structure/function (eg, frame-shifting mutations, stop mutation, and mutations in canonical splice sites, especially if these mutations are highly conserved) are prioritized. The disadvantage of the candidate-based approach is that it typically relies heavily on a priori knowledge of the pathophysiology of the disorder of interest so that variants in known molecular pathways can be prioritized; however, this is not always the case in Mendelian disorders, which are often found to result from novel, unsuspected pathways.

A final approach, the *extremes of the distribution strategy*, which is most useful for the study of quantitative complex traits or complex traits with rare presentations (such as very young individuals who suffer from myocardial infarction), involves performing WES on the extremes of the distribution (7).

For example, patients with very low cholesterol levels could be compared to controls with cholesterol levels that are within the normal range. The advantages of this approach are the ability to perform exome sequencing on a limited number of individuals (thereby keeping costs down) and enriching the cases to include those individuals most likely to harbor rare variants in coding regions that could lead to such extremes of a particular phenotype.

APPLICATION OF WES TO CARDIOVASCULAR DISEASE

There are several instructive examples of the application of WES to cardiovascular diseases and traits (23–30). For example, in a recent study of recurrent cardiac arrest in infants, Crotti et al performed exome sequencing on two unrelated probands with recurrent cardiac arrest (both of whom also had a markedly prolonged QT interval) and their healthy, unaffected parents using a combination of de novo and overlap strategies (25). By prioritizing rare variants, the authors found de novo variants in the calmodulin genes *CALM1* and *CALM2*. To determine whether these mutations were causal, the authors followed up on these results by performing candidate gene sequencing in 82 individuals with unexplained congenital long QT syndrome, among whom two additional infants with recurrent cardiac arrest harbored the heterozygous *CALM1* and *CALM2* mutations found in the two probands. Functional studies were also performed to determine the pathophysiological effect of the genetic mutations. The causal role of the variants was supported by the finding that recombinant mutant calmodulin demonstrated significantly reduced calcium-binding affinity.

Table 8.1 lists several example studies that have used a variety of the aforementioned WES strategies to identify causal variants for cardiovascular disorders. Each of these studies demonstrates both the successes and challenges of next-generation sequencing in the genetic dissection of cardiovascular diseases and traits.

NEXT-GENERATION SEQUENCING TECHNOLOGIES FOR CLINICAL DIAGNOSTICS

In addition to the use of next-generation sequencing tools in research studies, clinical application of these technologies has emerged, as demonstrated above in the case of the patient with X-linked inhibitor of apoptosis deficiency. Application of tools such as

TABLE 8.1   Examples of Whole-Exome-Sequencing Strategies Applied to Cardiovascular Diseases

Author	Disease	WES Strategy*	Main Findings	Notes
Musunuru et al. (26)	Familial combined hypolipidemia	Double-hit, linkage analysis	Two probands (siblings) from a family with familial combined hypolipidemia underwent WES. The two studied individuals shared only one gene (*ANGPTL3*) that contained novel variants (both within exon 1 of the gene). Linkage analysis using quantitative traits (LDL- and HDL-cholesterol levels) was also used to verify that the locus associated with cholesterol levels included *ANGPTL3*. Sanger sequencing of exon 1 in other family members revealed very low cholesterol levels in compound heterozygotes.	The double-hit strategy requires an autosomal recessive disorder and assumes that a single rare homozygous or two rare compound heterozygous mutations cause the disease of interest.
Norton et al. (27)	DCM	Linkage analysis	In a study of 48 individuals with DCM from 17 families, the authors found that truncating variants in *TTN*, the gene that encodes titin, were associated with DCM in 7/17 families.	The linkage analysis approach to WES analyses assumes that a fully penetrant mutation segregates with the disorder. Although the authors found that truncating *TTN* variants did segregate with DCM, two novel truncating *TTN* variants did not segregate with DCM, illustrating the challenge of determining whether variants identified by WES are truly pathogenic.
Theis et al. (28)	DCM	Homozygosity mapping, linkage analysis	In a single family (17 adult descendants of first cousins) studied by echocardiography, two female siblings were found to have DCM and an additional family member was found to have idiopathic LV enlargement. After linkage analysis mapped an autosomal recessive DCM locus to 7q21, WES was performed along with iterative bioinformatics to identify a homozygous missense mutation in *GATAD1*, which encodes GATA zinc finger domain-containing protein 1, as the cause of the autosomal recessive DCM.	GATAD1 regulates gene expression by binding to a histone modification site. Thus, the study by Theis et al. supports findings from experimental mouse models of genetic disruption of histone deacetylases, which have implicated epigenetic dysregulation as a cause of LV dysfunction. This study also shows the utility of homozygosity mapping in a single affected family with consanguinity for the study of an autosomal recessive trait.
Weeke et al. (29)	Drug-induced long QT syndrome	Overlap	WES was performed on 65 patients with drug-induced long QT syndrome and 148 drug-exposed control subjects, all of European descent. Rare variants in seven genes were identified through inspection of shared variants among cases. After replication studies, rare coding variants in the *KCNE1* and *ACN9* emerged as risk factors for drug-induced long QT syndrome.	The success of the overlap strategy requires: (a) multiple unrelated individuals with the disease phenotype or interest; (b) a small number of causative loci; and (c) accurate phenotyping to truly differentiate cases from controls and also to make sure that the cases have a homogeneous disorder.

TABLE 8.1 Examples of Whole-Exome-Sequencing Strategies Applied to Cardiovascular Diseases (*continued*)

Author	Disease	WES Strategy*	Main Findings	Notes
Zaidi et al. (30)	Congenital heart disease	de novo	Using WES of parent–offspring trios, the odds ratio of de novo mutations in 362 severe congenital heart disease cases was found to be 7.5 for damaging mutations when compared to 264 controls. De novo mutations in histone modifying (methylation and ubiquitination) genes were found to be especially common in patients with congenital heart disease.	The de novo strategy is most useful in sporadic cases of diseases that reduce fecundity. Once all inherited mutations are filtered out (after comparing exomes from the proband with parents), individuals typically have only 0–3 de novo mutations, which greatly reduces the number of possible causative variants.
Boczek et al. (23)	Long QT syndrome	Linkage analysis	WES was performed on three members (symptomatic index case, unaffected father, and affected maternal aunt) from a 15-member, multigenerational family with a history of autosomal dominant long QT syndrome. Affected members of the family had been previously tested negative for known genetic mutations that cause long QT syndrome. Using the overlap technique along with bioinformatics tools, novel mutations in the L-type calcium channel (*CACNA1C*) were identified.	This study was unique in its use of three systems biology ranking algorithms for the ranking of potential causative genes to narrow down the identified variants from WES.
Boyden et al. (24)	PHAII (hypertension, hyperkalemia, and metabolic acidosis)	Linkage analysis using GWAS data, WES, and targeted resequencing	A cohort of 52 PHAII kindreds (n = 126 affected individuals) were included in the study. Genome-wide genotyping of SNPs was performed in all individuals in all kindreds for linkage analysis. Index cases of 11 kindreds underwent WES to identify causative genes within the linkage peaks. Targeted resequencing of the most likely causative gene (*KLHL3*) and a related gene (*CUL3*) resulted in the identification of novel pathogenic variants associated with PHAII.	Despite the challenges of locus heterogeneity, mixed models of transmission, and frequent de novo mutations, the study by Boyden et al. was successful because the authors used a combination of techniques to identify novel genes for PHAII, providing new insight into the physiology of blood pressure regulation.

*Gilissen et al (9) provides an excellent review of disease identification strategies for exome-sequencing studies.
Abbreviations: DCM, dilated cardiomyopathy; HDL, high-density lipoprotein; LDL, low-density lipoprotein; PHAII, pseudohypoaldosteronism, type II; WES, whole-exome sequencing.

WES to the clinical realm has advantages and disadvantages, along with potential ethical issues.

In a provocative study on the clinical application of WGS, Dewey et al conducted an exploratory study of 12 adults who all underwent WGS (31). Several problems in the clinical interpretation of WGS were identified by this study. First, while WGS was capable of identifying previously described single nucleotide variants, insertion/deletion variants were not well covered or detected accurately. Review of the detected, possibly pathogenic variants by trained professionals required an enormous amount of time for each study participant, and there was only moderate agreement between professionals (kappa = 0.52, 95% confidence interval 0.40–0.64). Furthermore, after manual review, 69% of genetic variants initially categorized as disease-causing by automated query of mutation databases were reclassified as variants of uncertain or lesser significance. This important study demonstrated that WGS is most likely not ready at the present time for clinical use in the general population, given the low reproducibility of results and the uncertainty regarding clinically reportable findings.

On the other hand, for patients with unexplained Mendelian disorders, exome sequencing may be useful. In a study of 250 consecutive probands who underwent WES for clinical purposes, an underlying genetic defect (many of which occurred de novo) was identified in 25% of the study participants (32). In cases of Mendelian disorders, WES may be able to streamline identification of the causal variant, and, in some cases, may be more cost-effective than traditional genetic testing panels (33). However, the ethical issues of what to do with incidental (but medically actionable) findings in WES, and the ability to provide appropriate genetic counseling for patients who undergo WES, remain controversial areas. Nevertheless, unbiased WES can have unprecedented ability to increase diagnostic accuracy, a phenomenon known as reverse phenotyping (34). In these cases, causal variants identified by exome sequencing assist clinicians in identifying the correct diagnosis, which can have substantial impact on patient care.

FUTURE DIRECTIONS

Exome sequencing has proven to be quite useful for a large number of Mendelian disorders. However, the use of WES for the dissection of complex traits has proven to be more difficult. In some cases, the identification of rare variants through studies of Mendelian disorders has led to discoveries that can impact the population. The discovery that mutations in *PCSK9* could lead to very low levels of low-density lipoprotein cholesterol led to new biological insight of cholesterol metabolism and ultimately gave rise to a whole new class of lipid-lowering therapies (35–38). Nevertheless, despite scientific advances through the study of Mendelian disorders, it is expected that even more discoveries will be made through exome sequencing for common, complex traits.

In the future, through the continually decreasing costs of WES; the increasing availability of large cohorts of well-phenotyped individuals with cardiovascular diseases or traits; and improvements in analytic approaches, it is foreseeable that WES on a large scale will become more feasible, and therefore may allow for the "common disease, rare variant" hypothesis to be proven correct, thereby providing new molecular insight into the pathogenesis of common cardiovascular diseases and syndromes such as hypertension, heart failure, coronary artery disease, and atrial fibrillation. Exome-sequencing studies of complex traits will however still be challenging; rare, noncoding genetic mutations may play a major role in the pathogenesis of common diseases—therefore, it will take WGS to study these variants; it will remain challenging to combine results from heterogeneous sequencing platforms and analytic techniques across large population-based cohort studies; and just like results from GWAS, the identified variant only represents the tip of the iceberg—there will still be a need for functional studies, and cellular and model organ systems, to confirm the pathogenic role of the identified variant (21).

CONCLUSIONS

The future is bright for next-generation sequencing technologies, which are currently bringing about a paradigm shift in both clinical and research genetics and genomics. At the present time, exome sequencing may represent the sweet spot between GWAS and WGS in terms of its cost and focus on rare protein-coding variants. However, when designing research studies that use exome sequencing, carefully examining the available samples and inheritance patterns will be critical for choosing the best combination of strategies (both exome-sequencing and nonexome-sequencing techniques, in addition to bioinformatics) to increase the chances of successfully identifying causal variants. There are already several examples of cardiovascular diseases and traits that have benefitted from exome-sequencing approaches for the identification of causal variants and novel pathways. As sequencing technology continues to improve, and costs continue to fall, it is not implausible to believe that personalized diagnostics and targeted therapeutics will be within reach.

REFERENCES

1. Collins FS. Shattuck lecture—medical and societal consequences of the Human Genome Project. *N Engl J Med.* 1999;341(1):28–37.
2. Kruglyak L. The road to genome-wide association studies. *Nat Rev Genet.* 2008;9(4):314–318.
3. Feero WG, Guttmacher AE, Collins FS. Genomic medicine—an updated primer. *N Engl J Med.* 2010;362(21):2001–2011.
4. O'Donnell CJ, Nabel EG. Genomics of cardiovascular disease. *N Engl J Med.* 2011;365(22):2098–2109.
5. Manolio TA, Collins FS, Cox NJ, et al. Finding the missing heritability of complex diseases. *Nature.* 2009;461(7265):747–753.
6. Schork NJ, Murray SS, Frazer KA, Topol EJ. Common vs. rare allele hypotheses for complex diseases. *Curr Opin Genet Dev.* 2009;19(3):212–219.
7. Bamshad MJ, Ng SB, Bigham AW, et al. Exome sequencing as a tool for Mendelian disease gene discovery. *Nat Rev Genet.* 2011;12(11):745–755.
8. Gilissen C, Hoischen A, Brunner HG, Veltman JA. Unlocking Mendelian disease using exome sequencing. *Genome Biol.* 2011;12(9):228.
9. Gilissen C, Hoischen A, Brunner HG, Veltman JA. Disease gene identification strategies for exome sequencing. *Eur J Hum Genet.* 2012;20(5):490–497.
10. Majewski J, Schwartzentruber J, Lalonde E, Montpetit A, Jabado N. What can exome sequencing do for you? *J Med Genet.* 2011;48(9):580–589.
11. Ng SB, Turner EH, Robertson PD, et al. Targeted capture and massively parallel sequencing of 12 human exomes. *Nature.* 2009;461(7261):272–276.
12. Schuler BA, Prisco SZ, Jacob HJ. Using whole exome sequencing to walk from clinical practice to research and back again. *Circulation.* 2013;127(9):968–970.
13. Worthey EA, Mayer AN, Syverson GD, et al. Making a definitive diagnosis: successful clinical application of whole exome sequencing in a child with intractable inflammatory bowel disease. *Genet Med.* 2011;13(3):255–262.
14. Abecasis GR, Altshuler D, Auton A, et al. A map of human genome variation from population-scale sequencing. *Nature.* 2010;467(7319):1061–1073.
15. Teer JK, Mullikin JC. Exome sequencing: the sweet spot before whole genomes. *Hum Mol Genet.* 2010;19(R2):R145–R151.
16. Ward LD, Kellis M. Interpreting noncoding genetic variation in complex traits and human disease. *Nat Biotechnol.* 2012;30(11):1095–1106.
17. Schwartz K. Familial hypertrophic cardiomyopathy. Nonsense versus missense mutations. *Circulation.* 1995;91(12):2865–2867.
18. Schwartz K, Carrier L, Guicheney P, Komajda M. Molecular basis of familial cardiomyopathies. *Circulation.* 1995;91(2):532–540.
19. Remme CA, Wilde AA, Bezzina CR. Cardiac sodium channel overlap syndromes: different faces of SCN5A mutations. *Trends Cardiovasc Med.* 2008;18(3):78–87.
20. Stitziel NO, Kiezun A, Sunyaev S. Computational and statistical approaches to analyzing variants identified by exome sequencing. *Genome Biol.* 2011;12(9):227.
21. Do R, Kathiresan S, Abecasis GR. Exome sequencing and complex disease: practical aspects of rare variant association studies. *Hum Mol Genet.* 2012;21(R1):R1–R9.
22. Becker J, Semler O, Gilissen C, et al. Exome sequencing identifies truncating mutations in human SERPINF1 in autosomal-recessive osteogenesis imperfecta. *Am J Hum Genet.* 2011;88(3):362–371.
23. Boczek NJ, Best JM, Tester DJ, et al. Exome sequencing and systems biology converge to identify novel mutations in the L-type calcium channel, CACNA1C, linked to autosomal dominant long QT syndrome. *Circ Cardiovasc Genet.* 2013;6(3):279–289.
24. Boyden LM, Choi M, Choate KA, et al. Mutations in kelch-like 3 and cullin 3 cause hypertension and electrolyte abnormalities. *Nature.* 2012;482(7383):98–102.
25. Crotti L, Johnson CN, Graf E, et al. Calmodulin mutations associated with recurrent cardiac arrest in infants. *Circulation.* 2013;127(9):1009–1017.
26. Musunuru K, Pirruccello JP, Do R, et al. Exome sequencing, ANGPTL3 mutations, and familial combined hypolipidemia. *N Engl J Med.* 2010;363(23):2220–2227.
27. Norton N, Li D, Rampersaud E, et al. Exome sequencing and genome-wide linkage analysis in 17 families illustrate the complex contribution of TTN truncating variants to dilated cardiomyopathy. *Circ Cardiovasc Genet.* 2013;6(2):144–153.
28. Theis JL, Sharpe KM, Matsumoto ME, et al. Homozygosity mapping and exome sequencing reveal GATAD1 mutation in autosomal recessive dilated cardiomyopathy. *Circ Cardiovasc Genet.* 2011;4(6):585–594.
29. Weeke P, Mosley JD, Hanna D, et al. Exome sequencing implicates an increased burden of rare potassium channel variants in the risk of drug-induced long QT interval syndrome. *J Am Coll Cardiol.* 2014;63(14):1430–1437.
30. Zaidi S, Choi M, Wakimoto H, et al. De novo mutations in histone-modifying genes in congenital heart disease. *Nature.* 2013;498(7453):220–223.
31. Dewey FE, Grove ME, Pan C, et al. Clinical interpretation and implications of whole-genome sequencing. *JAMA.* 2014;311(10):1035–1045.
32. Yang Y, Muzny DM, Reid JG, et al. Clinical whole-exome sequencing for the diagnosis of mendelian disorders. *N Engl J Med.* 2013;369(16):1502–1511.
33. Jacob HJ. Next-generation sequencing for clinical diagnostics. *N Engl J Med.* 2013;369(16):1557–1558.
34. Schulze TG, McMahon FJ. Defining the phenotype in human genetic studies: forward genetics and reverse phenotyping. *Hum Hered.* 2004;58(3–4):131–138.
35. Cohen J, Pertsemlidis A, Kotowski IK, Graham R, Garcia CK, Hobbs HH. Low LDL cholesterol in individuals of African descent resulting from frequent nonsense mutations in PCSK9. *Nat Genet.* 2005;37(2):161–165.
36. Abifadel M, Varret M, Rabes JP, et al. Mutations in PCSK9 cause autosomal dominant hypercholesterolemia. *Nat Genet.* 2003;34(2):154–156.
37. Blom DJ, Hala T, Bolognese M, et al. A 52-week placebo-controlled trial of evolocumab in hyperlipidemia. *N Engl J Med.* 2014;370(19):1809–1819.
38. Roth EM, McKenney JM, Hanotin C, Asset G, Stein EA. Atorvastatin with or without an antibody to PCSK9 in primary hypercholesterolemia. *N Engl J Med.* 2012;367(20):1891–1900.

9

CHAPTER

Gene–Environment Interactions

Stella Aslibekyan

TAKE HOME POINTS

1. Gene–environment interactions is the umbrella term for a variety of scenarios that describe how environmental and genetic risk factors relate to disease susceptibility or progression and/or treatment response.
2. Despite the potential of gene–environment interaction studies to improve cardiovascular disease (CVD) risk prediction and personalized treatment approaches, few findings to date have been replicated or implemented in the clinic.
3. Research practices that are likely to yield clinically useful gene–environment interaction findings include ensuring adequate sample size in the design phase, appropriate confounder adjustment, reporting stratum-specific estimates, rigorous correction for multiple testing, replication in independent populations, and/or functional validation.

Although the role of genetic factors in the pathogenesis of cardiovascular disease (CVD) is supported by a large and constantly growing body of evidence, the identified genes account for only a small percentage of heritable risk (1), suggesting that the answer to the age-old "nature or nurture" question in the context of complex traits is "always both." Many of the genetic markers linked to CVD only confer risk in the presence of specific environmental factors, for example, lifestyle or medication use. As a result, the clinical utility of these genetic markers as diagnostic or risk stratification tools becomes rather limited if the patient's environment is not taken into account. Similarly, the effect of many environmental variables varies greatly among individuals, and such variation

is often mediated by genetics—a fact indirectly and anecdotally corroborated by the ubiquitous trope of a chain-smoking, junk-food-ingesting, couch potato relative or acquaintance who lives to become a centenarian (2). The etiology of CVD is complex, nonlinear, and multifactorial. Therefore, a nuanced understanding of interactions between genetic and environmental factors is essential for successfully translating genetic findings into clinical applications.

DEFINING INTERACTIONS

In 1946, J. B. S. Haldane published one of the first quantitative analyses of gene–environment interactions, describing the concept to be "exceedingly complex" (3). Since then, numerous studies and methodological developments have only added to Haldane's confusion, revealing interactions to be not a monolithic entity but rather an umbrella term for multiple disease scenarios. The modern definition of **gene–environment interaction** encompasses a variety of ways in which environmental risk factors can relate to disease and genetic susceptibility variants (4–6).

In the simplest case, the genetic mutation itself alters the level of a cardiovascular risk factor, for example, the *APOE* genotype and plasma low-density lipoprotein (LDL) cholesterol levels (7), leading to disease onset or progression (Figure 9.1A). Emerging findings from the field of epigenetics provide support for the reverse scenario, under which the environmental factor (eg, smoking) results in heritable changes to the genome (eg, through differential

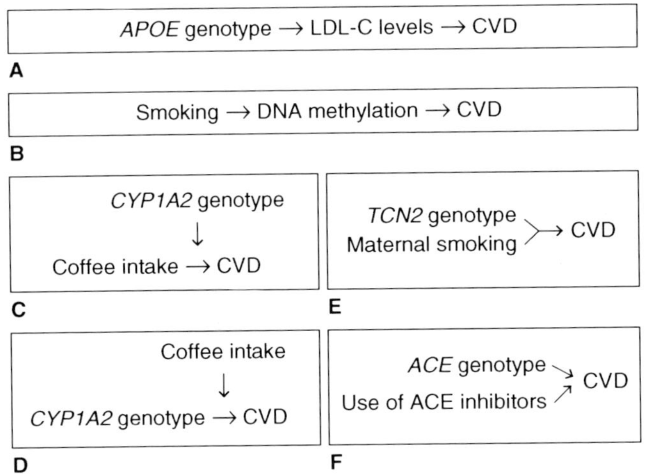

FIGURE 9.1 Examples of gene–environment interaction scenarios in cardiovascular disease (CVD).

DNA methylation), thus also affecting the risk of disease (Figure 9.1B). Alternatively, the genotype may amplify the risk conferred by a known environmental factor, for example, the *CYP1A2* genotype and caffeine consumption (8) in determining the risk of myocardial infarction (MI) or vice versa, with the environmental factor amplifying the effect of the genotype (Figure 9.1C, 9.1D). In other cases, the risk of disease does not change unless both the genetic and the environmental risk factors are present (eg, *TCN2* polymorphism and maternal smoking in congenital heart disease) (9) (Figure 9.1E). Finally, both genetic and environmental factors may influence disease risk independently, synergistically affecting disease risk in case of their co-occurrence (eg, *ACE* genotype and the use of angiotensin-converting-enzyme [ACE] inhibitors) (10) (Figure 9.1F). Although the above scenarios represent all possible models of interaction, the true causal picture may be further complicated by confounding (6) or higher-order relationships among variables, for example, gender–gene–environment interactions observed in models of high-density lipoprotein (HDL) cholesterol (11).

In epidemiological literature, discussion of interactions encompasses two easily confused but distinct concepts: **statistical interaction**, which has a precise mathematical definition but not necessarily etiological relevance, and **causal or biological interaction**, which represents the true underlying mechanism but may not always be captured by statistical models (12). Additionally, assuming an interaction is present, it can be classified as **negative or antagonistic**, in which the joint effect of two factors is less than a combination of their independent effects, and **positive or synergistic** otherwise (13). In both cases, statistical interaction refers to any deviation from conditional independence, and is observed when the effect of the genotype

on the risk of disease varies by level of environmental exposure, or vice versa (14). Statistical interaction is akin to circumstantial evidence, because its presence in the data is dependent on how the statistical model is specified. If the model is **multiplicative**, as is the case with any models estimating odds, risk, or hazards ratios, statistical interaction refers to a scenario where the joint effect of two variables is not equal to the product of their marginal effects. But if the model is **additive** and estimates risk differences rather than ratios, testing for statistical interaction implies testing whether the joint effect is the sum, rather than product, of the two marginal effects. In most cases, analyses using multiplicative versus additive models will reach conflicting conclusions on interactions. More specifically, in a sample that is large enough, it is not uncommon to see evidence of negative or no interaction on the multiplicative scale but positive interaction on the additive scale. Such inconsistencies do not exist when a positive interaction is observed on the multiplicative scale, as it necessarily implies positive interaction on the additive scale (15). A true biological interaction, however, should not be scale-dependent. In studies of gene–environment interactions, that leads to a perplexing question: Which of the approaches—additive or multiplicative—paints an accurate picture of the underlying etiology?

The consensus among methodologists is to trust the additive rather than the multiplicative model if the conclusions of the two approaches happen to diverge. It has been shown mathematically that in the absence of causal (or biological) interactions, the total risk of disease is equivalent to the sum and not the product of risks conferred by each of the participating factors (16). Therefore, a true mechanistic interaction is to be conceptualized as a deviation from additivity. Unfortunately, many statistical models used in cardiovascular epidemiology, most notably logistic and Cox regression models, are inherently multiplicative and therefore at first glance look unsuitable for estimating additive interactions. But if they are redefined in a specific way, for example, using a logarithmic transformation, they too can be valid tools for assessing biological interaction. Such modifications have been implemented in commonly available software like SAS and Microsoft Excel (17). While reporting additive interactions is especially relevant from a biological and public health standpoint, current best practices recommend reporting both estimates of additive and multiplicative interactions, as well as the corresponding confidence intervals and effect estimates of one factor (eg, gene) across strata of another (eg, environment), for transparency and ease of interpretation (18).

The final source of methodological confusion in the study of interactions lies in the nebulous distinction between the terms **interaction** and **effect (measure) modification**. While often used interchangeably, the two correspond to distinct epidemiological constructs (19). Interaction should be used in studies aimed at estimating the causal effects of two exposures, in this case both genotype and environment. Effect modification, or effect measure modification, on the other hand, is estimated in studies that are interested in estimating causal effects of one exposure across the strata of the other, but not both. In the first case, the model should adjust for two sets of confounders, one for the genetic exposure (usually ancestry) and one for the environmental exposure (age, sex, sociodemographic, and clinical variables). In the second case, however, it is only necessary to account for the set of confounders that correspond to the primary variable, that is, the one whose effect across different strata is being estimated. Interaction and effect modification are neither mutually exclusive nor always coincident. Most gene–environment studies warrant the use of the term "effect modification," because the effect of the environmental variable (eg, smoking) has already been well established and it is only the effect of the gene across its strata that is of interest. This terminological distinction matters for both analysis and interpretation of the findings, and therefore should be made early in the study design stage.

INVESTIGATING INTERACTIONS: STUDY DESIGN

Family-Based Studies

Most early studies of both genes and gene–environment interactions were done in families, leveraging pedigree information to infer inheritance of the alleles implicated in the disease process. In a classic manuscript on "inborn errors of metabolism," Archibald Garrod used family histories to surmise that the effect of the genetic variant could be "masked" by diet or comorbid conditions (20). More than a century after the publication of Garrod's paper, family-based studies remain a valid and relatively inexpensive tool in genetic epidemiology. Common family-based approaches include case–parent trios, case–control studies with related controls, and more general family pedigrees; each design comes with distinct strengths and limitations.

Twin studies represent a special case of family-based studies and can be used to distinguish between the overall genetic and environmental contributions to disease by comparing concordance rates between monozygotic and dizygotic pairs. Such studies rarely report on specific environmental risk factors that the twins may or may not share, precluding any true gene–environment interaction analysis (21). Even when the environmental risk factors are known, twin studies are not ideal for investigating interactions because their statistical power is quite limited (22). Also, standard twin data analysis methods are inappropriate as they assume genetic and nongenetic effects to be independent. However, recently developed methods enable researchers to circumvent that assumption, allowing for estimation of the variance component specifically attributed to the gene–environment interaction (23). Finally, while findings from twin studies may be useful for evaluating the overall contributions of genetic and environmental factors, they may be biased due to unmeasured epigenetic effects and should be interpreted with caution.

Case–parent and other types of pedigree studies conceptualize interactions as departures from Mendelian transmission (24). The case–parent design requires collecting genotypes on the case and his or her parents, and environmental data on the case only. Because the parents and the proband are perfectly matched on ethnic background, any potential bias due to population stratification is eliminated if the data are properly analyzed. However, unless the study also collected environmental information on the parents—a special challenge for late-onset diseases, for example, cardiovascular outcomes—such design cannot be used to estimate the main effect of the environmental variable, just that of the genotype and the gene–environment interaction. In many cases, it may be easier to obtain exposure data on an unaffected sibling or cousin, who serves as a matched control for the case. The **case–sibling** paradigm also controls for confounding by ethnicity and improves statistical power when the allele of interest is dominant (25) or rare (26). The case–parent design, on the other hand, is more efficient for studying recessive genes (25). Another way to reduce the required sample size involves restriction to cases whose parents are also affected, because it increases allele frequency in the sample. To avoid bias, such restriction must also be applied to controls (26).

The attractiveness of family-based studies lies in their immunity to unmeasured confounding by population stratification and admixture, ability to implement both linkage and association analyses, and relative efficiency for investigating interactions involving rare genetic variants. Nevertheless, family studies are limited by decreased power to detect genetic main effects, logistical difficulties associated with collecting exposure information from relatives, and limited generalizability due to oversampling of intact families (27). They are also prone to recall and survival bias, since environmental exposures are usually ascertained retrospectively (24). Additionally, as the price of directly analyzing the DNA sequence drops, family studies may no longer be as cost effective in the future (21).

Population-Based Studies

In population-based study designs, both genotype and phenotype information is collected on unrelated individuals. Evidence from **prospective cohort** studies, which measure both genotype and environmental exposures at baseline and follow individuals over time, is the current gold standard for assessing gene–environment interactions. Because prospective studies require a large number of individuals to protect against loss to follow-up, they are especially well-suited to investigating interactions in common diseases (eg, MI) or several outcomes at a time. Prospective studies collect information on exposure before disease occurs, so threats to validity from both recall bias and survival bias are minimal. Although confounding by population stratification is of concern, it is less likely if participants of all ethnic backgrounds are equally likely to stay in the study or be lost to follow-up (24). The high validity of prospective cohort findings makes them a good choice for confirmatory studies of already identified gene–environment interactions (28).

Unfortunately, like most gold standard measures, prospective cohort studies are expensive and effort-intensive. A good compromise can be achieved by using a **nested case–control** design within an existing prospective cohort, which selects a complete set of cases and compares their genetic and environmental risk factors with a set of unaffected individuals from the same study population. This approach preserves some of the desirable features of the prospective design, namely the reduced chance of biased exposure ascertainment, but gains efficiency by using only a subset of all participants. But results from such studies may be prone to selection bias if loss to follow-up is differential with respect to exposure and disease status, population stratification bias if retention rates vary by ethnicity, and exposure misclassification if environmental variables were not ascertained repeatedly over the course of the entire study and not only at baseline (21).

On the other end of the population-based study spectrum is the **case–control** design, where participants are selected into the study based on the outcome and thus both DNA and data on environmental exposures are obtained retrospectively. The case-control design, which is relatively inexpensive compared to the cohort studies, is especially suitable for exploratory rather than confirmatory studies of gene–environment interactions, or when the disease of interest is rare. If the controls are not representative of the population that gave rise to the cases, the results may suffer from selection bias. Population stratification represents one special case of selection bias, and may occur if controls systematically differ from the cases in their genetic ancestry. It may be addressed either by matching cases to controls based on ethnicity (29), but that may not be sufficient for recently admixed populations such as Hispanic Americans (24). Alternatively, if ancestry informative markers or whole-genome data are available, bias due to population stratification may be reduced by applying genomic control methods (30). But the main issue in assessing gene–environment interactions from case–control studies is the systematic misclassification of exposure, for example, situations when cases remember their exposures differently from controls (recall bias) or when a rapidly fatal disease makes it possible to ascertain exposure on only a fraction of cases (survival bias). Despite the threat that exposure misclassification poses to the validity of main effect estimates, it may still produce valid estimates of interaction parameters and is therefore not of concern if the research question is limited to gene–environment interactions per se (31).

In situations when the main effects of both genotype and the environmental factors have been well established and the research question only concerns their interaction, another attractive approach may be the **case-only** design. By recruiting only affected individuals, the case-only design reduces both costs and concerns regarding appropriate control selections. It assumes that genetic and environmental factors are independent in the population that gave rise to the cases, and conceptualizes differences in exposure among different genotype groups in cases as evidence of interaction (32). The independence assumption results in improved statistical power (and thus lower required sample size) over traditional case–control studies, but is not directly testable, making the case-only design especially prone to bias and confounding. Moreover, using the case-only method precludes estimation not only of the main effects, but also of the joint effect of the genes and environment, which is ostensibly more important from a public health standpoint (28). Overall, the case-only design is best suited for "first-pass" studies of gene–environment interactions with a clearly defined genomic region of interest.

Finally, when the environmental variable (eg, drug or dietary intervention) is the subject of a **clinical trial**, the resulting data could be used to estimate the effects of gene-intervention interactions. If the participants are randomized into the trial stratified by genotype, and the effect of the drug is found to be differential across the strata, such findings may provide important groundwork for personalized medicine (28).

However, for ethical reasons the trial design is not an appropriate way to investigate genetic interactions with noxious exposures, for example, environmental pollutants (33). Moreover, the costs of interventions that stratify on genotype are even more prohibitive than those of classical clinical trials, making this design a good choice only when preliminary evidence of gene–drug or gene–diet interactions is compelling.

Investigating Interactions: Analysis

In addition to the variety of study design paradigms, gene–environment interaction studies differ by genotyping strategy, ranging from candidate gene studies to pathway analyses to agnostic methods such as gene–environment-wide interaction studies (GEWISs) (34). The choice of genotyping strategy often dictates the methodology behind statistical analysis, with sample size and power being critical considerations in the study of gene–environment interactions.

Candidate gene approaches test a specific hypothesis that emerges from a priori biological understanding of the relevant pathophysiological mechanisms. The genes can either be selected based on their relation to a known environmental factor (eg, alcohol dehydrogenase polymorphisms and drinking behavior), or based on the gene's intrinsic importance to the disease of interest (eg, *LDLR*, which encodes an LDL receptor, and dietary lipid intake in the context of atherosclerosis). The simplest version of a candidate gene–environment interaction study involves a single genetic variant within a gene and an environmental factor. For example, a recent study focused on a promoter polymorphism that had previously been shown to affect expression and efficiency of one of the key enzymes in polyunsaturated fatty acid metabolism (35). The investigators found that variation at the promoter locus may modify the association between dietary intake of alpha-linolenic acid, a compound commonly found in plant seeds, and the risk of the metabolic syndrome. In many cases, however, the causal variant within the gene is not known, and single nucleotide polymorphism (SNP) methods may not be sufficient to establish whether a gene is involved in a particular disease process. As such, candidate gene–environment studies may conceptualize the genetic exposure as a risk score, a haplotype, or another aggregate measure of variation within a gene. The availability of HapMap data to impute unmeasured genetic variants makes haplotype analysis especially lucrative, as the existing methods can accommodate main as well as interaction effects (33).

Another strength of candidate gene analyses lies in their hypothesis-driven nature and reduced multiple testing burden, and thus a lower probability of false positive findings. It must be said, however, that despite their appeal, candidate gene studies have not identified many reproducible gene–environment interactions (36). In a 10-year review of candidate gene–environment studies in psychiatry, only 27% of replication studies produced statistically significant findings, compared to 96% of initial discovery studies (37). In addition to differences between discovery and replication study populations and challenges involving selecting the best candidate genes, an important culprit in nonreplication is insufficient statistical power, stemming from inadequate sample size of most candidate gene studies.

The limited success of candidate studies prompted the search for other analytic approaches to detecting gene–environment interactions. One promising alternative, which preserves the hypothesis-driven perspective, is **pathway-based** gene–environment interaction research. Using a hierarchical model framework, such studies allow estimation of overall effects of each biological pathway—for example, interactions between an environmental factor such as folic acid intake and variation in all the genes involved in folate metabolism (33). The advantages of hierarchical models include the biologically informed scope and the ability to distinguish between main effects and interactions. Compared to agnostic large-scale approaches like GEWIS, they also reduce the multiple testing burden and in some cases may improve statistical power. But like all the other pathway-based approaches, these studies are limited by their reliance on curated pathways like Gene Ontology or Ingenuity Pathways Analysis, which frequently lag behind in incorporating new biological knowledge and are thus biased toward already known genes. Findings from pathway-based analyses are also more challenging to interpret or develop into clinical applications, and often require follow-up with single-gene or SNP studies.

Recently, genome-wide approaches such as **GEWIS** have received considerable attention for their ability to efficiently and comprehensively detect loci involved in gene–environment interactions. GEWIS is an umbrella term for a variety of analytic methods that consider interactions on a global, genome-wide scale, and may be customized for specific research questions. Like genome-wide association studies (GWAS), GEWIS approaches are agnostic in nature and test a large number of interactions between environmental variables and genetic markers. As a result, they require stringent corrections for multiple comparisons, which often limit their power. Some GEWIS designs, such as the two-step approach proposed by Murcray et al (38), exploit the assumption of independence between genes and the environment to conduct a screening test of multiplicative interaction. The *P*-values from the screening step are then compared with alpha/M, where alpha is the predefined level of statistical significance (eg, 0.05) and M is the number of null hypothesis tests that were rejected during the first step. The statistical power of the two-step approach is higher than in standard methods, especially for lower M values (eg, 5,000 markers); however, as the number of interrogated genetic markers goes up to modern-day GWAS levels, the advantage disappears (39). In contrast, a proposed empirical Bayes procedure (40) exploits the gene–environment independent assumption for the actual test for interaction rather than the screening step, although some of the resulting power gain may be due to increased type 1 error (41) and should thus be used with caution.

Misspecification of the environmental effect and population stratification may seriously threaten the validity of GEWIS findings (42). The first problem may be addressed by using an approach that does not require specification of the environmental exposure, for example, a case-only design. But that strategy will still be invalid under two conditions: (a) if there is population stratification and (b) if the effect of the genotype on disease occurs through the exposure itself, for example, when a polymorphism in an alcohol-metabolizing enzyme that causes an adverse flushing reaction determines how much an individual chooses to drink and affects the risk of heart disease only through that behavior. The second problem, population stratification, may be addressed by conducting separate analyses in each ethnic group, but such subdivision of the study sample is likely to further limit statistical power and may only be feasible in large cohorts.

A different class of GEWIS strategies involves joint testing of genetic main effects and gene–environment interactions across the genome. As the goal of genome-wide studies is to identify novel etiologically relevant variants, it may not matter whether the observed association is with the gene directly or through its interaction with an environmental variable. Therefore, the null hypothesis in joint tests is that a given genetic variant has no effect on the outcome regardless of the level of the environmental risk factor. If a moderate-to-large interaction effect is present or if an interaction effect is in opposite direction to the main genetic effect, the joint approach is more powerful than the standard test, although both are subject to the multiple testing problem (43). In situations where the expected interaction effect is small or null, the marginal test may present a more powerful alternative (44). The joint test may also be extended to meta-analyze effects across studies, both generally (45) and when there is variability in definitions of environmental exposure across studies (46).

The emergence of the exposome concept, broadly defined as the composite measure of all environmental exposures throughout the life course, has offered new challenges and opportunities for gene–environment studies. While no **genome–exposome-wide interaction** searches have been published to date, considering genetic and environmental variation jointly and totally is a promising approach to identifying effects that may have been missed by conventional GWAS or GEWIS (47). Leveraging the whole exposome is likely to be useful for generating novel hypotheses, but the high probability of false-positive findings with such high-dimensional data certainly requires follow-up studies, preferably with replication in independent populations.

INVESTIGATING INTERACTIONS: STATISTICAL POWER

Statistical power is the key consideration in the design of gene–environment studies. The sample size required to detect an interaction effect is generally much greater than for main effects. For binary genetic and environmental exposures, the sample size required to see a 2×2 interaction is four times the sample size required to see a main effect of the same magnitude (48). That is a high burden, and because of cost and logistical challenges, most existing studies of gene–environment interactions are severely underpowered, often producing false, nonreplicable results. Calculating statistical power under realistic assumptions of effect is paramount to rigorous gene–environment studies, and may be done using freely available software such as Quanto (biostats .usc.edu/software) for case–control, case–parent, or case-only designs, and Power (http://dceg.cancer .gov/tools/design/power) for case-control and cohort designs (28). In the GEWIS context, the required sample sizes may be computed using formulas derived by Murcray et al (49). Alternatively, power and sample size may be estimated using statistical simulations.

GENE–ENVIRONMENT INTERACTIONS IN CARDIOVASCULAR DISEASE

Evidence supports the existence of gene–environment interactions in nearly all CVD conditions (50). In many cases, gene–environment interactions can account for seemingly inconsistent findings of genetic or environmental associations with disease across study populations. To date, only a few of the gene–environment interactions in cardiovascular pathophysiology have been replicated in independent populations.

For example, while genetic variants on chromosome 9p21 have been linked to the risk of MI, the risk tends to vary by more than 10% among different ethnic groups (51). The association between risk alleles located in the 9p21 region and MI also appears to vary by dietary pattern, with substantially lower effect estimates among individuals consuming a diet rich in raw vegetables and fruits. In other words, exposure to a healthier diet buffers the causative effect of the risk alleles in the 9p21 region and can account for some of the observed differences between study populations. This gene–diet interaction, first observed in a large-scale case–control INTERHEART study of MI, was subsequently replicated for the incidence of CVD in an independent prospective cohort (51). Similar nutritional buffering was observed for associations between well-characterized polymorphisms in *FTO* and *MC4R* and the risk of type 2 diabetes, with stricter adherence to the Mediterranean diet pattern counteracting the genetic predisposition conferred by risk alleles (52). As genetic testing for chronic disease alleles becomes more widespread, such protective interactions with dietary factors provide a useful public health framework, empowering individuals with higher risk genotypes to make healthier choices rather than resigning to the "our genes are our destiny" mentality.

In the clinical setting, pharmacogenetic effects represent an especially noteworthy type of gene–environment interaction, as they may pave the way to personalized treatment approaches based on the patient's genotype. The classic success story of cardiovascular pharmacogenetics is that of warfarin dosing. Warfarin, a powerful and widely used anticoagulant, is characterized by tremendous variability of response, with an effective daily dose ranging from 0.5 to 60 mg (53). Treatment protocols that were based solely on nongenetic covariates like the International Normalized Ratio (INR) had only a 69% success rate in predicting the correct maintenance dose (54). The discovery of biological mechanisms underlying warfarin metabolism and activity, involving variants in *CYP2C9* and *VKORC1* respectively, has greatly improved prediction across a variety of ethnic groups, resulting in a Food and Drug Administration (FDA)-mandated change in warfarin labeling in 2007 and the introduction of dosing tables in 2010 (55). Currently, clinical trials are ongoing to evaluate the effectiveness of pharmacogenetically guided dosing on anticoagulation control, risk of adverse events, and health care costs (56). It is noteworthy that even despite the high prognostic value of the *CYP2C9* and *VKORC1* findings, a large portion of variability in warfarin response—quantified as 40% among Caucasians and 60% among African Americans—remains unexplained (56). For other cardiovascular

drugs, for example, statins, gene–drug interactions explain even less variability, accounting for approximately 10% to 15% of differences in efficacy (57). However, genotypes in *SLCO1B1*, which encodes a transporter responsible for clearance of statins, may be a promising tool for identifying patients who are likely to suffer from statin-induced myopathies or exhibit noncompliance; that test is currently offered on personalized genomics platforms, for example, 23andMe (58).

Another special case of gene–environment interactions in CVD occurs through epigenetic mechanisms such as DNA methylation, histone modification, microRNAs, and other heritable changes to the genetic sequence. Epigenetic changes, which arise as a result of environmental exposures on the genome, have garnered considerable attention as both markers of disease and drug targets due to their clear biological link to the phenotype as well as reversibility. The environmental factors that can alter the epigenome are numerous, including diet, physical activity, smoking, alcohol intake, stress, and a range of environmental pollutants. For example, exposure to black carbon and particulate matter was shown to be associated with decreased methylation of the LINE-1 repetitive elements across the genome, which in turn is correlated with the prevalence of hypertension, ischemic heart disease, and stroke (59–61). The identification of epigenetic changes in CVD has led to interesting and promising findings. Specifically, supplementation with curcumin (turmeric), a low-cost nutritional agent that naturally inhibits histone acetylation, improved lipid profiles in healthy volunteers (62) and prevented ventricular hypertrophy in animal models of heart failure (63). However, unlike in the treatment of cancer, the development of effective cardiovascular epigenetic drugs is still in its infancy, as is identification of epigenetic markers that could be used in cardiovascular risk stratification and primary prevention (64).

THE PROMISE AND CHALLENGES OF GENE–ENVIRONMENT INTERACTION RESEARCH

Traditional studies of genetic and environmental contributions to CVD, in which these two risk factor groups are considered separately, consistently underestimate the population attributable risk of either exposure, potentially misinforming the setting of public health priorities (65). Studies of gene–environment interactions provide a more comprehensive and accurate picture of disease determinants, often offering additional insights into the underlying etiological mechanisms. In the presence of true biological interactions, incorporating synergy or antagonism among factors in statistical models can raise statistical power, resulting in identification of novel genetic risk markers. Moreover, incorporating gene–environment interactions into risk prediction models may improve prognostic accuracy, although simulation studies show that the resulting increase in discrimination ability is unlikely to be dramatic (66).

Currently available methods for investigating gene–environment interactions are constrained by numerous limitations, evidenced by the low rate of replication of findings across studies. Reasons for such failure to reproduce are manifold, but the chief one is the lack of statistical power, which results from small sample sizes and limited magnitude of interaction effects. Also, underpowered studies rarely adjust for multiple comparisons, resulting in many false-positives results and contributing to nonreplication. Environmental homogeneity is another factor that can hinder detection of gene–environment interactions. For example, studies of interactions between FTO polymorphisms and physical activity in cardiometabolic outcomes that were carried out in Western populations did not report a significant interaction (67), in contrast to results from a similar study conducted in India, where the range of physical activity is much larger. This discrepancy was also observed within the Indian study population itself: the interaction effect was only significant among participants from Trivandrum, a less modernized region with higher variability in physical activity, but not among individuals from New Delhi (68). This example illustrates the importance of exposure ascertainment in gene–environment studies, which are as prone to measurement error as nongenetic studies. Exposure measurement may be further complicated by high correlations among multiple factors, for example, diet and physical activity, which necessitate careful control for confounding (65). For these reasons, large sample sizes with prospectively collected exposure data provide the best setting for investigating interactions. Additionally, the varying approaches to exposure ascertainment often result in estimates that are challenging to combine across studies, suggesting the need for coordinated research efforts such as consortia.

Other challenges facing gene–environment interaction research lie in the identification of potential causal variants. Most of the replicated evidence for interaction comes from studies of already known candidate genes, leaving most of the available sequence information untapped. Although genome-wide methods have the potential to identify novel variants of interest, most commonly used genotyping arrays only include common polymorphisms, which are less likely to play a causal role in disease. Whole-exome and whole-genome approaches may remedy that by enabling discovery of less frequent mutations, although that is likely to only exacerbate the limitations on statistical power unless these variants are aggregated by region or as a genetic score. Several statistical methods for aggregating rare variants have now been extended to incorporate gene–environment interactions, with the sequence kernel association test (SKAT) showing the best overall power and computational efficiency in preliminary studies (69). However, these methods do not yet have the capacity to meta-analyze findings across different populations, creating the additional challenge of achieving adequate sample size within a single study.

SUMMARY

Gene–environment interaction studies are integral to understanding the pathogenesis and accurately estimating the public health burden of complex diseases, including cardiovascular pathologies. Accounting for interactions offers the opportunity to gain insights into etiological mechanisms, identify novel genetic and epigenetic risk markers, and ultimately develop tools for risk stratification and personalized treatment approaches. However, with a few notable exceptions such as warfarin dosing, most existing findings of gene–environment interactions in CVD have not been replicated due to study design constraints. Thus, clinical and public health applications of gene–environment interaction research are not likely to be immediate. In addition to development of novel statistical methods, several best practices are likely to accelerate this process: (a) considering interactions in the study design phase rather than in the analysis, thus ensuring adequate sample size to detect both main and interaction effects; (b) conducting tests of interactions on both additive and multiplicative scales; (c) reporting stratum-specific estimates of association as well as the corresponding confidence intervals; (d) collecting prospective detailed information on exposure and potentially relevant confounders; (e) rigorous adjustment for multiple testing; and (f) replication of findings in independent populations and/or functional validation in cell-based assays or animal models. By addressing the validity as well as the interpretability of findings, these research practices may yield results of both etiological and practical importance, bringing the clinical community closer to harnessing the promise of gene–environment interaction research.

REFERENCES

1. Manolio TA, Collins FS, Cox NJ, et al. Finding the missing heritability of complex diseases. *Nature.* 2009;461:747–753.
2. Rajpathak SN, Liu Y, Ben-David O, et al. Lifestyle factors of people with exceptional longevity. *J Am Geriatr Soc.* 2011;59:1509–1512.
3. Haldane JBS. The interaction of nature and nurture. *Ann Eugen.* 1946;13:197–205.
4. Ottman R. An epidemiologic approach to gene-environment interaction. *Genet Epidemiol.* 1990;7: 177–185.
5. Ottman R. Gene-environment interaction: definitions and study design. *Prev Med.* 1996;6:764–770.
6. Geneletti S, Gallo V, Porta M, et al. Assessing causal relationships in genomics: from Bradford-Hill criteria to complex gene-environment interactions and directed acyclic graphs. *Emerg Themes Epidemiol.* 2011;8:5.
7. Davignon J, Greg RE, Sing CF. Apolipoprotein E polymorphism and atherosclerosis. *Atherosclerosis.* 1988; 8:1–21.
8. Cornelis MC, El-Sohemy A, Kabagambe EK, Campos H. Coffee, CYP1A2 genotype, and risk of myocardial infarction. *JAMA.* 2006;295:1135–1141.
9. Hobbs CA, Cleves MA, Karim MA, et al. Maternal folate-related gene environment interactions and congenital heart defects. *Obstet Gynecol.* 2010;116(2, Pt 1): 316–322.
10. Knol MJ, VanderWeele TJ, Groenwold RHH, et al. Estimating measures of interaction on an additive scale for preventive exposures. *Eur J Epidemiol.* 2011; 26:433–438.
11. Corella D, Guillen M, Saiz C, et al. Environmental factors modulate the effect of the APOE genetic polymorphism on plasma lipid concentrations: ecogenetic studies in a Mediterranean Spanish population. *Metabolism.* 2001;8:936–944.
12. Clayton D, McKeigue PM. Epidemiological methods for studying genes and environmental factors in complex diseases. *Lancet.* 2001;358:1356–1360.
13. Rothman KJ. Synergy and antagonism in cause-effect relationships. *Am J Epidemiol.* 1974;99:385–388.
14. Institute of Medicine (US) Committee on Assessing Interactions Among Social, Behavioral, and Genetic Factors in Health; Hernandez LM, Blazer DG, eds. *Genes, Behavior, and the Social Environment: Moving Beyond the Nature/Nurture Debate.* Washington, DC: National Academies Press (US); 2006.
15. Koopman JS. Interaction between discrete causes. *Am J Epidemiol.* 1981;113:716–724.
16. Rothman KJ. *Epidemiology: An Introduction.* New York, NY: Oxford University Press; 2002.
17. Andersson T, Alfredsson L, Kallberg H, et al. Calculating measures of biological interaction. *Eur J Epidemiol.* 2005;20:575–579.
18. Knol MJ, VanderWeele TJ. Recommendations for presenting analyses of effect modification and interaction. *Int J Epidemiol.* 2012;41:514–520.
19. VanderWeele TJ. On the distinction between interaction and effect modification. *Epidemiology.* 2009; 20:863–871.
20. Garrod AE. The incidence of alkaptonuria: a study in chemical individuality. *Lancet.* 1902;ii:1616–1620.
21. Hunter DJ. Gene-environment interactions in human diseases. *Nat Rev Genet.* 2005;6:287–298.
22. Ottman R. Epidemiologic analysis of gene-environment interaction in twins. *Genet Epidemiol.* 1994;11: 75–86.
23. Wallace HM. A model of gene-gene and gene-environment interactions and its implications for targeting environmental interventions by genotype. *Theor Biol Med Model.* 2006;3:35.
24. Kraft P, Hunter DJ. Integrating epidemiology and genetic association: the challenge of gene-environment interaction. *Phil Trans R Soc B.* 2005;360:1609–1616.
25. Gauderman WJ. Sample size requirements for matched case-control studies of gene-environment interaction. *Stat Med.* 2002;21:35–50.
26. Gauderman WJ, Witte JS, Thomas DC. Family-based association studies. *J Natl Cancer Inst Monogr.* 1999; 26:31–37.
27. Liu C, Maity A, Lin X, et al. Design and analysis issues in gene and gene-environment studies. *Environ Health.* 2012;11:93.
28. Dempfle A, Scherag A, Hein R, et al. Gene-environment interactions for complex traits: definitions, methodological requirements, and challenges. *Eur J Hum Genet.* 2008;16:1164–1172.
29. Wacholder S, Rothman N, Caporaso N. Population stratification in epidemiologic studies of common genetic variants and cancer: quantification of bias. *J Natl Cancer Inst.* 2000;92:1151–1158.
30. Devlin B, Roeder K. Genomic control for association studies. *Biometrics.* 1999;55:997–1004.
31. Garcia-Closas M, Thompson WD, Robins JM. Differential misclassification and the assessment of gene-environment interactions in case-control studies. *Am J Epidemiol.* 1998;147:426–433.
32. Khoury M, Flanders WD. Nontraditional epidemiologic approaches in the analysis of gene-environment interactions: case-control studies with no controls! *Am J Epidemiol.* 1996;144:207–213.
33. Thomas D. Methods for investigating gene-environment interactions in candidate pathway and genome-wide association studies. *Annu Rev Public Health.* 2010;31:21–36.
34. Khoury M, Wacholder S. Invited commentary: from genome-wide association studies to gene-environment-wide interaction studies—challenges and opportunities. *Am J Epidemiol.* 2009;169:227–230.
35. Truong H, DiBello JR, Ruiz-Narvaez E, et al. Does genetic variation in the delta6-desaturase promoter modify the association between alpha-linolenic acid and the prevalence of metabolic syndrome? *Am J Clin Nutr.* 2009;89:920–925.
36. Mukherjee B, Ahn J, Gruber SB, Chatterjee N. Testing gene-environment interactions in large-scale case-control association studies: possible choices and comparisons. *Am J Epidemiol.* 2012;175:177–190.
37. Duncan LE, Keller MC. A critical review of the first 10 years of candidate gene-by-environment interaction research in psychiatry. *Am J Psychiatry.* 2011; 168:1041–1049.
38. Murcray CE, Lewinger JP, Gauderman WJ. Gene-environment interaction in genome-wide association studies. *Am J Epidemiol.* 2009;169:219–226.

39. Chatterjee N, Wacholder S. Invited commentary: efficient testing of gene-environment interaction. *Am J Epidemiol.* 2009;169:231–233.

40. Mukherjee B, Chatterjee N. Exploiting gene-environment independence for analysis of case-control studies: an empirical Bayes-type shrinkage estimator to trade-offs between bias and efficiency. *Biometrics.* 2008;64:685–694.

41. Thomas DC, Lewinger JP, Murcray CE, Gauderman WJ. Invited commentary: GE-Whiz! Ratcheting gene-environment studies up to the whole genome and the whole exposome. *Am J Epidemiol.* 2012;175:203–207.

42. Cornelis MC, Tchetgen EJ, Liang L, et al. Gene-environment interactions in genome-wide association studies: a comparative study of tests applied to empirical studies of type 2 diabetes. *Am J Epidemiol.* 2012;175:191–202.

43. Kraft P, Yen YC, Stram DO, et al. Exploiting gene-environment interaction to detect genetic associations. *Hum Hered.* 2007;63:111–119.

44. Aschard H, Lutz S, Maus B, et al. Challenges and opportunities in genome-wide environmental interaction (GWEI) studies. *Hum Genet.* 2012;131:1591–1613.

45. Manning AK, LaValley M, Liu CT, et al. Meta-analysis of gene-environment interaction: joint estimation of SNP and SNP x environment regression coefficients. *Genet Epidemiol.* 2011;35:11–18.

46. Aschard H, Hancock DB, London SJ, Kraft P. Genome-wide meta-analysis of joint tests for genetic and gene-environment interaction effects. *Hum Hered.* 2011;70:292–300.

47. Patel CJ, Bhattacharya J, Butte AJ. An environment-wide association study (EWAS) on type 2 diabetes mellitus. *PLoS ONE.* 2010;5:e10746.

48. Leon AC, Heo M. Sample sizes required to detect interactions between two binary fixed-effects in a mixed-effects linear regression model. *Comput Stat Data Anal.* 2009;53:603–608.

49. Murcray CE, Lewinger JP, Conti DV, et al. Sample size requirements to detect gene-environment interactions in genome-wide association studies. *Genet Epidemiol.* 2011;35:201–210.

50. Flowers E, Froelicher ES, Aouizerat BE. Gene-environment interactions in cardiovascular disease. *Eur J Cardiovasc Nurs.* 2012;11:472–478.

51. Do R, Xie C, Zhang X, et al. The effect of chromosome 9p21 variants on cardiovascular disease may be modified by dietary intake: Evidence from a case/control and a prospective study. *PLoS Med.* 2011;8:e1001106.

52. Saez-Tormo G, Pinto X, Munoz MA, et al. Associations of the FTO rs9939609 and the MC4R rs17782313 polymorphisms with type 2 diabetes are modulated by diet, being higher when adherence to the Mediterranean diet pattern is low. *Cardiovasc Diabetol.* 2012;11:137.

53. James AH, Britt RP, Risking CL, Thompson SG. Factors affecting the maintenance dose of warfarin. *J Clin Pathol.* 1992;45:704–706.

54. Aithal GP, Day CP, Kesteven PJL, Daly AK. Association of polymorphisms in the cytochrome p450 CYP2C9 with warfarin dose requirement and risk of bleeding complications. *Lancet.* 1999;353:717–719.

55. Jorgensen AL, Fitzgerald RJ, Oyee J, et al. Influence of CYP2C9 and VKORC1 on patient response to warfarin: a systematic review and meta-analysis. *PLoS ONE.* 2012;7:e44064.

56. Limdi NA. Warfarin pharmacogenetics: challenges and opportunities for clinical translation. *Front Pharmacol.* 2012;3:183.

57. Voora D, Ginsburg GS. Clinical application of cardiovascular pharmacogenetics. *J Am Coll Cardiol.* 2012;60:9–20.

58. Voora D, Shah SH, Spasojevic I, et al. The SLCO1B1*5 genetic variant is associate with statin-induced side effects. *J Am Coll Cardiol.* 2009;54:1609–1616.

59. Baccarelli A, Wright RO, Bollati V, et al. Rapid DNA methylation changes after exposure to traffic particles. *Am J Respir Crit Care Med.* 2009;179:572–578.

60. Baccarelli A, Tarantini L, Wright RO, et al. Repetitive element DNA methylation and circulating endothelial and inflammation markers in the VA normative aging study. *Epigenetics.* 2010;5:222–228.

61. Baccarelli A, Ghosh S. Environmental exposures, epigenetics and cardiovascular disease. *Curr Opin Clin Nutr Metab Care.* 2012;15:323–329.

62. Napoli C, Infante T, Casamassimi A. Maternal-fetal epigenetic interactions in the beginning of cardiovascular damage. *Cardiovasc Res.* 2011;92:367–374.

63. Sunagawa Y, Morimoto T, Wada H, et al. A natural p300-specific histone acetyltransferase inhibitor, curcumin, in addition to angiotensin-converting enzyme inhibitor, exerts beneficial effects on left ventricular systolic function after myocardial infarction in rats. *Circ J.* 2011;75:2151–2159.

64. Napoli C, Crudele V, Soricelli A, et al. Primary prevention of atherosclerosis: a clinical challenge for the reversal of epigenetic mechanisms? *Circulation.* 2012;125:2363–2373.

65. Ordovas JM, Tai ES. Why study gene-environment interactions? *Curr Opin Lipidol.* 2008;19:158–167.

66. Aschard H, Chen J, Cornelis MC, et al. Inclusion of gene-gene and gene-environment interactions unlikely to dramatically improve risk prediction for complex diseases. *Am J Hum Genet.* 2012;90:962–972.

67. Cornes BK, Lind PA, Medland SE, et al, Martin NG. Replication of the association of common rs9939609 variant of FTO with increased BMI in an Australian adult twin population but no evidence for gene by environment (G x E) interaction. *Int J Obes (Lond).* 2009;33:75–79.

68. Moore SC, Gunter MJ, Daniel CR, et al. Common genetic variants and central adiposity among Asian-Indians. *Obesity.* 2012;20:1902–1908.

69. Kazma R, Witte J. Hunting for rare genetic variants in complex disease? Do not forget about gene-environment interactions! American Society for Human Genetics Annual Meeting abstract, 2012.

Genetic Counseling

Kelly M. Bontempo

TAKE HOME POINTS

1. Genetic counseling is an essential aspect of cardiovascular genetics and genomics in clinical practice.
2. Effective genetic counseling is necessary for the proper interpretation of genetic testing, diagnosis of hereditary syndromes, education of patients and family members, and discussion of ethical aspects of genetic testing and screening.
3. With the advent of targeted therapies for genetic cardiovascular disorders, along with expansion of genomic analysis to whole genome sequencing, the need for genetic counseling has never been greater.

As with any diagnosis, identifying a genetic disorder in a patient should, at a minimum, involve helping them to understand the nature and implications of the disease. When a disorder is hereditary, there is an added responsibility to educate family members about their risks and available risk-reducing options. Knowledge of a heritable disease in a family can raise complex psychological and social issues. Therefore, the practice of clinical genetics must focus on the diagnosis and management of both medical and psychosocial aspects of heritable disease.

The diagnosis and management of hereditary conditions are particularly time-consuming and require extensive preparation and follow-up. With the rapid expansion of genetic knowledge and medical technology, the demand for genetic testing, education, and supportive counseling has been increasing. While clinical geneticists had provided these services fairly independently in the past, additional genetic practitioners have surfaced in order to assist physicians. Genetic counseling is a relatively young, multifaceted medical profession with services provided by master's-level, board-certified professionals trained in both medical genetics and psychosocial counseling. Genetic counselors act as diagnosticians, advocates, grief counselors, researchers, and health care professionals providing supportive care, education, resources, and referrals (1).

Established standards for genetic counseling services include obtaining a history that incorporates family and ethnic information, advising patients of the genetic risks to themselves and other family members, offering genetic testing or prenatal diagnosis when indicated, and discussing treatment or management options for reducing the risk of disease (2). Alongside obtaining a diagnosis or risk assessment, patients seek genetic counseling services for validation, to gain support, and to reduce their anxiety (3). Genetic counselors do not make decisions for the patient, but provide information and support in reaching a decision most appropriate for the patient and family. As individuals are assuming increasingly greater responsibility for their own health care decisions, genetic counselors can help patients arrive at decisions that are medically and ethically appropriate.

DEFINITION OF GENETIC COUNSELING

As the practice of genetic counseling evolved into an essential part of clinical care, national societies began to adopt definitions of the practice. In 1975, the American Society of Human Genetics (ASHG) proposed a definition of genetic counseling, which aimed to highlight principal features of the practice.

This definition aimed to acknowledge genetic counseling as a communication process that places the highest value on patient autonomy:

> Genetic counseling is a communication process that deals with the human problems associated with the occurrence or risk of occurrence of a genetic disorder in a family. This process involves an attempt by one or more appropriately trained persons to help the individual or family to: (1) comprehend the medical facts including the diagnosis, probable course of the disorder, and the available management, (2) appreciate the way heredity contributes to the disorder and the risk of recurrence in specified relatives, (3) understand the alternatives for dealing with the risk of recurrence, (4) choose a course of action which seems to them appropriate in their view of their risk, their family goals, and their ethical and religious standards and act in accordance with the decision, and (5) to make the best possible adjustment to the disorder in an affected family member and/or to the risk of recurrence of that disorder.
>
> —American Society of Human Genetics (4)

In the almost four decades since the ASHG definition was stated, our knowledge of medical genetics and genetic implications in health care has grown rapidly. Genetic counseling practice has expanded beyond the prenatal and pediatric settings, with many counselors now focusing on adult-onset conditions within a particular discipline, including, but not limited to, oncology, neurology, and cardiology. Genetic counselors also provide information for conditions that are not entirely genetic, including multifactorial disorders such as diabetes, teratogen exposure during pregnancy, ultrasound anomalies, and birth defects. To accommodate an evolving practice, the National Society of Genetic Counselors (NSGC) appointed a task force to develop an updated definition describing the process and goals of genetic counseling:

> Genetic counseling is the process of helping people understand and adapt to the medical, psychological and familial implications of genetic contributions to disease. This process integrates:
>
> - Interpretation of family and medical histories to assess the chance of disease, occurrence or recurrence.
> - Education about inheritance, testing, management, prevention, resources and research.
> - Counseling to promote informed choices and adaptation to the risk or condition.
>
> —National Society of Genetic Counselors (5)

Common indications for genetic counseling include prenatal diagnosis, previous child with congenital anomalies or intellectual disability, family history of a hereditary disease, unusual presentation or findings, repeated pregnancy loss or infertility, abnormal newborn screening or other forms of genetic screening (carrier or maternal serum screening), teratogen exposure, consanguinity, and pre- and posttest counseling, particularly for presymptomatic diagnoses or adult-onset conditions. While genetic counselors have traditionally practiced in the areas of prenatal diagnosis, pediatrics, cancer risk assessment, and the laboratory setting, with advances in genomic analysis, pharmacogenomics, and personalized medicine, practitioners should expect to see genetic counselors make a larger impact on the primary care setting in the foreseeable future.

THE POWER OF THE PEDIGREE AND FAMILY HEALTH HISTORY

Successful genetic counseling is dependent on collecting accurate and relevant medical information about an individual and his or her family members. Careful analysis of medical and family history can reveal the inheritance pattern of a particular condition, eliminate differentials to expose the likely diagnosis, identify at-risk relatives, and ultimately guide genetic testing and management decisions. Just as important, the gathering of family history at the beginning of a session provides an excellent opportunity to build rapport with a patient and establish a trusting relationship.

Conversations and observations during the acquisition of family history can reveal the emotional and social impact of a disorder on a family. Family members may express guilt, shame, anxiety, anger, or denial about their particular situation. For example, identification of an X-linked mode of inheritance can trigger feelings of guilt in a mother. A counselor's ability to recognize these emotions creates the opportunity to explore these feelings and ultimately divert the patient from maladaptive coping mechanisms such as self-blame, rationalizations, or other intellectualizations (6).

A detailed family history should be collected on all patients presenting for a genetics evaluation (6–8). Before beginning, it is helpful to explain to the patient that health information from the entire family is essential to providing the most appropriate medical care. For a genetics evaluation, the family history should include at a minimum a three-generation pedigree with information about the patient and his or her first-degree (parents, siblings, children), second-degree (half-siblings, aunts/uncles, nephews/nieces, grandparents), and third-degree relatives (first cousins, half-aunts/uncles, half-nieces/nephews). Some patients will be better historians than others, so the practitioner should not be discouraged if the information is limited. Essential information to record for individuals in the pedigree include gender, current age, age at diagnosis, age at death, cause of death, significant health problems, significant surgeries, pregnancy complications (miscarriage, stillbirth), infertility, distinguishing full relatives from half-relatives, and medically documented diagnoses from suspected diagnoses. It is also important to collect ethnic information, as certain ethnic groups are at increased risk for specific genetic disorders (eg, Amish, Ashkenazi Jews, and French Canadians). In some ethnic groups, consanguineous marriages are customary and preferred. Offspring of consanguineous relationships are at increased risk for autosomal recessive genetic disorders: the closer the relationship, the greater the risk.

Figures 10.1 and 10.2 summarize *Standardized Human Pedigree Nomenclature* set forth by the Pedigree Standardization Task Force of the NSGC (8). Using a pedigree to symbolize a patient's medical and family histories allows for readily accessible and succinctly summarized medical information on a single page. Developing skills in basic family history assessment will become key for all health care professions as genetic testing becomes further integrated into primary care and other nongenetic specialties (9). Having patients prepare family histories in advance of their appointment is a way to empower patients to take charge of their health, and

	Male	Female	Unspecified gender
Individual	□ b. 1986	○ 27yo	◇ 2 weeks
Affected	■	●	◆
Multiple individuals	3	*Females and unspecified gender are also denoted in the same way as males only using the circle and diamond.*	
Multiple individuals, number unknown	n		
Deceased	▱ d. 50		
Proband	↗□	The proband is the affected family member coming to medical attention, while a consultand is the individual seeking counseling/testing.	
Stillbirth	▱ SB 24 wk	Include gestational age and karyotype if known.	
Pregnancy	P 11 wk	Include gestational age and karyotype if known. If affected, use light shading.	

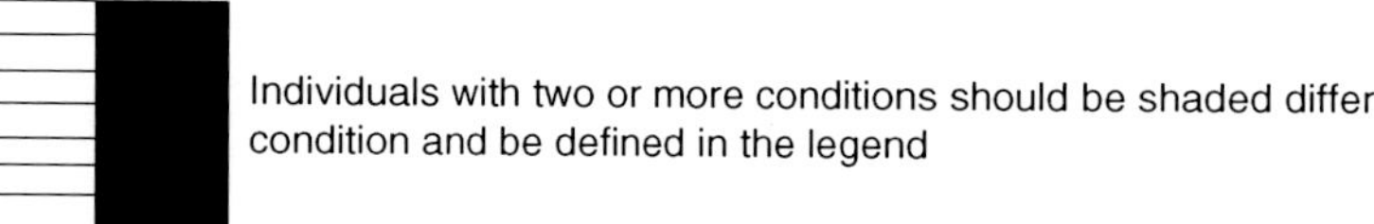

Individuals with two or more conditions should be shaded differently for each condition and be defined in the legend

FIGURE 10.1 Common pedigree symbols.

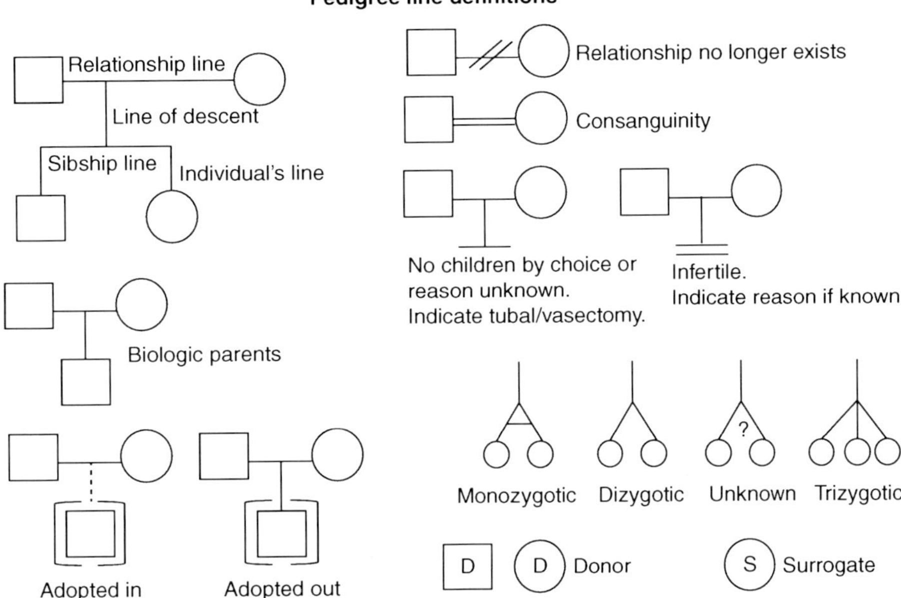

FIGURE 10.2 Common pedigree symbols, continued.

also allow the provider to spend more focused time in confirming and interpreting family history at the visit (10). Online tools are available to help accomplish this, including "My Family Health Portrait" put forth by the Surgeon General (familyhistory.hhs.gov /fhh-web/home.action).

Genetic practitioners are particularly skillful in pattern recognition. A general family history may prompt the counselor to ask more direct questions about specific individuals or characteristics of the family. For example, a 16-year-old female presented to a genetics clinic for evaluation of primary amenorrhea. While obtaining the family history, the patient's mother revealed that she had a parathyroid tumor removed a few years ago. The genetic counselor subsequently asked targeted questions about other family members and learned that an uncle had multiple peptic ulcers. While these medical concerns may appear seemingly unrelated to the family members, the counselor recognized the possibility of a hereditary cancer syndrome—multiple endocrine neoplasia type 1 (MEN1). While the family may or may not meet criteria for a clinical diagnosis or genetic testing, the suspicion alone can encourage family members to have basic screening for further evaluation or redirect further information gathering. Oftentimes during a genetic intake the individual being seen for assessment, or their family members, will be diagnosed with a condition with a possible genetic component outside of the condition for which the original referral was made. One study demonstrated that taking a family history in the primary care setting identified at least one condition with a possible genetic component in 40.3% of patients (11). Table 10.1 describes some of the "red flags" in a family history that suggest a disorder may have a genetic etiology.

TABLE 10.1   Red Flags in a Family History Suggestive of a Genetic Disorder

- Multiple closely related individuals affected with the same condition, particularly if the condition is rare
- Common disorder with earlier age of onset than typically expected (eg, premenopausal breast cancer, heart disease in individuals less than 40–50 years)
- Individual or couple with three or more pregnancy losses
- Medical issues in the offspring of consanguineous parents
- Sudden cardiac death in a person who seemed healthy
- Bilateral disease in paired organs (eg, eyes, kidneys, lungs, breasts)
- Associated cancers (eg, breast and ovarian, endometrial and colon)
- An individual with
  - Two or more medical conditions that could be due to a genetic disease (eg, hearing loss and renal disease, two primary cancers)
  - Two or more major birth anomalies
  - One major birth anomaly with two minor anomalies
  - Cleft lip and/or palate
  - Congenital heart defect
  - A medical condition and dysmorphic features
  - Developmental delay with dysmorphic features and/or birth anomalies
  - Developmental delay with other medical conditions
  - Loss of developmental milestones and/or progressive intellectual delay
  - Autism or pervasive developmental disorder
  - Progressive behavioral problems
  - Unexplained hypotonia, ataxia, or seizures
  - Progressive neurological condition, movement disorder, and/or muscle weakness
  - Unexplained cardiomyopathy
  - Hematologic condition with excessive bleeding or clotting
  - Unusual birthmarks, particularly if associated with seizures, learning disabilities, or dysmorphic features
  - Hair anomalies (eg, hirsutism, brittle, coarse, kinky, sparse, or absent)
  - Congenital or juvenile diabetes or blindness
  - Cataracts at a young age
  - Primary amenorrhea or primary adrenocortical insufficiency (male)
  - Ambiguous genitalia
  - Proportionate or disproportionate short stature with dysmorphic features and/or delayed or arrested puberty
  - Premature ovarian failure
  - Male with hypogonadism and/or significant gynecomastia and/or infertility
  - Congenital absence of the vas deferens
  - Oligospermia or azoospermia
- A fetus with
  - A major structural anomaly
  - Significant growth retardation
  - Multiple anomalies

Source: From Ref. (12). Bennett RL. *The Practical Guide to the Genetic Family History*. 2nd ed. New York, NY: Wiley-Blackwell; 2010.

RISK ASSESSMENT BEYOND BASIC MENDELIAN PRINCIPLE

Genetic risk assessment is an integral part of genetic counseling and testing. Genetic risk refers to the probability of carrying a disease-causing mutation or being affected with a particular genetic disease. Understanding, interpreting, and communicating risk is a fundamental component of genetic counseling. Risk information must be communicated clearly and empathetically, and practitioners should expect each patient to react to and interpret results differently. When a disorder has simple Mendelian inheritance, risk information for a particular family member is more easily determined. On the other hand, risk calculations can be far from straightforward with non-Mendelian patterns or complex presentations (Table 10.2). There are a number of factors that influence the interpretation of a pedigree and also affect the manifestation of a particular condition.

Germline mosaicism is a situation whereby only the gametes, egg or sperm cells, carry a genetic mutation and the mutation cannot be identified in the somatic cells of the individual. In theory, the possibility of germline mosaicism should be taken into consideration every time the parents of a child with an apparently de novo autosomal dominant or X-linked mutation are being counseled (12). Germline mosaicism can be assumed in cases where unaffected parents

TABLE 10.2 Patterns of Inheritance

Mendelian inheritance	Autosomal dominant Autosomal recessive X-linked dominant X-linked recessive
Non-Mendelian patterns	Mitochondrial Imprinting Multifactorial Sporadic, de novo
Complex presentations	Variable expressivity Reduced penetrance Age dependence Anticipation Sex-limited Sex-influenced Germline mosaicism Phenocopies Skewed X-inactivation

have two or more children with a dominant condition (when nonpaternity has been ruled out). Importantly, parents should first undergo careful evaluation and examination to ensure the neither shows evidence of somatic mosaicism, whereby the somatic cells of the body have more than one genotype. Certain genetic conditions are known to have increased incidence of germline mosaicism and therefore increased risk of recurrence. Examples of these include Duchenne muscular dystrophy (DMD; 7%–10%), facioscapulohumeral muscular dystrophy (10%), osteogenesis imperfecta (6%), retinoblastoma (5%), and tuberous sclerosis complex (2%–3%) (13–17). When providing recurrence risk, the literature should be consulted to establish if empiric risk data are available for any particular condition. And, if a pathogenic mutation has been identified, the option of prenatal or preimplantation diagnosis should be offered.

Penetrance describes the probability that a given gene mutation would manifest as disease. If a condition has complete penetrance, all individuals with a pathogenic mutation will have clinical symptoms. On the other hand, those with a reduced or incompletely penetrant condition may or may not show symptoms. For example, ectrodactyly–ectodermal dysplasia–cleft syndrome (EEC syndrome) is a genetic disorder characterized by ectrodactyly, ectodermal dysplasia, and orofacial clefts. Reports suggest that 93% to 98% of individuals with a mutation will manifest symptoms (18). On the other hand, hereditary hemochromatosis is a condition with reduced penetrance. In the general population of European descent, the expected homozygote frequency for the *HFE* C282Y mutation is approximately 1 in 80; however, this is not reflected in the number of patients presenting with

disease. Disease presentation in hereditary hemochromatosis is also age-related and sex-influenced. If a condition in a pedigree appears to have "skipped" a generation, reduced or incomplete penetrance should be considered.

EEC syndrome is also a good example of *variable expressivity*, which refers to the range of signs and symptoms that can occur in different people with the same genetic condition. Even individuals within the same family with the same genetic mutation can have tremendous variability. These intrafamilial differences are most likely due to other modifier genes, environmental and lifestyle factors. Some individuals with EEC syndrome, even family members with identical *TP63* mutations, may not manifest clefting or ectrodactyly (19). Individuals with mild clinical symptoms may be missed in a pedigree.

Anticipation is a phenomenon whereby the signs and symptoms of a genetic disease appear at an earlier age and tend to be more severe as the condition is passed from one generation to the next. Anticipation is typically observed in syndromes that result from a trinucleotide repeat expansion, such as in myotonic dystrophy type 1. Individuals with more than 100 CTG repeats in the *DMPK* gene manifest the classic form of the disease, and repeat size correlates with onset and increasing disease severity. Expansion of repeats in this form of myotonic dystrophy occurs with maternal transmission. A congenital form of the disease characterized by infantile hypotonia, respiratory deficits, and intellectual disability can occur when there are over 1,000 CTG repeats in the *DMPK* gene. An infant with congenital myotonic dystrophy type 1 may have a grandparent whose only sign of disease is early balding.

X-inactivation is the phenomenon by which one X chromosome is randomly inactivated in each somatic cell of a female as a method of dosage compensation, also known as lyonization. If the paternal X is inactivated more often than the maternal X or vice versa, the X-inactivation pattern is *skewed*, which may unmask recessive traits. For example, females who carry the X-linked DMD mutation may have mild symptoms of DMD, such as muscle weakness, myalgia, or cramps, while others may develop dilated cardiomyopathy (20). If the normal X is preferentially inactivated, females may exhibit more severe symptoms of the disease. Classic DMD presentation has been reported in women with Turner syndrome (monosomy X) and uniparental isodisomy of the X chromosome (21).

When several factors are involved in the pathogenesis of a given disorder, it is essential for a practitioner to have a thorough guide for calculating risk for non-Mendelian disorders or disorders with complex inheritance. The following books are recommended

TABLE 10.3 Bayesian Analysis for Mother of Son With Duchenne Muscular Dystrophy With Three Unaffected Sons and No Other Family History

	Carrier	Noncarrier
Prior probability	2/3	1/3 De novo rate
Conditional probability	1/2 × 1/2 × 1/2 = 1/8 The probability that she would have three unaffected sons under the assumption that she is a carrier	1 The probability that she would have three unaffected sons under the assumption that she is a noncarrier
Joint probability	1/12	1/3
Posterior probability	(1/12)/(1/12 + 1/3) = 1/5 (20%)	(1/3)/(1/12 + 1/3) = 4/5 (80%)

for instruction on genetic risk calculation and basic risk information for multiple genetic conditions: *Introduction to Risk Calculation in Genetic Counseling* (by Ian D. Young) and *Practical Genetic Counseling* (by Peter S. Harper).

Frequently in genetics practice, counselors employ Bayesian analysis to calculate conditional probabilities, and genetic risk can be modified using information obtained from the family history and pedigree, before genetic testing is considered. Take for example, a woman who has four sons, only one of whom is affected with DMD, a life-limiting X-linked recessive condition. The woman is interested in expanding her family, but has not undergone carrier screening for DMD. Therefore, we do not know if the woman is a carrier of DMD or if the mutation occurred for the first time in her son, that is, de novo. The woman is wondering what her chances are for being a carrier. With knowledge of three unaffected sons, and no other family history, her risk of being a carrier is low, although not zero. The two hypotheses in this situation are that the *mother is a carrier* and that *she is a noncarrier*. Because DMD is a life-limiting condition, the probability that she is a carrier is 2/3, and the probability that she is a noncarrier is 1/3 (the de novo rate for lethal X-linked conditions). In life-limiting or lethal X-linked conditions, the disease incidence is expected to decline over time because the affected person cannot reproduce; in other words, one out of three X chromosomes are lost with each generation (females have XX, males have XY). Since the incidence of DMD has stayed constant, there must be a 33% new mutation rate (1/3). The carrier rate is then deduced by simple subtraction (1 − 1/3 = 2/3).

However, the probability that the mother is a carrier is less than 2/3 because she has three sons who are unaffected. Bayesian analysis allows us to incorporate additional information about the family to modify the woman's a priori risk (2/3). Our goal is to determine the probability that the conditional information (three unaffected sons) would occur if the hypothesis (being a carrier) were true. Using this methodology, the prior probability is multiplied by the conditional probability to form the joint probability. The posterior probability for each hypothesis is the probability that each hypothesis is true after taking into account both prior and subsequent information (22). To calculate the posterior probability for each hypothesis, one simply divides the joint probability for that hypothesis by the sum of all of the joint probabilities (Table 10.3).

Using this methodology, the patient's risk of being a carrier can be modified from 2/3, or 66%, to 1/5, or 20%. The patient can, of course, have formal carrier testing if the mutation has been identified in the affected child. However, if the mutation cannot be identified or the patient cannot afford testing, Bayesian methodology can provide statistical risk information for the patient and can help guide reproductive decision making. Importantly, a practitioner must always consider rare occurrences, such as germline mosaicism. The inheritance pattern of the genetic disorder must also be taken into consideration. All professional recommendations agree that carrier testing of minors for X-linked recessive disease should ideally be deferred until the age of maturity (23). In our example, it is not possible to predict whether or not a female carrier of DMD will manifest any signs of the disorder, and current policy states that female fetuses should not be tested prenatally for carrier status (24). There are, however, special circumstances in which carrier testing may be indicated for at-risk female fetuses with chromosome abnormalities such as monosomy X or 45,X/46,XX mosaics.

Bayesian analysis can also be applied to genetic test results coupled with penetrance data for a particular condition. Take Huntington's disease (HD) for example, an autosomal dominant neurodegenerative

TABLE 10.4 Bayesian Analysis for Autosomal Dominant Huntington Disease (HD) Based on Penetrance Data for Clinically Unaffected At-Risk 40-Year Old

	Carrier	Noncarrier
Prior probability	1/2 Autosomal dominant condition with 50% risk of inheritance	1/2
Conditional probability	1 − 3/10 = 7/10 The probability that a patient would not show symptoms of HD by age 40 under the assumption that he or she is a carrier	1
Joint probability	7/20	1/2
Posterior probability	(7/20)/(7/20 + 1/2) = 7/17 (41%)	(1/2)/(7/20 + 1/2) = 10/17 (58%)

disorder caused by expanded triplet repeats in the *HTT* gene. Among those with an inherited susceptibility to HD, 30% have clinically detectable symptoms by the age of 40 (25). Using Bayesian analysis, a patient's a priori risk of 50% (the chance that the mutation was inherited from an affected parent) can be decreased to 41% if they have not shown symptoms by age 40 (Table 10.4). If they have not shown symptoms by age 70, their risk to have the condition is approximately 5%. While roughly 80% of at-risk patients choose not to proceed with genetic testing for HD, Bayesian analysis can provide exploration of their risk and perhaps provide some level of relief to an anxious patient who is not showing symptoms.

When assessing risk based on test results, it is essential that all medical providers understand the sensitivity, specificity, and detection rate for a particular genetic test. At present, no single genetic test detects 100% of mutations for a given condition. Bayesian analysis can also be used to provide a residual risk based on a negative test result. As an example, carrier screening for Pompe disease (a glycogen storage disorder that can result in cardiac abnormalities) has a detection rate by sequencing of up to 93%. If one member of a couple undergoes carrier screening for Pompe disease and is negative, their residual risk to be a carrier is reduced from 1% to 0.07%. If the other member of the couple is a known carrier, the couple's risk to have a child affected by this condition is reduced from 1:800 to 1:11,320. Additionally, a mutation cannot always be identified in families where there is a clear pattern of inheritance. For example, although approximately 50% of hypertrophic cardiomyopathy is familial, a disease-causing mutation is identified in only 60% to 70% of familial hypertrophic cardiomyopathy. Thus, at-risk relatives should follow screening guidelines when available if genetic testing is inconclusive.

PSYCHOSOCIAL COUNSELING

While genetic counselors are highly trained in medical genetics and risk assessment, what makes them particularly unique among health care providers is their additional training in psychosocial counseling. Frequently, there are competing priorities between the medical and psychological considerations when discussing genetic diseases, which requires a great deal of time exploring a patient's concerns, emotions, perceptions, reactions, coping mechanisms, cultural beliefs, and support systems. Genetic counselors create a supportive environment, which enhances the patient's ability to cope with the complexity of information, make difficult choices, and face the potential consequences of a diagnosis (6).

In addition to ongoing support in the clinic setting, genetic counselors are a particularly useful resource for identifying support groups, community agencies, advocacy groups, long-term counseling services, and for connecting families with the same condition.

GENETIC COUNSELORS IN CARDIOVASCULAR PRACTICE

Cardiovascular genetics is the genetics specialty that focuses on hereditary cardiovascular diseases. Individuals who are typically referred to a cardiogenetics clinic include those with a personal and/or family history of cardiomyopathy, aortopathy, cardiac amyloidosis, arrhythmia, familial coronary artery disease, muscular dystrophy, or congenital heart disease. This also extends to individuals with a genetic syndrome associated with cardiac disease (eg, Marfan syndrome) or those with an inherited metabolic disorder with cardiac manifestations (very long-chain

TABLE 10.5 Individuals Likely to Benefit From a Cardiogenetics Evaluation

- Preconception/prenatal counseling of adults with congenital heart disease
- Parents of a child with congenital heart disease
- Individuals with a strong family history of congenital heart disease or cardiac disease at a young age in one or more close relatives
- A pattern of cardiac disease being passed through generations in a family (ie, dilated or hypertrophic cardiomyopathy)
- A personal or family history of syncope
- Multiple relatives with related disorders such as stroke (particularly aged less than 60), high cholesterol, and high blood pressure (particularly aged less than 40)
- A history of sudden death in a family (particularly aged less than 50), including unexplained accidental death such as car accidents or drowning
- Sudden infant death syndrome (SIDS)
- Noncardiac features that suggest an underlying genetic syndrome:
  - Facial dysmorphism
  - Extracardiac malformations (eg, limb anomalies, congenital hearing loss)
  - Cognitive impairment
  - Behavioral and/or psychiatric disorders
  - Multisystem involvement (eg, coexistent renal, hepatic, hematologic, thyroid, immunologic, sensorineural, or vision abnormalities)
- Patients with high-risk anatomy to assess for:
  - DiGeorge syndrome/22q11.2 deletion (conotruncal defects, interrupted aortic arch, truncus arteriosus, ventricular septal defect, tetralogy of Fallot, aortic arch anomaly, discontinued branch pulmonary arteries)
  - Noonan syndrome (pulmonary valve stenosis)
  - Turner syndrome (female with coarctation of the aorta)
  - William syndrome (supravalvular aortic stenosis)

Source: Burchill L, Greenway S, Silversides CK, Mital S. Genetic counseling in the adult with congenital heart disease: what is the role? *Curr Cardiol Rep*. 2011;13(4):347–355 and Bennett RL. *The Practical Guide to the Genetic Family History*. 2nd ed. New York, NY: Wiley-Blackwell; 2010.

acyl-CoA dehydrogenase deficiency). Table 10.5 describes individuals who are likely to benefit from a cardiogenetics evaluation. Comprehensive evaluation of patients presenting to a cardiogenetics clinic should include detailed family history and risk assessment, examination for syndromic features, and when applicable, genetic testing and concomitant genetic counseling (26). Research has demonstrated that families with inherited cardiovascular disease prefer having genetic counseling as part of their care and have better psychological outcomes, including less worry and better adjustment to their diagnosis (27).

The understanding of the genetic basis of both syndromic and isolated congenital or late-onset cardiac disease has grown tremendously, while advances in cardiovascular care have increased the population of adults with heart disease who reach childbearing age (28). Genetic counselors function in several capacities within the cardiovascular setting (Figure 10.3). In addition to their traditional roles, they are available to discuss prenatal diagnosis or preimplantation genetic diagnosis, be a psychosocial resource for children and adults who may be devastated by physical restrictions, and participate in research. Counselors partake in a number of activities to help interpret test results, including bioinformatic prediction tools, revisiting variants of uncertain significance, family

investigations to derive pathogenicity, and may work in collaboration with laboratories that perform functional studies. Counselors often engage in psychosocial research to better understand the impact of

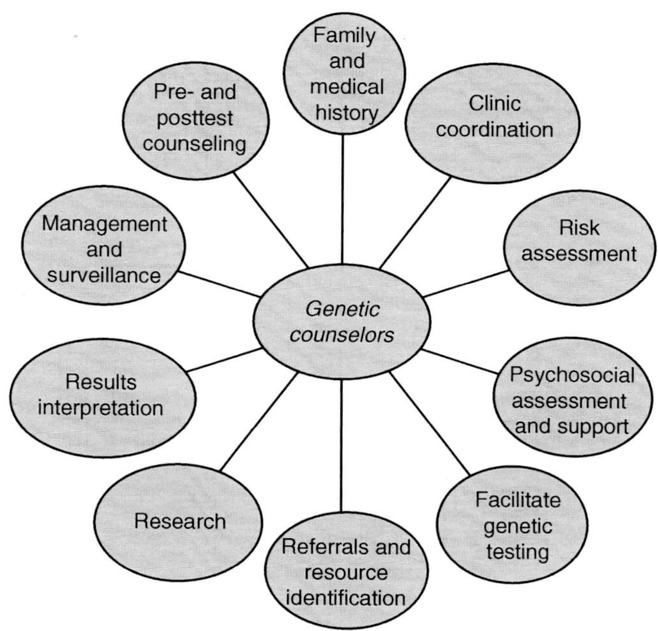

FIGURE 10.3 The role of genetic counselors in cardiovascular genetics and genomics.

TABLE 10.6 Examples of Extracardiac Findings in Genetic Syndromes

Unusual skin findings	Ehlers–Danlos Naxos Loeys–Dietz Hemochromatosis Fabry Dyslipoproteinemias
Blood disorders or immune deficiency	Barth DiGeorge Jacobson
Hearing loss	Jervell and Lange-Nielsen Multiple lentigines syndrome Mitochondrial disorders
Skeletal and joint issues	Ehlers–Danlos Holt–Oram Marfan Duane-Radial Ray Loeys–Dietz Klippel-Feil
Neuromuscular disease	Myotonic dystrophy Friedreich's ataxia Duchenne–Becker muscular dystrophies Emery–Dreifuss muscular dystrophy Limb-girdle muscular dystrophy Mitochondrial disorders
Unusually tall or short	Marfan Barth Homocystinuria Sotos Turner Noonan Ellis–van Creveld
Oral or dental anomalies	Andersen–Tawil DiGeorge Loeys–Dietz Marfan
Learning disabilities Cognitive impairment Developmental delay	Myotonic dystrophy Homocystinuria Noonan Lysosomal storage Danon Andersen–Tawil Timothy DiGeorge

a diagnosis, and try to identify predictors of worse outcomes to develop strategies that improve patient well-being (29). With knowledge of many genetic conditions with cardiac manifestations, counselors have the ability to triage patients who should be referred for a complete genetic evaluation with a clinical geneticist or other subspecialty based on medical history (Table 10.6).

CARDIOVASCULAR GENETIC TESTING

Perhaps the most well-known role of the cardiac genetic counselor is in facilitating genetic testing. Genetic counselors can interpret genetic test results and integrate those results into patient care. In a study conducted by ARUP Laboratories, it was determined that 30% of orders for complex gene tests within an 11-month period contained errors that originated with ordering physicians or their office staff (30). Physicians were reported to have ordered the wrong test or confused diseases with similar names. Since many treatment decisions are based on genetic test results, these errors have the potential to be quite harmful. For many genetic disorders, particularly cardiovascular ones, testing strategy is moving toward multigene panels and whole genome/exome analysis, both of which will require genetic practitioners to be members of multidisciplinary teams to accurately interpret and incorporate genetics and genomics into practice.

After a patient has been evaluated by a cardiovascular specialist, a genetic counselor identifies the most informative family member to test, determines the most useful and appropriate genetic test, discusses the benefits and limitations of testing, and ensures that the individual or family understands the implications of the possible results. Testing should begin with a clearly affected family member, preferably the most severely affected, increasing the likelihood of identifying a pathogenic mutation. Testing an unaffected individual is acceptable if there are no other options (all affected individuals are deceased or affected family members will not consent to testing), but the results need to be interpreted carefully. If the test is positive the result is straightforward, as this identifies the etiology of the disease. However, if the patient's result is negative, it could indicate several possibilities: there is no mutation in the gene(s) tested, the methodology could not detect the mutation, or the correct gene has not been tested or perhaps not yet discovered. In this situation, the patient cannot be assured they have not inherited the pathogenic mutation. If an affected relative is first found to have a mutation, other family members can then undergo testing for the familial mutation and accurate result interpretation can be provided.

A particularly bothersome result that poses a unique challenge in risk assessment is the identification of a variant of uncertain significance (VUS or VOUS). These variants typically lack adequate evidence to support pathogenicity, often because the variant has been found in both affected and unaffected individuals. Misinterpretation of information by the family or a provider can be especially harmful in the cardiovascular setting, and caution must be exercised when giving a VUS result. Testing other family members for an identified VUS is not recommended. Research studies may exist to help investigate the VUS further, such as tracking the variant within the family to determine if the variant is found in other individuals with cardiac disease. While the family may not benefit immediately, enough evidence from various sources may eventually warrant a change in classification.

Cardiogenetics is on the verge of a major transformation as reports of the clinical utility of whole exome/genome sequencing (WES/WGS) continue to surface. With the development of next-generation sequencing technologies, fairly rapid analysis of multiple genes can be performed, and eventually WES/WGS will replace smaller gene panels altogether. As an example, sudden cardiac death is often the first symptom of a cardiomyopathy or channelopathy. Testing via traditional methods would involve ordering one or more panels of several genes, which would consume time, money, and the potentially limited DNA sample. Testing for a single panel risks missing the pathogenic mutation, whereas ordering multiple panels may result in redundancy and wasted resources. WES/WGS would allow for testing all genes included on these panels as well as the possibility of new gene identification. The implementation of WES/WGS will require more time-intensive pre- and posttest "genomic counseling." Patients will need to understand the range of outcomes, particularly the high likelihood of multiple variants of uncertain significance, but also incidental findings beyond the aim of the study, which may or may not have a significant impact on their health. For example, there are over 30 known cancer susceptibility syndromes and other heritable conditions present in 0.2% of the population. The American College of Medical Genetics has published guidelines in 2013 that recommend the reporting of incidental findings for certain conditions. Additionally, The American College of Medical Genetics states that whole exome/genome informed consent counseling should be performed by a medical geneticist or genetic counselor (2013). Genetic practitioners will need to address patients' understanding of these implications, desire to be informed of this information, and when necessary, refer them to additional specialists.

PARTNERS IN PRACTICE

Genetic counselors possess the skills and expertise to guide patients through the intricacies of genetic and genomic medicine. With their unique understanding of the medical and psychosocial issues that surround chronic conditions, many counselors have joined nongenetic disciplines, and have become further removed from the traditional genetics team. The NSGC Code of Ethics specifically refers to the duty of genetic counselors to share their knowledge and provide mentorship for other health care providers. In the arena of cardiovascular health, genetic counselors can help cardiologists stay updated and knowledgeable on the latest advancements in genetic and genomic technology, further facilitating the application of complex genetic knowledge into patient and family care.

CONCLUSIONS

Genetic counseling is a critical aspect of cardiovascular genetics and genomics in clinical practice. Effective genetic counseling is essential for the interpretation of genetic testing, diagnosis of hereditary syndromes, education of patients and family members, and discussion of ethical aspects of genetic testing and screening. With the advent of targeted therapies for genetic cardiovascular disorders, along with expansion of genomic analysis to genome-wide sequencing, the need for genetic counseling has never been greater.

REFERENCES

1. Kessler S. *Genetic Counseling*. New York, NY: Academic Press; 1979.
2. Nussbaum RL, McInnes RR, Willard HF. *Thompson & Thompson Genetics in Medicine*. 7th edition. Saunders Elsevier; 2007.
3. Veach PM, LeRoy BS, Bartels DM. *Facilitating the Genetic Counseling Process: A Practice Manual*. New York, NY: Springer Verlag; 2003.
4. Epstein CJ, Childs B, Fraser FC, et al. Genetic Counseling. *Am J Hum Genet*. 1975;27:240–242.
5. Resta R, Biesecker BB, Bennett RL, et al. A new definition of Genetic Counseling: National Society of Genetic Counselors' Task Force report. *J Genet Couns*. 2006;15(2):77–83.
6. Uhlman WR, Schette JL, Yashar BM, eds. *A Guide to genetic counseling*. Hoboken, NJ: John Wiley and Sons, Inc; 2009.
7. Bennett RL. The family medical history. *Prim Care Clin Office Pract*. 2004;31:479–795.
8. Bennett RL, French KS, Resta RG, Doyle DL. Standardized human pedigree nomenclature: update and assessment of the recommendations of the National Society of Genetic Counselors. *J Genet Couns*. 2008; 17(5):424–433.
9. Shugar AL. The family history: an integral component of paediatric health assessment. *Paediatr Child Health*. 2003;8(1):33–35.
10. Bennett RL. The family medical history as a tool in preconception consultation. *J Community Genet*. 2012;3(3):175–183.
11. Rose P, Humm E, Hey K, et al. Family history taking and genetic counselling in primary care. *Fam Pract*. 1999;16:78–83.
12. Bennett RL. *The Practical Guide to the Genetic Family History*. 2nd ed. New York, NY: Wiley-Blackwell; 2010.
13. Young ID. *Introduction to Risk Calculation in Genetic Counseling*. 3rd edition. New York, NY: Oxford University Press; 2007:241.
14. Bakker E, Veenema H, Den Dunnen JT, et al. Germinal mosaicism increases the recurrence risk for 'new' Duchenne muscular dystrophy mutations. *J Med Genet*. 1989;26(9):553–559.
15. Kohler J, Rupilius B, Otto M, et al. Germline mosaicism in 4q35facio scapulo humeral muscular dystrophy (FSHD1A) occurring predominantly in oogenesis. *Hum Genet*. 1996;98(4):485–490.
16. Byers PH, Bonadio JF, Cohn DH, et al. Osteogenesis imperfecta: the molecular basis of clinical heterogeneity. *Ann NY Acad Sci*. 1988;543:117–128.
17. Sippel KC, Fraioli RE, Smith GD, et al. Frequency of somatic and germ-line mosaicism in retinoblastoma: implications for genetic counseling. *Am J Hum Genet*. 1998;62(3):610–619.
18. Rose VM, Au KS, Pollom G, et al. Germ-line mosaicism in tuberous sclerosis: how common? *Am J Hum Genet*. 1999;64(4):986–992.
19. Didier Lacombe. EEC syndrome. Orphanet. Retrieved March 2011 from http://www.orpha.net/consor/cgi -bin/OC_Exp.php?lng=EN&Expert=1896
20. Fryns JP, Legius E, Dereymaeker AM, Van den Berghe H. EEC syndrome without ectrodactyly: report of two new families. *J Med Genet*. 1990;27:165.
21. Hoogerwaard EM, van der Wouw PA, Wilde AA, et al. Cardiac involvement in carriers of Duchenne and Becker muscular dystrophy. *Neuromuscul Disord*. 1999b;9:347–351.
22. Quan F, Janas J, Toth-Fejel S, et al. Uniparental disomy of the entire X chromosome in a female with Duchenne muscular dystrophy. *Am J Hum Genet*. 1997;60(1):160–165.
23. Ogino S, Wilson RB. Bayesian analysis and risk assessment in genetic counseling and testing. *J Mol Diagn*. 2004;6(1):1–9.
24. Borry P, Evers-Kiebooms G, Cornel MC, et al. Genetic testing in asymptomatic minors: background considerations towards ESHG recommendations. *Eur J Hum Genet*. 2009;17:711–719.
25. Abbs S, Tuffery-Giraud S, Bakker E, et al. Best practice guidelines on molecular diagnostics in Duchenne/Becker muscular dystrophies. *Neuromuscul Disord*. 2010;20:422–427.
26. Harper PS, Newcombe RG. Age at onset and life table risks in genetic counselling for Huntington's disease. *J Med Genet*. 1992;29(4):239–242.
27. Sturm AC. Genetic testing in the contemporary diagnosis of cardiomyopathy. *Curr Heart Fail Rep*. 2013;10(1):63–72.
28. Ingles J, Lind JM, Phongsavan P, Semsarian C. Psychosocial impact of specialized cardiac genetic clinics for hypertrophic cardiomyopathy. *Genet Med*. 2008;10:117–120.
29. Marelli AJ, Mackie AS, Ionescu-Ittu R, et al. Congenital heart disease in the general population: changing prevalence and age distribution. *Circulation*. 2007;115:163–172.
30. Ingles J, Yeates L, Semsarian C. The emerging role of the cardiac genetic counselor. *Heart Rhythm*. 2011; 8(12):1958–1962.
31. Miller CE, Krautscheid P, Baldwin EE, et al. (2011) Value of Genetic Counselors in the Laboratory. Retrieved on January 10, 2014, from http://www .aruplab.com/files/resources/genetics/White-paper-1 -value-of-GCs-in-lab.pdf
32. Klitzman R, Chung W, Marder K. Attitudes and practices among internists concerning genetic testing. *J Genet Couns*. 2013;22(1):90–100.

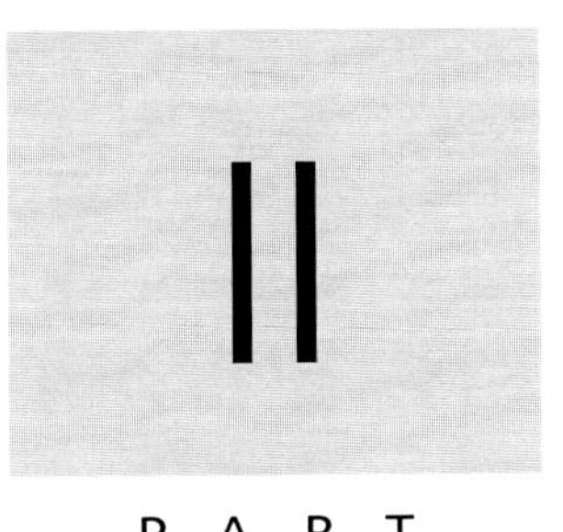

GENETICS OF CARDIOVASCULAR DISORDERS/TRAITS

Blood Pressure Genomics

Changwei Li and Tanika N. Kelly

TAKE HOME POINTS

1. The identification of genes involved in monogenic blood pressure (BP) conditions has provided important clues for the study of genes and pathways that may play a role in the complex BP phenotype.
2. Genome-wide association studies (GWAS) have identified common polymorphisms associated with BP traits. Current data suggest that those in the top decile of BP risk alleles have approximately 6 mmHg higher systolic blood pressure (SBP), 4 mmHg higher diastolic blood pressure (DBP), and twofold higher prevalence of hypertension, compared to those in the lowest decile.
3. The formation of GWAS meta-analysis consortia, the emergence of epigenetics, and the advent of next-generation sequencing (NGS) technology will likely provide further insight into the genetic determinants of hypertension.

CASE STUDY: SEVERE MONOGENIC HYPERTENSION—AN APPLICATION OF WHOLE-EXOME SEQUENCING

In a landmark study published in *Science* in 2011, Choi and colleagues examined 22 patients with severe hypertension, hypokalemia, high aldosterone to renin ratio, and aldosterone producing adenoma (APA) of the adrenal cortex (1). By performing whole-exome capture and Illumina sequencing of both the APA and blood of four patients (at greater than 150-fold coverage of each targeted base), the investigators identified two separate protein-altering mutations of the *KCNJ5* gene, which were confirmed in targeted sequencing study of the

APA–blood pairs of the remaining 18 patients. These findings strongly implicated a novel potassium channel gene in the etiology of a monogenic form of hypertension. Such remarkable results were achieved through the first application of whole-exome sequencing (WES) to the study of BP (1,2). Since this report, WES has identified a mutation to the *KLH3* gene resulting in another Mendelian BP disorder, familial hyperkalemic hypertension (3). WES (see Chapter 8) is now being applied to population-based studies, with investigators eagerly awaiting results, which will help to determine whether such an approach may identify rare variants influencing the complex BP phenotype (4).

Elevated blood pressure (BP) is a major global health challenge due to its high prevalence and associated increased risk of cardiovascular disease (CVD) and premature death (5–8). An estimated 978 million adults, or 28% of the world's adult population, had uncontrolled hypertension in 2008 (5). More alarming, conservative estimates indicate that the global burden of hypertension will increase to more than 1.5 billion by 2025 (6). As the most important modifiable risk factor for CVD and all-cause mortality, elevated BP was responsible for approximately 7.6 million deaths globally, or 13.5% of all deaths, in 2001 (7,8).

BP is influenced by both genomic and environmental factors, as well as their interactions (9–13). Although established early on as a heritable trait with many monogenic forms of BP dysregulation clearly described, our understanding of the genomic architecture of the complex BP phenotype was initially slow to progress (11). Early genome-wide linkage analyses, candidate gene studies, and GWAS were relatively unsuccessful in identifying reproducible loci related to BP (14–20). However, increased methodological stringency and the recent formations of large BP consortia have enabled important breakthroughs in hypertension genomic research. Through GWAS meta-analyses, numerous loci have now been robustly associated with BP in populations of European, Asian, and African ancestries (9–13). Although much of the heritability of BP still remains unexplained, there is renewed optimism as we turn our attention toward next-generation approaches for the discovery of novel genomic determinants of this complex trait.

MONOGENIC BP DISORDERS

Some of the earliest advancements in human BP genomics research involved the identification of genes responsible for severe, inherited forms of hypertension and hypotension. Although many physiological processes are responsible for the regulation of BP, the vast majority of genes identified for monogenic BP disorders play key roles in renal-sodium handling (Table 11.1) (3,21–62). Many genes have been shown to exert their effects by directly influencing sodium and water reabsorption in the nephron's distal tubule, leading to changes in plasma volume, cardiac output, and BP (63). For example, Liddle syndrome, which is characterized by severe early-onset hypertension, results from gain-of-function mutations to the epithelial sodium channel genes *SCNN1B* and *SCNN1G*, expressed in the nephron's distal tubule (64,65). Such gain-of-function mutations have been shown to increase renal epithelial sodium channel activity, directly increasing renal-sodium reabsorption and consequently BP (65). Other genes influence sodium reabsorption indirectly through their regulation of the renin–angiotensin–aldosterone system (RAAS) (31,66–69). For example, pseudohypoaldosteronism type 1, which is characterized by neonatal renal salt wasting and hypotension, can result from a loss-of-function mutation to the gene encoding the mineralocorticoid receptor (*NR3C2*) (56). Such a mutation limits the upregulation of epithelial sodium channel activity that typically results from the interaction of the mineralocorticoid receptor with aldosterone (56).

Identification of genes responsible for monogenic hypertension and hypotension disorders has provided valuable insights into the genomic mechanisms and biological pathways underlying BP regulation. Furthermore, such research has also provided important clues for the complex BP phenotype. For example, in comparison to the rare variants in genes responsible for monogenic BP disorders, investigators have postulated that common genetic variation in these genes may have more modest effects, contributing to the interindividual variation in the complex BP phenotype (70). As such, these genes have been the target of myriad candidate gene studies of BP and hypertension (71), and are considered very promising targets for follow-up when present at GWAS-identified loci (10).

GENOMICS OF BP AS A COMPLEX TRAIT

Heritability of BP

BP has long been established as a heritable trait, suggesting a significant contribution of genetic factors to this complex phenotype (72–77). The heritability of BP has been shown to range from about 30% to 60% in pedigree data to as high as 70% in twin studies (72,73). Longitudinal data from the Framingham Heart Study showed that 57% and 56% of interindividual variability in SBP and DBP, respectively, was due to genetic factors (74). Data from Nigerian families suggest heritabilities of 34% to 45% and 29% to 43% for systolic BP (SBP) and diastolic BP (DBP), respectively (75,76). Similarly, in the Chinese population, Gu et al estimated significant heritabilities of 31% and 32% for SBP and DBP, respectively (77).

Whole-Genome Linkage Analyses of BP

Given the widespread success of genome-wide linkage analyses in the identification of genes for Mendelian disorders, investigators were initially optimistic about using this approach to localize genomic regions harboring susceptibility loci for the complex BP phenotype. Numerous genome-wide linkage scans of SBP, DBP, or hypertension were subsequently conducted but with somewhat disappointing results. To date, approximately 34 quantitative trait loci (QTLs) for SBP, DBP, and hypertension phenotypes achieved a logarithm (base 10) of odds (LOD) score of 3.0 or higher (74,78–96). Among the identified loci, only three genomic regions, located on chromosomes 2, 3, and 7, achieved an LOD score of 3.0 or higher in two or more studies (78–83). However, linkage of BP phenotypes to chromosomes 3 and 7 may not represent true evidence of replication (78,79). For example, Koivukoski and colleagues linked 3p14.1-3q12.3 to BP phenotypes in a meta-analysis (80). This region was previously identified among Framingham Heart Study participants included in Koivukoski's study (81). Similarly, a meta-analysis by Rice and colleagues identified significant linkage of 7p21.3-7p15.3 to SBP (82), which was also identified in a study by Adeyemo and colleagues whose participants contributed to

TABLE 11.1 Monogenic Blood Pressure Disorders

Syndrome	Chr	Gene Location (BP)	Gene(s)	Inheritance	Mechanism	Reference
Hypertension Disorders						
Apparent mineralocorticoid excess	16	67464555-67471456	HSD11B2	Autosomal recessive	Inhibits conversion of cortisol to cortisone (↑ MR activity)	(21)
Autosomal dominant hypertension with brachydactyly	12	Unknown	HYT4	Autosomal dominant	Unknown	(22)
Congenital adrenal hyperplasia, due to 11-beta-hydroxylase deficiency	8	143953772-143961262	CYP11B1	Autosomal recessive	Accumulation of 11-deoxycorticosterone (↑ MR activity)	(23,24)
Congenital adrenal hyperplasia, due to 17-alpha-hydroxylase deficiency	10	104590288-104597290	CYP17A1	Autosomal recessive	Production of excessive corticosterone and deoxycorticosterone (↑ MR activity)	(25,26)
Familial glucocorticoid resistance	2	14441673-14447178	GR	Autosomal recessive or dominant	Overproduction of mineralocorticoids (↑ MR activity)	(27,28)
Familial hyperaldosteronism, type II		Unknown		Autosomal dominant	Unknown	(29)
Familial hyperaldosteronism, type III	11	128761251-128790930	KCNJ5	Autosomal dominant	Increased aldosterone production (↑ MR activity)	(30)
Glucocorticoid-remediable aldosteronism	8	143953772-143961262	CYP11B1	Autosomal dominant	Increased aldosterone production (↑ MR activity)	(31)
	8	143991975-143999259	CYP11B2			
Hypertension exacerbated by pregnancy	4	148999913-149365850	NR3C2	Autosomal dominant	Constitutive MR activity and altered receptor specificity (↑ MR activity)	(32)
Hypertension, hypercholesterolemia, and hypomagnesemia			tRNA	Mitochondrial inheritance	Loss of mitochondrial function	(33)
Insulin resistance and hypertension	3	12329349-12475855	PPARG	Autosomal dominant	Loss-of-function mutation results in insulin resistance and hypertension	(34)
Liddle syndrome	15	56119120-56285944	NEDD4	Autosomal dominant	Constitutive activation of the renal epithelial sodium channel	(35,36)
	16	23194036-23228204	SCNN1G			
	16	23289552-23392620	SCNN1B			

(continued)

TABLE 11.1 Monogenic Blood Pressure Disorders (*continued*)

Syndrome	Chr	Gene Location (BP)	Gene(s)	Inheritance	Mechanism	Reference
Pseudo-hypoaldosteronism type II	1	Unknown		Autosomal dominant	Dysregulation of sodium chloride transporter and renal epithelial sodium channel	(3,37–40)
	2	225334867-225450110	*CUL3*			
	5	136953189-137071779	*KLHL3*			
	12	861759-1020618	*WNK1*			
	17	40932696-40948954	*WNK4*			
Resistance to diastolic hypertension	5	169805168-169816681	*KCNMB1*	Unknown	Unknown	(41)
Hypotension Disorders						
Bartter syndrome	1	55464606-55476556	*BSND*	Autosomal recessive	Loss of function of the renal ion transporters, channels, and their subunits in the thick ascending loop of Henle	(42–46)
	1	16345370-16360545	*CLCNKA*			
	1	16370272-16383803	*CLCNKB*			
	11	128706210-128737268	*KCNJ1*			
	15	48483861-48596275	*SLC12A1*			
Congenital adrenal hyperplasia	8	38001167-38008783	*STAR*	Autosomal recessive	Mineralocorticoid deficiency ($\downarrow$ MR activity)	(47–52)
	6	32006042-32009447	*CYP21A2*			
	10	104590288-104597290	*CYP17A1*			
Corticosterone methyloxidase, type I deficiency	8	143991975-143999259	*CYP11B2*	Autosomal recessive	Impaired aldosterone biosynthesis ($\downarrow$ MR activity)	(53)
Corticosterone methyloxidase, type II deficiency	8	143991975-143999259	*CYP11B2*	Autosomal recessive	Aldosterone deficiency ($\downarrow$ MR activity)	(53)
Gitelman's syndrome	1	16370272-16383803	*CLCNKB*	Autosomal recessive	Loss of function of the renal thiazide-sensitive sodium chloride cotransporter	(54,55)
	16	56899119-56949762	*SLC12A3*			
Pseudo-hypoaldosteronism, type I	4	148999913-149365850	*NR3C2*	Autosomal dominant	Loss of MR function	(56–59)
	12	6456009-6486896	*SCNN1A*	Autosomal recessive	Loss of renal epithelial sodium channel activity	(60–62)
	16	23194036-23228204	*SCNN1G*			
	16	23289552-23392620	*SCNN1B*			

Abbreviations: Chr, chromosome; MR, mineralocorticoid receptor.

114

Rice's meta-analysis (83). The only truly independent evidence of significant linkage identified by more than one study was at 2q31-2q34. Hsueh and colleagues linked this region to DBP among Old Order Amish families (78), while Morrison and colleagues linked 2q34 to hypertension among African American families (79). The availability of genome-wide genotyping platforms led to a general shift away from this approach in favor of more powerful association methods.

Candidate Gene Studies of BP

Over 1,500 genes have been related to BP in human populations, with the vast majority derived from candidate gene studies (97). Based on a priori knowledge of biological function, candidate gene studies offer a powerful approach for detecting genetic variants that influence common complex traits, like BP. Despite their popularity, early candidate gene studies of BP were hampered by inconsistent findings, which to some extent may have reflected methodological limitations including small sample sizes, measurement error in the phenotype, inappropriate correction for multiple testing, and lack of verification in independent samples (98–100). Still some investigators have supported the continued use of candidate gene studies, noting that biologically relevant loci may be missed by GWAS, which use very stringent alpha thresholds for determining statistical significance (101).

Some of the more recent candidate gene studies have successfully identified genetic associations that are reproducible in independent samples (101–104). Such successes are likely the result of the employment of large sample sizes and appropriate correction for multiple testing. While the latter feature is clearly integral to a sound study design, it should be noted that the adjustment for multiple testing in the candidate gene study is somewhat relaxed compared to that of GWAS due to the fewer number of statistical tests typically conducted. For example, Johnson and colleagues examined approximately 7,000 single nucleotide polymorphisms (SNPs) from 30 genes known to encode antihypertensive drug targets with BP among nearly 30,000 participants from six studies in the Cohorts for Heart and Aging Research in Genomic Epidemiology (CHARGE) consortium. Using gene-based adjustment methods, they identified significant associations with *ADRB1* SNP rs1801253 and *AGT* SNP rs2004776 that were successfully replicated in large, independent samples (101). It is unlikely that these variants would have been identified by GWAS, as the *P* values achieved in the discovery stage CHARGE sample were greater than 1×10^{-6}, which is a cut point commonly used to determine which variants will be carried forward

in multistaged GWAS (10,12). Furthermore, more recent candidate gene studies have taken advantage of advances in high-throughput genotyping technology to identify gene variants related to BP. Such studies have utilized gene-centric arrays, which interrogate large numbers of variants in a multitude of genes and biological pathways (103,104). Using the Illumina HumanCVD BeadChip (San Diego, California), which genotypes approximately 50,000 SNPs from 2,000 genes demonstrated to associate with CVD-related traits, Johnson and colleagues identified BP-related SNPs in the *LSP1/TNNT3, MTHFR-NPPB, AGT, ATP2B1, NPR3, HFE, NOS3* and *SOX6* genes among a discovery stage sample of 25,118 participants and replication study of 59,349 participants (103). In summary, these findings demonstrate that despite their tarnished reputation, candidate gene studies may still play a role in our quest to discover variants related to BP. Candidate gene studies are likely to gain importance in the era of the fast-paced expansion of bioinformatic knowledge related to gene expression, function, and gene pathways. However, great care must be taken to avoid some of the early failings of gene-based approaches.

GWAS of BP

By interrogating a dense panel of SNPs covering the entire genome, GWAS represent an agnostic and powerful approach for the discovery of susceptibility loci for common complex traits. As such, there was initial enthusiasm at the prospect of using GWAS to identify novel BP-related variants. However, in contrast to GWAS for other CVD-related phenotypes (14,105,106), early GWAS failed to identify any associations with BP at a level of genome-wide significance (P less than 5×10^{-8}) (14–19). For example, in the Wellcome Trust Case Control Consortium (WTCCC), investigators used an Affymetrix GeneChip Human Mapping 500K Array Set (Santa Clara, California) to compare approximately 2,000 hypertension cases to 3,000 controls. In the WTCCC study, there were no signals that achieved even a suggestive association of P less than 5×10^{-7} (14), although the control group was likely to contain a number of hypertensive subjects. Similarly, when Levy and colleagues examined the continuous SBP and DBP phenotypes among approximately 1,300 participants of the Framingham Heart Study, the most significant associations for SBP and DBP were 1.7×10^{-6} and 3.3×10^{-6}, respectively (16). While a couple of the more recent GWAS have identified BP loci that meet conventional significance thresholds with evidence of replication (107,108), the failure of early GWAS created an impetus for the formation of consortia with the purpose of conducting GWAS meta-analyses in very large samples capable of detecting the modest effects of BP loci (9–12). A comprehensive summary of findings from these consortia are shown in Table 11.2.

GWAS Meta-Analyses in Populations of European Ancestry

In June 2009 two consortia, CHARGE and Global Blood Pressure Genetics (Global BPgen), reported findings of their large-scale GWAS meta-analyses. With discovery stage sample sizes of 29,136 and 34,133 participants in CHARGE and Global BPgen, respectively, they together identified 13 independent loci associated with BP at a level of genome-wide significance (P less than 5×10^{-8}) (9,10). These findings represented an important advance in BP genomics research, providing some of the first robust evidence of genetic association for the BP phenotype. Since the 2009 publications, additional BP GWAS meta-analyses have been conducted in populations of European ancestry, including two from the International Consortium of BP (ICBP), which is the largest GWAS meta-analysis of BP to date with a discovery stage sample of approximately 70,000 participants (11,13), and one from the HYPERGENES Project with a smaller sample size of 1,865 hypertension cases and 1,750 controls (109). In total, these studies have identified 38 loci robustly associated with BP traits (Table 11.2).

Although inference of causal genes and variants based on GWAS signals alone is difficult due to regional linkage disequilibrium (LD) structure, findings from CHARGE, Global BPgen, ICBP, and HYPERGENES have provided association evidence for some biological candidate genes previously suspected to influence BP (110–113). For example, meta-analysis of CHARGE and Global BPgen findings revealed an association of SBP with intronic marker rs1004467 ($P = 1.28 \times 10^{-13}$) of the *CYP17A1* gene, which is responsible for a monogenic form of hypertension (Table 11.1) (9,113). Similarly in the GWAS meta-analysis by Global BPgen, Newton-Cheh and colleagues identified a strong signal for SBP at 1p36. The most significant SNP at that locus was rs17367504 ($P = 7 \times 10^{-24}$), an intronic variant of the *MTHFR* gene, which has been implicated in BP due to its role in regulating homocysteine, a biomarker linked to endothelial dysfunction and hypertension (112). Several other relevant biological candidates are also present at this locus including *NPPA* and *NPPB*, which encode natriuretic peptides, RAAS component *AGTRAP*, and ion channel *CLCN6* (10). A final example includes one from the recent HYPERGENES GWAS meta-analysis (109). Salvi and colleagues identified novel *NOS3* variant rs3918226 associated with hypertension susceptibility ($P = 6.20 \times 10^{-16}$). *NOS3* encodes a protein, which catalyzes the synthesis of nitric oxide by vascular endothelium, which lowers BP by increasing vasodilation (109).

While GWAS meta-analysis results have provided association evidence for some genes with known biological relevance, the majority of loci identified have not been previously implicated in studies of BP regulation in human populations. For example, the *ATP2B1* gene at the 12q21 locus achieved genome-wide significance for SBP, DBP, and mean arterial pressure (MAP) in GWAS meta-analyses conducted by CHARGE, Global BPgen, and ICBP (9–11,13). However, *ATPB1* was never examined in previous candidate gene studies of BP. While post hoc investigation into the potential biological plausibility of *ATP2B1* revealed a previous experiment demonstrating increased mRNA expression in the spontaneously hypertensive rat (114), this gene had never been deemed a high priority for study in human populations using nonagnostic approaches. While some genes at implicated loci, like that of *ATP2B1*, have demonstrated plausibility for association with BP based on our current knowledge, other loci discovered by GWAS meta-analyses have provided completely novel insights into BP regulation. For example, the *SH2B3* locus achieved genome-wide significance for SBP, DBP, and MAP in GWAS meta-analyses by CHARGE, Global BPgen, and ICBP (9–11,13). *SH2B3* had been shown previously to exert an effect on cytokine sensitivity in studies of knockout mice and was associated with auto-immune conditions in human populations (9). Based on these studies, Levy and colleagues speculated that immune response pathways may influence BP by mechanisms not previously appreciated (9).

GWAS Meta-Analyses in Populations of Non-European Ancestry

In June 2011, Kato and colleagues published findings from the Asian Genetic Epidemiology Network (AGEN), a large consortium of GWAS conducted in East Asian populations (12). With GWAS data from nearly 20,000 East Asian participants and follow-up genotyping in an additional 30,000, AGEN identified five novel loci that achieved P less than 5×10^{-8} for association with SBP and/or DBP phenotypes (Table 11.2) (12). Similar to findings of GWAS meta-analyses in European populations, some of the newly identified loci harbored genes with apparent biological relevance to BP (eg, the natriuretic peptide receptor gene *NPR3* at 5p13), while the genomic mechanisms at other loci were unclear (eg, the lead SNP at 2p24 was over 250 kb away from and in no LD with the nearest genes). Of particular importance, the AGEN meta-analysis replicated seven of the 13 loci that had been identified previously by the CHARGE and Global BPgen consortia,

(*text continues on page 121*)

TABLE 11.2 Genetic Variants That Achieved P Less Than 5×10^{-8} in Previous GWAS Meta-Analyses, According to Their One Megabase Position

Chr Region	Lead SNP	Position	Nearest Gene	Functional Relevance	Associated Phenotype(s)	Identifying Consortium
1p36	rs880315	10719453	*CASZ1*[a]	Intron	DBP	AGEN
	rs17367504	11785365	*MTHFR*[a]	Intron	SBP, DBP, HTN, MAP	Global BPgen, AGEN[b], ICBP[d,e]
1p13	rs17030613	112971190	*CAPZA1*[a]	Intron	DBP	AGEN[b]
	rs2932538	113018066	*MOV10*	Near promoter	SBP, DBP	ICBP[d]
2q24	rs16849225	164615066	*FIGN*	Intergenic	SBP	AGEN[b]
	rs13002573	164623454	*FIGN*	Intergenic	PP	ICBP[e]
	rs1446468	164671732	*FIGN*	Intergenic	SBP, DBP, MAP	ICBP[e]
3p24	rs13082711	27512913	*SLC4A7*	Intergenic	DBP, MAP	ICBP[d,e]
3p22	rs3774372	41852418	*ULK4*[a]	Missense	DBP	ICBP[d]
	rs9815354	41887655	*ULK4*[a]	Intron	DBP	CHARGE, AGEN[b]
	rs1717027	41962924	*ULK4*[a]	Intron	DBP	COGENT
3p21	rs319690	47902488	*MAP4*[a]	Intron	SBP, DBP, MAP	ICBP[e]
3q26	rs419076	170583580	*MECOM*[a]	Intron	SBP, DBP, MAP	ICBP[d,e]
	rs1343040	170668987	*MECOM*[a]	Intron	MAP	ICBP[e]
4q12	rs871606	54494002	*CHIC2*	Intergenic	PP	ICBP[e]
4q21	rs13149993	81377569	*FGF5*	Intergenic	MAP	ICBP[e]
	rs1458038	81383747	*FGF5*	Intergenic	SBP, DBP, MAP	ICBP[d,e]
	rs16998073	81403365	*FGF5*	Intergenic	SBP, DBP, HTN, MAP	Global BPgen, AGEN[b], ICBP[e]
4q24	rs13107325	103407732	*SLC39A8*[a]	Missense	SBP, DBP, MAP	ICBP[d,e]
4q25	rs6825911	111601087	*ENPEP*	Intergenic	DBP	AGEN[b]
4q32	rs13139571	156864963	*GUCY1A3*[a]	Intron	DBP	ICBP[d,e]
5p13	rs1173756	32825609	*NPR3*[a]	3′ UTR	PP	ICBP[e]
	rs1173766	32840285	*NPR3*	Intergenic	SBP	AGEN[b]
	rs1173771	32850785	*NPR3*	Intergenic	SBP, DBP, HTN. MAP, PP	ICBP[d,e]

(continued)

TABLE 11.2 Genetic Variants That Achieved *P* Less Than 5×10^{-8} in Previous GWAS Meta-Analyses, According to Their One Megabase Position (*continued*)

Chr Region	Lead SNP	Position	Nearest Gene	Functional Relevance	Associated Phenotype(s)	Identifying Consortium
5q33	rs9313772	157737035	*EBF1*	Intergenic	MAP	ICBP[e]
	rs11953630	157777980	*EBF1*	Intergenic	SBP, DBP, MAP	ICBP[d,e]
6p22	rs1799945	26199158	*HFE*[a]	Missense	SBP, DBP, HTN, MAP	ICBP[d,e]
	rs198846	26215442	*HIST1H1T*	Near 3' UTR	MAP	ICBP[e]
6p21	rs805303	31724345	*BAG6*[a]	Intron	SBP, DBP, HTN	ICBP[d]
6q22.33	rs13209747	127157147	*RSPO3*	Intergenic	SBP, DBP	COGENT
6q25.1	rs17080102	151046463	*PLEKHG1*[a]	Intron	SBP, DBP	COGENT
7p15.2	rs17428471	27304392	*EVX*	Intergenic	SBP, DBP	COGENT
7q22	rs17477177	106199094	*PIK3CG*	Intergenic	SBP, PP	ICBP[e]
7q36	rs3918226	150321109	*NOS3*	Near promoter	HTN	HYPERGENES
8q24	rs2071518	120504993	*NOV*[a]	3' UTR	PP	ICBP[e]
10p12	rs4373814	18459978	*CACNB2*[a]	Intron	SBP, DBP	ICBP[d]
	rs1813353	18747454	*CACNB2*[a]	Intron	SBP, DBP, HTN, MAP	ICBP[d,e]
	rs11014166	18748804	*CACNB2*[a]	Intron	DBP, MAP	CHARGE, AGEN[b], ICBP[e]
	rs12258967	18767965	*CACNB2*[a]	Intron	MAP	ICBP[e]
10q21	rs4590817	63137559	*C10orf107*[a]	Intron	SBP, DBP, HTN, MAP	ICBP[d,e]
	rs1530440	63194597	*C10orf107*[a]	Intron	DBP, MAP	Global BPgen, AGEN[b], ICBP[e]
10q23	rs9663362	95885167	*PLCE1*[a]	Intron	PP	ICBP[e]
	rs932764	95885930	*PLCE1*[a]	Intron	SBP, HTN	ICBP[d,e]
10q24	rs1004467	104584497	*CYP17A1*[a]	Intron	SBP, MAP, PP	CHARGE, ICBP[e]
	rs3824755	104585839	*CYP17A1*[a]	Intron	PP	AGEN[b], ICBP[e]
	rs11191548	104836168	*NT5C2*	Intergenic	SBP, DBP, HTN, MAP, PP	Global BPgen, AGEN[b], ICBP[d,e]
	rs11191593	104929205	*NT5C2*[a]	Intron	MAP	ICBP[e]
10q25	rs2782980	115771517	*ADRB1*	Intergenic	MAP	ICBP[e]
11p15	rs7129220	10307114	*ADM*	Intergenic	SBP	ICBP[d,e]

TABLE 11.2 Genetic Variants That Achieved *P* Less Than 5×10^{-8} in Previous GWAS Meta-Analyses, According to Their One Megabase Position (*continued*)

Chr Region	Lead SNP	Position	Nearest Gene	Functional Relevance	Associated Phenotype(s)	Identifying Consortium
11p15.1	rs1401454	16206759	*SOX6*[a]	Intron	DBP	COGENT
11p15	rs381815	16858844	*PLEKHA7*[a]	Intron	SBP, DBP, MAP	CHARGE, AGEN[b], ICBP[d,e]
11q22	rs633185	100098748	*ARHGAP42*[a]	Intron	SBP, DBP, HTN, MAP	ICBP[d,e]
	rs604723	100115756	*ARHGAP42*[a]	Intron	MAP	ICBP[e]
11q24	rs11222084	129778440	*ADAMTS-8*	Intergenic	PP	ICBP[e]
12q21	rs4842666	88465680	*POC1B*	Intergenic	SBP	CHARGE
	rs11105328	88466521	*POC1B*	Intergenic	SBP	CHARGE
	rs2681472	88533090	*ATP2B1*[a]	Intron	SBP, DBP, HTN, MAP, PP	CHARGE, AGEN[b], ICBP[e]
	rs2681492	88537220	*ATP2B1*[a]	Intron	SBP, DBP, MAP, PP	CHARGE, ICBP[e]
	rs11105354	88550654	*ATP2B1*[a]	Intron	SBP, HTN	CHARGE
	rs12579302	88574634	*ATP2B1*	Near promoter	SBP, HTN	CHARGE
	rs17249754	88584717	*ATP2B1*	Intergenic	SBP, DBP, HTN, MAP, PP	CHARGE, AGEN[b,c], ICBP[e]
	rs11105364	88593407	*ATP2B1*	Intergenic	SBP, HTN	CHARGE
	rs11105368	88598572	*ATP2B1*	Intergenic	SBP, HTN	CHARGE
	rs11105378	88614872	*ATP2B1*	Intergenic	SBP, HTN	CHARGE
	rs12230074	88614998	*ATP2B1*	Intergenic	SBP, HTN	CHARGE
12q24	rs3184504	110368991	*SH2B3*[a]	Missense	SBP, DBP, MAP	CHARGE, ICBP[d,e]
	rs4766578	110388754	*ATXN*[a]	Intron	DBP	CHARGE
	rs10774625	110394602	*ATXN*[a]	Intron	DBP	CHARGE
	rs653178	110492139	*ATXN2*[a]	Intron	DBP, MAP	CHARGE, Global BPgen, ICBP[e]
	rs671	110726149	*ALDH2*[a]	Missense	SBP, DBP	AGEN[b]
	rs11066132	110952589	*NAA25*[a]	Intron	SBP, DBP	AGEN[b]
	rs2074356	111129784	*HECTD4*[a]	Intron	SBP, DBP	AGEN[b]
	rs11066280	111302166	*HECTD4*[a]	Intron	SBP, DBP	AGEN[b]
12q24	rs2384550	113837114	*TBX3*	Intergenic	DBP	CHARGE, AGEN[b]
	rs10850411	113872179	*TBX3*	Intergenic	DBP	ICBP[d,e]
	rs35444	114036820	*TBX3*	Intergenic	DBP	AGEN[b]

(continued)

TABLE 11.2 Genetic Variants That Achieved *P* Less Than 5 × 10⁻⁸ in Previous GWAS Meta-Analyses, According to Their One Megabase Position (*continued*)

Chr Region	Lead SNP	Position	Nearest Gene	Functional Relevance	Associated Phenotype(s)	Identifying Consortium
15q24	rs1378942	72864420	*CSK*[a]	Intron	SBP, DBP, HTN, MAP	Global BPgen, AGEN[b], ICBP[d,e]
	rs6495122	72912698	*ULK3*[a]	Intron	DBP, MAP	CHARGE, ICBP[e]
15q26	rs2521501	89238392	*FES*[a]	Intron	SBP, DBP, MAP	ICBP[d,e]
17q21	rs12946454	40563647	*ACBD4*	Near promoter	SBP	Global BPgen, AGEN[b]
17q21	rs8069437	42261948	*WNT3*	Intergenic	PP	ICBP[e]
	rs17608766	42368270	*GOSR2*[a]	Intron	SBP, PP	ICBP[d,e]
17q21	rs12940887	44757806	*ZNF652*[a]	Intron	SBP, DBP	ICBP[d,e]
	rs16948048	44795465	*ZNF652*	Near promoter	DBP	Global BPgen, AGEN[b]
18p11	rs8096897	13428905	*c18orf1*[a]	Intron	SBP	CHARGE
20p12	rs1327235	10917030	*JAG1*	Intergenic	SBP, DBP, MAP	ICBP[d,e]
20q13	rs6026748	57179210	*ZNF831*	Intergenic	MAP	ICBP[e]
	rs6015450	57184512	*ZNF831*	Intergenic	SBP, DBP, HTN, MAP	ICBP[d,e]

[a]Corresponding variant is located within this gene; [b]AGEN publication by Kato and colleagues (5). [c]AGEN publication by Kelly and colleagues (115). [d]ICBP publication by Ehret and colleagues (11). [e]ICBP publication by Wain and colleagues (13).

Abbreviations: AGEN, Asian Genetic Epidemiology Network; CHARGE, Cohorts for Heart and Aging Research in Genomic Epidemiology; Chr, chromosome; DBP, diastolic blood pressure; Global BPgen, Global Blood Pressure Genetics; HTN, hypertension; ICBP, International Consortium of Blood Pressure; MAP, mean arterial pressure; PP, pulse pressure; SBP, systolic blood pressure.

including four at a level of genome-wide significance (12). Similarly, a subsequent AGEN meta-analysis of MAP and PP phenotypes revealed transethnic replication of five loci previously implicated in the GWAS meta-analysis of MAP and PP conducted in European populations (13,115).

The first GWAS meta-analysis of BP in samples of African ancestry was published by Franceschini and colleagues in September 2013 (116). The Continental Origins and Genetic Epidemiology Network (COGENT), a consortium of 19 studies conducted in 29,378 individuals of African ancestry (116), identified five loci at a level of genome-wide significance. Franceschini et al employed a transethnic meta-analysis approach, which included locus discovery by the COGENT BP GWAS meta-analysis and replication in samples of African (N = 10,000), European (N = 69,385), and East Asian ancestry (N = 19,608) (Table 11.2) (116). Among the five loci achieving genome-wide significance, three were novel, while two had been previously reported in European and East Asian samples (116). These findings indicate that the physiological effects of many common polymorphisms may be generalizable across populations with diverse genetic backgrounds.

Of the eight GWAS meta-analyses of BP phenotypes, only three were conducted in populations of primarily non-European ancestry (12). This is particularly unfortunate since novel genomic mechanisms may be discovered in unique populations due to differences in allele frequencies or factors that interact with genes to influence BP. Furthermore, due to differences in LD structure across populations, the identification of loci demonstrating transethnic replication may help to localize signals for future sequencing and functional study. Thus, the investigation of genomic factors influencing BP in populations with differing genetic backgrounds should continue to be pursued. Findings from these studies will be essential to enhancing our understanding of the molecular mechanisms underlying BP regulation.

THE FUTURE OF BP GENOMICS RESEARCH

To date, most identified SNPs have displayed modest effect sizes and together have explained a limited proportion of the heritability of BP (70,117). While genetic factors are theorized to explain roughly 30% to 50% of the interindividual variation in BP (74–76,78), it was estimated that the currently identified common variants explain only about 0.9% of such variability (11). As we look toward the future, new approaches are being sought to help explain the "missing heritability" of BP. On the horizon are global GWAS meta-analyses, examination of novel BP phenotypes, research of gene–gene and gene–environment interactions (including epigenetic studies [see Chapter 5]), and, as we move beyond GWAS, NGS studies. While setbacks are likely to occur as we continue to move forward, there is optimism that such work will make headway in our quest to better understand the genomic architecture of BP.

Global GWAS Meta-Analyses

GWAS meta-analyses have already made important strides in advancing hypertension genomic research, identifying 43 loci robustly associated with BP phenotypes. Examination of the cumulative effects of these loci using genetic risk scoring methods has been impressive (11,118). For example, Ehret and colleagues demonstrated that individuals in the top decile of BP risk alleles had on average 6 mmHg higher SBP, 4 mmHg higher DBP, and twofold higher prevalence of hypertension compared to those in the lowest decile (11). These findings were confirmed by Fava and colleagues, who also showed that genetic risk score predicted longitudinal changes in SBP and DBP as well as hypertension incidence (118). Despite these promising findings, currently identified loci explain only about 0.9% of the estimated heritability of BP (11). It is theorized that up to 2.2% of the interindividual variation in BP may be explained by common genetic variants (11). Very large sample sizes will be required to detect the remaining SNPs. Mega consortia are now being formed that include genetically diverse samples from around the world. By substantially increasing sample sizes, these studies will have power to detect additional BP loci (119,120). Furthermore, such research will present an outstanding opportunity to refine genomic signals in the search for causal variants by leveraging LD structure across populations (119,120). In undertaking these studies, investigators will likely encounter new challenges, such as how to appropriately account for the genetic heterogeneity that exists between ethnically diverse samples while maximizing study power (119,120). However, novel insights into other phenotypes, such as serum proteins, have already been identified by global GWAS meta-analysis approaches (121). It is likely that BP will soon follow suit.

Novel BP Phenotypes

Hypertension is a heterogeneous phenotype, influenced by numerous underlying biological pathways (71). Because reducing phenotype heterogeneity can increase power to detect genomic loci (122,123), the investigation of distinct physiological hypertension subtypes may facilitate important progress in the search for BP genes and variants. A nice example of work already being conducted in this area is that to better understand the genomic underpinnings of salt-sensitive hypertension (124,127). He and colleagues recently published results from their GWAS of salt-sensitivity phenotypes (124). Their study identified four novel loci, which achieved genome-wide significance with relatively small discovery and replication phase samples of 1,881 and 698 participants, respectively, providing empirical evidence of the efficiency of this approach. In addition to potential gains in statistical power, examination of hypertension subtypes is relevant from a public health and clinical perspective, potentially allowing for the optimization of prevention and therapeutic efforts through a better understanding of an individual's hypertension pathophysiology.

Furthermore, individual BP measures vary substantially over days, months, and years (126–128). Minimizing the variability of BP measurements can increase statistical power to detect genetic variants. For example, Ganesh and colleagues carried out a GWAS of long-term average SBP, DBP, MAP, and pulse pressure (PP) values. They identified four novel loci, which achieved genome-wide significance and estimated a 20% power advantage for detecting BP variants compared to studies using BP measured at a single visit (129). Similarly, increased statistical power may be gained by reducing BP variability due to environmental factors such as dietary sodium or potassium intake (124). As future studies focus more heavily on low frequency and rare variants, with effects that may be even more difficult to detect, utilizing average BP measures or other methods that minimize BP variability may provide a good strategy to further delineate the genomic architecture of BP.

Gene–Gene and Gene–Environment Interaction

Given the commonly accepted belief that complex traits like BP are influenced by the interaction of genetic and environmental factors, it has been suggested that research of such interactions could help to explain some of the missing heritability of these traits (70,130,131) (see Chapter 9). Still, there is a paucity of data from GWAS examining how genes interact with each other and with environmental factors to influence BP. Since current methods for detecting interactions lack statistical power, investigators may be hesitant to undertake such analyses, especially in light of the early difficulties of BP GWASs in identifying simple single marker associations (132). However, before moving completely beyond GWAS, it may be worthwhile to leverage data from existing large consortia to explore the interactions between genes and environmental factors on BP.

In the discussion of gene–environment interaction, it would be remiss to ignore the emerging field of epigenetics. It is hypothesized that one of the mechanisms by which environmental factors interact with genes to modify their effects is through epigenetic modifications that include DNA methylation, histone modification, and alteration of microRNA expression (101). Studies have already shown a loss of global genomic methylation content among hypertension patients, as well as hypermethylation of the *HSD11B2* gene (133,134). Further research in this area will be critical to untangle the complex web of genetic and environmental factors that act together to determine an individual's BP.

Next-Generation Sequencing Studies

Cohen and colleagues achieved early success identifying rare variants with large influence on lipid phenotypes, prompting investigators to turn their attention toward sequencing studies to help clarify the role of rare genetic variants in the complex BP phenotype (135–138). Already there is some suggestion that rare variants could help to explain the missing heritability of BP. For example, Ji and colleagues reported that carriers of rare functional mutations in three renal-salt-handling genes (*SLC12A3*, *SLC12A1*, and *KCNJ1*) had significantly reduced SBP (mean reduction = 9.0 mmHg, *P* = .0002) and DBP (mean reduction = 5.0 mmHg, *P* = .003) compared to noncarriers (139). Similarly, Rao et al resequenced a locus of the *CHGA* gene and discovered a Gly364Ser amino acid substitution that decreased DBP by approximately 5 mmHg (140). While these previous studies have sequenced just a limited number of genes, the advent of NGS technology has made it plausible to deeply sequence large stretches of DNA, whole exomes, or even the entire genome in large, population-based studies (141). As such, the National Heart, Lung, and Blood Institute sponsored an initiative to identify low frequency and rare variants, which may contribute to heart, lung, and blood disorders by conducting whole-exome sequencing in ongoing population-based studies (4). With much of the sequencing completed and cataloged in the database of Genotypes and Phenotypes (dbGaP), results from the BP working group are eagerly anticipated (142).

CONCLUSIONS

Although the genomic mechanisms underlying BP regulation have yet to be fully elucidated, important advances in the field have been made. Identification of genes involved in monogenic BP conditions has provided important clues for the study of genes and pathways that may play a role in the complex BP phenotype. Although initially slow to progress, genetic association studies seem to have finally delivered on their promise to identify common polymorphisms associated with this trait. While it is true that much of the heritability of BP remains unexplained, the variants robustly identified by previous GWAS meta-analyses already show nonnegligible associations with BP and its comorbid conditions. Current data suggest that those in the top decile of BP risk alleles have on average 5.8 mmHg higher SBP, 3.7 mmHg higher DBP, twofold higher prevalence of hypertension, 44% increased risk of stroke, and 43% higher prevalence of coronary artery disease compared to those in the lowest decile (11). Furthermore, with the formation of global GWAS meta-analysis consortia, the emergence of epigenetics, and the advent of NGS technology, the future for BP genomics research is bright. Investigators are optimistic that the coming years will offer a clearer picture of the genomic architecture of BP. Such insights could eventually be used to identify individuals at high risk for hypertension who may benefit most from primary prevention efforts. In addition, findings may be used to develop novel gene-based therapies for the treatment of hypertension (143). Such advancements will have important public health and clinical implications, helping to curb the growing CVD epidemic at a national and global level (71,144).

REFERENCES

1. Choi M, Scholl UI, Yue P, et al. K+ channel mutations in adrenal aldosterone-producing adenomas and hereditary hypertension. *Science*. 2011;331(6018): 768–772.
2. Zennaro M-C, Jeunemaitre X. Mutations in KCNJ5 gene cause hyperaldosteronism. *Circ Res*. 2011; 108(12):1417–1418.
3. Louis-Dit-Picard H, Barc J, Trujillano D, et al. KLHL3 mutations cause familial hyperkalemic hypertension by impairing ion transport in the distal nephron. *Nat Genet*. 2012;44(4):456–460, S1–S3.
4. Tennessen JA, Bigham AW, O'Connor TD, et al. Evolution and functional impact of rare coding variation from deep sequencing of human exomes. *Science*. 2012;337(6090):64–69.
5. Danaei G, Finucane MM, Lin JK, et al. National, regional, and global trends in systolic blood pressure since 1980: systematic analysis of health examination surveys and epidemiological studies with 786 country-years and 5.4 million participants. *Lancet*. 2011;377(9765):568–577.
6. Kearney PM, Whelton M, Reynolds K, et al. Global burden of hypertension: analysis of worldwide data. *Lancet*. 2005;365(9455):217–223.
7. Lopez AD, Mathers CD, Ezzati M, et al. Global and regional burden of disease and risk factors, 2001: systematic analysis of population health data. *Lancet*. 2006;367(9524):1747–1757.
8. Lim SS, Vos T, Flaxman AD, et al. A comparative risk assessment of burden of disease and injury attributable to 67 risk factors and risk factor clusters in 21 regions, 1990-2010: a systematic analysis for the Global Burden of Disease Study 2010. *Lancet*. 2012;380(9859):2224–2260.
9. Levy D, Ehret GB, Rice K, et al. Genome-wide association study of blood pressure and hypertension. *Nat Genet*. 2009;41(6):677–687.
10. Newton-Cheh C, Johnson T, Gateva V, et al. Genome-wide association study identifies eight loci associated with blood pressure. *Nat Genet*. 2009;41(6):666–676.
11. Ehret GB, Munroe PB, Rice KM, et al. Genetic variants in novel pathways influence blood pressure and cardiovascular disease risk. *Nature*. 2011; 478(7367):103–109.
12. Kato N, Takeuchi F, Tabara Y, et al. Meta-analysis identifies five novel loci associated with blood pressure in East Asians. *Nat Genet*. 2011;43:531–538.
13. Wain LV, Verwoert GC, O'Reilly PF, et al. Genome-wide association study identifies six new loci influencing pulse pressure and mean arterial pressure. *Nat Genet*. 2011;43(10):1005–1011.
14. Wellcome Trust Case Control Consortium. Genome-wide association study of 14,000 cases of seven common diseases and 3,000 shared controls. *Nature*. 2007; 447(7145):661–678.
15. Saxena R, Voight BF, Lyssenko V, et al. Genome-wide association analysis identifies loci for type 2 diabetes and triglyceride levels. *Science*. 2007; 316(5829):1331–1336.
16. Levy D, Larson MG, Benjamin EJ, et al. Framingham Heart Study 100K Project: genome-wide associations for blood pressure and arterial stiffness. *BMC Med Genet*. 2007;8(Suppl 1):S3.
17. Kato N, Miyata T, Tabara Y, et al. High-density association study and nomination of susceptibility genes for hypertension in the Japanese National Project. *Hum Mol Genet*. 2008;17(4):617–627.
18. Sabatti C, Service SK, Hartikainen A-L, et al. Genome-wide association analysis of metabolic traits in a birth cohort from a founder population. *Nat Genet*. 2009;41(1):35–46.
19. Wang Y, O'Connell JR, McArdle PF, et al. From the cover: whole-genome association study identifies STK39 as a hypertension susceptibility gene. *Proc Natl Acad Sci USA*. 2009;106(1):226–231.
20. Harrap SB. Where are all the blood-pressure genes? *Lancet*. 2003;361(9375):2149–2151.
21. Ferrari P. The role of 11beta-hydroxysteroid dehydrogenase type 2 in human hypertension. *Biochim Biophys Acta*. 2010;1802(12):1178–1187.
22. Nagai T, Nishimura G, Kato R, et al. Del(12) (p11.21p12.2) associated with an asphyxiating thoracic dystrophy or chondroectodermal dysplasia-like syndrome. *Am J Medl Genet*. 1995;55(1):16–18.
23. Helmberg A, Ausserer B, Kofler R. Frame shift by insertion of 2 basepairs in codon 394 of CYP11B1 causes congenital adrenal hyperplasia due to steroid 11 beta-hydroxylase deficiency. *J Clin Endocrinol Metab*. 1992;75(5):1278–1281.

24. White PC, Dupont J, New MI, et al. A mutation in CYP11B1 (Arg-448----His) associated with steroid 11 beta-hydroxylase deficiency in Jews of Moroccan origin. *J Clin Invest.* 1991;87(5):1664–1667.

25. Biglieri EG, Kater CE. 17 alpha-hydroxylation deficiency. *Endocrinol Metab Clin North Am.* 1991;20(2):257–268.

26. Kagimoto M, Winter JS, Kagimoto K, et al. Structural characterization of normal and mutant human steroid 17 alpha-hydroxylase genes: molecular basis of one example of combined 17 alpha-hydroxylase/17,20 lyase deficiency. *Mol Endocrinol.* 1988;2(6):564–570.

27. Trebble P, Matthews L, Blaikley J, et al. Familial glucocorticoid resistance caused by a novel frameshift glucocorticoid receptor mutation. *J Clin Endocrinol Metab.* 2010;95(12):E490–E499.

28. van Rossum EF, Lamberts SW. Glucocorticoid resistance syndrome: a diagnostic and therapeutic approach. *Best Pract Res Clin Endocrinol Metab.* 2006;20(4):611–626.

29. Mulatero P. A new form of hereditary primary aldosteronism: familial hyperaldosteronism type III. *J Clin Endocrinol Metab.* 2008;93(8):2972–2974.

30. Monticone S, Hattangady NG, Nishimoto K, et al. Effect of KCNJ5 mutations on gene expression in aldosterone-producing adenomas and adrenocortical cells. *J Clin Endocrinol Metab.* 2012;97(8):E1567–E1572.

31. Halperin F, Dluhy RG. Glucocorticoid-remediable aldosteronism. *Endocrinol Metab Clin North Am.* 2011;40(2):333–341.

32. Geller DS, Farhi A, Pinkerton N, et al. Activating mineralocorticoid receptor mutation in hypertension exacerbated by pregnancy. *Science.* 2000;289(5476):119–123.

33. Wilson FH, Hariri A, Farhi A, et al. A cluster of metabolic defects caused by mutation in a mitochondrial tRNA. *Science.* 2004;306(5699):1190–1194.

34. Barroso I, Gurnell M, Crowley VE, et al. Dominant negative mutations in human PPARgamma associated with severe insulin resistance, diabetes mellitus and hypertension. *Nature.* 1999;402(6764):880–883.

35. Hansson JH, Schild L, Lu Y, et al. A de novo missense mutation of the beta subunit of the epithelial sodium channel causes hypertension and Liddle syndrome, identifying a proline-rich segment critical for regulation of channel activity. *Proc Natl Acad Sci USA.* 1995;92(25):11495–11499.

36. Abriel H, Loffing J, Rebhun JF, et al. Defective regulation of the epithelial Na+ channel by Nedd4 in Liddle's syndrome. *J Clin Invest.* 1999;103(5):667–673.

37. Brautbar N, Levi J, Rosler A, et al. Familial hyperkalemia, hypertension, and hyporeninemia with normal aldosterone levels. A tubular defect in potassium handling. *Arch Intern Med.* 1978;138(4):607–610

38. Farfel Z, Iaina A, Rosenthal T, et al. Familial hyperpotassemia and hypertension accompanied by normal plasma aldosterone levels: possible hereditary cell membrane defect. *Arch Intern Med.* 1978;138(12):1828–1832.

39. Wilson FH, Disse-Nicodeme S, Choate KA, et al. Human hypertension caused by mutations in WNK kinases. *Science.* 2001;293(5532):1107–1112.

40. Boyden LM, Choi M, Choate KA, et al. Mutations in kelch-like 3 and cullin 3 cause hypertension and electrolyte abnormalities. *Nature.* 2012;482(7383):98–102.

41. Fernandez-Fernandez JM, Tomas M, Vazquez E, et al. Gain-of-function mutation in the KCNMB1 potassium channel subunit is associated with low prevalence of diastolic hypertension. *J Clin Invest.* 2004;113(7):1032–1039.

42. Simon DB, Karet FE, Hamdan JM, et al. Bartter's syndrome, hypokalaemic alkalosis with hypercalciuria, is caused by mutations in the Na-K-2Cl cotransporter NKCC2. *Nat Genet.* 1996;13(2):183–188.

43. Simon DB, Bindra RS, Mansfield TA, et al. Mutations in the chloride channel gene, CLCNKB, cause Bartter's syndrome type III. *Nat Genet.* 1997;17(2):171–178.

44. Birkenhager R, Otto E, Schurmann MJ, et al. Mutation of BSND causes Bartter syndrome with sensorineural deafness and kidney failure. *Nat Genet.* 2001;29(3):310–314.

45. Schlingmann KP, Konrad M, Jeck N, et al. Salt wasting and deafness resulting from mutations in two chloride channels. *N Engl J Med.* 2004;350(13):1314–1319.

46. Nozu K, Inagaki T, Fu XJ, et al. Molecular analysis of digenic inheritance in Bartter syndrome with sensorineural deafness. *J Med Genet.* 2008;45(3):182–186.

47. Bose HS, Sugawara T, Strauss JF, Miller WL; International Congenital Lipoid Adrenal Hyperplasia Consortium. The pathophysiology and genetics of congenital lipoid adrenal hyperplasia. *N Engl J Med.* 1996;335(25):1870–1878.

48. Lin D, Gitelman SE, Saenger P, Miller WL. Normal genes for the cholesterol side chain cleavage enzyme, P450scc, in congenital lipoid adrenal hyperplasia. *J Clin Invest.* 1991;88(6):1955–1962.

49. Balraj P, Lim PG, Sidek H, et al. Mutational characterization of congenital adrenal hyperplasia due to 21-hydroxylase deficiency in Malaysia. *J Endocrinol Invest.* 2013;36(6):366–374.

50. Abbaszadegan MR, Hassani S, Vakili R, et al. Two novel mutations in CYP11B1 and modeling the consequent alterations of the translated protein in classic congenital adrenal hyperplasia patients. *Endocrine.* 2013;44(1):212–219.

51. Han B, Liu W, Zuo CL, et al. Identifying a novel mutation of CYP17A1 gene from five Chinese 17alpha-hydroxylase/17, 20-lyase deficiency patients. *Gene.* 2013;516(2):345–350.

52. White PC, Bachega TA. Congenital adrenal hyperplasia due to 21 hydroxylase deficiency: from birth to adulthood. *Semin Reprod Med.* 2012;30(5):400–409.

53. Portrat-Doyen S, Tourniaire J, Richard O, et al. Isolated aldosterone synthase deficiency caused by simultaneous E198D and V386A mutations in the CYP11B2 gene. *J Clin Endocrinol Metab.* 1998;83(11):4156–4161.

54. Ng HY, Lin SH, Hsu CY, et al. Hypokalemic paralysis due to Gitelman syndrome: a family study. *Neurology.* 2006;67(6):1080–1082.

55. Jeck N, Konrad M, Hess M, et al. The diuretic- and Bartter-like salt-losing tubulopathies. *Nephrol Dial Transplant.* 2000;15(Suppl 6):19–20.

56. Geller DS, Rodriguez-Soriano J, Vallo Boado A, et al. Mutations in the mineralocorticoid receptor gene cause autosomal dominant pseudohypoaldosteronism type I. *Nat Genet.* 1998;19(3):279–281.

AF RISK PREDICTION

Recently, several instruments for AF risk prediction have been developed on the basis of clinical risk factors for AF (23–25). The risk prediction tools were developed predominantly in low-risk community-dwelling individuals of European ancestry, though the algorithms also have been examined in African Americans.

Data suggest that biochemical risk factors, such as brain natriuretic peptides, may further enhance AF risk prediction beyond clinical risk factors (26). The role of assessing genetic factors to enhance AF risk prediction has not been fully explored. We previously observed that discrimination of AF risk improved slightly beyond age and sex alone when considering multiple susceptibility genotypes at the top AF susceptibility locus on chromosome 4q25 (9). The addition of genotype information for the top risk variants at the two most significantly associated AF susceptibility loci on chromosomes 4q25 and 16q22 did not enhance prediction in one Swedish examination (27). In contrast, consideration of genotypes at AF susceptibility signals demonstrated an improvement in discrimination beyond a clinical risk factor model in the Women's Genome Health Study (28).

Whether future genetic risk prediction efforts that incorporate larger numbers of variants and variants associated with larger risks of AF can meaningfully influence risk stratification efforts remains to be determined. Furthermore, the optimal populations in which to deploy AF risk prediction instruments have yet to be defined, in part because therapies for AF prevention have not been systematically explored. To date, the utility of AF risk estimation based on clinical risk factors or genetics has not been formally tested in the clinic.

FUTURE DIRECTIONS

Missing Heritability of AF

Candidate gene studies and GWAS have revealed common genetic variants with small to modest effect sizes in AF cohorts and the general population (29–32). However, a large proportion of the heritability of AF cannot be explained by genetic variants reported to date (29–33). Despite high heritability estimates of a variety of phenotypes, GWAS published to date have left us with a sizeable gap in explaining the genetic contribution to variation in quantitative traits and diseases (29,30). The issue of missing heritability is of major importance in the future of complex genetics of AF (Figure 13.2) (30,34).

One potential source for the missing heritability of AF lies in an unknown number of yet-to-be-identified

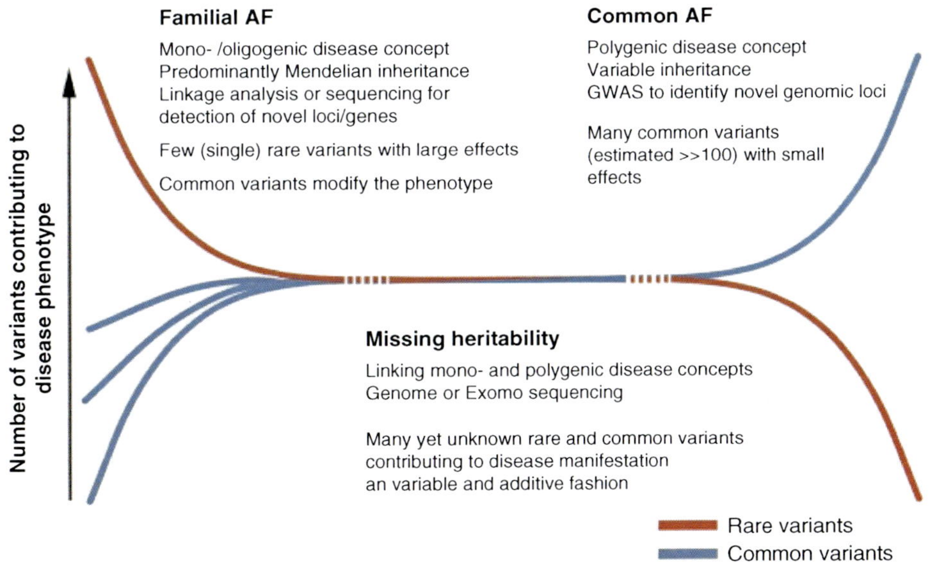

FIGURE 13.2 GWAS, Mendelian families, and the missing heritability of AF. The figure illustrates three situations typically encountered in current genetic studies. The blue and red lines symbolize the number of rare and common variants contributing to a trait's heritability. High-frequency variants that usually confer only small effect sizes can be identified by GWAS; low-frequency variants often exert strong effects and are detected by sequencing. The missing heritability may be found in the middle, where many variants with intermediate frequency explain the phenotypic variability. In GWAS situations, undetected rare variants account for missing heritability; in Mendelian AF situations, rare variant patterns are modified by common variants.

Source: Adapted with permission from Ref. (30). Sinner MF, Ellinor PT, Meitinger T, et al. Genome-wide association studies of atrial fibrillation: past, present, and future. *Cardiovasc Res*. 2011;89(4):701–709.

common variants. Second potential sources are less common variants, which currently are not sufficiently covered by GWAS. It has been noted that there often is an inverse relation between the frequency of the variant and effect size; that is, rare variants tend to have larger effect sizes and common variants often have small to modest effect sizes. The most probable scenario is that combinations of common and rare variants, in addition to other disease mechanisms, such as gene–gene and gene–environment interactions, copy number variants, and DNA methylation, combine to further explain the heritability of AF (30,33,35).

GWAS of AF—Future Directions

Although the recent GWAS findings have identified promising novel molecular pathways for AF, the translation into clinical application has not yet happened. Over the next years, a wide variety of basic, genomic, and clinical studies are needed to translate genetic findings into clinical applications. Specific challenges include addressing the missing heritability of AF, determining the molecular mechanisms through which the identified susceptibility loci lead to AF, and relating such variations to AF risk prediction and clinical outcomes. So far, a measurable value of the genetic findings for AF risk prediction or medical care of AF patients has not been demonstrated (30).

The most direct way to identify additional loci for AF is to simply increase the number of individuals studied. Variants that are detected only in larger samples will typically have smaller effect sizes. With a limited effect size, these SNPs may still provide useful information about molecular pathways leading to AF; whether such variants will prove useful for risk prediction or assessing clinical outcomes remains to be determined (33,35).

Another extension of GWAS for AF is to dissect the genetic architecture at identified loci in more detail. The utility of this approach has been demonstrated at the 4q25 locus. As an example of the complexity of the genetics of common diseases, two additional independent signals were identified at the 4q25 locus after adjusting for the initial, genome-wide significant finding (22). Similarly, it is likely that there are additional, as yet unidentified signals, at the other reported loci for AF. As a consequence, it is very likely that a multitude rather than a low, circumscribed number of variants will explain a fraction of the heritability of AF. Future sequencing efforts at the 4q25 and other GWAS loci for AF might help to identify further independent signals, and potentially will detect the true functional variants rather than SNPs tagging them. The advancement of next-generation sequencing technologies will substantially improve the detection of rare and structural variation and thus further help to expedite the discovery of genetic factors (30).

Future studies also will likely be directed at GWAS for AF in diverse populations. To date, all published GWAS for AF have been conducted in individuals of European or Japanese descent. It remains unclear if AF in Africans and other races/ethnicities have similar or distinct genomic etiologies for AF. Using the combined information from multiple ancestries may be a powerful tool for narrowing genetic loci and identifying causative SNPs (30).

"OMICS" OF AF

Research in several fields, including epigenomics, transcriptomics, proteomics, and metabolomics is beginning to shed light on the interplay between genetic variation and risk for AF. Research in the expansive fields of "omics" is emerging rapidly. Although such endeavors will elucidate novel pathways in the initiation and maintenance of AF, "omics" research is still in its infancy.

Epigenomics is the field of genetics that investigates factors that determine stable modifications of gene expression that can be transmitted through cell division, such as methylation and histone remodeling (see also Chapter 5). Epigenetic phenomena may alter gene expression, thereby influencing the potential impact of changes in the DNA sequence (29,36). Such epigenetic changes may be heritable, and the global changes are incorporated in the *epigenome*. The *transcriptome* is the unique set of genes expressed or transcribed under specific conditions. The resulting diversity of proteins produced by a cell or organ can now be characterized by high-throughput gel electrophoresis or mass spectrometry as the *proteome*, and the metabolic processing profiles determined by separation and detection methods (eg, gas chromatography or mass spectrometry), the *metabolome*. The *phenome* is determined by genetic and environmental interactions through measurable structural and functional features. *Systems biology* is the study to integrate these diverse disciplines (Figure 13.3) (29,37,38).

Transcriptomics of AF

Role of mRNA

Gene expression is mediated by both environmental and genetic factors (39,40). Recent studies on large-scale transcriptional profiling have revealed the complexity of transcriptional regulation (41,42). More than 9,000 genes were expressed in the human

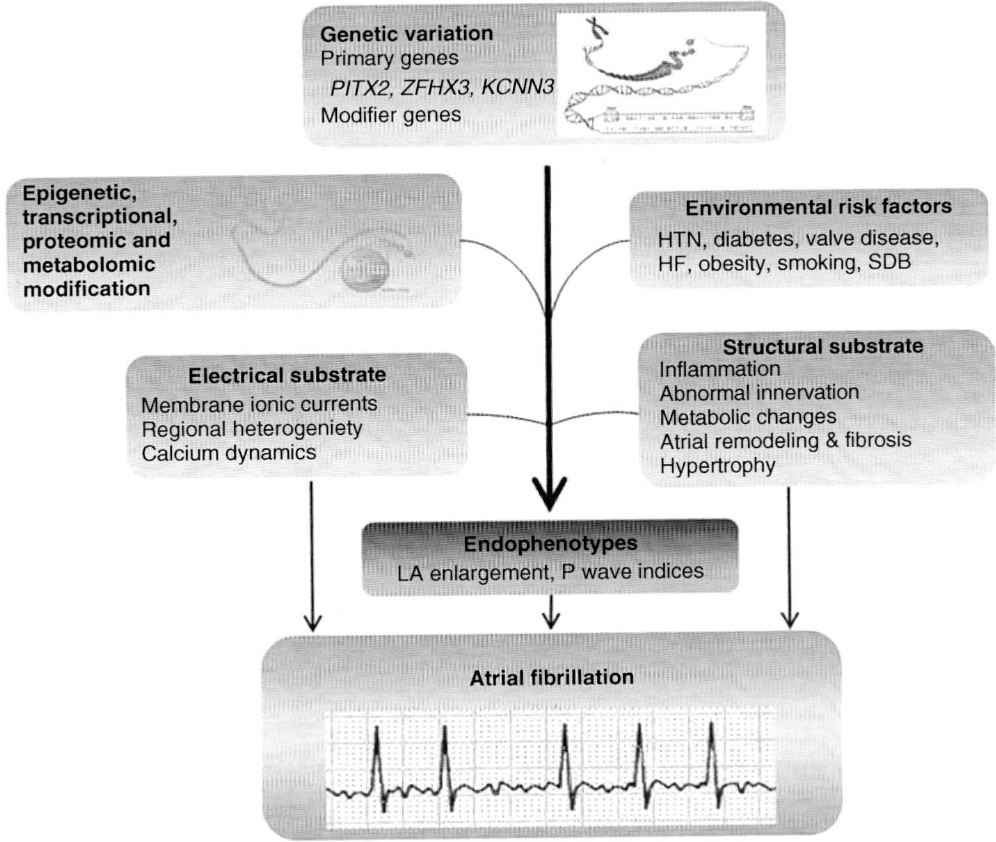

FIGURE 13.3 Systems biology approach to AF. Genetic loci provide targets for epigenomic, transcriptomic, proteomic, and metabolomic investigations. Epidemiological and clinical investigations identify risk factors. In turn, systems biology translates epidemiological and genomic pathways, providing insight into electrical and structural substrates. Systems biology in turn may yield novel hypotheses for risk factors, target intermediate endophenotypes for genetic dissection, and elucidate upstream targets for therapeutic intervention.

Abbreviations: HF, heart failure; HTN, hypertension; LA, left atrial; SDB, sleep disordered breathing.

Source: Adapted with permission from Ref. (29). Magnani JW, Rienstra M, Lin H, et al. Atrial fibrillation: current knowledge and future directions in epidemiology and genomics. *Circulation*. 2011;124(18):1982–1993.

atria (43). Various studies have demonstrated that AF has profound effects on gene expression (44–47), such as natriuretic peptides (48), connexin40 (49), signal-regulated kinases, and angiotensin-converting enzymes (ACEs) (50). Many genes involved in myocardial fibrosis also vary in their expression (51–54). A number of differentially expressed genes were in the ion-channel family, which plays an essential role in the pathophysiology of AF initiation, perpetuation, and adaptation (47,55,56). AF also alters ion-channel expression in patients with valvular heart disease (57). The expression of calcium homeostasis proteins was different with long-term persistent AF but not paroxysmal AF, although it might represent a secondary rather than a primary phenomenon (58).

Gene expression also provides an important way to elucidate the effects of genetic variations on various diseases. Given that most of the identified disease-related loci reside in intronic or even intergenic regions (59), their functional relation to disease

pathobiology remains largely unknown. The identification of genetic variations that mediate gene expression could provide valuable insights into disease etiology. Thousands of expression quantitative trait loci have been identified (60). Many of them were associated with at least one common cardiovascular risk factor (61). Interestingly, a large proportion of expression quantitative trait loci are located outside of corresponding gene regions (60). A recent study on human atria also found that one AF top locus, 10q22, was actually associated with *MYOZ1* expression, which is located 5 kb downstream of the genetic variant, suggesting that *MYOZ1* might be a candidate gene for AF (62).

Challenges

Gene expression varies in different tissues or cell types (63). Hundreds of genes are differentially expressed even between left and right atria (62,64,65). The majority of expression quantitative trait loci may be tissue specific, reflecting the difference in tissue-specific function (60,66). The human left atrium is the ideal tissue for the study of AF. However, invasive left atrial specimen collection is usually unfeasible except from highly selected samples (eg, patients with valvular heart disease undergoing invasive procedure). Future advances in stem cell technologies offers a promising solution to study gene expression in different stem cell derived tissues.

Gene expression is traditionally measured by the real-time polymerase chain reaction (PCR), which is labor intensive and low throughput. Microarray enables a genome-wide profiling of gene expression efficiently, and thus has gained popularity in the last decade. Due to the inherent limitation of cross-hybridization of microarrays, a further validation is usually required to exclude potential nonspecific signals and further confirm the results (51,67). Recent advances in ribonucleic acid (RNA) sequencing technologies are able to provide digital and accurate measurement of transcript abundance (68). Deep sequencing is particularly useful to discover novel and low-abundance or wide-dynamic-range transcripts (69). As the sequencing cost continues to drop, RNA sequencing will become a more popular tool for transcriptomic profiling.

Role of MicroRNA

MicroRNAs (miRNAs) are small, single-stranded, noncoding RNAs that have recently gained recognition as key gene regulatory factors in cardiovascular development and disease (70–73) (see Chapter 6). Unlike transcription factors, which regulate gene expression by acting on the 5′-flanking region of genes, miRNAs primarily silence gene expression at the posttranscriptional level by acting on the 3′-untranslated region of genes (74,75). miRNAs have been shown to influence cardiac excitability and arrhythmogenesis, processes implicated in susceptibility to AF (76,77). A group of miRNAs that are able to regulate the genes encoding cardiac ion channels/ion transporters/Ca²⁺-handling proteins and other relevant genes have been identified (77–83). Some of these miRNAs have been shown to be involved in AF, and some are considered to have the potential to regulate AF based on their target genes (Table 13.2).

MicroRNAs as Biomarkers

MiRNAs can be detected circulating in blood plasma or serum as a result of cellular damage or secretion (73,84). In contrast to intracellular miRNAs and other extracellular disease mediators, circulating miRNAs have characteristics potentially useful as markers of

TABLE 13.2 Target Genes of miRNAs With Potential Roles in AF

miRNA	Target Genes/Proteins				Function	Possible Role in AF
Upregulation	**Repression**					
	CC	CA	CR	Other		
miR-328	KCNJ12	CACNA1C CACNB2 CACNB1	KCND3		ICaL reduction APD shortening	A-ER (78)
miR-223	SLC8A1				Ca²⁺ handling via NCX1	A-ER (77)
Downregulation	**Activation**					
	CC	CA	CR	Other		
miR-1	KCNJ2 GJA1	HCN2		Hsp60 Hsp70	Ik₁ increase, conduction slowing, enhanced automaticity, proapoptosis	A-ER (79,80) Ectopic activity A-SR
miR-26	KCNJ2				Ik₁ increase, APD shortening	A-ER (81)
miR-133	KCNH2	HCN2 HCN4	KCNQ1	TGF-beta 1 TGFBRII CTGF Caspase 9	APD prolongation, enhanced automaticity, antifibrosis Matrix remodeling Antiapoptosis	A-SR (82) Ectopic activity?
miR-590				TGF-beta 1 TGFBRII	Antifibrosis	A-SR (82)
miR-208	GJA5			THRAP1	Conduction slowing Pro-fibrosis	A-ER (83) A-SR?

Question marks indicate speculations yet to be experimentally verified.

Abbreviations: A-ER, atrial electrical remodeling; APD, action potential duration; A-SR, atrial structural remodeling; CA, cardiac automaticity; CC, cardiac conduction; CR, cardiac repolarization.

prognostic or diagnostic importance in AF because they are stable and easily detectable; miRNAs comprise nucleic acids, and their sequences can be amplified (84). Circulating miRNAs have been associated with cardiovascular risk factors and coronary artery disease (85). Circulating levels of miR-150 have been associated with AF (86).

MicroRNAs: Future Implications

AntagomiRs are engineered oligonucleotides that can silence miRNAs (87–89). MiRNAs might also be modifiable using vectors to express target sites and thereby consume (or "soak up") miRNAs that enhance AF vulnerability (75). As miRNA research advances, a thorough understanding of the broad functions of any particular miRNA will be necessary to avoid unappreciated or even harmful "off-target" effects. Importantly, although individual miRNAs have been associated with AF, it remains unknown how networks of miRNAs interact to influence AF susceptibility in human populations. It is likely that multiple miRNAs work to maintain homeostatic networks and that exposure of cardiomyocytes to stress states may alter levels of multiple related miRNA–mRNA coexpression pairs. In addition, miRNA–mRNA expression networks must be defined in specific type (ie, AF vs atrial flutter), patterns (eg, new onset, paroxysmal, persistent, or permanent) and populations (eg, postoperative AF) of AF in order to understand the gene regulatory systems involved in these distinct settings. A better understanding of miRNA–mRNA coexpression networks is also necessary since a therapeutic "antagomiR cocktail" including several synergistic miRNAs might be effective in targeting key regulatory pathways (29,30,75).

Proteomics and Metabolomics of AF

Pathological cardiac remodeling, a process underlying AF, involves not just structural and electrophysiological change, but also metabolic alterations. Upregulation of gene transcripts involved in cardiomyocyte metabolism, including glycolysis and ketone processing, as well as heat shock and cytoskeletal proteins is seen in the atria of individuals with permanent AF (47,89–91). Mitochondrial protein expression products also are altered in the setting of AF and may serve as important biomarkers of diagnostic or prognostic importance in AF (90).

Although levels of inflammatory proteins (eg, C-reactive protein, interleukin-6) and natriuretic peptides (eg, B-type natriuretic peptide) (26) are associated with AF and may enhance AF risk prediction, few studies have used proteomics to investigate whether or not protein–protein interactions or post-translational modifications relate to AF risk.

SYSTEMS BIOLOGY—APPLICATION TO AF

Accumulating evidence indicates that complex diseases such as AF are influenced by the interactions of multiple genes. Like other complex phenotypes, the mechanisms underlying AF are incompletely captured using single-gene-based approaches (92). Furthermore, biological systems comprise circuitries of interacting components such as proteins, nucleic acids, and other small molecules that operate in concert to create complicated molecular networks (93). A two-hit model of AF vulnerability has been promulgated and suggests that exposure to a risk factor, for example, hypertension, alters gene expression (perhaps via pressure sensors) and unmasks AF vulnerability in susceptible individuals (eg, those with poorly functioning calcium channels) (32,94).

FROM THE BENCH TO THE BEDSIDE: INTEGRATING GENETIC INFORMATION INTO CLINICAL PRACTICE

Ultimately, it remains to be established whether or not associations between genetic and "omic" variations and AF will remain research tools or will be useful to clinicians. In order for "omic" information to be useful to the clinician, it must provide actionable intelligence about AF prevention, risk, response to AF treatments, or risk for complications (eg, stroke) (30). The integration of genetic information into clinical practice also has implications for the development of new AF treatments as well as AF prevention by identifying upstream gene regulatory networks responsible for generating a substrate vulnerable to AF. A potential important strategy for preventing the complications of AF is to develop effective therapies to prevent its occurrence, or to identify individuals at high risk for the occurrence of AF in whom to institute more intensive AF screening measures.

In general, there are two directions for the development of AF treatments. One is the design of drugs based on target mechanism (ie, enhanced automaticity vs macro-reentry). The other is modulation of atrial-specific ion-channel proteins using atrial-selective drugs (32). Potential targets for atrial-selective agents include I_{Kur} and I_{KACh} (95). Genetic profiling also may help identify individuals with AF from enhanced automaticity versus others with macro-reentry and thereby classify patients relatively more or less likely to benefit from a particular agent. Genetic information may also be useful in identifying individuals more likely to progress quickly from paroxysmal to persistent AF. As conceptualized in Figure 13.4, a genotype-based

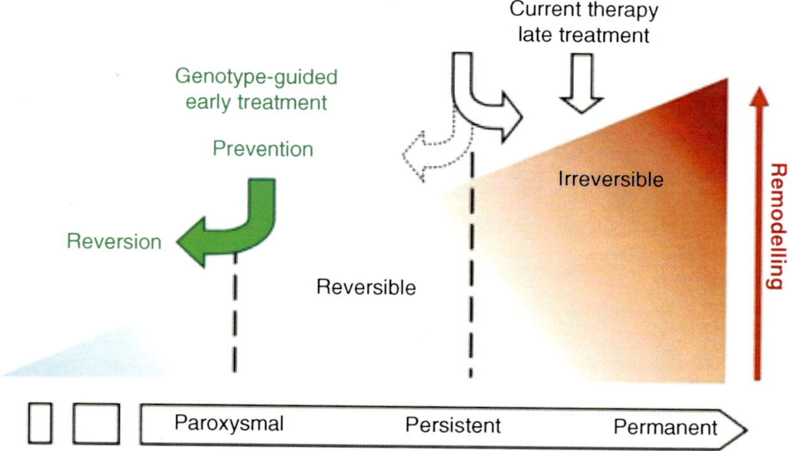

FIGURE 13.4 Schematic representation of the time course of AF and potential opportunities for genotype-guided therapies. Current pharmacological and procedural treatments for AF are initiated after the onset of the arrhythmia and in many cases after sustained periods of AF. Ultimately, genetic information may be useful in identifying high-risk patients, or those patients more likely to respond to current therapies. An early, genotype-guided treatment might thus help to prevent or ameliorate progression of AF. In the transition from paroxysmal to persistent to permanent AF, many factors contribute. Hypertension, ischemic heart disease, and heart failure contribute particularly to cardiac remodeling and thus to the transition of AF.

Source: Adapted with permission from Ref. (30). Sinner MF, Ellinor PT, Meitinger T, et al. Genome-wide association studies of atrial fibrillation: past, present, and future. *Cardiovasc Res.* 2011;89(4):701–709.

approach might enable providers to identify at-risk individuals before irreversible pathological cardiac remodeling has occurred (30).

Personalized Medicine

Individuals with particular genetic polymorphisms might exhibit different therapeutic or toxic responses to antiarrhythmic medications. Sodium channel blockers may unmask an abnormal ECG phenotype in patients with loss-of-function *SCN5A* mutations, indicating an increased susceptibility to the proarrhythmic effects of blocking sodium channels (1). Individuals carrying the R558 allele are similarly susceptible (96). Therefore, AF families with *SCN5A* mutations and nonfamilial AF patients who are R558 allele carriers might be advised to avoid using class I antiarrhythmics in the management of AF. Although a recent investigation associated certain ACE DD/ID genotypes with antiarrhythmic drug therapy failure, replication of these findings is necessary, and the current level of evidence supporting ACE DD/ID genotyping for prediction of antiarrhythmic drug response is weak (97).

Warfarin is important in the management of AF to prevent embolic events. However, its use is limited by its narrow therapeutic range and the marked interindividual variation in response (98,99). Cytochrome P450 2C9 (CYP2C9) is the major isoform of the hepatic cytochrome P450 enzymes; it modulates the physiological effect of warfarin. The *CYP2C9*2*

(rs1799853) and *CYP2C9*3* (rs1057910) polymorphisms can lead to decreased enzymatic activity; therefore, patients with at least one *CYP2C9* variant have between a 1.4- and 3.6-fold increased risk for supratherapeutic international normalized ratios and a delay before they reach stable dosing (98). In addition, variation in the vitamin K epoxide reductase complex, subunit 1 (*VKORC1*), has been correlated with variations in warfarin dosage (99). Testing for these variant alleles might allow for more personalized dosing of warfarin in the management of AF (Figure 13.5). However, based on the recently published European Pharmacogenetics of Anticoagulant Therapy (EU-PACT) trials (100,101) and the Clarification of Optimal Anticoagulation through Genetics (COAG) trial (102), incorporation of genotype-based guidance of warfarin therapy into clinical practice cannot be recommended at present.

Polymorphisms on chromosome 4q25, including rs2200733 and rs10033464, were associated with increased risk of AF recurrence following catheter ablation (103). This result may assist proposals regarding a new approach to the treatment of AF using catheter ablation. Of course, the previously mentioned paradigms are merely hypothesis generating, and will require prospective and randomized assessment. Nevertheless, these studies serve as a proof-of-concept that systems-biology-based approaches might be useful in tailoring not just medical, but also interventional AF therapeutics.

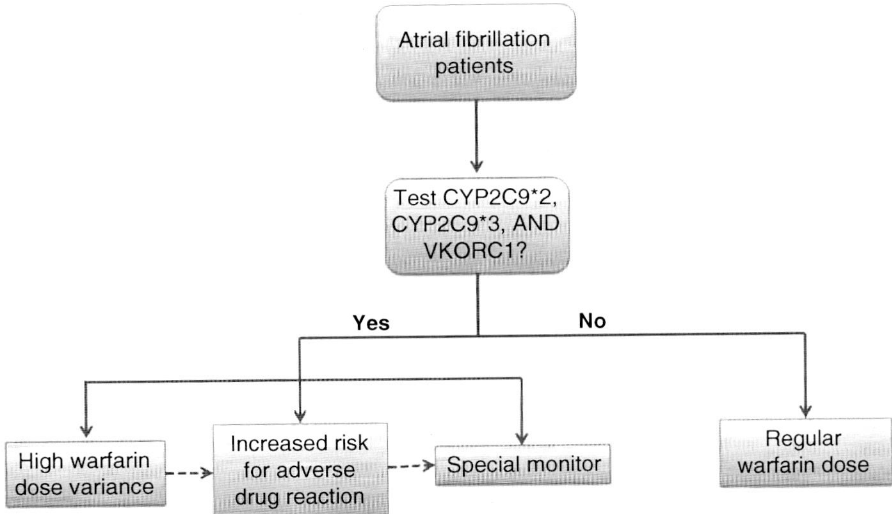

FIGURE 13.5 The pharmacogenetics of warfarin therapy for atrial fibrillation.

Source: Modified and adapted with permission from Ref. (32). Xiao J, Liang D, Chen YH. The genetics of atrial fibrillation: from the bench to the bedside. *Annu Rev Genomics Hum Genet*. 2011;12:73–96.

CONCLUSIONS

AF is a heritable condition, and at least nine susceptibility loci have been identified by GWAS. However, the identified loci explain only a small proportion of the heritability of AF. Future work will focus on larger scale GWAS to identify additional associated variants. GWAS, as well as new genotyping and functional studies, will help define the mechanisms by which AF-associated SNPs lead to disease and how they ultimately relate genotypes to clinical outcomes and treatments. "Omic" studies, still in early development and largely confined to animal studies and small, selected cohorts with AF, provide new avenues for identifying cellular and developmental pathways for AF. We envision that systems biology will identify phenotypic traits, integrate genetic information, and utilize "omics" platforms to provide new insights into the pathogenesis of AF as well as aid in the development of new AF risk scores, AF prevention, and guide the design and application of AF screening and treatments.

REFERENCES

1. Ellinor PT, Nam EG, Shea MA, et al. Cardiac sodium channel mutation in atrial fibrillation. *Heart Rhythm*. 2008;5:99–105.
2. Das S, Makino S, Melman YF, et al. Mutation in the S3 segment of KCNQ1 results in familial lone atrial fibrillation. *Heart Rhythm*. 2009;6(8):1146–1153.
3. Levy RL. Paroxysmal auricular fibrillation and flutter without signs of organic cardiac disease in two brothers. *J Mt Sinai Hosp*. 1942;8:765–770.
4. Wolff L. Familial auricular fibrillation. *New Eng J Med*. 1943;229:396–398.
5. Chen YH, Xu SJ, Bendahhou S, et al. KCNQ1 gain-of-function mutation in familial atrial fibrillation. *Science*. 2003;299:251–254.
6. Ellinor PT, Petrov-Kondratov VI, Zakharova E, et al. Potassium channel gene mutations rarely cause atrial fibrillation. *BMC Med Genet*. 2006;7:70.
7. Hodgson-Zingman DM, Karst ML, Zingman LV, et al. Atrial natriuretic peptide frameshift mutation in familial atrial fibrillation. *N Engl J Med*. 2008;359(2):158–165.
8. Fox CS, Parise H, D'Agostino RB, Sr, et al. Parental atrial fibrillation as a risk factor for atrial fibrillation in offspring. *JAMA*. 2004;291:2851–2855.
9. Lubitz SA, Yin X, Fontes JD, et al. Association between familial atrial fibrillation and risk of new-onset atrial fibrillation. *JAMA*. 2010;304:2263–2269.
10. Arnar DO, Thorvaldsson S, Manolio TA, et al. Familial aggregation of atrial fibrillation in Iceland. *Eur Heart J*. 2006;27:708–712.
11. Christophersen IE, Ravn LS, Budtz-Joergensen E, et al. Familial aggregation of atrial fibrillation: a study in Danish twins. *Circ Arrhythm Electrophysiol*. 2009;2:378–383.
12. Ellinor PT, Yoerger DM, Ruskin JN, MacRae CA. Familial aggregation in lone atrial fibrillation. *Hum Genet*. 2005;118:179–184.
13. Marcus GM, Smith LM, Vittinghoff E, et al. A first-degree family history in lone atrial fibrillation patients. *Heart Rhythm*. 2008;5:826–830.
14. Gudbjartsson DF, Arnar DO, Helgadottir A, et al. Variants conferring risk of atrial fibrillation on chromosome 4q25. *Nature*. 2007;448:353–357.
15. Mommersteeg MT, Brown NA, Prall OW, et al. Pitx2c and Nkx2-5 are required for the formation and identity of the pulmonary myocardium. *Circ Res*. 2007;101:902–909.
16. Mommersteeg MT, Hoogaars WM, Prall OW, et al. Molecular pathway for the localized formation of the sinoatrial node. *Circ Res*. 2007;100:354–362.

17. Gudbjartsson DF, Holm H, Gretarsdottir S, et al. A sequence variant in ZFHX3 on 16q22 associates with atrial fibrillation and ischemic stroke. *Nat Genet.* 2009;41:876–878.

18. Benjamin EJ, Rice KM, Arking DE, et al. Variants in ZFHX3 are associated with atrial fibrillation in individuals of European ancestry. *Nat Genet.* 2009;41:879–881.

19. Ellinor PT, Lunetta KL, Albert CM, et al. Meta-analysis identifies six new susceptibility loci for atrial fibrillation. *Nat Genet.* 2012;44:670–675.

20. Ellinor PT, Lunetta KL, Glazer NL, et al. Common variants in KCNN3 are associated with lone atrial fibrillation. *Nat Genet.* 2010;42:240–244.

21. Schnabel RB, Kerr KF, Lubitz SA, et al. Large-scale candidate gene analysis in whites and African Americans identifies IL6R polymorphism in relation to atrial fibrillation: the National Heart, Lung, and Blood Institute's Candidate Gene Association Resource (CARe) project. *Circ Cardiovasc Genet.* 2011;4:557–564.

22. Lubitz SA, Sinner MF, Lunetta KL, et al. Independent susceptibility markers for atrial fibrillation on chromosome 4q25. *Circulation.* 2010;122:976–984.

23. Schnabel RB, Sullivan LM, Levy D, et al. Development of a risk score for atrial fibrillation (Framingham Heart Study): a community-based cohort study. *Lancet.* 2009;373:739–745.

24. Schnabel RB, Aspelund T, Li G, et al. Validation of an atrial fibrillation risk algorithm in whites and African Americans. *Arch Intern Med.* 2010;170:1909–1917.

25. Alonso A, Krijthe BP, Aspelund T, et al. Simple risk model predicts incidence of atrial gibrillation in a racially and geographically diverse population: the CHARGE-AF consortium. *J Am Heart Assoc.* 2013;2:e000102.

26. Schnabel RB, Larson MG, Yamamoto JF, et al. Relations of biomarkers of distinct pathophysiological pathways and atrial fibrillation incidence in the community. Circulation. 2010;121:200–207.

27. Smith JG, Newton-Cheh C, Almgren P, et al. Genetic polymorphisms for estimating risk of atrial fibrillation in the general population: a prospective study. *Arch Intern Med.* 2012;172:742–744.

28. Everett BM, Cook NR, Conen D, et al. Novel genetic markers improve measures of atrial fibrillation risk prediction. *Eur Heart J.* 2013;34:2243–2251.

29. Magnani JW, Rienstra M, Lin H, et al. Atrial fibrillation: current knowledge and future directions in epidemiology and genomics. *Circulation.* 2011;124(18):1982–1993.

30. Sinner MF, Ellinor PT, Meitinger T, et al. Genome-wide association studies of atrial fibrillation: past, present, and future. *Cardiovasc Res.* 2011;89(4):701–709.

31. Lubitz SA, Ellinor PT. Personalized medicine and atrial fibrillation: will it ever happen? *BMC Med.* 2012;10:155.

32. Xiao J, Liang D, Chen YH. The genetics of atrial fibrillation: from the bench to the bedside. *Annu Rev Genomics Hum Genet.* 2011;12:73–96.

33. Goldstein DB. Common genetic variation and human traits. *N Engl J Med.* 2009;360(17):1696–1698.

34. Manolio TA, Collins FS, Cox NJ, et al. Finding the missing heritability of complex diseases. *Nature.* 2009;461(7265):747–753.

35. Gibson G. Hints of hidden heritability in GWAS. *Nat Genet.* 2010;42(7):558–560.

36. Baccarelli A, Rienstra M, Benjamin EJ. Cardiovascular epigenetics: basic concepts and results from animal and human studies. *Circ Cardiovasc Genet.* 2010;3(6):567–573.

37. Piran S, Liu P, Morales A, Hershberger RE. Where genome meets phenome: rationale for integrating genetic and protein biomarkers in the diagnosis and management of dilated cardiomyopathy and heart failure. *J Am Coll Cardiol.* 2012;60(4):283–289.

38. De Souza AI, Camm AJ. Proteomics of atrial fibrillation. *Circ Arrhythm Electrophysiol.* 2012;5(5):1036–1043.

39. Emilsson V, Thorleifsson G, Zhang B, et al. Genetics of gene expression and its effect on disease. *Nature.* 2008;452:423–428.

40. Dixon AL, Liang L, Moffatt MF, et al. A genome-wide association study of global gene expression. *Nat Genet.* 2007;39:1202–1207.

41. Djebali S, Davis CA, Merkel A, et al. Landscape of transcription in human cells. *Nature.* 2012;489:101–108.

42. Lappalainen T, Sammeth M, Friedlander MR, et al. Transcriptome and genome sequencing uncovers functional variation in humans. *Nature.* 2013;501:506–511.

43. Putt ME, Hannenhalli S, Lu Y, et al. Evidence for coregulation of myocardial gene expression by mef2 and nfat in human heart failure. *Circ Cardiovasc Genet.* 2009;2:212–219.

44. Cervero J, Segura V, Macias A, et al. Atrial fibrillation in pigs induces left atrial endocardial transcriptional remodelling. *Thromb Haemost.* 2012;108:742–749.

45. Thijssen VL, van der Velden HM, van Ankeren EP, et al. Analysis of altered gene expression during sustained atrial fibrillation in the goat. *Cardiovasc Res.* 2002;54:427–437.

46. Heerdt PM, Kant R, Hu Z, et al. Transcriptomic analysis reveals atrial kcne1 down-regulation following lung lobectomy. *J Mol Cell Cardiol.* 2012;53:350–353.

47. Barth AS, Merk S, Arnoldi E, et al. Reprogramming of the human atrial transcriptome in permanent atrial fibrillation: expression of a ventricular-like genomic signature. *Circ Res.* 2005;96:1022–1029.

48. Mace LC, Yermalitskaya LV, Yi Y, et al. Transcriptional remodeling of rapidly stimulated hl-1 atrial myocytes exhibits concordance with human atrial fibrillation. *J Mol Cell Cardiol.* 2009;47:485–492.

49. Dupont E, Ko Y, Rothery S, et al. The gap-junctional protein connexin40 is elevated in patients susceptible to postoperative atrial fibrillation. *Circulation.* 2001;103:842–849.

50. Goette A, Staack T, Rocken C, et al. Increased expression of extracellular signal-regulated kinase and angiotensin-converting enzyme in human atria during atrial fibrillation. *J Am Coll Cardiol.* 2000;35:1669–1677.

51. Adam O, Lavall D, Theobald K, et al. Rac1-induced connective tissue growth factor regulates connexin 43 and n-cadherin expression in atrial fibrillation. *J Am Coll Cardiol.* 2010;55:469–480.

52. He X, Gao X, Peng L, et al. Atrial fibrillation induces myocardial fibrosis through angiotensin ii type 1 receptor-specific arkadia-mediated downregulation of smad7. *Circ.Res.* 2011;108:164–175.

53. Burstein B, Nattel S. Atrial fibrosis: mechanisms and clinical relevance in atrial fibrillation. *J Am Coll Cardiol.* 2008;51:802–809.

54. Burstein B, Qi XY, Yeh YH, et al. Atrial cardiomyocyte tachycardia alters cardiac fibroblast function: a novel consideration in atrial remodeling. *Cardiovasc. Res.* 2007;76:442–452.

55. Van Gelder IC, Brundel BJ, Henning RH, et al. Alterations in gene expression of proteins involved in the calcium handling in patients with atrial fibrillation. *J Cardiovasc Electrophysiol.* 1999;10:552–560.

56. Nattel S. New ideas about atrial fibrillation 50 years on. *Nature.* 2002;415:219–226.

57. Gaborit N, Steenman M, Lamirault G, et al. Human atrial ion channel and transporter subunit gene-expression remodeling associated with valvular heart disease and atrial fibrillation. *Circulation.* 2005; 112:471–481.

58. Brundel BJ, van Gelder IC, Henning RH, et al. Gene expression of proteins influencing the calcium homeostasis in patients with persistent and paroxysmal atrial fibrillation. *Cardiovasc Res.* 1999;42:443–454.

59. Hindorff LA, Sethupathy P, Junkins HA, et al. Potential etiologic and functional implications of genome-wide association loci for human diseases and traits. *Proc Natl Acad Sci USA.* 2009;106:9362–9367.

60. Schadt EE, Molony C, Chudin E, et al. Mapping the genetic architecture of gene expression in human liver. *PLoS Biol.* 2008;6:e107.

61. Zeller T, Wild P, Szymczak S, et al. Genetics and beyond—the transcriptome of human monocytes and disease susceptibility. *PLoS One.* 2010;5:e10693.

62. Lin H, Dolmatova EV, Morley MP, et al. Gene expression and genetic variation in human atria. *Heart Rhythm.* 2013;11(2):266–271.

63. Debey S, Schoenbeck U, Hellmich M, et al. Comparison of different isolation techniques prior gene expression profiling of blood derived cells: impact on physiological responses, on overall expression and the role of different cell types. *Pharmacogenomics J.* 2004;4:193–207.

64. Kahr PC, Piccini I, Fabritz L, et al. Systematic analysis of gene expression differences between left and right atria in different mouse strains and in human atrial tissue. *PLoS One.* 2011;6:e26389.

65. Hsu J, Hanna P, Van Wagoner DR, et al. Whole genome expression differences in human left and right atria ascertained by RNA sequencing. *Circ Cardiovasc Genet.* 2012;5:327–335.

66. Powell JE, Henders AK, McRae AF, et al. Genetic control of gene expression in whole blood and lymphoblastoid cell lines is largely independent. *Genome Res.* 2012;22:456–466.

67. Skopek P, Hynie S, Chottova-Dvorakova M, et al. Effects of acute stressors on the expression of oxytocin receptor mRNA in hearts of rats with different activity of HPA axis. *Neuro Endocrinol Lett.* 2012; 33:124–132.

68. Mortazavi A, Williams BA, McCue K, Schaeffer L, Wold B. Mapping and quantifying mammalian transcriptomes by RNA-seq. *Nat Methods.* 2008;5: 621–628.

69. Matkovich SJ, Zhang Y, Van Booven DJ, Dorn GW, 2nd. Deep mRNA sequencing for in vivo functional analysis of cardiac transcriptional regulators: application to galphaq. *Circ Res.* 2010;106:1459–1467.

70. Small EM, Frost RJ, Olson EN. MicroRNAs add a new dimension to cardiovascular disease. *Circulation.* 121(8):1022–1032.

71. Small EM, Olson EN. Pervasive roles of microRNAs in cardiovascular biology. *Nature.* 469(7330):336–342.

72. van Rooij E, Olson EN. MicroRNAs: powerful new regulators of heart disease and provocative therapeutic targets. *J Clin Invest.* 2007;117(9):2369–2376.

73. McManus DD, Ambros V. Circulating MicroRNAs in cardiovascular disease. *Circulation.* 2011;124(18): 1908–1910.

74. Wang Z, Lu Y, Yang B. MicroRNAs and atrial fibrillation: new fundamentals. *Cardiovasc Res.* 2011; 89(4):710–721.

75. Kim GH. MicroRNA regulation of cardiac conduction and arrhythmias. *Transl Res.* 2013;161(5):381–392.

76. Latronico MV, Catalucci D, Condorelli G. Emerging role of microRNAs in cardiovascular biology. *Circ Res.* 2007;101(12):1225–1236.

77. Luo X, Zhang H, Xiao J, Wang Z. Regulation of human cardiac ion channel genes by microRNAs: theoretical perspective and pathophysiological implications. *Cell Physiol Biochem.* 2010;25(6):571–586.

78. Lu Y, Zhang Y, Wang N, et al. MicroRNA-328 contributes to adverse electrical remodeling in atrial fibrillation. *Circulation.* 2010;122(23):2378–2387.

79. Zhao Y, Ransom JF, Li A, et al. Dysregulation of cardiogenesis, cardiac conduction, and cell cycle in mice lacking miRNA-1-2. *Cell.* 2007;129(2):303–317.

80. Girmatsion Z, Biliczki P, Bonauer A, et al. Changes in microRNA-1 expression and IK1 up-regulation in human atrial fibrillation. *Heart Rhythm.* 2009; 6(12):1802–1809.

81. Luo X, Pan Z, Shan H, et al. MicroRNA-26 governs profibrillatory inward-rectifier potassium current changes in atrial fibrillation. *J Clin Invest.* 2013; 123(5):1939–1951.

82. Shan H, Zhang Y, Lu Y, et al. Downregulation of miR-133 and miR-590 contributes to nicotine-induced atrial remodelling in canines. *Cardiovasc Res.* 2009;83(3):465–472.

83. Callis TE, Pandya K, Seok HY, et al. MicroRNA-208a is a regulator of cardiac hypertrophy and conduction in mice. *J Clin Invest.* 2009;119(9):2772–2786.

84. Wang K, Zhang S, Marzolf B, et al. Circulating microRNAs, potential biomarkers for drug-induced liver injury. *Proc Natl Acad Sci USA.* 2009;106: 4402–4407.

85. Fichtlscherer S, De Rosa S, Fox H, et al. Circulating microRNAs in patients with coronary artery disease. *Circ Res.* 107:677–684.

86. Liu Z, Zhou C, Liu Y, et al. The expression levels of plasma microRNAs in atrial fibrillation patients. *PLoS One.* 2012;7(9):e44906.

87. Krützfeldt J, Rajewsky N, Braich R, et al. Silencing of microRNAs in vivo with 'antagomirs.' *Nature.* 2005;438(7068):685–689.

88. van Rooij E, Olson EN. MicroRNA therapeutics for cardiovascular disease: opportunities and obstacles. *Nat Rev Drug Discov.* 2012;11(11):860–872.

89. van Rooij E. The art of microRNA research. *Circ Res.* 2011;108(2):219–234.

90. De Souza AI, Camm AJ. Proteomics of atrial fibrillation. *Circ Arrhythm Electrophysiol.* 2012;5(5): 1036–1043.

91. Mayr M, Yusuf S, Weir G, et al. Combined metabolomic and proteomic analysis of human atrial fibrillation. *J Am Coll Cardiol.* 2008;51(5):585–594.

92. Bodmer W, Bonilla C. Common and rare variants in multifactorial susceptibility to common diseases. *Nat Genet.* 2008;40(6):695–701.

93. Jeong H, Tombor B, Albert R, et al. The large-scale organization of metabolic networks. *Nature.* 2000; 407:651–654.

94. Otway R, Vandenberg JI, Guo G, et al. Stretch-sensitive KCNQ1 mutation: a link between genetic and environmental factors in the pathogenesis of atrial fibrillation? *J Am Coll Cardiol.* 2007;49:578–586.

95. Ehrlich JR, Biliczki P, Hohnloser SH, Nattel S. Atrial-selective approaches for the treatment of atrial fibrillation. *J Am Coll Cardiol.* 2008;51:787–792.

96. Chen LY, Ballew JD, Herron KJ, et al. A common polymorphism in SCN5A is associated with lone atrial fibrillation. *Clin Pharmacol Ther.* 2007;81:35–41.

97. Darbar D, Motsinger AA, Ritchie MD, et al. Polymorphism modulates symptomatic response to anti-arrhythmic drug therapy in patients with lone atrial fibrillation. *Heart Rhythm.* 2007;4(6):743–749.

98. Higashi MK, Veenstra DL, Kondo LM, et al. Association between CYP2C9 genetic variants and anticoagulation-related outcomes during warfarin therapy. *JAMA.* 2008;287:1690–1698.

99. Rieder MJ, Reiner AP, Gage BF, et al. Effect of VKORC1 haplotypes on transcriptional regulation and warfarin dose. *N Engl J Med.* 2005;352:2285–2293.

100. Pirmohamed M, Burnside G, Eriksson N, et al.; EU-PACT Group. A randomized trial of genotype-guided dosing of warfarin. *N Engl J Med.* 2013; 369(24):2294–2303.

101. Verhoef TI, Ragia G, de Boer A, et al.; EU-PACT Group. A randomized trial of genotype-guided dosing of acenocoumarol and phenprocoumon. *N Engl J Med.* 2013;369(24):2304–2312.

102. Kimmel SE, French B, Kasner SE, et al.; COAG Investigators. A pharmacogenetic versus a clinical algorithm for warfarin dosing. *N Engl J Med.* 2013; 369(24):2283–2293.

103. Husser D, Adams V, Piorkowski C, et al. Chromosome 4q25 variants and atrial fibrillation recurrence after catheter ablation. *J Am Coll Cardiol.* 2010; 55:747–753.

57. Sartorato P, Lapeyraque AL, Armanini D, et al. Different inactivating mutations of the mineralocorticoid receptor in fourteen families affected by type I pseudohypoaldosteronism. *J Clin Endocrinol Metab.* 2003;88(6):2508–2517.

58. Viemann M, Peter M, Lopez-Siguero JP, et al. Evidence for genetic heterogeneity of pseudohypoaldosteronism type 1: identification of a novel mutation in the human mineralocorticoid receptor in one sporadic case and no mutations in two autosomal dominant kindreds. *J Clin Endocrinol Metab.* 2001; 86(5):2056–2059.

59. Tajima T, Kitagawa H, Yokoya S, et al. A novel missense mutation of mineralocorticoid receptor gene in one Japanese family with a renal form of pseudohypoaldosteronism type 1. *J Clin Endocrinol Metab.* 2000;85(12):4690–4694.

60. Arai K, Zachman K, Shibasaki T, Chrousos GP. Polymorphisms of amiloride-sensitive sodium channel subunits in five sporadic cases of pseudohypoaldosteronism: do they have pathologic potential? *J Clin Endocrinol Metab.* 1999;84(7):2434–2437.

61. Strautnieks SS, Thompson RJ, Gardiner RM, Chung E. A novel splice-site mutation in the gamma subunit of the epithelial sodium channel gene in three pseudohypoaldosteronism type 1 families. *Nat Genet.* 1996;13(2):248–250.

62. Chang SS, Grunder S, Hanukoglu A, et al. Mutations in subunits of the epithelial sodium channel cause salt wasting with hyperkalaemic acidosis, pseudohypoaldosteronism type 1. *Nat Genet.* 1996;12(3):248–253.

63. Lifton RP. Molecular genetics of human blood pressure variation. *Science.* 1996;272(5262):676–680.

64. Shimkets RA, Lifton RP, Canessa CM. The activity of the epithelial sodium channel is regulated by clathrin-mediated endocytosis. *J Biol Chem.* 1997; 272(41):25537–25541.

65. Snyder PM, Price MP, McDonald FJ, et al. Mechanism by which Liddle's syndrome mutations increase activity of a human epithelial Na+ channel. *Cell.* 1995; 83(6):969–978.

66. Ehret GB, Caulfield MJ. Genes for blood pressure: an opportunity to understand hypertension. *Eur Heart J.* 2013;34(13):951–961.

67. Williams SS. Advances in genetic hypertension. *Curr Opin Pediatr.* 2007;19(2):192–198.

68. Vehaskari VM. Heritable forms of hypertension. *Pediatr Nephrol.* 2009;24(10):1929–1937.

69. Hassan-Smith Z, Stewart PM. Inherited forms of mineralocorticoid hypertension. *Curr Opin Endocrinol Diabetes Obes.* 2011;18(3):177–185.

70. Manolio TA, Collins FS, Cox NJ, et al. Finding the missing heritability of complex diseases. *Nature.* 2009;461(7265):747–753.

71. Marteau JB, Zaiou M, Siest G, Visvikis-Siest S. Genetic determinants of blood pressure regulation. *J Hypertens.* 2005;23(12):2127–2143.

72. Cheng LS, Livshits G, Carmelli D, et al. Segregation analysis reveals a major gene effect controlling systolic blood pressure and BMI in an Israeli population. *Hum Biol.* 1998;70(1):59–75.

73. Perusse L, Moll PP, Sing CF. Evidence that a single gene with gender- and age-dependent effects influences systolic blood pressure determination in a population-based sample. *Am J Hum Genet.* 1991;49(1):94–105.

74. Levy D, DeStefano AL, Larson MG, et al. Evidence for a gene influencing blood pressure on chromosome 17. Genome scan linkage results for longitudinal blood pressure phenotypes in subjects from the Framingham Heart Study. *Hypertension.* 2000;36(4): 477–483.

75. Adeyemo AA, Omotade OO, Rotimi CN, et al. Heritability of blood pressure in Nigerian families. *J Hypertens.* 2002;20(5):859–863.

76. Rotimi CN, Cooper RS, Cao G, et al. Maximum-likelihood generalized heritability estimate for blood pressure in Nigerian families. *Hypertension.* 1999;33(3): 874–878.

77. Gu D, Rice T, Wang S, et al. Heritability of blood pressure responses to dietary sodium and potassium intake in a Chinese population. *Hypertension.* 2007;50(1):116–122.

78. Hsueh WC, Mitchell BD, Schneider JL, et al. QTL influencing blood pressure maps to the region of PPH1 on chromosome 2q31-34 in Old Order Amish. *Circulation.* 2000;101(24):2810–2816.

79. Morrison AC, Cooper R, Hunt S, et al. Genome scan for hypertension in nonobese African Americans: the National Heart, Lung, and Blood Institute Family Blood Pressure Program. *Am J Hypertens.* 2004;17(9):834–838.

80. Koivukoski L, Fisher SA, Kanninen T, et al. Meta-analysis of genome-wide scans for hypertension and blood pressure in Caucasians shows evidence of susceptibility regions on chromosomes 2 and 3. *Hum Mol Genet.* 2004;13(19):2325–2332.

81. Palmer LJ, Scurrah KJ, Tobin M, et al. Genome-wide linkage analysis of longitudinal phenotypes using sigma2A random effects (SSARs) fitted by Gibbs sampling. *BMC Genet.* 2003;(4 Suppl 1): S12.

82. Rice T, Cooper RS, Wu X, et al. Meta-analysis of genome-wide scans for blood pressure in African American and Nigerian samples. The National Heart, Lung, and Blood Institute GeneLink Project. *Am J Hypertens.* 2006;19(3):270–274.

83. Adeyemo A, Luke A, Wu X, et al. Genetic effects on blood pressure localized to chromosomes 6 and 7. *J Hypertens.* 2005;23(7):1367–1373.

84. Jacobs KB, Gray-McGuire C, Cartier KC, Elston RC. Genome-wide linkage scan for genes affecting longitudinal trends in systolic blood pressure. *BMC Genet.* 2003;4(Suppl 1):S82.

85. Mocci E, Concas MP, Fanciulli M, et al. Microsatellites and SNPs linkage analysis in a Sardinian genetic isolate confirms several essential hypertension loci previously identified in different populations. *BMC Medical Genetics.* 2009;10:81.

86. Simino J, Shi G, Kume R, et al. Five blood pressure loci identified by an updated genome-wide linkage scan: meta-analysis of the Family Blood Pressure Program. *Am J Hypertens.* 2011;24(3):347–354.

87. Gong M, Zhang H, Schulz H, et al. Genome-wide linkage reveals a locus for human essential (primary) hypertension on chromosome 12p. *Hum Mol Genet.* 2003;12(11):1273–1277.

88. Harrap SB, Wong ZYH, Stebbing M, et al. Blood pressure QTLs identified by genome-wide linkage analysis and dependence on associated phenotypes. *Physiol Genomics.* 2002;8(2):99–105.

89. Hoffmann K, Planitz C, Ruschendorf F, et al. A novel locus for arterial hypertension on chromosome 1p36 maps to a metabolic syndrome trait cluster in the Sorbs, a Slavic population isolate in Germany. *J Hypertens.* 2009;27(5):983–990.

90. Atwood LD, Samollow PB, Hixson JE, et al. Genome-wide linkage analysis of blood pressure in Mexican Americans. *Genet Epidemiol.* 2001;20(3):373–382.

91. Zhu DL, Wang HY, Xiong MM, et al. Linkage of hypertension to chromosome 2q14-q23 in Chinese families. *J Hypertens.* 2001;19(1):55–61.

92. Perola M, Kainulainen K, Pajukanta P, et al. Genome-wide scan of predisposing loci for increased diastolic blood pressure in Finnish siblings. *J Hypertens.* 2000; 18(11):1579–1585.

93. McArdle PF, Dytch H, O'Connell JR, et al. Homozygosity by descent mapping of blood pressure in the Old Order Amish: evidence for sex specific genetic architecture. *BMC Genet.* 2007;8:66.

94. Puppala S, Coletta DK, Schneider J, et al. Genome-wide linkage screen for systolic blood pressure in the Veterans Administration Genetic Epidemiology Study (VAGES) of Mexican-Americans and confirmation of a major susceptibility locus on chromosome 6q14.1. *Human Hered.* 2011;71(1):1–10.

95. Krushkal J, Ferrell R, Mockrin SC, et al. Genome-wide linkage analyses of systolic blood pressure using highly discordant siblings. *Circulation.* 1999; 99(11):1407–1410.

96. Caulfield M, Munroe P, Pembroke J, et al. Genome-wide mapping of human loci for essential hypertension. *Lancet.* 2003;361(9375):2118–2123.

97. Yu W, Gwinn M, Clyne M, et al. A navigator for human genome epidemiology. *Nat Genet.* 2008;40(2): 124–125.

98. Conen D, Cheng S, Steiner LL, et al. Association of 77 polymorphisms in 52 candidate genes with blood pressure progression and incident hypertension: the Women's Genome Health Study. *J Hypertens.* 2009; 27(3):476–483.

99. Zintzaras E, Kitsios G, Stefanidis I. Endothelial NO synthase gene polymorphisms and hypertension: a meta-analysis. *Hypertension.* 2006;48(4);700–710.

100. Basson J, Simino J, Rao DC. Between candidate genes and whole genomes: time for alternative approaches in blood pressure genetics. *Curr Hypertens Rep.* 2012; 14(1):46–61.

101. Johnson AD, Newton-Cheh C, Chasman DI, et al. Association of hypertension drug target genes with blood pressure and hypertension in 86 588 individuals. *Hypertension.* 2011;57:903–910.

102. Takeuchi F, Yamamoto K, Katsuya T, et al. Reevaluation of the association of seven candidate genes with blood pressure and hypertension: a replication study and meta-analysis with a larger sample size. *Hypertens Res.* 2012;35(8):825–831.

103. Johnson T, Gaunt TR, Newhouse SJ, et al. Blood pressure loci identified with a gene-centric array. *Am J Hum Genet.* 2011;89(6):688–700.

104. Ganesh SK, Tragante V, Guo W, et al. Loci influencing blood pressure identified using a cardiovascular gene-centric array. *Hum Mol Genet.* 2013;22(8):1663–1678.

105. Samani NJ, Erdmann J, Hall AS, et al. Genome-wide association analysis of coronary artery disease. *N Engl J Med.* 2007;357(5):443–453.

106. Frayling TM, Timpson NJ, Weedon MN, et al. A common variant in the FTO gene is associated with body mass index and predisposes to childhood and adult obesity. *Science.* 2007;316(5826):889–894.

107. Hiura Y, Tabara Y, Kokubo Y, et al. A genome-wide association study of hypertension-related phenotypes in a Japanese population. *Circ J.* 2010;74(11):2353–2359.

108. Padmanabhan S, Melander O, Johnson T, et al. Genome-wide association study of blood pressure extremes identifies variant near UMOD associated with hypertension. *PLoS Genet.* 2010;6(10):e1001177.

109. Salvi E, Kutalik Z, Glorioso N, et al. Genomewide association study using a high-density single nucleotide polymorphism array and case-control design identifies a novel essential hypertension susceptibility locus in the promoter region of endothelial NO synthase. *Hypertension.* 2012;59(2):248–255.

110. Conen D, Glynn RJ, Buring JE, et al. Natriuretic peptide precursor a gene polymorphisms and risk of blood pressure progression and incident hypertension. *Hypertension.* 2007;50(6):1114–1119.

111. Kosuge K, Soma M, Nakayama T, et al. A novel variable number of tandem repeat of the natriuretic peptide precursor B gene's 5'-flanking region is associated with essential hypertension among Japanese females. *Int J Med Sci.* 2007;4(3):146–152.

112. Qian X, Lu Z, Tan M, et al. A meta-analysis of association between C677T polymorphism in the methylenetetrahydrofolate reductase gene and hypertension. *Eur J Hum Genet.* 2007;15(12):1239–1245.

113. White PC. Disorders of aldosterone biosynthesis and action. *N Engl J Med.* 1994;331(4):250–258.

114. Monteith GR, Kable EP, Kuo TH, Roufogalis BD. Elevated plasma membrane and sarcoplasmic reticulum Ca2+ pump mRNA levels in cultured aortic smooth muscle cells from spontaneously hypertensive rats. *Biochem Biophys Res Commun.* 1997;230(2):344–346.

115. Kelly TN, Takeuchi F, Tabara Y, et al. Genome-wide association study meta-analysis reveals transethnic replication of mean arterial and pulse pressure loci. *Hypertension.* 2013;62(5):853–859.

116. Franceschini N, Fox E, Zhang Z, et al. Genome-wide association analysis of blood-pressure traits in African-ancestry individuals reveals common associated genes in African and non-African populations. *Am J Hum Genet.* 2013;93(3):545–554.

117. Maher B. Personal genomes: the case of the missing heritability. *Nature.* 2008;456(7218):18–21.

118. Fava C, Sjogren M, Montagnana M, et al. Prediction of blood pressure changes over time and incidence of hypertension by a genetic risk score in Swedes. *Hypertension.* 2013;61(2):319–326.

119. Wang X, Chua H-X, Chen P, et al. Comparing methods for performing trans-ethnic meta-analysis of genome-wide association studies. *Hum Mol Genet.* 2013; 22(11):2303–2311.

120. Morris AP. Transethnic meta-analysis of genomewide association studies. *Genet Epidemiol.* 2011;35(8): 809–822.

121. Franceschini N, van Rooij FJA, Prins BP, et al. Discovery and fine mapping of serum protein loci through transethnic meta-analysis. *Am J Hum Genet.* 2012;91(4):744–753.

122. Manchia M, Cullis J, Turecki G, et al. The impact of phenotypic and genetic heterogeneity on results of genome wide association studies of complex diseases. *PLoS One.* 2013;8(10):e76295.

123. Traylor M, Bevan S, Rothwell PM, et al. Using phenotypic heterogeneity to increase the power of genome-wide association studies: application to age at onset of ischaemic stroke subphenotypes. *Genet Epidemiol.* 2013;37(5):495–503.

124. He J, Kelly TN, Zhao Q, Li H, et al. Genome-wide association study identifies 8 novel loci associated with blood pressure responses to interventions in Han Chinese. *Circ Cardiovasc Genet.* 2013;6(6):598–607.

125. Citterio L, Simonini M, Zagato L, et al. Genes involved in vasoconstriction and vasodilation system affect salt-sensitive hypertension. *PLoS One.* 2011; 6(5):e19620.

126. Thomas C, Wood GC, Langer RD, Stewart WF. Elevated blood pressure in primary care varies in relation to circadian and seasonal changes. *J Hum Hypertens.* 2008;22(11):755–760.

127. Minami J, Kawano Y, Ishimitsu T, et al. Seasonal variations in office, home and 24 h ambulatory blood pressure in patients with essential hypertension. *J Hypertens.* 1996;14(12):1421–1425.

128. Coca A. Circadian rhythm and blood pressure control: physiological and pathophysiological factors. *J Hypertens Suppl.* 1994;12(5):S13–S21.

129. Ganesh SK, Ehret GB, Chakravarti AobotC-BaIc. *Genome-Wide Association Analyses of Long-Term Blood Pressure Traits.* In: American Society of Human Genetics Annual Meeting 2012; San Francisco; 2012.

130. Murcray CE, Lewinger JP, Gauderman WJ. Gene-environment interaction in genome-wide association studies. *Am J Epidemiol.* 2009;169(2):219–226.

131. Bloom JS, Ehrenreich IM, Loo WT, et al. Finding the sources of missing heritability in a yeast cross. *Nature.* 2013;494(7436):234–237.

132. Murcray CE, Lewinger JP, Conti DV, et al. Sample size requirements to detect gene-environment interactions in genome-wide association studies. *Genet Epidemiol.* 2011;35(3):201–210.

133. Friso S, Pizzolo F, Choi S-W, et al. Epigenetic control of 11 beta-hydroxysteroid dehydrogenase 2 gene promoter is related to human hypertension. *Atherosclerosis.* 2008;199(2):323–327.

134. Smolarek I, Wyszko E, Barciszewska AM, et al. Global DNA methylation changes in blood of patients with essential hypertension. *Med Sci Monit.* 2010;16(3):CR149–CR155.

135. Cohen J, Pertsemlidis A, Kotowski IK, et al. Low LDL cholesterol in individuals of African descent resulting from frequent nonsense mutations in PCSK9. *Nat Genet.* 2005;37(2):161–165.

136. Cohen JC, Kiss RS, Pertsemlidis A, et al. Multiple rare alleles contribute to low plasma levels of HDL cholesterol. *Science.* 2004;305(5685):869–872.

137. Cohen JC, Pertsemlidis A, Fahmi S, et al. Multiple rare variants in NPC1L1 associated with reduced sterol absorption and plasma low-density lipoprotein levels. *Proc Natl Acad Sci USA.* 2006;103(6):1810–1815.

138. Cohen JC, Boerwinkle E, Mosley TH, Jr, Hobbs HH. Sequence variations in PCSK9, low LDL, and protection against coronary heart disease. *N Engl J Med.* 2006;354(12):1264–1272.

139. Ji W, Foo JN, O'Roak BJ, et al. Rare independent mutations in renal salt handling genes contribute to blood pressure variation. *Nat Genet.* 2008; 40(5):592–599.

140. Rao F, Wen G, Gayen JR, et al. Catecholamine release-inhibitory peptide catestatin (chromogranin A(352-372)): naturally occurring amino acid variant Gly364Ser causes profound changes in human autonomic activity and alters risk for hypertension. *Circulation.* 2007;115(17):2271–2281.

141. Metzker ML. Sequencing technologies–the next generation. *Nat Rev Genet.* 2010;11(1):31–46.

142. Mailman MD, Feolo M, Jin Y, et al. The NCBI dbGaP database of genotypes and phenotypes. *Nat Genet.* 2007;39(10):1181–1186.

143. Lander ES. Initial impact of the sequencing of the human genome. *Nature.* 2011;470(7333):187–197.

144. Brugts JJ, Isaacs A, de Maat MP, et al. A pharmaco-genetic analysis of determinants of hypertension and blood pressure response to angiotensin-converting enzyme inhibitor therapy in patients with vascular disease and healthy individuals. *J Hypertens.* 2011;29(3):509–519.

12

C H A P T E R

Genetics of Electrocardiographic Traits

Andrew J. Sauer and Sanjiv J. Shah

TAKE HOME POINTS

1. Virtually all electrocardiographic (ECG) traits have a genetic component, and understanding the genetics of ECG traits can provide insight into electrophysiology and the molecular pathogenesis of arrhythmias.
2. The most commonly described gene associated with sodium channel protein abnormalities is the *SCN5A* gene; variants in this gene can result in a variety of abnormal ECG traits, including abnormal or delayed depolarization, conduction delay, deranged repolarization, and/or also pacemaker dysfunction.
3. Genome-wide association studies (GWAS) have provided tremendous insight into ECG traits relevant to both the general population and those with congenital electrophysiological syndromes; for example, variants in *NOS1AP* (which encodes a protein that participates in neuronal nitric oxide synthase regulation) have been associated with both a prolonged QT interval in the general population and with an increased risk for cardiac events in patients with long QT syndrome (LQTS).

CASE PRESENTATION

A 32-year-old woman of European descent is hospitalized after a sudden cardiac arrest. The patient had a witnessed collapse in an airport and an automatic external defibrillator was readily available. She was shocked once and regained consciousness. Upon presentation to the emergency department, a 12-lead electrocardiogram (ECG)

revealed sinus rhythm at a rate of 60 beats per minute, PR interval of 180 ms, QRS interval of 80 ms, and QT interval of 470 ms. Subsequent evaluation including echocardiography and coronary angiography did not reveal any abnormalities. The patient was given a preliminary diagnosis of idiopathic ventricular fibrillation.

However, upon taking a careful history, her outpatient cardiologist discovers that she had recently enrolled in a methadone clinic for a history of opiate dependence. The patient also describes experiencing palpitations and symptoms of lightheadedness since starting the methadone. On review of her family history, she states her family is generally healthy with no history of unexplained death. However, one of her siblings has a long history of a seizure disorder that manifests as "drop attacks." Analysis of her sister's ECG reveals a QT interval of 500 ms. DNA sequence analysis reveals a heterozygous missense mutation of a coding region of the *KCNH2* allele resulting in an alanine-to-valine amino acid substitution at position 1116 (designated A1116V), which is known to be highly conserved among homologous sequences of several species. This rare mutation is shared by the patient, the patient's mother, and the patient's clinically affected sister (Figure 12.1).

Further testing of the family members reveals that the patient's father, affected sister, and unaffected brother carry a polymorphism (K897T) involving the *KCNH2* allele that is known to be common (present in approximately 33% of individuals of European descent) and is associated with a modest increase in the QT interval. Of note, the patient does not have the K897T polymorphism. Given the patient's genetic information, the patient's cardiologist is debating whether or not the patient should receive an implantable defibrillator in her case.

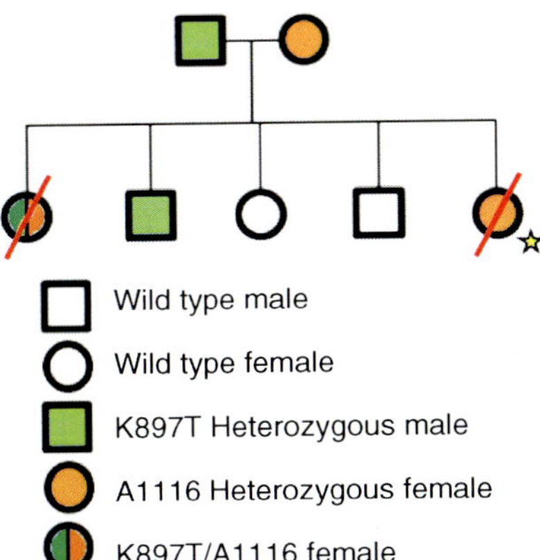

FIGURE 12.1 Case presentation pedigree demonstrating QT inheritance pattern. The yellow star denotes the patient described in the text. The red line represents a clinically affected family member.

Since the first sequencing of the human genome, there has been an explosion in knowledge of the genetic determinants of electrocardiographic (ECG) traits. Recently published genome-wide association studies (GWAS) comparing readily identified single nucleotide polymorphisms (SNPs) harbored near specific loci with easily measurable continuous ECG variables have advanced the understanding of the genetics of cardiac conduction, depolarization, and repolarization. While GWAS has proven to be an important tool for discovering novel loci that may further explain ECG genetics, sample sizes have become increasingly large while discovered effect sizes associated with common variants have become increasingly small (eg, often less than 10 ms of ECG interval effect size associated with each minor allele). Moreover, the identified common variants are not always replicated in other studies, thereby reducing the validity of the associations.

The undertaking of the study of complex genetics of ECG traits can be daunting, especially when one considers that the human genome contains 3.2 billion nucleotides, ordered in a seemingly random fashion, with a fascinatingly well-orchestrated access to the protein machinery leading to gene expression manifest as protein synthesis. It is remarkable that cardiac electrical activity is well-preserved in the vast majority of individuals considering the fact that a single nucleotide alteration, addition, or deletion may result in a significant change in the gene expression and final protein function in a large variety of pathways involved in cardiac conduction. Furthermore, understanding the genetics of ECG traits requires acknowledgment of the many known and unknown mechanisms that govern the ultimate manifestation of ECG phenotypes (Figure 12.2). The contemporary focus of human studies of complex ECG traits is on DNA sequence variants among individuals within populations (1). However, such variants have thus far failed to account for all of the variation in ECG traits expected from heritability studies. Other factors such as noncoding ribonucleic acids (RNAs), splice variants, epigenetics, microRNAs, and posttranslational modifications of encoded proteins, along with numerous environmental factors, likely account for the "unexplained heritability" of ECG traits. Thus,

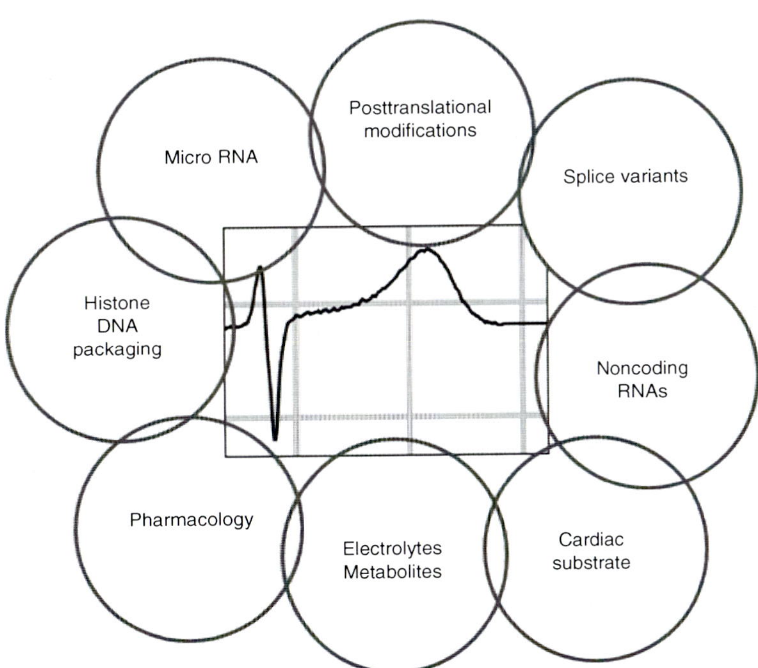

FIGURE 12.2 Multiple potential contributors to electrocardiographic traits. Common or rare genetic variants can demonstrate variable electrocardiographic trait expression due to several modifying factors as shown.

so far we have merely scratched the surface as we attempt to understand ECG trait genetics. However, it is important to review what we have learned thus far in order to have a better understanding of where we are going in the greater quest to fully understand the genetics of complex ECG traits.

GENETIC ARCHITECTURE OF COMPLEX ECG TRAITS

Allelic frequency within a population is known to range from the rare variant (often a mutation isolated to a single family or few families, known to have a large influence on an ECG phenotype) to the common variant (known to have a modest independent influence on the ECG phenotype). Natural selection has eliminated many of the mutations with strong effects from the population gene pool since humans with phenotypic expression of extreme derangements in cardiac electrophysiology would be less likely to survive to reproductive age. When investigating complex phenotypes such as ECG traits, the a priori knowledge of a potential gene involved in the expression of a trait may drive the hypothesis testing, a pathway for discovery known as the candidate gene approach (1). Alternatively, one could perform an unbiased survey of the entire genome attempting to associate a common nucleotide polymorphism with a particular phenotype (such as QT interval), as in GWAS.

Either way, most genetic studies examining ECG genetics are designed to test either a common variant–common disease (CV–CD) or a rare variant–common disease (RV–CD) hypothesis (2). The former presupposes that it is the cumulative contribution of a large number of common variants exerting modest effects that, in total, makes up the phenotypic trait. The latter assumes it is actually fewer rare variants with much larger individual effect sizes, which determine complex heritability. Regardless of CV–CD or RV–CD hypothesis, virtually all agree that a single variant rarely produces a complex phenotype independently. This has been well-exemplified in familial LQTS studies, for instance, in which the proband typically has a more markedly pronounced phenotypic expression than other mutation carriers observed within the family (3). Such an observation argues for the existence of additional alleles or strong environmental factors within the family that either accentuate or attenuate the phenotype despite the Mendelian mutation inheritance pattern. In reality, it is logical to hypothesize it is actually a combination of rare, uncommon, and common alleles with a heterogeneous continuum of effect sizes that contribute to the heritability of complex ECG traits (Figure 12.3).

ION-CHANNEL CANDIDATE GENES ASSOCIATED WITH ECG TRAITS

Prior to the GWAS revolution, most studies describing the genetics of ECG traits involved studying families with inherited channelopathies such as the LQTS, the short QT syndrome (SQTS), Brugada syndrome, catecholaminergic polymorphic ventricular tachycardia (CPVT), and idiopathic atrial fibrillation.

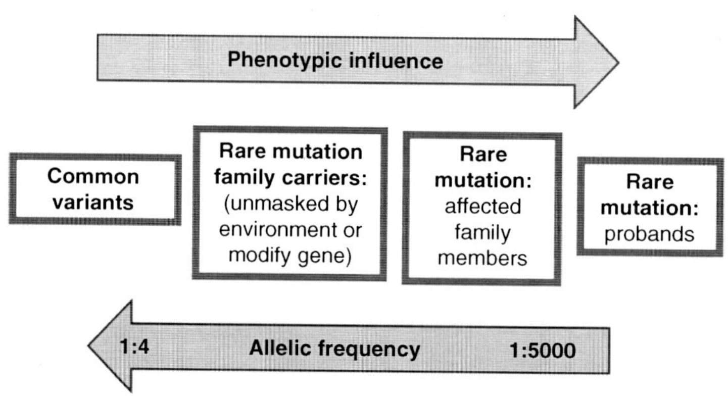

FIGURE 12.3 The spectrum of allelic frequencies and phenotypic expression. Since patients with severely abnormal ECG phenotypes (such as a severely prolonged QT or QRS interval) are less likely to survive to reproductive age, most of these rare mutations are restricted to individual families arising from spontaneous mutations and comprise a very small proportion of the population. Common variants, on the other hand, may be identified in as many as 30% to 40% of individuals within a population but are associated with more modest ECG trait alterations and minimal to no clinical consequence allowing for these variants to remain highly conserved. Combinations of uncommon and common genetic variants often manifest as an intermediate phenotype. Thus, ECG traits are ultimately a reflection of a spectrum of alleles with variable expression, interacting with numerous modifiers, manifesting as a spectrum of phenotypes.

TABLE 12.1   A Reference List of Candidate Genes Associated With Cardiac Ion Channelopathies

Gene	Chromosome Locus	Protein
Sodium		
SCN5A[a]	3p21-p24	Sodium channel alpha subunit ($Na_v1.5$)
SCN10A[a]	3p21	Sodium channel alpha subunit ($Na_v1.8$)
SCN4B	11q23.3	Sodium channel beta 4 subunit
SCN1B	19q13	Sodium channel beta 1 subunit
SCN3B	11q24.1	Sodium channel beta 3 subunit
Potassium		
KCNQ1[a]	11p15.5	I_{Ks} potassium channel alpha subunit (K_vLQT1)
KCNH2[a]	7q35-36	I_{Kr} potassium channel alpha subunit (HERG)
KCNE1[a]	21q22.1	I_{Ks} potassium channel beta subunit (MinK)
KCNE2	21q22.1	I_{Kr} potassium channel beta subunit (MiRP1)
KCNJ2[a]	17q23	I_{K1} late inward-rectifying channel (Kir2.1)
KCNJ8	12p11.23	ATP-sensitive potassium channel (Kir6.1)
KCND3[a]	1p13.2	I_{to} early voltage-gated potassium channel (Kv4.3)
Calcium		
CACNA1C	12p13.3	Voltage-gated L-type channel ($Ca_v1.2$)
RYR2	1q42.1-q43	Ryanodine receptor 2 (excitation–contraction)
CASQ2[a]	1p13.3	Calsequestrin 2 (calcium SR sequestration)
CALM1	14q32.11	Calmodulin 1 (excitation–contraction coupling)
PLN[a]	6q	Phospholamban (calcium SR sequestration)

[a]Associated with ECG traits in both family studies and genome-wide association studies.

Abbreviations: ATP, adenosine triphosphate; SR, sarcoplasmic reticulum.

Traditionally considered monogenic, these disorders formed the original models for the foundational understanding of the role and function of cardiac ion channels, which are now known to be heterogeneously distributed throughout the electrically active myocardial cells. The candidate genes (identified in pre-GWAS family studies) and their resultant phenotypes can be categorized in a variety of ways but one common method for appreciating their role is to discuss their influence on a specific ion-channel protein and the consequent downstream effect on ion current (Table 12.1 and Figure 12.4) (4).

Sodium Channel Gene Abnormalities

Undoubtedly, the most commonly described gene associated with sodium (Na^+) channel protein abnormalities is the *SCN5A* gene (encoding the $Na_v1.5$ protein). This gene represents a suitable example of how the specific type of mutation significantly impacts upon the ECG phenotype expression. Sodium channelopathies have been associated with abnormal or delayed depolarization, conduction delay, deranged repolarization, and also pacemaker dysfunction. In some cases, an *SCN5A* mutation can lead to the channel protein failing to close properly after initial depolarization leading to a Na^+ current "leak," resulting in prolonged repolarization, which is what is typically described in LQTS patients with this mutation. Patients with this particular gene mutation may have a characteristic ECG phenotype in which the T-wave onset is extremely delayed and the morphology may be peaked (5).

The *SCN5A* gene has also been implicated with the Brugada syndrome, first described in 1992, in which a rare mutation resulting in loss of Na^+ channel protein function leads to slowing of Phase 0 and also a reduction in the propagation of depolarization to neighboring myocytes which can lead to slowed cardiac conduction (6). These effects on the action potential and on heterogeneous conduction from one myocyte to the next lead to the characteristic but sometimes variable phenotype of a prolonged QRS duration (including a "pseudo" right bundle branch block morphology) and right precordial lead ST-segment elevation. The ECG pattern associated with this mutation may be heavily influenced by environmental modifiers such as temperature, pharmacological therapy, and autonomic modulation.

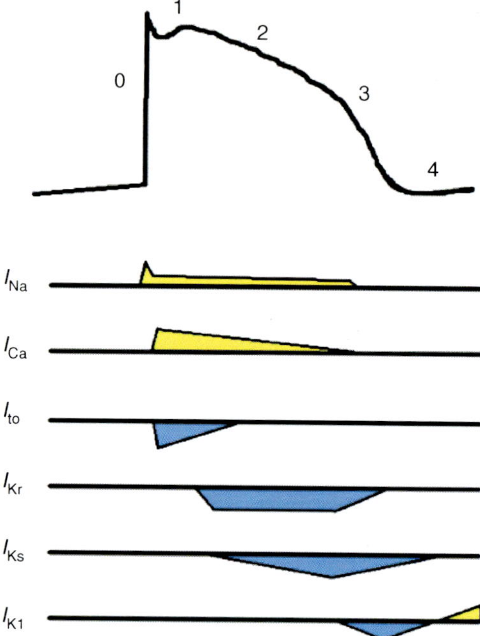

FIGURE 12.4 The monophasic action potential and the individual contribution of ion-channel currents. The myocardial cell is negatively polarized prior to the action potential. Phase 0 represents the rapid influx of positive ions (predominantly sodium) causing rapid depolarization of the cell (I_{Na}). Phase I is a "plateau" period, when outward potassium currents (I_{to}) are balanced by inward calcium currents (I_{Ca}) leading to a zero net change in cell polarization while the increase in cytosolic calcium interacts with myofilaments to prompt myocardial contraction. Phase II and Phase III represent early and late repolarization stages when additional outward potassium currents (I_{Kr} and I_{Ks}) exceed the diminishing influx of calcium ions allowing for repolarization of the cell and myocardial relaxation. Phase IV is characterized by a very slow depolarizing pacemaking current (I_{K1}) until the sodium channels responsible for Phase 0 are reactivated at a set depolarizing threshold.

Blue, efflux; yellow, myocardial cell influx of ion current.

Abbreviations: I_{Na}, sodium ion current; I_{Ca}, calcium ion current; I_{to}, early outward potassium current; I_{Kr}, rapid rectifying outward potassium current; I_{Ks}, slow rectifying outward potassium current; I_{K1}, very late potassium current.

Potassium Channel Genes

Phase III of the action potential, also known as repolarization, is a complex and delicate process, involving an intricate balance of competing inward currents (Na⁺ and calcium [Ca²⁺]) and outward currents (potassium [K⁺]) across the myocardial cell membrane. The model disease processes LQTS and SQTS have provided some of the initial framework for identifying candidate genes associated with repolarization abnormalities. The majority of repolarization phenotypes are secondary to either a gain-of-function or loss-of-function mutation affecting the K⁺ channel proteins. The most important K⁺ channels are those identified with rapid rectifying (I_{Kr}) and slow rectifying (I_{Ks}) current. The most common genes associated with I_{Kr} and I_{Ks} are *KCNH2* (encoding a channel protein known as Human Ether-a-go-go Related Gene, or HERG protein) and *KCNQ1* (encoding KvLQT1 channel protein), respectively, and the usual mutation effect is due to either abnormal protein trafficking to the cell membrane or abnormal ion-channel pore function causing the channel to either increase or decrease ion current (7).

The *KCNE1* gene encodes a regulatory protein (minK) for *KCNQ1* (KvLQT1) and the KCNE2 gene encodes a regulatory protein (MiRP1) for *KCNH2* (HERG) so it is not surprising that mutations involving coding regions for these regulatory proteins can have significant influence on rapid rectifying (I_{Kr}) and slow rectifying (I_{Ks}) currents. Mutations involving *KCNQ1* and *KCNH2* causing prolonged repolarization can also have stereotyped T-wave morphologies with *KCNQ1* mutations manifest on the ECG as a broad-based T-wave and *KCNH2* as a bifid or asymmetric T-wave (5). Less commonly reported mutations have been described involving the *KCNJ2* gene, which encodes the Kir2.1 protein involved in outward potassium current (I_{K1}), the terminal component of Phase III repolarization. Finally, *KCND3* which encodes the voltage-gated potassium channel Kv4.3 and is a known contributor to the Phase I transient outward current I_{to}, contributes to the early component of repolarization, and mutations involving this gene have also been identified in patients with Brugada syndrome (8).

Calcium Channel Genes

Ca²⁺ current into the myocardial cell is an important component of excitation–contraction coupling prompting myofilament activation and contraction. Ca²⁺ sequestration into the sarcoplasmic reticulum (SR) is an adenosine triphosphate (ATP)-requiring process important for ending the plateau phase (Phase II) of the action potential and allowing for repolarization while simultaneously inducing myofilament relaxation and the transition to diastole. Thus, any genes encoding proteins involved with Ca²⁺ currents into the cell or calcium sequestration into the SR will invariably impact ECG traits, particularly those involving heart rate (due to adrenergic regulation of Ca²⁺ handling) and repolarization (due to Ca²⁺ current contribution to Phase II plateau phase). The initial Ca²⁺ entry through the cell membrane involves L-type Ca²⁺ channels. The gene *CACNA1C* encodes the protein Cav1.2, which is responsible for the voltage-dependent L-type Ca²⁺ current and is responsive to adrenergic regulation (9). The gene *RYR2* encodes the ryanodine receptor channel protein, which is responsible for Ca^{2+} release from the SR into the sarcoplasma and also plays a major role in excitation–contraction coupling. Alterations in ryanodine receptor channel protein can have a downstream effect on repolarization, influencing both refractoriness as well as pacemaker function (10). Two genes, *CASQ2* and *SERCA2A*, are responsible for encoding additionally important proteins involved in calcium sequestration into the SR—calsequestrin and the SR Ca^{2+} ATPase, respectively (11).

LOCI ASSOCIATED WITH THE PR INTERVAL

The phenotypic trait of the ECG PR interval is a measurement of the period of time encompassing the onset of sinus node depolarization and including the propagation of the depolarization wave through the atria and the atrioventricular (AV) node, with the onset of the QRS complex representing initiation of ventricular depolarization. Thus, any inherited derangement of the action potential involving the sinus node, atria, AV node, or His–Purkinje system can influence the PR interval phenotype. Initial twin studies from the 1980s suggested a 34% heritability of the PR interval (12). Prior to GWAS, an association was demonstrated between *SCN5A* and conduction defects (13). Three landmark GWAS have since been recently published (two of which involved predominantly European ancestry populations and one exclusively involved African American subjects) examining the relationship between PR interval and SNPs noted to be in linkage disequilibrium (LD) with candidate gene loci (14–16). Table 12.2 provides a selective list of those minor allelic variants from each

loci (some loci harboring numerous associated variants not represented in the table) having the strongest association with PR interval.

The two variants that were found (in at least two different studies) to have the strongest association with PR interval appeared to be harbored within known candidate gene loci, specifically *SCN10A* and *SCN5A*, both encoding sodium channel proteins. Both of these genes are found on chromosome 3 and are in close proximity to each other. *SCN10A* is known to encode $Na_v1.8$ (a sodium channel found in both sensory nerves and cardiac tissue) and was shown to be associated with ECG conduction abnormalities, including increased risk of heart block, among individuals of European and Asian Indian ancestry (17). *SCN10A* and *SCN5A* variants are common in the population and are independently associated with a 3 to 4 ms increase in the PR interval.

Four additional loci were observed to have an association with the PR interval in at least two GWASs—specifically *CAV1*, *ARHGAP24*, *TBX5*, and *MEIS1*. Again, the effect size for each allele was small (ranging from 1 to 2 ms) and the mechanism underlying the associations remains unclear. However, each of the loci may have a conceivable impact on the distribution, frequency, or function of existing ion channels such as $Na_v1.5$ (encoded by *SCN5A*) either as a transcription factor or by impacting the cell construct such as the cytoskeleton or the cell membrane interface. Finally, it is worth noting that one variant near the locus *ITGA9* was associated with a modest effect (approximately 3 ms) on the PR interval but only within individuals of African ancestry. The *ITGA9* gene is known to encode a regulatory protein that modulates the function of $Na_v1.5$ cardiac sodium channels.

TABLE 12.2   Select Genes Within Loci Associated With the PR Interval and Their Proposed Function

Locus	Proposed Protein Function	Allele Frequency	P Value
SCN10A[a]	Cardiac Na^+ ($Na_v1.8$) channel	0.40	2×10^{-76}
SCN5A[a]	Cardiac Na^+ ($Na_v1.5$) channel	0.15	6×10^{-26}
CAV1[a]	Caveolins regulating plasma membrane	0.40	4×10^{-28}
ARHGAP24[a]	Cytoskeletal organization	0.31	6×10^{-20}
TBX5[a]	Conduction tissue transcription factor	0.28	3×10^{-17}
SOX5	Cardiac transcription factor	0.15	3×10^{-13}
NKX2-5	Cardiac homeobox transcription factor	0.40	9×10^{-13}
ITGA9[b]	Possible $Na_v1.5$ regulatory protein	0.18	4×10^{-11}
MEIS1[a]	Cardiac homeobox transcription factor	0.39	5×10^{-11}
WNT11	Cardiogenesis signaling protein	0.32	3×10^{-08}
MYH6	Augments myocardial contraction	0.34	2×10^{-05}

[a]Association with PR interval validated in more than one study.
[b]Observed only in individuals of African descent.

LOCI ASSOCIATED WITH THE QRS INTERVAL

The electrophysiological properties within the His–Purkinje system and the ventricular myocardium are represented electrocardiographically by the QRS interval. Prolongation of the QRS interval is an important independent predictor of cardiomyopathy, incident heart failure, sudden cardiac death, and mortality. Conduction velocity is determined by the magnitude of depolarizing inward currents flowing through voltage-gated sodium channels and also the extent of intercellular communication via gap junction–connexin coupling, which is variably influenced by myocardial cell and tissue architecture. Studies involving twins reared apart and together demonstrated a 30% to 60% heritability of the QRS interval phenotype (18). Initial candidate gene studies and GWAS involving a relatively small number of participants observed prolongation of the QRS interval implicating mutations of the cardiac Na⁺ channel encoding *SCN5A* and *SCN10A* genes (17,19). Subsequently, GWAS involving larger populations have further validated and expanded the list of candidate loci associated with the QRS interval phenotype (14,20,21). Table 12.3 provides a selective list of those minor allelic variants from each loci (some loci having numerous associated variants not represented in the table) having the strongest association with QRS interval. Again, each allelic variant identified by GWAS is found frequently in the population (approximately 15% to 40% minor allele frequency [MAF]) but imparts a very modest independent association with QRS duration (generally less than 1 ms). Not surprisingly, a consistent observation within these population studies has been the strong association between QRS interval and SNPs in LD with the *SCN5A* and *SCN10A* genes. Of particular interest is that the *SCN10A* gene has been shown to be preferentially expressed in the mouse His–Purkinje system when compared with the ventricular myocardium.

It has also become better appreciated that the Na⁺ channel can reside in distinctly different pools of interacting proteins that modulate $Na_v1.5$ activity and Na⁺ ion current (22). For instance, $Na_v1.5$ is particularly calcium ion sensitive, so it makes sense that genes involved with Ca^{2+} regulation (*PLN, PRKCA, CASQ2,* and *STRN*) would be in LD with SNPs associated with QRS interval. Also, in order to develop and maintain the integrity of the tissue involved with impulse conduction, transcription factors play a critical role. The transcription factor genes *TBX3, TBX5, TBX20, HAND1,* and *NFIA* have all been identified in loci associated with QRS duration. What appears to be more apparent since GWAS is the concept that Na⁺ channel function plays a central role in depolarization and conduction velocity but there is a complex interplay involving a variety of modulating Ca^{2+}-handling proteins, transcription factors, cyclin-dependent kinase inhibitors, gap junction proteins, and undoubtedly a variety of other pathways.

TABLE 12.3 Select Genes Within Loci Associated With the QRS Interval and Their Proposed Function

Locus	Proposed Protein Function	Allele Frequency	*P* Value
SCN10A[b]	Cardiac Na⁺ ($Na_v1.8$) channel	0.41	1×10^{-28}
CDKN1A[b]	Cyclin-dependent kinase inhibitor	0.25	3×10^{-27}
SCN5A[b]	Cardiac Na⁺ ($Na_v1.5$) channel	0.21	6×10^{-22}
PLN	Calcium-handling phospholamban protein	0.49	1×10^{-18}
NFIA[a]	Cardiac transcription factor	0.46	5×10^{-18}
HAND1	Cardiac transcription factor	0.36	7×10^{-14}
TBX20	Cardiac transcription factor	0.18	1×10^{-13}
SIPA1L1	Cell signaling, embryonic development	0.27	1×10^{-10}
TBX5[a]	Cardiac conduction transcription factor	0.29	1×10^{-10}
TBX3	Cardiac conduction transcription factor	0.27	3×10^{-10}
STRN	Ca^{2+}-calmodulin-binding protein	0.21	2×10^{-09}
PRKCA	Protein kinase modulates phospholamban	0.43	1×10^{-08}
CASQ2	Modulates ryanodine receptor	0.29	2×10^{-08}
DKK1[a]	Modulates connexin43 intercellular coupling	0.25	3×10^{-08}

[a]Association with QRS interval validated in more than one study.
[b]Association with QRS interval validated in at least three population studies.

LOCI ASSOCIATED WITH THE QT INTERVAL

The QT interval remains one of the most studied ECG traits as it has been associated with morbidity and mortality in large population-based studies (23). The QT interval phenotype is known to be influenced by underlying structural or coronary heart disease, electrolyte imbalance, and QT-prolonging pharmacological agents. Moreover, there is a well-demonstrated age- and sex-based modulation of the phenotype when comparing childhood, adolescent, and adult populations (24,25). Females have an increased QT duration when compared to males. When compared to healthy men, healthy adult women were found to have significantly lower *KCNQ1* and *KCNH2* mRNA levels (26). This difference in gene expression, likely modulated by sex hormones to some extent, may partially account for the sex differences in phenotype. However, males with cardiovascular disease were noted to have decreased mRNA expression of *KCNQ1* and *KCNH2* when compared to healthy males, suggesting that pathological processes can also feedback to the genetic transcription machinery regardless of genetic architecture.

Unlike other ECG traits, besides having a known genetic basis, the QT interval phenotype also has numerous identified environmental modifiers. Therefore, most agree with the theory that the QT phenotype and its consequences are the cumulative result of multiple modifiable factors (such as ischemia, drug therapy, and so on) interacting with a complex fixed genetic substrate (including rare variants with strong effects and common variants with modest effects) to cause some degree of reduced repolarization reserve that translates to the prolonged QT interval phenotype (27).

International studies of families with the rare inherited LQTS (prevalence estimated at 1:2000 to 1:5000) have established that approximately 95% of the familial mutations involve coding regions of *KCNQ1*, *KCNH2*, or *SCN5A*. However, given the multiple redundant cellular pathways to repolarization, a single mutation is most likely not enough to establish the phenotypic of QT prolongation in all individuals. Early observations of LQTS families demonstrated that approximately one-third of the family members with no phenotypic ECG evidence of LQTS actually carried the same familial mutation as the clinically affected proband (3). Furthermore, the phenotype can be effectively "unmasked" by exposing family members with known mutations to environmental modifiers such as epinephrine (in the case of *KCNQ1* mutation) or HERG-altering drug such as erythromycin (in the case of *KCNH2* mutation). Not surprisingly, one study demonstrated that 10% to 15% of patients presumed to have drug-induced QT prolongation actually had an identifiable mutation in the coding regions of *KCNQ1*, *KCNH2*, or *SCN5A*.

Certainly the variable phenotypic expression seen among family members with a common mutation could be justified by differences in environmental modifiers such as age, sex, comorbidities, or drug exposure. However, further investigation into probands has made a more convincing case that differences in genetic predisposition may provide an alternative explanation. For instance, one family study examined a proband presenting with ventricular fibrillation and ECG demonstrating a prolonged QT interval (in the setting of QT-prolonging drug) that was found to have a rare *KCNH2* mutation shared within the family in an autosomal dominance pattern (28). The researchers also discovered the proband carried a common variant on the nonmutant *KCNH2* allele that was not shared by the other family members with the rare mutation, illustrating the combined phenotypic effect of a rare *KCNH2* mutation, a common *KCNH2* variant, and culprit drug exposure cumulatively contributing to the severe reduction in repolarization reserve noted in the proband. Similar experiments and findings have been reported involving probands within families with rare and common variants involving the *KCNQ1* and *SCN5A* mutations (29,30).

The search for common allelic variants contributing to the complex phenotype of QT interval has included some very large GWAS populations that have served to add to the list of potential candidate genes underlying repolarization reserve and QT duration. In some of the earliest studies looking at smaller populations, a common variant was identified in the *NOS1AP* locus that was significantly associated with the QT interval (31). The gene *NOS1AP* is known to encode a protein important for neuronal nitric oxide synthase regulation but its role in modulating cardiac repolarization remains poorly understood. Within the past few years, three very large GWAS involving a combined total of 42,632 participants (two studies in individuals of European ancestry and one study in individuals of African ancestry) have identified common variants associated with the QT interval (32–34). Table 12.4 provides a selective list of those minor allelic variants from each locus with the strongest association with QT interval (some loci harboring numerous associated variants not represented in the table). Once again, the associated change in QT interval ranges from 1 to 3 ms for each identified allelic variant, exemplifying the modest individual effect on the QT interval phenotype. *NOS1AP* variants have also been associated with cardiac events in patients with congenital LQTS, demonstrating the impact of common variants on a disease due to rare mutations.

The powerful independent association of *NOS1AP* with QT interval is noteworthy, particularly since

TABLE 12.4 Select Genes Within Loci Associated With the QT Interval and Their Proposed Function

Locus	Proposed Protein Function	Allele Frequency	P Value
NOS1AP[a]	Nitric oxide synthase 1	0.26	2×10^{-78}
CNOT1	RNA transcription	0.24	3×10^{-25}
PLN	Calcium-handling phospholamban protein	0.49	5×10^{-22}
KCNQ1[b]	Alpha subunit, slow inward-rectifying K$^+$ channel (I_{Ks})	0.21	3×10^{-17}
KCNH2[b]	Alpha subunit, rapid inward-rectifying K$^+$ channel (I_{Kr})	0.22	5×10^{-17}
RNF207	Ring finger protein	0.28	1×10^{-17}
ATP1B1[a]	Beta-subunit, Na/K-ATPase	0.13	1×10^{-15}
LITAF	Tumor necrosis factor	0.49	5×10^{-15}
SCN5A[b]	Cardiac Na$^+$ (Na$_v$1.5) channel	0.34	1×10^{-14}
KCNJ2[b]	Late inward-rectifying K$^+$ channel Kir2.1 (I_{K1})	0.35	6×10^{-12}
LIG3	DNA ligase III	0.46	6×10^{-12}
KCNE1[b]	Beta subunit slow inward-rectifying K$^+$ channel (I_{Ks})	0.01	2×10^{-08}

[a]Association with QT interval validated in individuals of African descent.
[b]Gene loci encoding known cardiac ion channels associated with QT interval in long QT syndrome studies.

this is not a locus known to harbor a DNA segment directly coding for a cardiac ion channel. That said, a variety of known cardiac ion-channel loci (*KCNQ1*, *KCNH2*, *SCN5A*, *KCNJ2*, *KCNE1*) contained common allelic variants associated with QT interval, which is to be expected given the multiple previously identified family mutations involving these genes. Finally, *NOS1AP* and *ATP1B1* variants are unique in that their association with QT interval is preserved throughout all of the major population-based studies. In particular, nitric oxide pathways (as they relate to *NOS1AP*) and energy metabolism (as they relate to *ATP1B1*) are pathways modulated significantly by underlying pathophysiology such as heart failure, ischemia, and ventricular hypertrophy. The preserved genome-wide association with QT interval among a racially diverse population involving these two novel loci is unlikely a coincidence and should prompt considerable further investigation into the underlying mechanism.

LOCI ASSOCIATED WITH EARLY REPOLARIZATION

The term early repolarization pattern (ERP) generally refers to the presence of J-waves (QRS–ST junction or J-point elevation) coinciding with ST-segment elevations on the ECG. ERP has been defined as J-wave elevation of at least 1 mV in at least two contiguous inferior or lateral leads. It should be reinforced that ERP remains a common ECG trait (a population prevalence of 1%–13%) with an increased frequency found among African Americans and athletes. For decades, since it was first described by Grant et al in 1951, the general consensus among the cardiovascular community had been that ERP is a normal variant and was believed to be benign (35). However, when two studies published in 2008 demonstrated an association between the ERP phenotype and both sudden cardiac death and idiopathic ventricular fibrillation (36,37), the interest in a potential pathophysiological mechanism underlying ERP emerged. In addition, some ERP phenotypes appear to be less benign than others, particularly those with horizontal or downsloping ST-segments (38).

Thus, the heritability of ERP has become a topic of burgeoning interest. Within the Framingham and Health 2000 Survey populations, ERP prevalence ranged from 3% to 6% and was associated with male sex and younger age. But most notably, siblings of individuals with ERP had an increased ERP prevalence of 11.6% and an increased probability of ERP (odds ratio 2.22, $P = .047$), suggesting a significant genetic contribution to the ERP phenotype (39). Mutations involving genes encoding Na$^+$, K$^+$, and Ca^{2+} ion channels have been associated with the ERP phenotype, including *SCN5A*, *KCNJ8*, *ABCC9*, *CACNA1C*, *CACNB2b*, and *CACNA2D1* (40). Sometimes there is gene-specific phenotype variability or overlap, in which a specific mutation (*KCNJ8*, for example) may manifest as a Brugada-like ECG phenotype pattern in one patient and an ERP pattern in a different patient (41). There has also been an interest in identifying common variants associated with ERP but these efforts have been less conclusive than GWAS analyses involving other more common ECG traits. The largest population study investigating this question included a discovery cohort of 7,482 individuals from the Framingham Heart Study, the Health 2000 Study, and the KORA F4 Study (42). A horizontal or descending ST-segment morphology was observed in 71% of the 452 individuals demonstrating ERP. Using logistic regression to demonstrate strength of association with ERP, a total of eight SNPs were discovered in LD with eight different candidate genes. However, the authors' replication cohort involving 7,151 individuals was unable to reproduce the associations between any of the discovered SNPs and ERP. The authors concluded that a lack of power may have explained these results, particularly since less than 10% of the nearly 15,000 participants had ERP. Furthermore, a significant proportion of the individuals with ERP had the upsloping "benign" ST-segment phenotype, which may represent a different genetic substrate than those with horizontal or downsloping ST-segments.

One particular ERP-associated SNP locus, *KCND3* (encoding the voltage-gated potassium channel $K_v4.3$ underlying the transient outward current I_{to}), may represent a promising candidate gene as it is known to be important for the early phase of repolarization and Phase I of the cardiac action potential. *KCND3* mutations have also been implicated in rare cases of LQTS and Brugada syndromes, in which repolarization kinetics are also markedly deranged. Regardless, the inability to replicate this association, along with the other discovery alleles, warrants further exploration in future GWAS populations. Until additional studies are able to validate previous hypothesis-generating observations, the genetic architecture of the ERP phenotype may remain poorly elucidated.

LOCI ASSOCIATED WITH HEART RATE (RR INTERVAL)

The resting heart rate, measurable on the ECG as the RR interval, represents an important complex trait since the RR interval and heart rate variability have important implications for prognosis. The increase in sympathetic tone and/or the loss in parasympathetic tone are known adverse contributors to elevation in resting heart rate (and decrease in heart rate variability) but the heritability of the RR interval is only recently becoming better elucidated. Based on twin studies, heart rate heritability is estimated to comprise 55% to 77% of the trait (43). The heart rate phenotype is in some ways more challenging to characterize because it reflects both sinoatrial function

and also encompasses alterations in the previously described PR, QRS, and QT intervals. For this reason, there have been few candidate gene studies able to provide insight into the RR phenotype. Much of what is being discovered about RR interval heritability comes from contemporary GWAS, which have included over 200,000 individuals in sum (44–47). Similar to other complex ECG traits influenced by common variants, for all GWAS-identified alleles in common SNPs (MAF greater than 10%) significantly associated with the RR interval, the estimated genetic contribution to the phenotype is less than 1% and usually less than 10 ms in effect size.

Two distinct loci (well-replicated by multiple GWAS involving ethnically diverse populations) have consistently been associated with the RR interval. The locus *GJA1* encoding a connexin family protein Cx43 has had the strongest association with resting heart rate in populations of both European (45) and African (47) ancestry. This protein is a major structural component of gap junctions, which are integral to the electrical coupling of cardiomyocytes. Of note, this gene has also been associated with other ECG traits in a few studies, including the QRS and QT intervals as well as a decreased risk of atrial fibrillation (46). Also, connexin43 is expressed most abundantly in the atria, which may partially explain its importance for sinus node function. Another locus highly associated with RR interval, *MYH6* (and nearby *MYH7* in one study) is known to encode a cardiac microRNA (miR-208a) that probably plays a regulatory role in cardiac conduction. Specifically, miR-208a is involved with regulation of connexin40 expression, another gap junction protein important for sinus node function. The loss of gap junction units (which could be a direct result of mutations involving connexin genes) has been associated with a sick sinus syndrome phenotype (bradycardia and sinus node exit block) in experimental models (48). *MYH6* also encodes a myosin heavy chain subunit important for the sarcomere and the cardiac contractile system. While not well-replicated in multiple studies, this same locus has also been associated with the PR interval—again, reinforcing the complexity of the RR interval as an ECG trait.

CASE RESOLUTION

The patient described earlier is an adult female who presents after sudden cardiac arrest. The ECG is notable for a corrected QT interval of 470 ms, which is in the borderline range for making the diagnosis of LQTS. The QT interval is also influenced by the use of methadone and the patient should be transitioned to an alternative narcotic. The patient does carry the *KCNH2* mutation but she does not harbor the K897T polymorphism shared by her affected sister and other members of her family (Figure 12.1), so her phenotype is likely less severe than her sister's. While it would not be unreasonable to implant an implantable cardioverter-defibrillator (ICD) based on her history of sudden cardiac arrest, it would also be appropriate in her case to hold the methadone and start a beta-blocker since this intervention alone is associated with a greater than 60% relative risk reduction in adverse events (syncope or sudden death) for adult female patients with a *KCNH2* mutation. If her QTc normalizes and she has no subsequent events while avoiding QT-prolonging drugs, the ICD could be safely deferred.

CONCLUSIONS

In recent years, there has been a wealth of studies on the genetics of complex ECG traits. While it is clear now that virtually every ECG trait has a variable genetic contribution, we have also grown to appreciate the role for numerous modifiers of genetic effects. The discoveries from both Mendelian studies and GWAS have elucidated candidate genes with direct and indirect cardiac electrophysiological effects. The important next stage of investigation will require exploring each of these loci to understand the mechanisms underlying the genotype–phenotype associations. Finally, the role of noncoding RNAs, splice variants, epigenetics, microRNAs, and posttranslational modifications of encoded proteins remains to be determined. Further understanding the genetics of ECG traits will continue to enhance our understanding of molecular electrophysiology and will hopefully provide new avenues for therapeutics–diagnostics ("theranostics")—personalized medicines that can be tailored to the individual patient based on genotype and ECG phenotype.

REFERENCES

1. Marian AJ, Belmont J. Strategic approaches to unraveling genetic causes of cardiovascular diseases. *Circ Res.* 2011;108:1252–1269.
2. Schork NJ, Murray SS, Frazer KA, Topol EJ. Common vs. rare allele hypotheses for complex diseases. *Curr Opin Genet Dev.* 2009;19:212–219.
3. Priori SG, Napolitano C, Schwartz PJ. Low penetrance in the long-qt syndrome: clinical impact. *Circulation.* 1999;99:529–533.
4. Schwartz PJ, Ackerman MJ, George AL, Jr, Wilde AA. Impact of genetics on the clinical management of channelopathies. *J Am Coll Cardiol.* 2013;62:169–180.
5. Moss AJ, Zareba W, Benhorin J, et al. ECG t-wave patterns in genetically distinct forms of the hereditary long qt syndrome. *Circulation.* 1995;92:2929–2934
6. Wilde AAM, Bezzina CR. Genetics of cardiac arrhythmias. *Heart.* 2005;91:1352–1358.
7. Goldenberg I, Moss AJ. Long qt syndrome. *J Am Coll Cardiol.* 2008;51:2291–2300
8. Delpón E, Cordeiro JM, Núñez L, et al. Functional effects of kcne3 mutation and its role in the development of brugada syndrome. *Circ Arrhythm Electrophysiol.* 2008;1:209–218.
9. Yang L, Katchman AN, Samad T, et al. B-adrenergic regulation of the L-type Ca2+ channel does not require phosphorylation of alpha1c ser1700. *Circ Res.* 2013;113(7):871–880.
10. Bround MJ, Asghari P, Wambolt RB, et al. Cardiac ryanodine receptors control heart rate and rhythmicity in adult mice. *Cardiovasc Res.* 2012;96:372–380.
11. Lyon AR, Bannister ML, Collins T, et al. Serca2a gene transfer decreases sarcoplasmic reticulum calcium leak and reduces ventricular arrhythmias in a model of chronic heart failure. *Circ Arrhythm Electrophysiol.* 2011;4:362–372.
12. Havlik RJ, Garrison RJ, Fabsitz R, Feinleib M. Variability of heart rate, p-r, qrs and q-t durations in twins. *J Electrocardiol.* 1980;13:45–48.
13. Schott JJ, Alshinawi C, Kyndt F, et al. Cardiac conduction defects associate with mutations in scn5a. *Nat Genet.* 1999;23:20–21.
14. Holm H, Gudbjartsson DF, Arnar DO, et al. Several common variants modulate heart rate, pr interval and qrs duration. *Nat Genet.* 2010;42:117–122.
15. Pfeufer A, van Noord C, Marciante KD, et al. Genome-wide association study of pr interval. *Nat Genet.* 2010;42:153–159.
16. Butler AM, Yin X, Evans DS, et al. Novel loci associated with pr interval in a genome-wide association study of 10 African American cohorts. *Circ Cardiovasc Genet.* 2012;5:639–646.
17. Chambers JC, Zhao J, Terracciano CMN, et al. Genetic variation in scn10a influences cardiac conduction. *Nat Genet.* 2010;42:149–152.
18. Hanson B, Tuna N, Bouchard T, et al. Genetic factors in the electrocardiogram and heart rate of twins reared apart and together. *Am J Cardiol.* 1989;63:606–609.
19. Bezzina CR, Shimizu W, Yang P, et al. Common sodium channel promoter haplotype in asian subjects underlies variability in cardiac conduction. *Circulation.* 2006;113:338–344.
20. Sotoodehnia N, Isaacs A, de Bakker PIW, et al. Common variants in 22 loci are associated with qrs duration and cardiac ventricular conduction. *Nat Genet.* 2010;42:1068–1076.
21. Ritchie MD, Denny JC, Zuvich RL, et al. Genome- and phenome-wide analyses of cardiac conduction identifies markers of arrhythmia risk. *Circulation.* 2013;127:1377–1385.
22. Shy D, Gillet L, Abriel H. Cardiac sodium channel nav1.5 distribution in myocytes via interacting proteins: the multiple pool model. *Biochim Biophys Acta.* 2013;1833:886–894.
23. Sauer AJ, Newton-Cheh C. Clinical and genetic determinants of torsade de pointes risk. *Circulation.* 2012;125:1684–1694.
24. Hobbs JB, Peterson DR, Moss AJ, et al. Risk of aborted cardiac arrest or sudden cardiac death during adolescence in the long-qt syndrome. *JAMA.* 2006;296:1249–1254.

25. Sauer AJ, Moss AJ, McNitt S, et al. Long qt syndrome in adults. *J Am Coll Cardiol.* 2007;49:329–337.

26. Moric-Janiszewska E, Glogowska-Ligus J, Paul-Samojedny M, et al. Age-and sex-dependent mrna expression of kcnq1 and herg in patients with long qt syndrome type 1 and 2. *Arch Med Sci.* 2011;7:941–947.

27. Roden DM. Taking the "idio" out of "idiosyncratic": predicting torsades de pointes. *Pacing Clin Electrophysiol.* 1998;21:1029–1034.

28. Crotti L, Lundquist AL, Insolia R, et al. Kcnh2-k897t is a genetic modifier of latent congenital long-qt syndrome. *Circulation.* 2005;112:1251–1258.

29. Terrenoire C, Wang K, Chan Tung KW, et al. Induced pluripotent stem cells used to reveal drug actions in a long qt syndrome family with complex genetics. *J Gen Physiol.* 2013;141:61–72.

30. Crotti L, Monti MC, Insolia R, et al. Nos1ap is a genetic modifier of the long-qt syndrome. *Circulation.* 2009;120:1657–1663.

31. Arking DE, Pfeufer A, Post W, et al. A common genetic variant in the nos1 regulator nos1ap modulates cardiac repolarization. *Nat Genet.* 2006;38:644–651.

32. Pfeufer A, Sanna S, Arking DE, et al. Common variants at ten loci modulate the qt interval duration in the qtscd study. *Nat Genet.* 2009;41:407–414.

33. Newton-Cheh C, Eijgelsheim M, Rice KM, et al. Common variants at ten loci influence qt interval duration in the qtgen study. *Nat Genet.* 2009;41:399–406

34. Smith JG, Avery CL, Evans DS, et al. Impact of ancestry and common genetic variants on qt interval in african americans. *Circ Cardiovasc Genet.* 2012;5:647–655.

35. Grant RP, Estes EH, Doyle JT. Spatial vector electrocardiography: The clinical characteristics of s-t and t vectors. *Circulation.* 1951;3:182–197.

36. Haïssaguerre M, Derval N, Sacher F, et al. Sudden cardiac arrest associated with early repolarization. *N Engl J Med.* 2008;358:2016–2023.

37. Rosso R, Kogan E, Belhassen B, et al. J-point elevation in survivors of primary ventricular fibrillation and matched control subjects: incidence and clinical significance. *J Am Coll Cardiol.* 2008;52:1231–1238.

38. Tikkanen JT, Junttila MJ, Anttonen O, et al. Early repolarization: electrocardiographic phenotypes associated with favorable long-term outcome. *Circulation.* 2011;123:2666–2673.

39. Noseworthy PA, Tikkanen JT, Porthan K, et al. The early repolarization pattern in the general population: clinical correlates and heritability. *J Am Coll Cardiol.* 2011;57:2284–2289.

40. Antzelevitch C. Genetic, molecular and cellular mechanisms underlying the j wave syndromes. *Circ J.* 2012;76:1054–1065.

41. Barajas-Martinez H, Hu D, Ferrer T, et al. Molecular genetic and functional association of brugada and early repolarization syndromes with s422l missense mutation in kcnj8. *Heart Rhythm.* 2012;9:548–555.

42. Sinner MF, Porthan K, Noseworthy PA, et al. A meta-analysis of genome-wide association studies of the electrocardiographic early repolarization pattern. *Heart Rhythm.* 2012;9:1627–1634.

43. Russell MW, Law I, Sholinsky P, Fabsitz RR. Heritability of ECG measurements in adult male twins. *J Electrocardiol.* 1998;30:64–68.

44. Cho YS, Go MJ, Kim YJ, et al. A large-scale genome-wide association study of asian populations uncovers genetic factors influencing eight quantitative traits. *Nat Genet.* 2009;41:527–534.

45. Eijgelsheim M, Newton-Cheh C, Sotoodehnia N, et al. Genome-wide association analysis identifies multiple loci related to resting heart rate. *Hum Mol Genet.* 2010;19:3885–3894.

46. den Hoed M, Eijgelsheim M, Esko T, et al. Identification of heart rate-associated loci and their effects on cardiac conduction and rhythm disorders. *Nat Genet.* 2013;45:621–631.

47. Deo R, Nalls MA, Avery CL, et al. Common genetic variation near the connexin-43 gene is associated with resting heart rate in african americans: a genome-wide association study of 13,372 participants. *Heart Rhythm.* 2013;10:401–408.

48. Hagendorff A, Schumacher B, Kirchhoff S, et al. Conduction disturbances and increased atrial vulnerability in connexin40-deficient mice analyzed by transesophageal stimulation. *Circulation.* 1999;99:1508–1515.

Atrial Fibrillation Genetics and Genomics

Amir Y. Shaikh, Steven A. Lubitz, Honghuang Lin, Emelia J. Benjamin, and David D. McManus

TAKE HOME POINTS

1. Atrial fibrillation (AF) is a heritable condition, and at least nine susceptibility loci have been identified by genome-wide association studies (GWAS); however, the identified loci explain only a small proportion of the heritability of AF.
2. Of the AF susceptibility loci identified by GWAS, the 4q25 locus (sentinel single nucleotide polymorphism [SNP] rs6817105, nearest gene *PITX2*) has the strongest known association with AF ($P = 1.8 \times 10^{-74}$; relative risk 1.64). *PITX2* encodes a transcription factor that plays a critical role in cardiogenesis—specifically in determining cardiac left- and right-sidedness and in the formation of the interatrial septum, SA node, and pulmonary vein myocytes.
3. "Omic" studies, still in early development and largely confined to animal studies and small, selected cohorts with AF, will most likely provide new avenues for identifying cellular and developmental pathways for AF, especially when combined together in a systemic biology approach.

CASE STUDY

A 34-year-old, previously healthy woman presented to her primary care physician with palpitations. She stated that she was in her usual state of health until 3 to 4 weeks prior to her appointment when she developed palpitations while jogging in her neighborhood. She felt slightly short of breath during the palpitations and felt an "irregular heart beat." These symptoms lasted a few minutes and then resolved. Over the past 3 to 4 weeks, her symptoms became more frequent and longer, lasting a few hours at a time. She denied lightheadedness, dizziness, chest pain, syncope, leg swelling, or any other symptoms.

The patient had no prior medical problems and maintained an active lifestyle, exercising regularly 3 to 4 times per week. She worked as a manager of a restaurant and had two small children. She denied alcohol, tobacco, or illicit drug use. Her family history was notable for paroxysmal AF in two older brothers (ages 38 and 42 at the time of AF diagnosis). Her mother was 68 years old and had a stroke a few years ago and was found to have AF at that time.

On physical examination, the patient had a normal blood pressure (118/74 mmHg) with a heart rate of 108 bpm. She appeared comfortable and in no distress. Her jugular venous pressure was normal and her lungs were clear to auscultation bilaterally. On cardiac examination, she had a normal point of maximal impulse, a tachycardic, irregularly irregular rhythm, and no murmurs. There were no extra heart sounds. Her abdominal, extremity, skin, and neurological exams were benign. Laboratory testing did not reveal any abnormalities, and a pregnancy test was negative.

A 12-lead electrocardiogram demonstrated AF with a heart rate of 117 bpm (Figure 13.1). An echocardiogram was subsequently performed. It showed normal

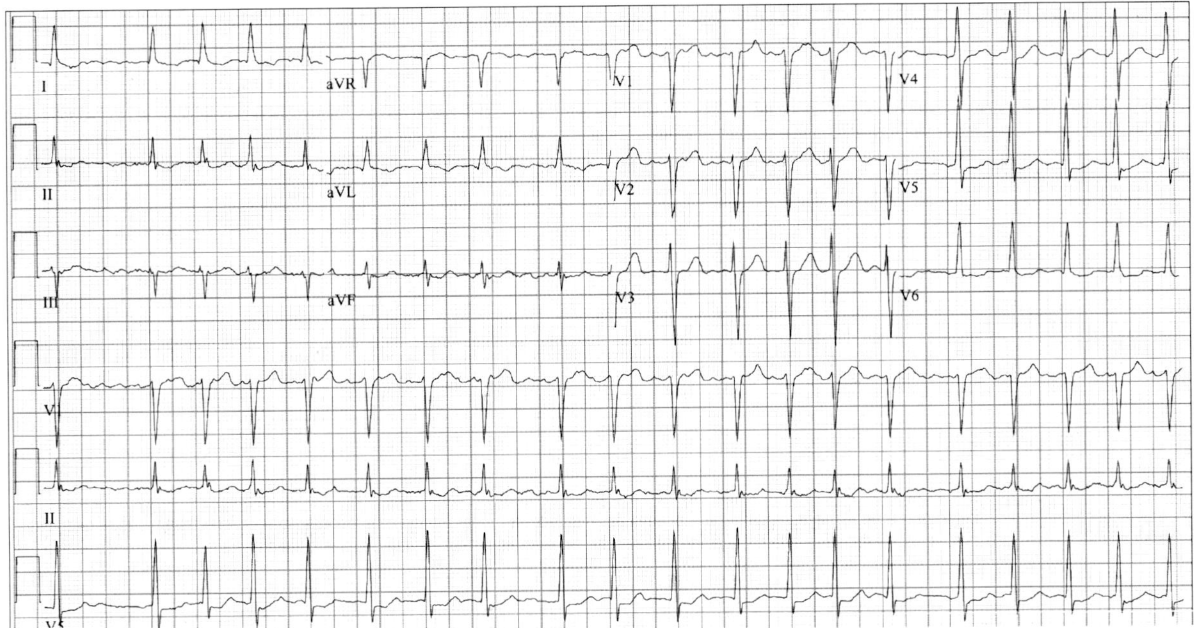

FIGURE 13.1 Electrocardiogram of the patient presented in the case study demonstrating atrial fibrillation.

cardiac chamber dimensions except for borderline left atrial enlargement. There was no evidence of left ventricular hypertrophy, and no significant valvular disease. The patient was diagnosed with lone AF and treated with aspirin and a beta-blocker. A 30-day event monitor showed that the patient spontaneously cardioverted to normal sinus rhythm 2 days following her appointment; over the remainder of the monitoring period she developed two brief episodes of paroxysmal AF.

Familial AF is rare. However, experimental studies of families with hereditary AF have identified monogenic disorders (primarily of cardiac ion channels) that result in atrial arrhythmias. For example, Ellinor and colleagues screened 57 probands with a family history of AF for mutations in the coding region of the cardiac sodium channel gene *SCN5A* (1). These investigators identified a single mutation (N1986K) that was present in one of the families studied. In order to further explore the functional significance of this mutation, the mutant N1986K sodium channel was expressed in *Xenopus* oocytes and compared to wild-type oocytes. These experiments revealed a hyperpolarizing shift in steady-stage inactivation of the N1986K mutant sodium channel. In another study, Das and colleagues studied a family containing multiple members with early-onset lone AF and found a serine-to-proline (S209P) mutation in the *KCNQ1* gene, which resulted in a gain-of-function effect with increased I_{Ks} current (2).

While genetic mutations as a cause of AF are rare, studies of familial AF like the ones described demonstrate the ability to: (a) dissect monogenic disorders in familial AF; and (b) study the functional electrophysiological consequences of the identified genetic mutations. In the current era of exome sequencing and whole-genome sequencing, genetic studies of families with AF (such as

the case study) will most likely become more commonplace. These studies should provide additional insight into the molecular pathogenesis of atrial arrhythmias, and will be complementary to the variety of "omics" technologies currently being used to explore the underlying molecular basis and pathophysiology of more common forms of AF.

Atrial fibrillation (AF) is the most common sustained arrhythmia in humans, and is associated with considerable morbidity, mortality, and health care costs. While the familial clustering of AF has been long recognized, advances in genetics and genomics have allowed for the increased understanding of both rare (monogenic) forms of AF and common (polygenic) forms of AF. Here we describe the heritability and genetic basis for AF; candidate gene and genome-wide association studies of AF; the problem of "missing heritability" in AF; future directions in the study of AF, including additional "omics" of AF such as transcriptomics, proteomics, and systems biology; and finally, the integration of genetic information into the clinical practice of AF.

AF HERITABILITY AND GENETIC BASIS FOR AF

Familial clustering of AF has been reported since the 1940s (3,4), though such families were typically regarded as rare occurrences. Indeed, genetic mapping in families with Mendelian AF led to the identification of mutations in genes such as cardiac expressed ion channels or in genes affecting myocyte repolarization (5–7).

The observation that genes involved in cardiac conduction and repolarization might be involved in AF pathogenesis prompted candidate gene association studies. However, the results of widespread candidate gene screening efforts and association testing in larger AF study samples revealed such mutations were rare (6). Furthermore, observed associations in candidate gene association studies were generally not replicable. In aggregate, the results of linkage analysis and early candidate gene resequencing efforts implied that variants identified in families with a high prevalence of AF were likely to be private to those specific families.

Novel epidemiological insights over the past decade led to the realization that widespread heritability underlying more typical forms of AF exists. In 2004, investigators from the community-based Framingham Heart Study reported that a parental history of AF was associated with an increased risk of AF in offspring, irrespective of clinical factors such as hypertension, diabetes, and heart disease (8). Later work in the Framingham Heart Study demonstrated that those with a first-degree relative with AF have about a 40% increased risk of AF, and that the risk of AF is increased twofold if the relative develops AF by 65 years of age (9). Familial AF remained associated with AF risk after adjustment for established clinical AF risk factors and led to a small improvement in the ability to discriminate AF risk.

Similar observations were made by investigators in Iceland who leveraged genealogical data to demonstrate that AF risk decreases as the degree of relatedness to a relative with AF decreases (10). Observations in twin studies from Denmark also support a heritable component underlying AF (11), as did additional studies in individuals with lone AF (12,13).

Observations of a widespread heritable component underlying AF prompted the search for common genetic variation underlying AF. The first GWAS of AF was performed in Icelanders, and demonstrated an AF susceptibility locus on chromosome 4q25 (14). The variants associated with AF at this locus resided in a noncoding region about 150 kilobases upstream of *PITX2*, a transcription factor important for determining cardiac left- and right-sidedness, pulmonary vein myocyte specification, and suppressing the default formation of a sinus node in the left atrium (15,16).

Since the seminal Icelandic AF GWAS, four others as well as a large-scale candidate gene approach have been conducted for AF in individuals of European ancestry. Currently, a total of ten AF susceptibility loci have been associated with AF—nine identified by GWAS and one by candidate gene study. The described loci implicate cardiac expressed ion channels, transcription factors involved in cardiac and pulmonary development, inflammation, and yet other pathways that have yet to be characterized (Table 13.1) (14,17–21). Moreover, fine mapping at the chromosome 4q25 locus in a sample of almost 6,000 individuals with AF and about 30,000 without AF identified multiple susceptibility variants at the 4q25 locus that associate with a substantial gradient of AF risk (22). A sixfold increased risk of AF was observed in the 1% of individuals homozygous for AF risk alleles at just three genetic variants at the locus relative to those with the average combination of AF risk alleles (22).

The GWAS findings implicate that whereas Mendelian forms of AF exist, more common forms of AF observed in the community reflect a complex trait with polygenic susceptibility. Even in individuals with the less common early-onset forms of AF, evidence supports a polygenic contribution. Furthermore, common AF susceptibility variants identified recently associate with a substantial gradient of AF risk.

TABLE 13.1 Common Variant AF Susceptibility Loci Identified in Large-Scale Candidate Gene or Genome-Wide Association Studies

Chromosomal Locus	Candidate Gene	Approach	Reference
1q21	*IL6R*	Candidate gene	(21)
1q21	*KCNN3*	GWAS	(19,20)
1q24	*PRRX1*	GWAS	(19)
4q25	*PITX2*	GWAS	(14,19)
7q31	*CAV1*	GWAS	(19)
9q22	*C9orf3*	GWAS	(19)
10q22	*SYNPO2L/MYOZ1*	GWAS	(19)
14q23	*SYNE2*	GWAS	(19)
15q24	*HCN4*	GWAS	(19)
16q22	*ZFHX3*	GWAS	(17–19)

Abbreviation: GWAS, genome-wide association study.

14

Inherited Ventricular Arrhythmias

Brittany Weber and Rajat Deo

TAKE HOME POINTS

1. The inherited arrhythmic disorders constitute approximately one-half of unexplained cardiac arrests and include the congenital long QT syndrome (LQTS), Brugada syndrome (BrS), catecholaminergic polymorphic ventricular tachycardia (CPVT), and short QT syndrome (SQTS). Although heterogeneous, these disorders are marked by the dysregulation of critical cardiac ion channels important for the genesis, propagation, and repolarization of the electrical signal.
2. The discovery of the genetic mutations underlying inherited arrhythmic disorders and the rapidly advancing understanding of the correlations between genotype and phenotype have significantly impacted the diagnosis of these disorders and the ability to identify and treat clinically silent family members who inherit the genetic abnormality.
3. The electrocardiogram (ECG) is a critical tool as it is often the "portal of entry" to an arrhythmic disorder; however, the ECGs in the postarrest phase of resuscitated sudden cardiac death (SCD) may demonstrate long QT or early repolarization changes due to metabolic disturbances and cardiac stunning; as a result, they may not be accurate for diagnosing specific conditions. Further, the absence of findings on one ECG is not infrequent and a disorder should not be "ruled out" based on a normal ECG. Nevertheless, cardiovascular clinicians should be well aware of the ECG abnormalities that provide clues to the presence of inherited arrhythmic disorders.

CASE STUDY

A 16-year-old boy presents to his pediatrician with symptoms of recurrent exercise-induced palpitations and lightheadedness and occasional syncope. The patient states that for the past few years, he has had multiple episodes of palpitations and lightheadedness when he tries to play sports. He has also had two to three episodes of syncope over the past few years during extreme exertion or emotional outbursts. He regained consciousness shortly after these episodes and they seemed to improve with rest and hydration. However, his symptoms are now becoming more frequent, and they are impeding his ability to play varsity sports at his high school.

Upon further questioning, the patient states that his father died suddenly at the age of 36, but his mother is alive and healthy. He has no siblings. He denies any alcohol or illicit drug use. Physical examination, electrocardiography, and echocardiography are all normal. In order to reproduce his symptoms and determine whether exercise-induced arrhythmias are present, the patient undergoes exercise treadmill testing on a Bruce protocol. During Stage 4 of the exercise protocol, he develops palpitations and lightheadedness. The ECG at the time of his symptoms demonstrates long runs of polymorphic ventricular tachycardia (VT). At times, bidirectional VT (beat-to-beat alteration in the frontal QRS axis) is present in the exercise ECG tracing.

After the exercise test, the patient is started on propranolol with gradual uptitration of the dose. He has greatly improved symptoms and no longer suffers from exercise- or emotion-induced lightheadedness, palpitations, or syncope. The patient subsequently undergoes genetic counseling and genetic testing, revealing a mutation in the *RYR2* gene. The patient was diagnosed with CPVT. His mother wondered whether an implantable cardioverter-defibrillator (ICD) was indicated, and whether her son should be allowed to participate in sports. The patient's physicians explained the risks and benefits of ICD implantation and participation in sports. The decision was made to defer ICD implantation and to avoid competitive sports, but to reevaluate after 6 months of beta-blocker therapy and repeated exercise treadmill testing.

Sudden cardiac death (SCD) refers to an unexpected death from a cardiovascular cause in a person with or without pre-existing heart disease. Most cases are associated with a witnessed collapse, death occurring within 1 hour of an acute change in clinical symptoms, or unexpected death that occurred within the previous 24 hours (1,2). SCD is a leading cause of mortality worldwide, and its incidence in the United States ranges between 180,000 and 450,000 cases annually. As such, SCD represents a large fraction of total mortality and a substantial health burden. The most common electrophysiological mechanism for SCD is ventricular fibrillation (VF) and the most prevalent underlying pathology is coronary heart disease (CHD) demonstrated by both clinical and autopsy studies (3). Epidemiological data have also demonstrated that 5% to 10% of total SCD occurs in the absence of CHD or structural heart disease, and this smaller subset often includes children and younger adults (4), suggesting an underlying genetic cause of SCD in these individuals.

The inheritable arrhythmic disorders (IADs) constitute approximately one-half of unexplained cardiac arrests (5) and include the congenital LQTS, BrS, CPVT, and SQTS. Although heterogeneous, these disorders are marked by the dysregulation of critical cardiac ion channels important for the genesis, propagation, and repolarization of the electrical signal. The discovery of these genetic mutations and the rapidly advancing understanding of the correlations between genotype and phenotype have significantly impacted the diagnosis and ability to identify and treat clinically silent family members who inherit the genetic abnormality.

LONG QT SYNDROME (LQTS)

The LQTS is a heritable arrhythmic disease that is characterized by a prolonged QT interval and may trigger ventricular arrhythmias resulting in SCD. Individuals with LQTS have a structurally normal heart and may experience syncope and cardiac arrest. These conditions are often triggered by adrenergic stress; however, approximately 10% to 15% of patients experience symptoms at rest (6). Long QT disease-causing mutations have been identified in approximately one-third of infants with a prolonged QT interval and translate to a prevalence of approximately 1:2000 apparently healthy live births (7).

Genetics of the LQTS

Most cases of the LQTS are inherited through an autosomal dominant pattern. Experimental studies have implicated 13 genetic mutations of the LQTS. These abnormalities are known to occur in several different genes including ion channels or proteins such as chaperons and modulators that regulate ion channels. Mutations lead to QT interval prolongation either by impairing repolarizing currents (loss of function) or by increasing depolarizing currents (gain of function). Greater than 80% of genetically positive long QT cases have LQT1, LQT2, or LQT3 genotypes with mutations that involve *KCNQ1*, *KCNH2*, and *SCN5A*, respectively. In approximately 15% to 20% of LQTS cases, genetic testing is negative. Some genetic variants of the LQTS are associated with disease in other organs. For example, Jervell and Lange-Nielsen syndrome, which is autosomal recessive, is characterized by QT prolongation and congenital deafness. Andersen–Tawil syndrome is a QT-prolonging disorder that is associated with facial dysmorphisms and hypokalemic periodic paralysis. Timothy syndrome demonstrates marked QT prolongation, syndactyly, paroxysmal hypoglycemia, atrioventricular block, congenital heart defects, developmental disorders, reduced immune response, and life-threatening arrhythmias (8–10).

The three main genes, *KCNQ1* (LQT1), *KCNH2* (LQT2), and *SCN5A* (LQT3) account for over 90% of all genotype-positive cases whereas the additional minor genes only contribute approximately 5%. LQT1 is the most common form of LQTS (30%) and results from a mutation in the *KCNQ1* gene. This gene encodes the alpha subunit of the potassium channel Kv7.1, which forms the adrenergic-sensitive, I_{Ks} repolarizing current (11). Mutations in *KCNQ1* either reduce the I_{ks} channel density by a loss-of-function mutation or result in a dominant negative effect by interfering with the function of the wild-type form. Two different *KCNQ1* isoforms are expressed: the normal *KCNQ1* subunit (isoform 1) and a second alternatively spliced subunit (isoform 2). Isoform 2 has been shown to produce a dominant negative effect on isoform 1 subunits, and the balance between these isoforms is thought to stabilize the slow component of the cardiac delayed rectifier potassium current (12).

Over 250 predominantly missense mutations have been described in *KCNQ1* (13). Mutations in the *KCNQ1* gene may also have overlapping clinical phenotypes such as atrial fibrillation or sick sinus syndrome (14,15). Alternatively, a gain-of-function mutation in *KCNQ1* can lead to SQTS and is discussed in the section following.

LQT2 is the second most common mutation in LQTS (25%–30%) and is secondary to loss-of-function mutations in *KCNH2* (HERG). This gene encodes the alpha subunit of the I_{Kr} repolarizing potassium channel (16). More than 300 LQT2 mutations have been described. Homozygous mutations are an exception and result in a severe, early-onset phenotype (13,17). In addition, it appears that some patients with QT prolongation in the setting of a specific drug or metabolic disturbance such as hypokalemia or hypomagnesemia may have an underlying genetic susceptibility. This condition reflects acquired LQTS and may be associated with less severe mutations in *KCNH2* (18). Some studies have suggested that these patients are often sensitive to serum potassium levels, which can be modulated with oral potassium supplements (19).

The third subtype, LQT3 (5%–10%), is a result of mutations that disrupt fast inactivation of the cardiac sodium channel *SCN5A* (20). This gain-of-function phenotype leads to the inward sodium current to persist abnormally during the plateau of the cardiac action potential, prolonging the QT interval (21). Mutations in this gene have also been implicated in causing the BrS, atrial and ventricular conduction disorders, and dilated cardiomyopathy.

Genetic Modifiers

Recent evidence has also highlighted the role of common inherited modifiers in LQTS. The high prevalence of concealed LQTS and low penetrance raises the question of whether other factors play a role. In LQT2 patients, coinheritance of a common polymorphism on the opposite allele such as *KCNH2*-K897T, which has a 30% carrier frequency in the Caucasian population, and an underlying *KCNH2*-A116V carrier mutation significantly exaggerated the repolarizing potassium current. This exemplifies how the inheritance of the same mutation in family members can lead to phenotypic heterogeneity based on a common modifier gene (22). Further, in a study that investigated an unusually large South African LQT1 founder population, *NOS1AP* variants were significantly associated with the occurrence of symptoms, clinical severity, and with greater likelihood of having a QT interval in the top 40% of values among all mutation carriers (23).

Clinical Manifestations

The most concerning manifestations of the LQTS are arrhythmic events due to torsades de pointes VT. This arrhythmia can result in syncope, cardiac arrest, and SCD. Early work demonstrated that the clinical manifestations of the LQTS may differ based on the specific genotype. Patients with LQT1 gene mutations have a reduction in myocardial repolarization that is more apparent at higher heart rates. Emotional or physical stress are conditions known to activate adrenergic signaling pathways and may be a clinical clue in the history (24). Patients with the LQT2 subtype can also experience syncope or SCD during periods of stress. Unique to this subtype, however, is the association between sudden loud noises (eg, an alarm clock) and arrhythmic events (25). Patients with LQT3 often experience symptoms at rest and during sleep. In addition to the characteristic QT prolongation, bradycardia and sinus pauses may also occur implicating a role for the sodium channel in the sinoatrial pacemaker (24). Atrial arrhythmias and particularly atrial fibrillation are also more common among LQTS patients (26,27). Despite these commonly described associations and clinical scenarios, it is critical to remember that activity or rest may trigger arrhythmic events in any of the LQT subtypes (Figure 14.1). This information should be used in conjunction with the other diagnostic data when evaluating patients suspected of having a mutation that prolongs the QT interval.

Diagnosis

The diagnosis of LQTS is complex, associated with a constellation of findings, and is primarily made in childhood. The mean age of onset is 12 years and earlier onset is often a marker of a more severe form of the disease (28). The most important tool in the clinical evaluation remains the clinical history. Common presentations of LQTS include palpitations, presyncope, syncope, and cardiac arrest. For every new patient, a clear and detailed personal, family, and medication history should be obtained. An understanding of whether symptoms occur during swimming, running, startling noises (eg, an alarm clock, a loud horn, a ringing phone), strong emotional situations (eg, anger, crying, test taking), and at rest should be obtained. In addition, a detailed family history should include any history of SCD, which may have occurred through drowning, sudden infant death syndrome, and automobile accidents. In women of childbearing age, a detailed peripartum history should be obtained as the risk for cardiac events is reduced during pregnancy but is increased significantly during the 9-month period postpartum (29). The physical exam and echocardiogram often

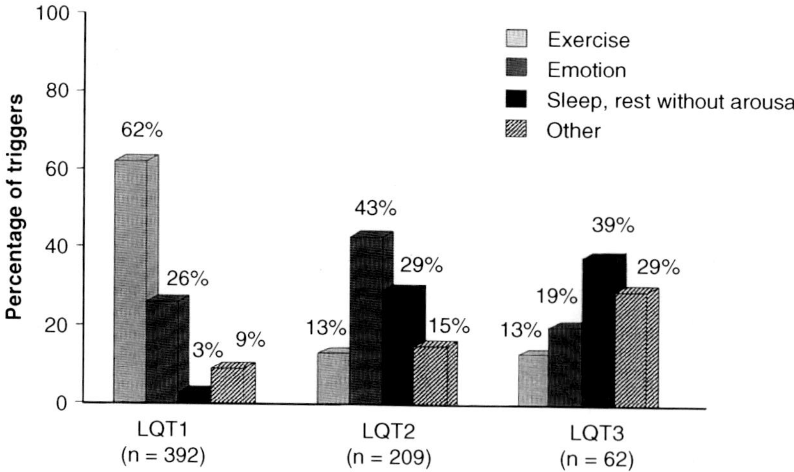

FIGURE 14.1 Triggers for cardiac events according to three genotypes. Numbers in parentheses indicate the number of triggers, not number of patients.

Source: From Ref. (24). Schwartz PJ, Priori SG, Spazzolini C, et al. Genotype-phenotype correlation in the long-QT syndrome: gene-specific triggers for life-threatening arrhythmias. *Circulation*. 2001;103(1):89–95.

show no abnormalities in patients with LQTS but may be helpful in ruling out other causes of symptoms or arrhythmias.

The diagnosis of LQTS is based on the ECG and the measurement of the heart rate corrected QT interval (QT_c). The corrected QT interval is calculated using Bazett's formula: $QT_c = QT/\sqrt{RR}$. Normally, the QT_c should not exceed 450 ms in males and 460 ms in post-pubertal females. LQTS is often characterized by a QT_c that exceeds the 99th percentile, greater than 470 ms for men and greater than 480 ms for woman. However, the ECG is not static and a single ECG may not manifest stereotypical features. Even among patients with a definitive diagnosis of LQTS, studies have found that the QT_c interval is variable with a mean of 47 ± 40 ms on repetitive ECGs (30). In addition, patients affected by congenital LQTS show frequent abnormalities in the T-wave morphology such as diphasic T-waves, notches, low amplitude, or very slow onset (31). It is recommended to measure the QT interval in tracings where the RR interval (or heart rate) is constant for at least 10 to 20 beats. The evaluation of the QT interval duration also requires review at different heart rates. Since Bazett's formula is not linear at fast or slow heart rates, general recommendations have been to exclude tracings with a heart rate greater than 100 beats per minute or less than 50 beats per minute.

Approximately 25% of patients with LQTS confirmed by the presence of a LQTS gene mutation may have a QT_c interval in the normal range (32,33). These patients may have a paradoxical QT response or QT prolongation that occurs with exertion or beta-adrenergic stimulation. Diagnostic testing may include provocative maneuvers with exercise or catecholamine infusion (34,35). QT prolongation on the ECG may be especially notable after exercise or epinephrine challenge in the LQT1 subtype (34). Routine use of provocative testing in clinical practice, however, requires additional studies.

Genetic Testing

Genetic testing can be helpful when clinical and ECG findings are consistent with the LQTS. In this scenario, identification of the genetic subtype can help to risk stratify and provide understanding on the risk of a cardiac event while on beta-blocker therapy. LQT1 patients on beta-blocker therapy have a lower risk of cardiac events than LQT2 or LQT3 patients (28) (Figure 14.2). Genetic testing can also be useful in identifying silent gene carriers. Variable penetrance, which describes the phenotypic heterogeneity among similar mutation carriers, is commonly observed in this syndrome. As a result, family members of affected individuals may not manifest the same degree of QT_c prolongation or clinical symptoms. Testing of specific mutations in family members can provide a conclusive diagnosis of the LQTS. Genetic testing can also be considered in asymptomatic patients with a prolonged QT interval (greater than or equal to 480 ms) after excluding the presence of other conditions that may increase repolarization such as the use of medications and electrolyte disturbances (36).

Genetic testing also poses certain challenges. Studies have demonstrated that approximately 4% of healthy Caucasians and 6% to 8% of African Americans harbor a rare, amino acid-altering (ie, nonsynonymous) genetic variant in one of the three major LQTS

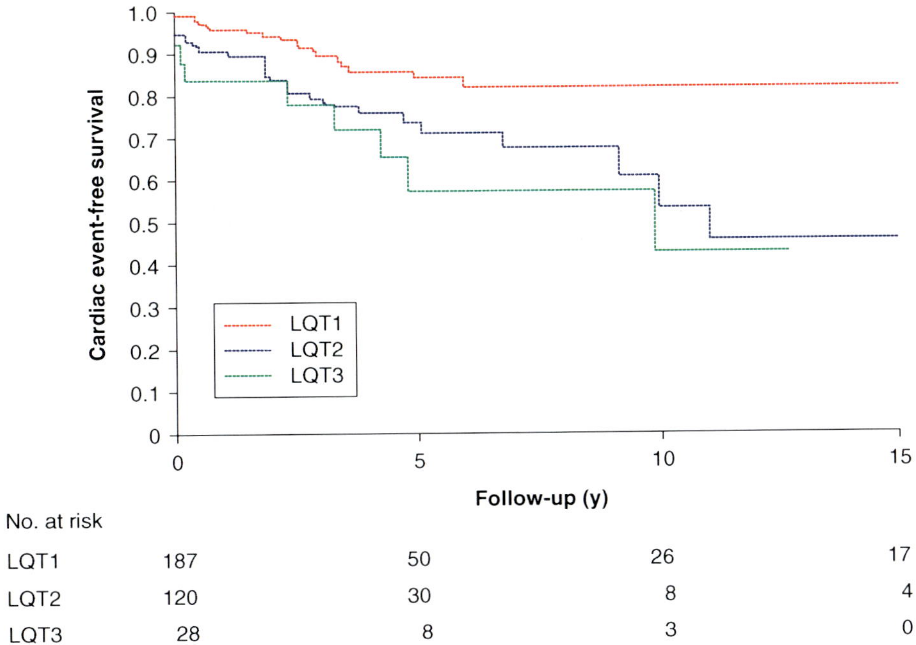

FIGURE 14.2 Cardiac events in LQTS patients receiving beta-blocker therapy. Kaplan–Meier analysis of cumulative cardiac-event-free survival in genotyped LQTS patients receiving beta-blockers according to the genetic variant of the disease (log-rank *P* less than .001). Events were defined as syncope, cardiac arrest, and SCD.

Source: From Ref. (28). Priori SG, Napolitano C, Schwartz PJ, et al. Association of long QT syndrome loci and cardiac events among patients treated with beta-blockers. *JAMA*. 2004;292(11):1341–1344.

genes (37). As a result, genetic testing should be considered in situations where there is reasonable clinical suspicion or to screen family members for a known mutation. In addition, the yield of genetic testing decreases to 14% for borderline cases (38). Testing may also identify variants of unknown significance (VUS), which are defined by a mutation that may be pathogenic or benign. Finally, it is critical to remember that a negative genetic test in a patient with a high clinical suspicion does not rule out the disease, especially given that 20% of patients remain genetically elusive.

Risk Stratification

Risk assessment has traditionally involved the evaluation of clinical and ECG factors. Patients with a QTc greater than 500 ms are at high risk of cardiac events (33,39). A history of syncope at any age is another risk factor for cardiac arrest or SCD (39,40). Children who have experienced syncope or a cardiac arrest before 7 years of age are at particularly high risk and may not be protected adequately with beta-blocker therapy (33,39). Prior studies have also revealed that a family history of SCD did not portend a higher likelihood of SCD. The genetic architecture is also important in the risk assessment of these patients. Potassium channel mutations in both LQT1 and LQT2 are more severe when the transmembrane segments are involved and

less severe when the N- or C-termini are affected (41–43) (Figures 14.3A and 14.3B).

Lower risk patients can also be identified. Concealed mutation-positive patients are at a lower risk for arrhythmic events than those with prolonged QT intervals. Specifically, the risk for an arrhythmic event in this group has been estimated to be between 5% and 10% between birth and age 40 in the absence of therapy (32,33). These individuals should undergo routine ECG screening and avoid QT-prolonging drugs.

The role of gender and genotype in LQTS is not entirely clear. Early epidemiological studies noted that 60% to 70% of patients with LQTS are women (44). One appealing hypothesis is the role sex hormones play in cardiovascular physiology. In support of this relationship is the increased risk of cardiac events that occurs in the 9-month postpartum period in female patients. Testosterone is known to shorten the QT interval in boys after puberty and may offer protection postpuberty. These findings support data that demonstrate that male sex is associated with a three-fold increase risk in life-threatening cardiac events prepuberty, and postpuberty the opposite trend is observed. However, this association is predominantly seen in LQT2 patients. In the LQT3 genotype, men have an increased risk across the life span. Although these data suggest that the role of sex hormones may be genotype specific, the mechanisms are not entirely defined (29,33,45).

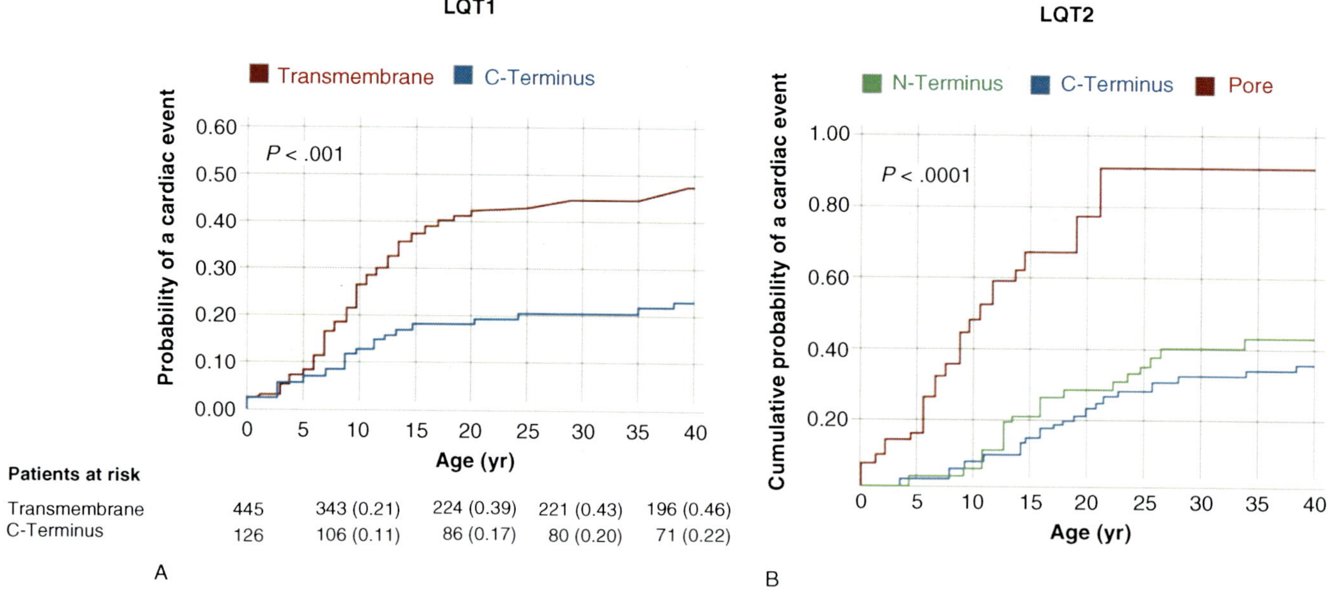

Patients at risk					
Transmembrane	445	343 (0.21)	224 (0.39)	221 (0.43)	196 (0.46)
C-Terminus	126	106 (0.11)	86 (0.17)	80 (0.20)	71 (0.22)

FIGURE 14.3 Mutation location further defines LQTS risk. For LQT1 and LQT2 patients, there is a significantly higher risk of cardiac events when mutations are located in the transmembrane or pore region. Shown is a Kaplan–Meier estimate of the cumulative probability of a first cardiac event by location in LQT1 patients (A) (42) and LQT2 patients (B) (41). The first cardiac event is defined as time to syncope, aborted cardiac arrest, or sudden death, whichever occurred first. *Source*: From Refs. (41,42).

Treatment Strategies

The management approach to patients with the LQTS requires a careful assessment of the overall risk for an arrhythmic event. In all patients suspected of carrying a mutation that involves cardiac repolarization/QT interval, measures to modify lifestyle by avoiding QT-prolonging medications should be performed. A comprehensive source of medications that result in QT prolongation can be found at www.qtdrugs.org.

Patients with LQT1 should avoid excessive exercise and especially swimming. Swimming is particularly dangerous in this subtype as 99% of arrhythmic episodes associated with swimming occur in LQT1 patients (46). Participation in competitive sports may be safer in non-LQT1 subtypes and a borderline phenotype; however, more definitive evidence is still required (47). Patients who participate in sports should not have a history of syncope or cardiopulmonary symptoms or family history of SCD. These individuals require a comprehensive clinical evaluation, adequate medical treatment for LQTS, and require the presence of automatic external defibrillators (AEDs) and personnel trained in basic life support (36,47). Similarly, patients with LQT2 should have minimal exposure to startling noises. Simple interventions such as removal of telephones and alarm clocks in their bedrooms should be advised (46).

Beta-blocker therapy remains the mainline therapy for LQTS patients (28,48,49). Unless there is a contraindication such as active asthma (28,49), beta-blockers are utilized to manage the spectrum of

LQT patients including those with a genetic diagnosis and normal QT_c. Long acting beta-blockers such as nadolol or sustained-release propranolol should be preferred as these medications can be given once or twice a day with avoidance of wide fluctuations in blood levels. In the largest multicenter study published to date, the efficacy of commonly used beta-blockers (propranolol, metoprolol, and nadolol) was compared in the management of LQTS. Propranolol and nadolol were significantly more effective than metoprolol in preventing breakthrough cardiac events (ie, syncope, aborted cardiac arrest, appropriate ICD shock, and SCD) in symptomatic patients (Figure 14.4). Further, symptomatic patients treated with metoprolol were four times more likely to have a breakthrough cardiac event than those treated with propranolol and nadolol. Propranolol was also superior to both nadolol and metoprolol in shortening cardiac repolarization time (50). While studies are not available to define the most effective dosage, full dosing for age and weight, if tolerated, is recommended. Propranolol is the most widely used drug, and a typical dose is 2 to 3 mg/kg per day, which can be increased up to 4 mg/kg in more malignant cases (51). Atenolol is another beta-blocker with limited data; however, it appears to be less effective than propranolol (52). For patients with LQT3, Na-channel specific medications such as mexiletine and flecainide have been investigated as additive therapies to beta-blockers. Recent data demonstrate that the response is dependent on the location of the mutation as over 80 mutations for LQT3 have

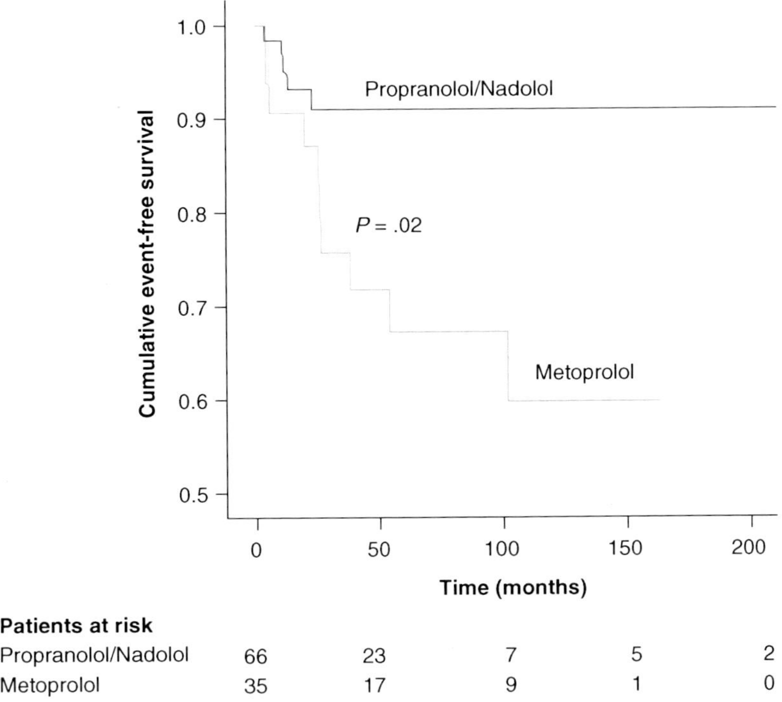

FIGURE 14.4 Differences in beta-blockers for the management of LQT1 and LQT2. Kaplan–Meier estimates of event-free survival of symptomatic patients initiated on different beta-blockers are depicted. An event was defined as syncope, cardiac arrest, appropriate ICD shock, or SCD that occurred while receiving beta-blockers, The cumulative event-free survival of symptomatic patients initiated on metoprolol (n = 35) (light gray line) was significantly different (*P* = .02) from that of patients initiated on propranolol and nadolol combined (n = 66) (dark gray line).

Source: From Ref. (50). Chockalingam P, Crotti L, Girardengo G, et al. Not all beta-blockers are equal in the management of long QT syndrome types 1 and 2: higher recurrence of events under metoprolol. *J Am Coll Cardiol*. 2012;60(20):2092–2099.

already been described (53). Thus, understanding the use of these medications could be a potential area in the future.

ICD therapy is indicated in patients who are resuscitated from cardiac arrest (54). In addition, ICD therapy is often indicated in LQTS patients who have had a syncopal event while taking beta-blocker therapy. Prophylactic implantation should be considered in very high-risk patients such as those who are symptomatic with two or more gene mutations, including those with the Jervell and Lange-Nielsen variant (55). The decision to implant is often challenging as ICD placement can result in complications including inappropriate shocks, infection, and lead failure. A decision to have an ICD implanted should be made only after careful consideration of the risk of SCD, the short and longer term risks of ICD implantation, and the values and preferences of the patient.

Finally, left sympathetic denervation, based on its antifibrillatory effect, is an alternative therapy for high-risk patients who fail mainstream therapy. This therapy is only indicated in specific situations such as patients with appropriate recurrent VF-terminating ICD shocks, breakthrough cardiac events while on adequate drug therapy, failure to tolerate beta-blocker therapy, or high-risk young patients where

primary drug therapy may not be sufficiently protective and with the hopes to serve as a bridge to the eventual implant of an ICD (46,56,57).

BRUGADA SYNDROME (BrS)

BrS can be a highly lethal disorder that is responsible for 4% of all SCDs and up to 20% of SCD events in patients with structurally normal hearts (58). It was first described by Pedro and Josep Brugada in the early 1990s when they identified eight patients resuscitated from a cardiac arrest with an abnormal ECG (59). The prevalence is estimated to be 5 in 10,000, although this figure may be an underestimate given the transient and dynamic nature of the ECG (58). In addition, the prevalence of malignant ventricular arrhythmias varies from 5% at 2 years in asymptomatic patients to 45% in patients with a previous cardiac arrest (60).

Genetics of Brugada Syndrome

BrS is an autosomal dominant disorder. Twelve genes have been implicated in this disease, and mutations are detected in approximately 30% of suspected patients.

The most common genes implicated in BrS include *SCN5A* and *CACN1AC* (61,62). Loss-of-function mutations in *SCN5A*, which encodes the cardiac sodium channel, have been identified in less than one-third of clinically diagnosed BrS patients. In an international registry of approximately 2,000 patients with suspected BrS and 1,300 healthy volunteers, genotyping of all 27 exons of *SCN5A* revealed 293 mutations with an overall 21% of BrS probands harboring a mutation compared to a 2% to 5% rate of rare variants in healthy controls (63). Gain-of-function mutations in the *SCN5A* gene cause LQT3. As a result, these two disorders are considered allelic diseases or disorders that result from different mutations in the same gene. The different mutations in *SCN5A* result in different gating and conductance changes of the sodium channel, which ultimately results in loss of function or gain of function of the sodium channel. Loss-of-function mutations in the *CACNA1C* and *CACNB2* genes, encoding for the alpha and beta subunits of the cardiac calcium channel, are the second most frequently associated mutation. These mutations lead to a decrease of the I_{ca} current and can result in a combined BrS and SQTS (64). Despite the documented genetic inheritance, it is estimated that the overall disease penetrance based on studies from several small BrS families is approximately 16% (65).

Clinical Manifestations

BrS is significantly more prevalent in men compared with women. Unlike LQTS, BrS typically manifests during adulthood with an average age of 41 years. However, there is a wide range with the youngest recorded patient at 2 days old and the oldest at 84 years. VF and SCD most often occur at rest or during sleep, which is likely secondary to circadian variation of sympathovagal balance, hormones, and other metabolic factors (58,66). Ventricular arrhythmias are often triggered by fevers and large meals, the latter being associated with glucose-induced insulin secretion that may elevate the ST-segment (67,68). A febrile state may unmask BrS secondary to premature inactivation of the sodium channels as a function of temperature (69). In addition, up to 20% of patients will demonstrate supraventricular arrhythmias (70). Importantly, atrial arrhythmias have been shown to be a risk factor for VF inducibility and thus have important prognostic implications (71).

Diagnosis

Similar to the other inherited arrhythmic diseases, the first step involves a detailed clinical and family history. A family history of SCD in young men during sleep may be present and direct the workup for BrS. The ECG is a key to diagnosis, and documentation of a type I BrS pattern in the right precordial leads (V1–V3) is required for diagnosis (Figure 14.5). Other disorders that can mimic a type I BrS pattern in a patient suspected to have BrS include LVH, atypical RBBB, early repolarization, arrhythmogenic right ventricular dysplasia, hypothermia, acute myocardial ischemia, pericarditis, pulmonary embolism, Prinzmetal

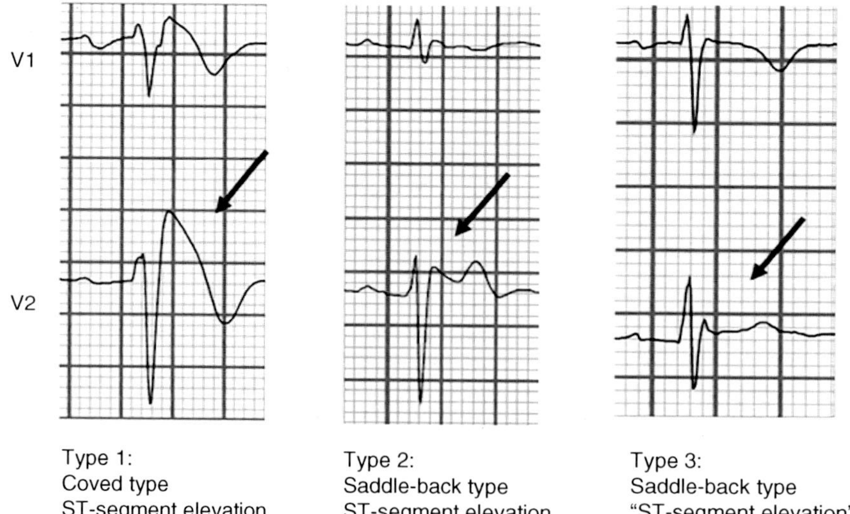

FIGURE 14.5 ECG characteristics of BrS. A type I BrS pattern on ECG is the only diagnostic ECG in BrS and is characterized by a prominent coved ST-segment with ST-segment or J-point elevation greater than 2 mm and a negative T-wave. Type II BrS (saddle-back, elevation greater than 2.5 mm) and Type III (coved or saddle-back with ST elevation less than 2.5 mm) are both considered to be suggestive but not diagnostic of BrS.

Source: From Ref. (82). Mizusawa Y, Wilde AAM. Brugada syndrome. *Circ Arrhythm Electrophysiol.* 2012;5(3):606–616.

angina, aortic dissection, metabolic disturbances (eg, hyperkalemia), and other various nervous system abnormalities (58). A type I BrS pattern on ECG is characterized by a prominent coved ST-segment with ST-segment or J-point elevation greater than 2 mm and a negative T-wave. A right bundle branch block pattern may also be observed but is not required for diagnosis. In addition, ECG patterns known as type II BrS (saddle-back, ST elevation greater than 2.5 mm) or type III (coved or saddle-back with ST elevation less than 2.5 mm) are considered to be suggestive but not diagnostic (72). Identification of a type II or type III pattern, however, should prompt medical attention as the ECG is dynamic in BrS patients, and all three patterns may coexist in a patient at different time periods. As a result, prolonged electrocardiographic monitoring through Holter and event monitoring is important. Further, pharmacological testing by administering sodium channel blockade (flecainide, ajmaline, procainamide, disopyramide, propafenone, or pilsicainide) can play an important role in the diagnosis of BrS (58). A drug challenge is only positive when the ECG reveals a type I BrS pattern (73,74).

The diagnosis for BrS requires a type I ECG pattern and one of the following clinical scenarios: positive family history including SCD in a family member younger than 45 years or an ECG type I BrS pattern in other members; syncope, seizures, or nocturnal agonal respiration; or history of documented ventricular arrhythmias (73).

Genetic Testing

Genetic testing for BrS can be considered for individuals with a high clinical suspicion based on resting ECG, provocative drug challenge, and the patient's clinical and family history. Mutation-specific genetic testing is recommended for family members after the identification of a BrS-mutation in an index patient. Genetic testing is not recommended for an isolated type II or type III BrS ECG pattern. *SCN5A* loss-of-function mutations represent the most common BrS genotype, estimated to represent approximately 15% to 30% and accounts for the vast majority of genotype-positive cases. However, a negative genetic test does not rule out the disease and in fact 65% of cases will remain genetically elusive (63,75). Based on these findings, genetic testing does not play a major role in risk stratification at present.

Risk Stratification

Risk stratification for BrS is based on clinical parameters. A history of syncope and the presence of a spontaneous type I ECG is considered a marker of a higher risk of SCD (76–79). Other clinical variables associated with VF include the presence of atrial fibrillation. Atrial fibrillation is present in nearly half of all BrS patients and is associated with a higher incidence of syncope and VF (70,80). A type I ECG pattern that is present during pharmacological testing represents a lower risk phenotype to experience either syncope or SCD. Also, neither a positive family history nor an *SCN5A* mutation has proven to be a risk marker for SCD (76,77). Assessing the inducibility of ventricular arrhythmias using electrophysiology studies is a controversial topic based on conflicting studies. Several recent studies on different series of patients indicate that the electrophysiology study should not be used for the purposes of risk stratification (78,81).

Management

The primary treatment for BrS remains ICD implantation. Implantation is recommended among patients who have survived a cardiac arrest or with spontaneous VT even in the absence of syncope (36,78). ICD implantation can also be useful in patients with a spontaneous type I ECG who have a history of syncope. Implantation is not indicated in asymptomatic BrS patients, who have a family history of SCD even when the ECG demonstrates a drug-induced type I ECG pattern (82).

There are no drugs currently approved for the primary prevention of SCD in BrS. Quinidine may be an effective antiarrhythmic drug and is currently undergoing evaluation in a clinical trial for the prevention of SCD (83). Both quinidine and isoproterenol have been used to treat the rare complication of VT storm that can be seen in BrS.

All patients should be advised to avoid drugs that could induce a BrS type pattern on the ECG or trigger an arrhythmia (www.brugadadrugs.org). Patients should also be counseled to take antipyretic medications during episodes of fever.

CATECHOLAMINERGIC POLYMORPHIC VT (CPVT)

CPVT is one of the most lethal arrhythmic disorders and is characterized by adrenergically mediated arrhythmias. Unlike LQTS, CPVT is associated with a higher penetrance, and the majority of patients will be symptomatic before the age of 40 without medical therapy (84,85) (Figure 14.6). The estimated prevalence of this disease is 1 in 10,000 (86). The mean age of symptom onset is approximately 8 to 12 years, and symptoms are typically observed in the setting of emotional stress or exertion.

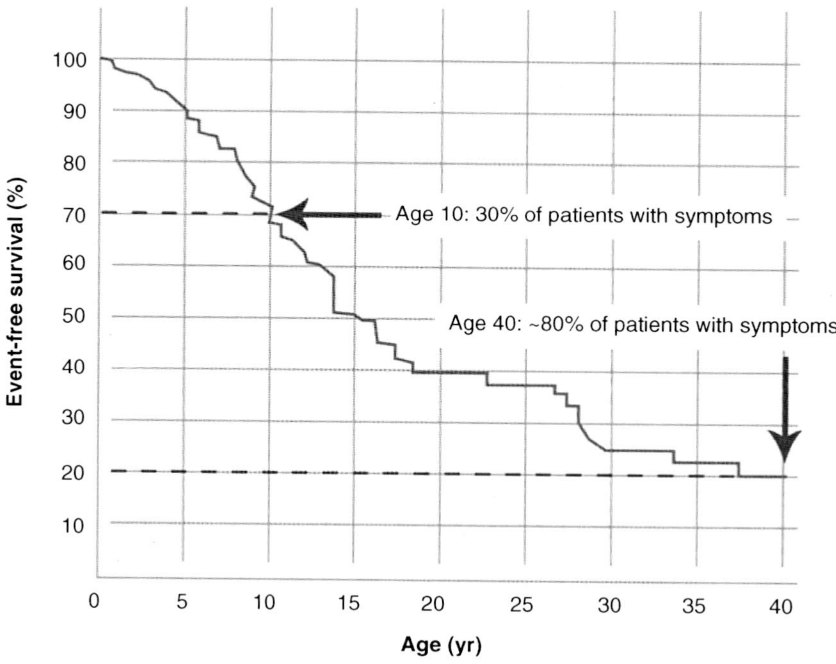

FIGURE 14.6   The natural history of CPVT. Survival analysis shows time to first cardiac event defined as syncope, VF, or sudden death in the absence of beta-blocker therapy in 119 patients.

Source: From Ref. (85). Napolitano C, Priori SG. Diagnosis and treatment of catecholaminergic polymorphic ventricular tachycardia. *Heart Rhythm*. 2007;4(5):675–678.

Genetics of CPVT

The two genes that have been implicated in causing CPVT include *RyR2*, which encodes the cardiac ryanodine receptor (87), and *CASQ2*, which encodes for calsequestrin (88,89). Mutations in *RyR2* account for approximately 65% of cases and are transmitted in an autosomal dominant pattern. The cardiac ryanodine receptors control the release of calcium from the sarcoplasmic reticulum (SR) during excitation–contraction coupling. Mutations in this protein result in the release of more calcium from the SR during the diastolic phase of the cardiac cycle. This mechanism can result in the development of delayed afterdepolarizations, a well-described mechanism of arrhythmogenesis (90). The rare, recessive form of CPVT is caused by mutations on the *CASQ2* gene and accounts for 3% to 5% of patients (38). Calsequestrin controls SR calcium by binding free calcium ions. As a result, mutations in this protein cause calcium leakage from the SR either by loss of the binding capacity or by altering *RyR2* function.

Clinical Manifestations

CPVT manifests with stress-induced syncope and cardiac arrest (84,87,91). The mean age of symptom onset is 12 years (92), and sudden cardiac arrest is the initial manifestation of the disease in approximately 30% of patients (38). Most patients present with cardiac symptoms during exercise, and it is rare to have an event at rest. In addition, as mentioned, the penetrance of the disease is high, and most patients with CPVT develop symptoms. Approximately 80% of patients not treated with a beta-blocker will become symptomatic before the age of 40.

Diagnosis

The ECG of patients with CPVT is usually normal and abnormalities are often only revealed under exercise or provocative studies (91,93). Exercise stress testing tends to elicit a bidirectional VT or polymorphic VT in the majority of patients. In particular, exercise often evokes supraventricular arrhythmias and atrial fibrillation (94), which progress toward the development of a bidirectional polymorphic VT (38). Monomorphic VT is much less frequently seen in this disorder.

Clinical features that suggest a diagnosis of CPVT include: the identification of exercise-induced polymorphic VT or bidirectional VT; history of syncope elicited by an acute emotion or physical activity; a history of exercise or emotion-related palpitations or dizziness; and the absence of structural heart disease. When the disease is observed in familial cases, there often is family history of juvenile SCD during adrenergic stress that may help the diagnosis.

Genetic Testing

Genetic testing in CPVT is highly relevant as 60% to 70% of suspected cases yield a positive mutation (95). As a result, comprehensive genetic testing for mutations in either *RyR2* or *CASQ2* can be useful for patients with a high clinical suspicion. Mutation-specific genetic testing is recommended for all first-degree family members after the identification of a positive index case (75). Given that SCD may be the initial manifestation of CPVT, identification and testing of family members can have important prognostic implications including the initiation of beta-blocker therapy. The yield of genetic testing in patients with the typical bidirectional VT is high (65%) and decreases to less than 15% for patients with a non-typical ECG pattern. Genotyping does not help with risk stratification of this disorder.

Risk Stratification

Limited markers of risk have been identified in CPVT. A diagnosis that occurs in childhood is associated with an adverse outcome (36). Further, the lack of appropriate beta-blocker therapy is also a predictor for future arrhythmic events. Finally, the presence of ventricular ectopy during exercise testing is a marker for arrhythmic events (92).

Management

CPVT patients should be instructed to avoid vigorous exercise. The mainstay of treatment is beta-blocker therapy and is recommended in patients with spontaneous or documented ventricular arrhythmias with stress. Although there are no studies comparing the differential effect of beta-blockers, the consensus recommendations are nadolol (1–2.5 mg/kg/day) or propranolol (2–4 mg/kg/day) (86). However, even with maximal doses of beta-blocker therapy, the incidence of recurrence is still up to 30% to 40% (84). An ICD should be implanted in CPVT patients who survive a cardiac arrest, and have a history of syncope or VT despite beta-blocker therapy. ICD is not currently recommended as primary prevention in an asymptomatic patient (36). In addition, beta-blockers may be effective in asymptomatic patients when a diagnosis of CPVT is made on genetic testing.

A clinician should also be aware that ICD implantation in a young patient still has limitations such as the risk of device-related events and risk of VT storm over time, as well as the technical difficultly of implanting in a young child. Thus, similar to LQTS, sympathetic denervation has also been suggested to be an alternative therapy when beta-blocker therapy fails (96). Careful discussion of the pros and cons of management should be conducted with family members. Finally, flecainide may be considered in patients with a diagnosis of CPVT who manifest recurrent syncope or VT while on a beta-blocker. Ongoing clinical trials are currently evaluating the efficacy of flecainide as an additive therapy to beta-blockers in the setting of recurring cardiac symptoms (38,97).

SHORT QT SYNDROME (SQTS)

SQTS is a relatively recently identified inherited channelopathy. It was first described in 2000 with the identification of a SQTS family with a markedly short QT interval that was associated with atrial fibrillation or SCD in individuals with a structurally normal heart (98). Given the recent description of this clinical entity, the true prevalence of SQTS is not yet known.

Genetics of SQTS

SQTS has an autosomal dominant mode of inheritance. Given the opposite clinical phenotype as LQTS, investigators evaluated whether the same genes implicated in the pathogenesis of LQTS could be involved in SQTS. Three LQTS mutations, *KCNQ1* (LQT1), *KCNH2* (LQT2), and *KCNJ2* (LQT7) have been identified in affected patients. However, instead of loss-of-function mutations, in vitro studies revealed a gain of function of the respective ionic channels, which effectively increased the repolarizing potassium current and shortened repolarization (99,100). Thus far, mutations in five different genes have been described, which are labeled SQT1-5 based on the chronology of discovery (98). Very few patients have been diagnosed and genotyped; therefore, the yield of genetic testing is currently unknown. Mutation-specific testing is recommended for family members after the identification of a genotype-positive index case (75).

Clinical Manifestations and Diagnosis

Given the rarity of this condition and no more than 60 cases reported over the past decade, we have a limited understanding of the clinical progression of this disease. As described by earlier reports, patients with SQTS can present with either atrial fibrillation or cardiac arrest. The diagnosis of SQTS has represented a challenge and should be strongly suspected in young individuals with a short QT interval on ECG in the presence of other clinical symptoms such as lone atrial fibrillation, VF, or a strong family history of arrhythmic events or SCD. The spectrum of the QT interval generally ranges from 220 to 360 ms. More recently, consensus panel recommendations indicate that a corrected QT less than or equal to 330 ms should be used for the diagnosis of SQTS (36). This QT_c is two standard deviations below the lower limits seen in general population studies (99–101). In addition, a diagnosis of SQTS can be made in the presence of a QT_c less than 360 ms and the presence of a known mutation, history of cardiac arrest in the absence of structural heart disease, family history of SQTS, or a family history of SCD in an individual less than 40 years of age.

Management

No official guidelines exist on management. An ICD is recommended as first-line treatment for secondary prevention of SCD in patients with prior sudden cardiac arrest and for SQTS patients with documented VT. ICD implantation can also be considered in asymptomatic SQTS patients with a family history of SCD (36). No clear-cut pharmacological therapy exists, although limited data have suggested that quinidine may be the most effective (102).

CONCLUSIONS

Although the various arrhythmic disorders described here each have key characteristics and nuances, the common theme in the approach to a patient suspicious for an inherited arrhythmic disease is similar. The first step for any patient is always a thoughtful and extensive history, thorough family history and review of the medication profile, and physical exam. In addition, the ECG is a critical tool as it is often the "portal of entry" to an arrhythmic disorder. ECGs in the postarrest phase may demonstrate long QT or early repolarization changes due to metabolic disturbances and cardiac stunning; as a result, they may not be accurate for diagnosing specific conditions. Further, the absence of findings on one ECG is not infrequent, and a disorder should not be "ruled out" based on a normal ECG. In addition to the ECG, Holter monitoring, echocardiography, and exercise testing are typical tests that are often evaluated.

As discussed throughout this chapter, genetic testing has emerged as another option that aids with diagnosis and screening family members. This method may be particularly valuable in identifying family members at an increased risk of SCD who may warrant additional diagnostic testing and treatment. Among patients who have survived a cardiac arrest, genetic testing is recommended when a diagnosis is suspected based on clinical data. Similarly, postmortem genetic testing can be considered in the setting of autopsy-negative sudden death cases especially if

circumstantial evidence points toward a clinical diagnosis of an inherited arrhythmic disorder. Mutation-specific genetic testing can then be recommended for family members. Despite the increasing availability of commercial genetic testing, future studies will need to address the management of patients with genetic variants and no clinical abnormalities.

This chapter has focused on the inherited arrhythmic disorders with a clinically relevant focus. Understanding the genetic bases for these disorders and the genotype–phenotype correlations has transformed our ability to care for these patients and family members.

REFERENCES

1. Deo R, Albert CM. Epidemiology and genetics of sudden cardiac death. *Circulation.* 2012;125(4):620–637.
2. John RM, Tedrow UB, Koplan BA, et al. Ventricular arrhythmias and sudden cardiac death. *Lancet.* 2012; 380(9852):1520–1529.
3. Arking DE, Sotoodehnia N. The genetics of sudden cardiac death. *Annu Rev Genomics Hum Genet.* 2012;13:223–239.
4. Zipes DP, Camm AJ, Borggrefe M, et al. ACC/AHA/ESC 2006 guidelines for management of patients with ventricular arrhythmias and the prevention of sudden cardiac death—executive summary: a report of the American College of Cardiology/American Heart Association Task Force and the European Society of Cardiology Committee for Practice Guidelines (Writing Committee to Develop Guidelines for Management of Patients with Ventricular Arrhythmias and the Prevention of Sudden Cardiac Death) Developed in collaboration with the European Heart Rhythm Association and the Heart Rhythm Society. *Eur Heart J.* 2006;27(17):2099–2140.
5. Krahn AD, Healey JS, Chauhan V, et al. Systematic assessment of patients with unexplained cardiac arrest: Cardiac Arrest Survivors With Preserved Ejection Fraction Registry (CASPER). *Circulation.* 2009;120(4):278–285.
6. Moss AJ. Long QT syndrome. *JAMA.* 2003;289(16): 2041–2044.
7. Schwartz PJ, Stramba-Badiale M, Crotti L, et al. Prevalence of the congenital long-QT syndrome. *Circulation.* 2009;120(18):1761–1767.
8. Ward OC. A new familial cardiac syndrome in children. *J Ir Med Assoc.* 1964;54:103–106.
9. Jervell A, Lange-Nielsen F. Congenital deaf-mutism, functional heart disease with prolongation of the Q-T interval and sudden death. *Am Heart J.* 1957;54(1): 59–68.
10. Priori SG, Napolitano C. Genetics of channelopathies and clinical implications. *Hurst's The Heart.* 13th ed. New York, NY: McGraw Hill; 2011:897–910.
11. Wang Q, Curran ME, Splawski I, et al. Positional cloning of a novel potassium channel gene: KVLQT1 mutations cause cardiac arrhythmias. *Nat Genet.* 1996;12(1):17–23.

12. Thomas D, Wimmer A-B, Karle CA, et al. Dominant-negative I(Ks) suppression by KCNQ1-deltaF339 potassium channels linked to Romano-Ward syndrome. *Cardiovasc Res.* 2005;67(3):487–497.

13. Schulze-Bahr E. Long QT Syndromes: Genetic Basis. In: Priori SG, ed. *Long QT Syndrome, Card Electrophysiol Clin 4.* Philadelphia, PA. WB Saunders Co; 2012:1–16.

14. Bartos DC, Duchatelet S, Burgess DE, et al. R231C mutation in KCNQ1 causes long QT syndrome type 1 and familial atrial fibrillation. *Heart Rhythm.* 2011; 8(1):48–55.

15. Lundby A, Ravn LS, Svendsen JH, et al. KCNQ1 mutation Q147R is associated with atrial fibrillation and prolonged QT interval. *Heart Rhythm.* 2007; 4(12):1532–1541.

16. Curran ME, Splawski I, Timothy KW, et al. A molecular basis for cardiac arrhythmia: HERG mutations cause long QT syndrome. *Cell.* 1995;80(5):795–803.

17. Biliczki P, Girmatsion Z, Brandes RP, et al. Trafficking-deficient long QT syndrome mutation KCNQ1-T587M confers severe clinical phenotype by impairment of KCNH2 membrane localization: evidence for clinically significant IKr-IKs alpha-subunit interaction. *Heart Rhythm.* 2009;6(12):1792–1801.

18. Sanguinetti MC, Jiang C, Curran ME, Keating MT. A mechanistic link between an inherited and an acquired cardiac arrhythmia: HERG encodes the IKr potassium channel. *Cell.* 1995;81(2):299–307.

19. Etheridge SP, Compton SJ, Tristani-Firouzi M, Mason JW. A new oral therapy for long QT syndrome: long-term oral potassium improves repolarization in patients with HERG mutations. *J Am Coll Cardiol.* 2003;42(10):1777–1782.

20. Wang Q, Shen J, Splawski I, et al. SCN5A mutations associated with an inherited cardiac arrhythmia, long QT syndrome. *Cell.* 1995;80(5):805–811.

21. Bennett PB, Yazawa K, Makita N, George AL Jr. Molecular mechanism for an inherited cardiac arrhythmia. *Nature.* 1995;376(6542):683–685.

22. Crotti L, Lundquist AL, Insolia R, et al. KCNH2-K897T is a genetic modifier of latent congenital long-QT syndrome. *Circulation.* 2005 30;112(9):1251–1258.

23. Crotti L, Monti MC, Insolia R, et al. NOS1AP is a genetic modifier of the long-QT syndrome. *Circulation.* 2009;120(17):1657–1663.

24. Schwartz PJ, Priori SG, Spazzolini C, et al. Genotype-phenotype correlation in the long-QT syndrome: gene-specific triggers for life-threatening arrhythmias. *Circulation.* 2001;103(1):89–95.

25. Wilde AA, Jongbloed RJ, Doevendans PA, et al. Auditory stimuli as a trigger for arrhythmic events differentiate HERG-related (LQTS2) patients from KVLQT1-related patients (LQTS1). *J Am Coll Cardiol.* 1999;33(2):327–332.

26. Zellerhoff S, Pistulli R, Mönnig G, et al. Atrial Arrhythmias in long-QT syndrome under daily life conditions: a nested case control study. *J Cardiovasc Electrophysiol.* 2009;20(4):401–407.

27. Johnson JN, Tester DJ, Perry J, et al. Prevalence of early-onset atrial fibrillation in congenital long QT syndrome. *Heart Rhythm.* 2008;5(5):704–709.

28. Priori SG, Napolitano C, Schwartz PJ, et al. Association of long QT syndrome loci and cardiac events among patients treated with beta-blockers. *JAMA.* 2004;292(11):1341–1344.

29. Seth R, Moss AJ, McNitt S, et al. Long QT syndrome and pregnancy. *J Am Coll Cardiol.* 2007;49(10): 1092–1098.

30. Goldenberg I, Mathew J, Moss AJ, et al. Corrected QT variability in serial electrocardiograms in long QT syndrome: the importance of the maximum corrected QT for risk stratification. *J Am Coll Cardiol.* 2006;48(5):1047–1052.

31. Zhang L, Timothy KW, Vincent GM, et al. Spectrum of ST-T-wave patterns and repolarization parameters in congenital long-QT syndrome: ECG findings identify genotypes. *Circulation.* 2000;102(23):2849–2855.

32. Goldenberg I, Horr S, Moss AJ, et al. Risk for life-threatening cardiac events in patients with genotype-confirmed long-QT syndrome and normal-range corrected QT intervals. *J Am Coll Cardiol.* 2011; 57(1):51–59.

33. Priori SG, Schwartz PJ, Napolitano C, et al. Risk stratification in the long-QT syndrome. *N Engl J Med.* 2003;348(19):1866–1874.

34. Vyas H, Hejlik J, Ackerman MJ. Epinephrine QT stress testing in the evaluation of congenital long-QT syndrome: diagnostic accuracy of the paradoxical QT response. *Circulation.* 2006;113(11):1385–1392.

35. Obeyesekere MN, Klein GJ, Modi S, et al. how to perform and interpret provocative testing for the diagnosis of Brugada syndrome, long-QT syndrome, and catecholaminergic polymorphic ventricular tachycardia. *Circ Arrhythm Electrophysiol.* 2011;4(6): 958–964.

36. Priori SG, Wilde AA, Horie M, et al. HRS/EHRA/APHRS expert consensus statement on the diagnosis and management of patients with inherited primary arrhythmia syndromes: document endorsed by HRS, EHRA, and APHRS in May 2013 and by ACCF, AHA, PACES, and AEPC in June 2013. *Heart Rhythm.* 2013;10(12):1932–1963.

37. Giudicessi JR, Ackerman MJ. Determinants of incomplete penetrance and variable expressivity in heritable cardiac arrhythmia syndromes. *Transl Res.* 2013;161(1):1–14.

38. Cerrone M, Cummings S, Alansari T, Priori SG. A clinical approach to inherited arrhythmias. *Circ Cardiovasc Genet.* 2012;5(5):581–590.

39. Goldenberg I, Moss AJ, Peterson DR, et al. Risk factors for aborted cardiac arrest and sudden cardiac death in children with the congenital long-QT syndrome. *Circulation.* 2008;117(17):2184–2191.

40. Goldenberg I, Moss AJ, Bradley J, et al. Long-QT syndrome after age 40. *Circulation.* 2008;117(17): 2192–2201.

41. Moss AJ, Zareba W, Kaufman ES, et al. Increased risk of arrhythmic events in long-QT syndrome with mutations in the pore region of the human ether-a-go-go-related gene potassium channel. *Circulation.* 2002;105(7):794–799.

42. Moss AJ, Shimizu W, Wilde AAM, et al. Clinical aspects of type-1 long-QT syndrome by location, coding type, and biophysical function of mutations involving the KCNQ1 gene. *Circulation.* 2007;115(19): 2481–2489.

43. Shimizu W, Moss AJ, Wilde AAM, et al. Genotype-phenotype aspects of type 2 long QT syndrome. *J Am Coll Cardiol.* 2009;54(22):2052–2062.

44. Moss AJ, Schwartz PJ, Crampton RS, et al. The long QT syndrome. Prospective longitudinal study of 328 families. *Circulation*. 1991;84(3):1136–1144.

45. Locati EH, Zareba W, Moss AJ, et al. Age- and sex-related differences in clinical manifestations in patients with congenital long-QT syndrome: findings from the International LQTS Registry. *Circulation*. 1998;97(22):2237–2244.

46. Schwartz PJ, Ackerman MJ. The long QT syndrome: a transatlantic clinical approach to diagnosis and therapy. *Eur Heart J*. 2013;34(40):3109–3116.

47. Johnson JN, Ackerman MJ. Competitive sports participation in athletes with congenital long QT syndrome. *JAMA*. 2012;308(8):764–765.

48. Schwartz PJ, Periti M, Malliani A. The long Q-T syndrome. *Am Heart J*. 1975;89(3):378–390.

49. Schwartz PJ, Spazzolini C, Crotti L. All LQT3 patients need an ICD: true or false? *Heart Rhythm*. 2009;6(1):113–120.

50. Chockalingam P, Crotti L, Girardengo G, et al. Not all beta-blockers are equal in the management of long QT syndrome types 1 and 2: higher recurrence of events under metoprolol. *J Am Coll Cardiol*. 2012;60(20):2092–2099.

51. Schwartz PJ, Crotti L, Insolia R. Long-QT syndrome: from genetics to management. *Circ Arrhythm Electrophysiol*. 2012;5(4):868–877.

52. Chatrath R, Bell CM, Ackerman MJ. Beta-blocker therapy failures in symptomatic probands with genotyped long-QT syndrome. *Pediatr Cardiol*. 2004;25(5):459–465.

53. Benhorin J, Taub R, Goldmit M, et al. Effects of flecainide in patients with new SCN5A mutation: mutation-specific therapy for long-QT syndrome? *Circulation*. 2000;101(14):1698–1706.

54. Jons C, Moss AJ, Goldenberg I, et al. Risk of fatal arrhythmic events in long QT syndrome patients after syncope. *J Am Coll Cardiol*. 2010;55(8):783–788.

55. Schwartz PJ, Spazzolini C, Crotti L, et al. The Jervell and Lange-Nielsen syndrome: natural history, molecular basis, and clinical outcome. *Circulation*. 2006;113(6):783–790.

56. Collura CA, Johnson JN, Moir C, Ackerman MJ. Left cardiac sympathetic denervation for the treatment of long QT syndrome and catecholaminergic polymorphic ventricular tachycardia using video-assisted thoracic surgery. *Heart Rhythm*. 2009;6(6):752–759.

57. Schwartz PJ, Priori SG, Cerrone M, et al. Left cardiac sympathetic denervation in the management of high-risk patients affected by the long-QT syndrome. *Circulation*. 2004;109(15):1826–1833.

58. Antzelevitch C, Brugada P, Borggrefe M, et al. Brugada syndrome: report of the second consensus conference: endorsed by the Heart Rhythm Society and the European Heart Rhythm Association. *Circulation*. 2005;111(5):659–670.

59. Brugada P, Brugada J. Right bundle branch block, persistent ST segment elevation and sudden cardiac death: a distinct clinical and electrocardiographic syndrome. A multicenter report. *J Am Coll Cardiol*. 1992;20(6):1391–1396.

60. Brugada P, Brugada R, Mont L, et al. Natural history of Brugada syndrome: the prognostic value of programmed electrical stimulation of the heart. *J Cardiovasc Electrophysiol*. 2003;14(5):455–457.

61. Chen Q, Kirsch GE, Zhang D, et al. Genetic basis and molecular mechanism for idiopathic ventricular fibrillation. *Nature*. 1998;392(6673):293–296.

62. Burashnikov E, Pfeiffer R, Barajas-Martinez H, et al. Mutations in the cardiac L-type calcium channel associated with inherited J-wave syndromes and sudden cardiac death. *Heart Rhythm*. 2010;7(12):1872–1882.

63. Kapplinger JD, Tester DJ, Alders M, et al. An international compendium of mutations in the SCN5A-encoded cardiac sodium channel in patients referred for Brugada syndrome genetic testing. *Heart Rhythm*. 2010;7(1):33–46.

64. Antzelevitch C, Pollevick GD, Cordeiro JM, et al. Loss-of-function mutations in the cardiac calcium channel underlie a new clinical entity characterized by ST-segment elevation, short QT intervals, and sudden cardiac death. *Circulation*. 2007;115(4):442–449.

65. Priori SG, Napolitano C, Gasparini M, et al. Clinical and genetic heterogeneity of right bundle branch block and ST-segment elevation syndrome: a prospective evaluation of 52 families. *Circulation*. 2000;102(20):2509–2515.

66. Kasanuki H, Ohnishi S, Ohtuka M, et al. Idiopathic ventricular fibrillation induced with vagal activity in patients without obvious heart disease. *Circulation*. 1997;95(9):2277–2285.

67. Nishizaki M, Sakurada H, Mizusawa Y, et al. Influence of meals on variations of ST segment elevation in patients with Brugada syndrome. *J Cardiovasc Electrophysiol*. 2008;19(1):62–68.

68. Napolitano C, Bloise R, Monteforte N, Priori SG. Sudden cardiac death and genetic ion channelopathies: long QT, Brugada, short QT, catecholaminergic polymorphic ventricular tachycardia, and idiopathic ventricular fibrillation. *Circulation*. 2012;125(16):2027–2034.

69. Dumaine R, Towbin JA, Brugada P, et al. Ionic mechanisms responsible for the electrocardiographic phenotype of the Brugada syndrome are temperature dependent. *Circ Res*. 1999;85(9):803–809.

70. Morita H, Kusano-Fukushima K, Nagase S, et al. Atrial fibrillation and atrial vulnerability in patients with Brugada syndrome. *J Am Coll Cardiol*. 2002;40(8):1437–1444.

71. Bordachar P, Reuter S, Garrigue S, et al. Incidence, clinical implications and prognosis of atrial arrhythmias in Brugada syndrome. *Eur Heart J*. 2004;25(10):879–884.

72. Gussak I, Antzelevitch C, Bjerregaard P, et al. The Brugada syndrome: clinical, electrophysiologic and genetic aspects. *J Am Coll Cardiol*. 1999;33(1):5–15.

73. Berne P, Brugada J. Brugada syndrome 2012. *Circ J*. 2012;76(7):1563–1571.

74. Hong K, Brugada J, Oliva A, et al. Value of electrocardiographic parameters and ajmaline test in the diagnosis of Brugada syndrome caused by SCN5A mutations. *Circulation*. 2004;110(19):3023–3027.

75. Ackerman MJ, Priori SG, Willems S, et al. HRS/EHRA expert consensus statement on the state of genetic testing for the channelopathies and cardiomyopathies this document was developed as a partnership between the Heart Rhythm Society (HRS) and the European Heart Rhythm Association (EHRA). *Heart Rhythm*. 2011;8(8):1308–1339.

76. Priori SG, Napolitano C, Gasparini M, et al. Natural history of Brugada syndrome: insights for risk stratification and management. *Circulation.* 2002; 105(11):1342–1347.

77. Gehi AK, Duong TD, Metz LD, et al. Risk stratification of individuals with the Brugada electrocardiogram: a meta-analysis. *J Cardiovasc Electrophysiol.* 2006;17(6):577–583.

78. Probst V, Veltmann C, Eckardt L, et al. Long-term prognosis of patients diagnosed with Brugada syndrome: Results from the FINGER Brugada Syndrome Registry. *Circulation.* 2010;121(5):635–643.

79. Brugada J, Brugada R, Antzelevitch C, et al. Long-term follow-up of individuals with the electrocardiographic pattern of right bundle-branch block and ST-segment elevation in precordial leads V1 to V3. *Circulation.* 2002;105(1):73–78.

80. Kusano KF, Taniyama M, Nakamura K, et al. Atrial fibrillation in patients with Brugada syndrome relationships of gene mutation, electrophysiology, and clinical backgrounds. *J Am Coll Cardiol.* 2008; 51(12):1169–1175.

81. Priori SG, Gasparini M, Napolitano C, et al. Risk stratification in Brugada syndrome: results of the PRELUDE (PRogrammed ELectrical stimUlation preDictive valuE) registry. *J Am Coll Cardiol.* 2012;59(1):37–45.

82. Mizusawa Y, Wilde AAM. Brugada syndrome. *Circ Arrhythm Electrophysiol.* 2012;5(3):606–616.

83. Belhassen B, Glick A, Viskin S. Efficacy of quinidine in high-risk patients with Brugada syndrome. *Circulation.* 2004;110(13):1731–1737.

84. Priori SG, Napolitano C, Memmi M, et al. Clinical and molecular characterization of patients with catecholaminergic polymorphic ventricular tachycardia. *Circulation.* 2002;106(1):69–74.

85. Napolitano C, Priori SG. Diagnosis and treatment of catecholaminergic polymorphic ventricular tachycardia. *Heart Rhythm.* 2007;4(5):675–678.

86. Napolitano C, Priori SG, Bloise R. Catecholaminergic polymorphic ventricular tachycardia. *Gene Reviews.* Seattle, WA: University of Washington; 2004.

87. Priori SG, Napolitano C, Tiso N, et al. Mutations in the cardiac ryanodine receptor gene (hRyR2) underlie catecholaminergic polymorphic ventricular tachycardia. *Circulation.* 2001;103(2):196–200.

88. Lahat H, Pras E, Olender T, et al. A missense mutation in a highly conserved region of CASQ2 is associated with autosomal recessive catecholamine-induced polymorphic ventricular tachycardia in Bedouin families from Israel. *Am J Hum Genet.* 2001;69(6):1378–1384.

89. Knollmann BC, Chopra N, Hlaing T, et al. Casq2 deletion causes sarcoplasmic reticulum volume increase, premature Ca2+ release, and catecholaminergic polymorphic ventricular tachycardia. *J Clin Invest.* 2006;116(9):2510–2520.

90. Liu N, Colombi B, Memmi M, et al. Arrhythmogenesis in catecholaminergic polymorphic ventricular tachycardia: insights from a RyR2 R4496C knock-in mouse model. *Circ Res.* 2006;99(3):292–298.

91. Leenhardt A, Lucet V, Denjoy I, et al. Catecholaminergic polymorphic ventricular tachycardia in children. A 7-year follow-up of 21 patients. *Circulation.* 1995;91(5):1512–1519.

92. Hayashi M, Denjoy I, Extramiana F, et al. Incidence and risk factors of arrhythmic events in catecholaminergic polymorphic ventricular tachycardia. *Circulation.* 2009;119(18):2426–2434.

93. Postma A, Denjoy I, Kamblock J, et al. Catecholaminergic polymorphic ventricular tachycardia: RYR2 mutations, bradycardia, and follow up of the patients. *J Med Genet.* 2005;42(11):863–870.

94. Fisher JD, Krikler D, Hallidie-Smith KA. Familial polymorphic ventricular arrhythmias: a quarter century of successful medical treatment based on serial exercise-pharmacologic testing. *J Am Coll Cardiol.* 1999;34(7):2015–2022.

95. Bai R, Napolitano C, Bloise R, et al. Yield of genetic screening in inherited cardiac channelopathies: how to prioritize access to genetic testing. *Circ Arrhythm Electrophysiol.* 2009;2(1):6–15.

96. Wilde AAM, Bhuiyan ZA, Crotti L, et al. Left cardiac sympathetic denervation for catecholaminergic polymorphic ventricular tachycardia. *N Engl J Med.* 2008;358(19):2024–2029.

97. Watanabe H, Chopra N, Laver D, et al. Flecainide prevents catecholaminergic polymorphic ventricular tachycardia in mice and humans. *Nat Med.* 2009; 15(4):380–383.

98. Gussak I, Brugada P, Brugada J, et al. Idiopathic short QT interval:a new clinical syndrome? *Cardiology.* 2000;94(2):99–102.

99. Funada A, Hayashi K, Ino H, et al. Assessment of QT intervals and prevalence of short QT syndrome in Japan. *Clin Cardiol.* 2008;31(6):270–274.

100. Mason JW, Ramseth DJ, Chanter DO, et al. Electrocardiographic reference ranges derived from 79,743 ambulatory subjects. *J Electrocardiol.* 2007; 40(3):228–234.

101. Kobza R, Roos M, Niggli B, et al. Prevalence of long and short QT in a young population of 41,767 predominantly male Swiss conscripts. *Heart Rhythm.* 2009;6(5):652–657.

102. Wolpert C, Schimpf R, Giustetto C, et al. Further insights into the effect of quinidine in short QT syndrome caused by a mutation in HERG. *J Cardiovasc Electrophysiol.* 2005;16(1):54–58.

Genetics of Cardiac Structure and Function

Sanjiv J. Shah, Sadiya S. Khan, and Donna K. Arnett

TAKE HOME POINTS

1. Abnormalities in cardiac structure and function are important intermediate phenotypes that mediate the transition from risk factors (such as hypertension, obesity, and diabetes) to heart failure. Several of these traits have been found to be heritable, and therefore likely have a significant genetic basis.
2. To determine the genetic basis of heart failure, it will most likely be more fruitful to study the genetics of cardiac structure and function phenotypes rather than heart failure per se because of the heterogeneity of the heart failure syndrome.
3. Determining the genetic basis of traits such as left ventricular (LV) mass has proven quite challenging in the linkage and genome-wide association study (GWAS) eras; advances in next-generation sequencing technologies, epigenomics, and gene–environment interaction, as well as improvements in phenotyping (eg, cardiac magnetic resonance [CMR]), will hopefully improve our understanding of the genetics of left ventricular hypertrophy (LVH).
4. Compared with common variants with small effects, it is more likely that rare variants with large effects contribute to cardiac structural and functional phenotypes such as LV mass and LVH.

CASE STUDIES

Three male African American patients, ranging in age from 50 to 60 years old, were referred by their internists to the echocardiography laboratory for evaluation of dyspnea. The first patient had no significant past medical history or family history of heart disease. However, he was found to have moderate LVH, with an interventricular septal wall thickness of 1.5 cm and posterior wall thickness of 1.6 cm.

The second patient had a history of severe, uncontrolled systemic hypertension. His echocardiogram was notable for mild diastolic dysfunction but was otherwise normal, with normal wall thickness and no valvular abnormalities.

The third patient had mild systemic hypertension, controlled on just one medication (chlorthalidone); however, his echocardiogram was notable for moderate concentric LVH, with septal and posterior wall thicknesses of 1.4 and 1.5 cm, respectively. On further questioning, the third patient revealed that both hypertension and "thick hearts" were common in his family; he had two siblings both with mild hypertension but with LVH. He also admitted to being a heavy smoker.

These three patients are examples of the heterogeneity of LVH in response to varying levels of afterload. The first patient has no history of hypertension or family history of hypertrophic cardiomyopathy; nevertheless, LVH is present. The second patient has a history of severe hypertension but despite the increased afterload, LV wall thickness has not increased. The final patient has moderate concentric LVH despite only mild hypertension, though based on the family history, appears to have a heritable form of LVH.

All three patients entered a research study on the genetic determinants of LVH. Interestingly, the first patient was found to have a single-gene mutation in a sarcomere protein gene (*MYBPC3*), thought to be a de novo mutation given the lack of family history of LVH or hypertrophic cardiomyopathy. No known genetic risk factors for LVH were detected in the second patient. The third patient, however, did have a variant in *NCAM1* (rs1436109, located in intron 1). *NCAM1* encodes neural cell adhesion molecule-1 (NCAM1), and rs1436019 has been associated with increased posterior wall thickness and relative wall thickness in hypertensive African American families (1). In this particular patient, the single-nucleotide polymorphism (SNP) and cigarette smoking together could have resulted in amplification of LVH in the setting of mild hypertension.

These three case studies demonstrate the variability of LVH, which has been found to be a heritable trait and is therefore likely influenced by both genetic and environmental factors. Both rare and common variants likely influence the complex trait of cardiac hypertrophy.

Adverse cardiac remodeling, such as LV dilation or LVH is a major risk factor for the development of coronary heart disease, heart failure, stroke, and death (2–7). Although certain comorbidities influence the development of cardiac remodeling, there is considerable interindividual variation in the development of abnormal cardiac structure. For example, hypertension, obesity, and diabetes are important determinants of LVH, but the presence and extent of these risk factors fail to identify many individuals with LVH, suggesting a genetic component (8). Genetic factors are thought to influence cardiac remodeling, supported by evidence of significant heritability of various parameters of LV structure, especially LV mass (9–16).

As early as 1969, electrocardiographic LVH was identified as a major determinant of cardiovascular death (17), and more sensitive measures of LVH have subsequently confirmed its importance in cardiovascular and stroke risk (18,19). Although effective treatments exist for other major cardiovascular risk factors (eg, hypercholesterolemia, hypertension), there is no exclusive treatment for LVH that is independent of lowering blood pressure (BP). Understanding the molecular pathogenesis of traits such as LVH through genetic studies may therefore allow the ability to develop therapies aimed at directly targeting the myocardium.

Several studies have investigated genetic determinants of cardiac structure using linkage, candidate gene, and GWAS approaches (1,20–26). As detailed below, the largest GWAS of cardiac structure and function conducted thus far (EchoGen) found genetic loci associated with LV dilation and hypertrophy, but these loci only explained 1% to 3% of the variance in these traits. EchoGen highlights the challenges in identifying the genetic determinants of cardiac structure and function, which have proven to be elusive complex phenotypes.

Here we discuss several aspects of the genetics of cardiac structure and function: (a) reasons to study genetics of cardiac structure/function; (b) challenges of accurately phenotyping cardiac structure/function; (c) heritability of cardiac structural and functional traits; (d) results from genetic studies of cardiac structure/function traits, particularly LV mass; and (e) future directions.

WHY STUDY THE GENETICS OF CARDIAC STRUCTURE AND FUNCTION?

There are several reasons to study the genetics of cardiac structure/function traits. First, as stated above, there is considerable interindividual variation in the development of abnormalities in cardiac structure and function in response to risk factors such as hypertension, which suggests the presence of genetic risk factors for the development of cardiac remodeling and dysfunction. Second, many cardiac structural and functional traits have been found to be heritable, which also supports the presence of a genetic component. Third, abnormalities in cardiac structure and function represent intermediate phenotypes that dictate the transition from hypertension to heart failure. Given the heterogeneity of the heart failure syndrome, it is likely that determining the genetics of cardiac structure/function will provide greater insight into heart failure pathogenesis compared to genetic studies of heart failure per se. Finally, indices of cardiac structure and function are risk factors for a wide variety of adverse cardiovascular endpoints; it is therefore hoped that understanding the molecular basis for traits such as LVH will allow for the development of therapies that ultimately will decrease adverse cardiovascular events in the population.

LVH is a common condition in the United States, both in the general population (occurring in 16% of whites and 33%–43% of African Americans (27)) and in hypertensives (occurring in 22%–60% of hypertensives (28)). Similar rates have been reported elsewhere (29,30). LVH is an independent risk factor for a variety of cardiovascular endpoints including stroke, myocardial infarction, heart failure, and all-cause mortality (4,5,31–33). LVH has been shown to be a better predictor of mortality than coronary artery disease and LV ejection fraction in some populations (34).

Increased LV mass has historically been considered an adaptive response to hypertension; variation in BP, however, accounts for only a modest fraction of the observed variance in LV mass (15,35). In fact, mounting evidence suggests that increased LV mass can precede the onset of hypertension in normotensives (36–38). It is widely accepted that genetic factors influence LV mass and other structural cardiac phenotypes.

As outlined below, heritability estimates have been as large as 0.59 for LV mass (16) and the heritabilities of other structural and functional echocardiographic phenotypes, such as wall thickness, left ventricular internal dimension (LVID), and mitral annular velocities, are also significant (10,39). Linkage (21,40,41), candidate gene association (42,43), and GWAS (1,23,24,26) have reported suggestive and statistically significant correlations between genetic loci and LVH and other structural and functional echocardiographic phenotypes. A pilot exome study in the Hypertension Genetic Epidemiology Network (HyperGEN) (44), described in detail below, also points to the potential role that rare genetic variation plays in these important phenotypes.

As stated above, studying the genetics of cardiac structure and function is likely to be more fruitful than studying the genetics of the heart failure syndrome itself. For example, consider two patients with heart failure, one with severe LV dilation and severely reduced LV ejection fraction, and the other with a small LV, severe concentric LVH, and preserved LV ejection fraction. It is unlikely that these two patients, both of whom have the heart failure syndrome, have a shared genetic risk factor for heart failure. Rather, patients with dilated cardiomyopathy and concentric LVH have different genetic risk factors, which can only be identified by studying the genetics of cardiac structure and function.

CHALLENGES OF ACCURATE PHENOTYPING OF CARDIAC STRUCTURE AND FUNCTION

Unlike some complex cardiovascular traits, such as cholesterol levels, accurate phenotyping of cardiac structure and function can be quite challenging because of the inability to directly measure the traits of interest. Furthermore, unlike genetic studies of other organs and tissues, large-scale access to suitable cardiac tissue (ie, cardiac biopsy specimens) is also not feasible. Several noninvasive modalities can be used to phenotype cardiac structure and function, including electrocardiography (ECG), echocardiography, and CMR, each of which has its strengths and limitations. For example, ECG QRS voltage is a marker of LVH, which has been found to be a heritable trait (14). ECG is inexpensive, relatively simple to perform accurately, and can be used in very large, population-based studies. In addition, measurement of QRS magnitude and duration is also highly reproducible, with less intra- and interobserver variability compared to

echocardiographic estimation of LV mass. These are all reasons that may explain the findings of Mayosi et al who demonstrated that ECG measures of LVH are more heritable compared to echocardiographic measures of LVH (14). However, ECG markers of LVH cannot distinguish between LV-enlargement-induced increases in LV mass and concentric LVH, which are likely to have different genetic determinants. Furthermore, some ECG markers of LVH such as the Cornell product include QRS duration, which can lead to potential incorrect associations of genetic variants with "LVH" when in fact the identified variants influence ventricular depolarization (45).

Thus far, most investigators have relied on 1-dimensional (1D), M-mode indices to estimate LV mass for studies on the heritability and genetics of LV mass and LVH. However, 1D-based LV mass estimation formulas, such as the Penn method (46) (LV mass = $1.04 \times [(\text{LVID} + \text{IVS} + \text{PW})^3 - (\text{LVID})^3] - 13.6$) can lead to large errors in LV mass in distorted hearts (due to inappropriate modeling of the ventricle) and in normal hearts (since small errors in measurement are magnified due to the cubic nature of the formula). Two-dimensional (2D) methods (such as the area-length and truncated ellipsoid methods (47)) and 3-dimensional (3D) methods (real-time 3D echocardiography or CMR) increase phenotypic precision (Figure 15.1), which may be very important for studies of LVH genetics. Reduction in phenotypic precision could have reduced the ability to detect genetic loci influencing cardiac structural traits.

CMR is a standard of reference for noninvasive measurement of cardiac structure (48). Compared with M-mode and 2D echocardiography, CMR is more reproducible, uses a 3D technique, has superior image quality, offers precise endocardial border definition, and is not dependent on sonographer technique or acoustic windows (48,49). Indeed, Busjahn et al found very high estimates of heritability of LV mass ($h^2 = 0.82$), much higher than echocardiography-based estimates, among 25 healthy twin pairs (50). These authors reasoned that the increased phenotypic precision of CMR allows for the study of fewer numbers of subjects and may allow genetic assessment of cardiac structures, which is not possible with conventional echocardiography.

Advances in imaging techniques, along with the ability to estimate load-independent indices of cardiac function, may also allow for novel insight into the genetics of cardiac mechanics. Tissue Doppler imaging and speckle-tracking echocardiography (Figure 15.2) provide the ability to examine regional myocardial deformation for determination of tissue velocities and strain, respectively (51,52). These measures of intrinsic myocardial function are not fully load-independent, but are increasingly being

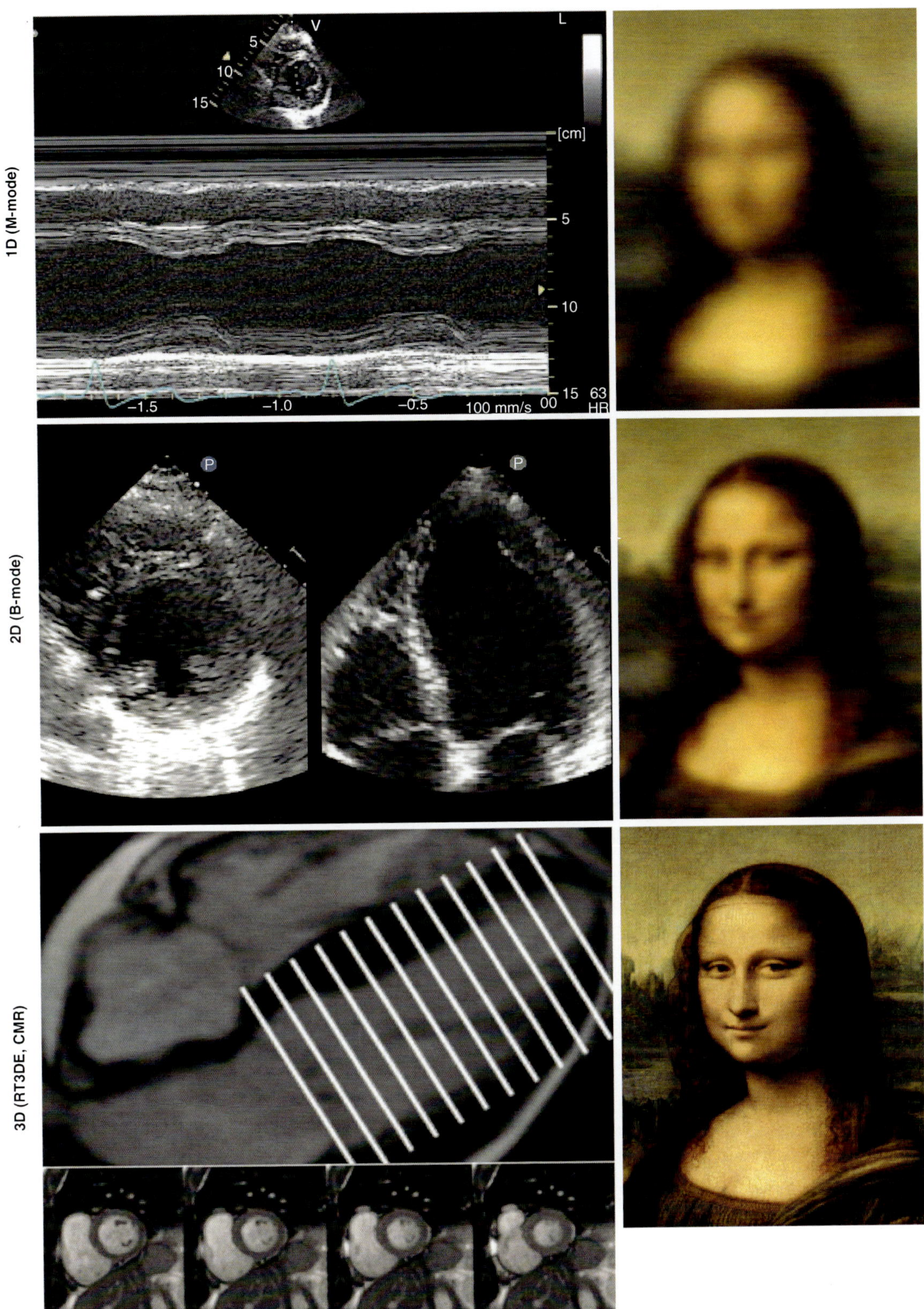

FIGURE 15.1 Comparison of methods for phenotyping left ventricular mass. The majority of genetic studies of left ventricular (LV) mass have used 1-dimensional (1D) methods (ie, M-mode echocardiography-based measurements [top panel]); however, this method results in only a "fuzzy" estimate of the true LV mass, symbolized by the unfocused photograph of da Vinci's Mona Lisa on the right. The use of 2-dimensional (2D) echocardiography (middle panel), improves upon 1D methods by increasing the number of data points used to create the LV mass estimate; however, it too is imperfect because it still relies on geometric assumptions. The 3-dimensional (3D) methods, real-time 3D echocardiography and cardiac MRI, allow for 3D volumetric assessment of LV mass, and therefore are the most accurate methods for estimation of LV mass. The bottom panel displays an example of cardiac MRI technique for estimation of LV mass, symbolized by the most focused photograph on the right.

Abbreviations: CMR, cardiac magnetic resonance; RT3DE, real-time 3D echocardiography.

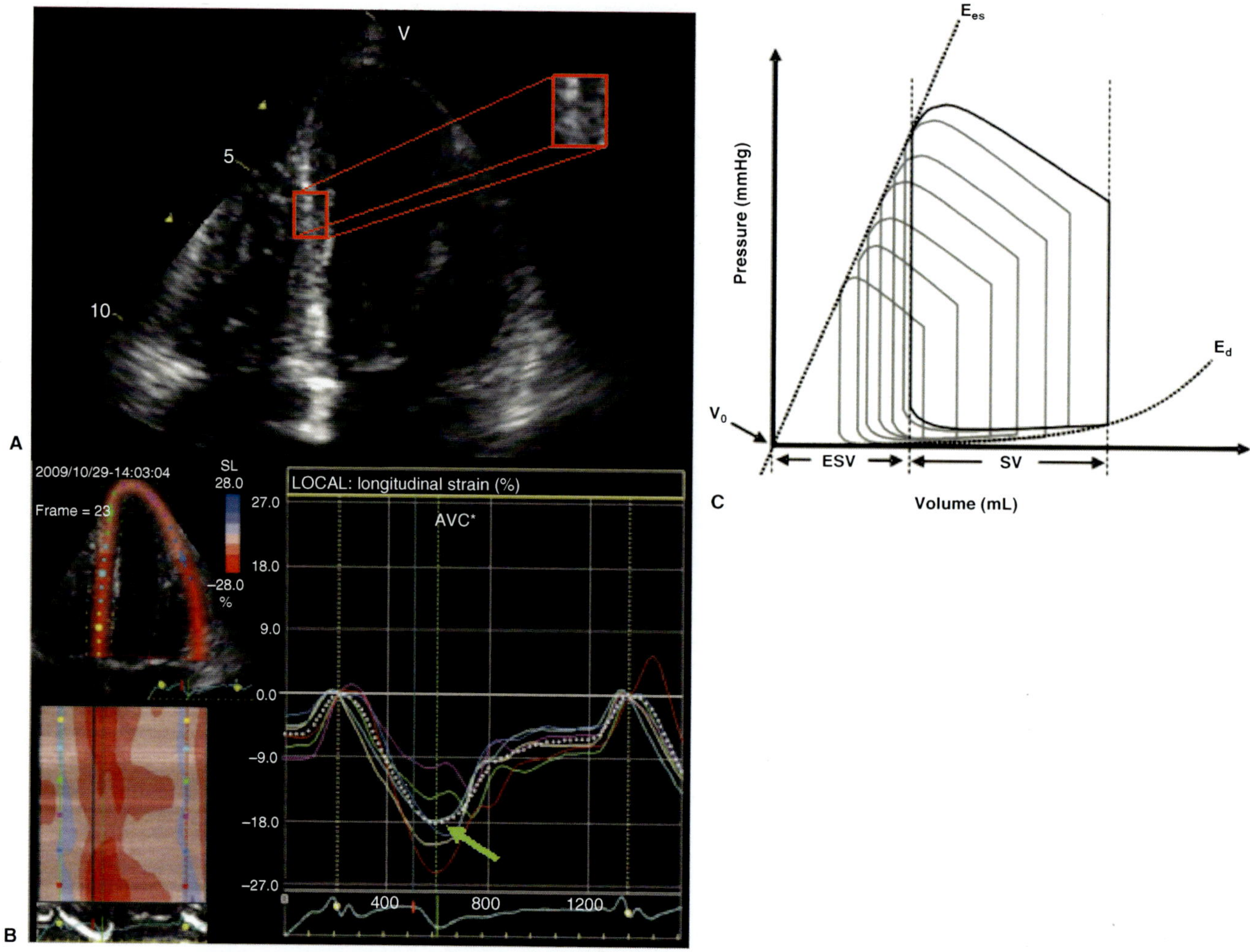

FIGURE 15.2 Advanced phenotyping techniques for cardiac function: speckle-tracking echocardiography and pressure–volume analysis. Speckle-tracking echocardiography and pressure–volume analysis are two examples of advanced quantification techniques for the evaluation of cardiac function. (A) Speckle-tracking echocardiography involves the analysis of speckle artifacts, which are randomly generated by reflections, refraction, and scattering of ultrasound beams. These speckles serve as natural acoustic markers for tagging myocardial motion throughout the cardiac cycle. A cluster of speckles (called a "kernel") is defined and followed frame to frame, thereby allowing calculation of Lagrangian strain (the ratio of change in length of a myocardial segment divided by its original length [$\Delta L/L_0$], multiplied by 100). (B) The parasternal short axis and apical 4-, 3-, and 2-chamber views are each divided into six segments and regional strains (color lines) and averaged strain (white dotted line) are graphed throughout the cardiac cycle. The peak average strain in systole is termed "global strain," with the direction of strain (longitudinal, circumferential, or radial) noted. Thus, the green arrow in the middle panel represents peak systolic global longitudinal strain. (C) Invasive pressure–volume analysis involves plotting pressure vs. volume throughout the cardiac cycle. Preload reduction maneuvers (such as occlusion of the inferior vena cava) result in successively smaller pressure–volume loops, which allows for the calculation of the load-independent measures such as the slopes of the end-systolic and end-diastolic pressure–volume relationships (end-systolic elastance and end-diastolic elastance, respectively). Single-beat, noninvasive methods are now available for estimation of pressure–volume relationships, thereby allowing for use of these measures in large-scale genetic studies.

Abbreviations: E_d, end-diastolic elastance; E_{es}, end-systolic elastance; ESV, end-systolic volume; SV, stroke volume; V_0, LV volume at the x-axis intercept of the end-systolic pressure–volume relationship.

recognized as important markers of subclinical myocardial dysfunction that provide insight into the health of myocytes (53). Thus, these indices are especially attractive as potential traits for genetic studies.

Pressure–volume analysis of cardiac function (Figure 15.2) is a way to generate load-independent indices of systolic and diastolic ventricular function (54). However, measuring the slope of the end-systolic pressure–volume relationship (end-systolic elastance) and end-diastolic pressure–volume relationship (end-diastolic elastance) has traditionally required invasive, conductance-catheter measurements with preload reduction maneuvers, thus rendering genetic studies of pressure–volume indices in humans virtually impossible. Fortunately, it is possible to estimate pressure–volume indices noninvasively using single-beat methods (55–57). While not yet examined in large-scale genetic studies, these load-independent indices of cardiac structure and function may allow for understanding the genetic determinants of ventricular systolic and diastolic chamber stiffness.

HERITABILITY OF INDICES OF CARDIAC STRUCTURE AND FUNCTION

Several studies have examined the heritability of indices of cardiac structure and function. Most of these studies have focused on LV mass and LVH (10–12,14–16,39,41,50,58–63), though the heritability of LV dimensions (10,58,61,64), left atrial size (65,66), and Doppler and tissue Doppler velocities (39,41,61) has also been shown. Most of the studies of heritability of LV mass have used M-mode echocardiography-based estimation of LV mass, with variable adjustment for body size and other determinants of LVH (Table 15.1). These studies have shown that LV mass is a heritable trait with heritabilities ranging from 0.15 to 0.59. In a population-based study, Schunkert and colleagues demonstrated that siblings of participants with LVH were more likely to have increased septal and posterior wall thickness, relative wall thickness, concentric remodeling, and concentric

LVH (but not eccentric LVH) compared to matched controls.

The most comprehensive heritability study of cardiac structure and function published thus far is by Jin and colleagues (39). These investigators used comprehensive echocardiography, including Doppler and tissue Doppler imaging, to evaluate the heritability of a wide range of echocardiographic measures of cardiac structure and function in 52 families ($n = 459$ individuals). Besides showing that LV mass and LV dimensions were heritable, Jin et al also found moderate heritability of tissue Doppler indices ($h^2 = 0.36$ [$P = .003$] and 0.53 [$P < .001$] for early [e′] and atrial [a′] diastolic tissue velocities, respectively, after adjusting for age, sex, height, weight, systolic BP, and heart rate) and ejection fraction ($h^2 = 0.48$ [adjusted $P < .001$]).

Overall, these studies demonstrate that indices of cardiac structure and function are heritable traits, thereby providing rationale for studying the genetic determinants of these traits.

GENETICS OF CARDIAC STRUCTURE AND FUNCTION: PUBLISHED STUDIES

As mentioned above, several studies have investigated the genetic determinants of cardiac structure and function using linkage, candidate gene, and GWAS approaches (1,11,21–26,40–42,44,59,62,67–77). Examples of linkage, candidate gene, and GWASs published in the GWAS era (2007–present) are displayed in Table 15.2. Here we will focus on the results of a study of rare, single-gene mutations in the Framingham

TABLE 15.1 Heritability Studies of Left Ventricular Mass

Study Population	N	LV Mass Method*	Heritability	Covariates	P**
Chinese Family Study (11)	1,145	Penn method; not indexed	0.15	SBP	.0009
Framingham Heart Study (15)	2,624	Penn method; not indexed	0.24–0.32	Age, ht, wt, BP	–
Twin Study (European) (16)	752	Penn method; not indexed	0.59	None	<.05
Juo et al. (Hispanic Caribbean) (13)	623	Penn method; indexed to BSA	0.50	Age, sex	<.001
Bella et al (Native American) (89)	1,373	Penn method; not indexed	0.27	Age, sex, center	<.001
Mayosi et al (European) (14)	955	ASE method; not indexed	0.23–0.41	None	<.001
Swan et al (European) (60)	220	Penn method; not indexed	0.53	Age, sex, BP, wt	.006
Jin et al (European) (39)	459	ASE method; not indexed	0.23–0.48	Age, sex, ht, wt, SBP, heart rate	<.001

*Penn method = 1.04 × ([LVIDd + PWTd + IVSTd]³ – [LVIDd]³) – 13.6; ASE method = 0.8 × (1.04 × ([LVIDd + PWTd + IVSTd]³ – [LVIDd]³)) + 0.6.
**P values shown are those for the most minimally adjusted model.

Abbreviations: ASE, American Society of Echocardiography; BP, blood pressure; BSA, body surface area; ht, height; IVSTd, interventricular septal wall thickness at end diastole; LVIDd, left ventricular internal diameter at end diastole; LV mass, left ventricular mass; PWTd, posterior wall thickness at end diastole; SBP, systolic blood pressure; wt, weight.

TABLE 15.2 Selected Studies Identifying Genetic Risk Factors for Indices of Cardiac Structure and Function

Author (Year)	Quantitative Phenotype	Study Design	Main Findings	Notes
Parry et al (2013) (25)	LVH	GWAS	Genotype data was available from 973 cases with LVH and 1,443 non-LVH controls from the Genetics of Diabetes Audit and Research Database in Tayside, Scotland. Two SNPs previously associated with LVH were significant: rs17132261 (OR 2.03, P = .02) and rs2292462 (OR 0.82, P = 2.3 × 10^{-3}). Meta-analysis resulted in an additional SNP reaching genome-wide significance: rs4966014 (P = 1.35 × 10^{-8}).	This study was performed exclusively in patients with diabetes (Go-DARTS study) with replication in a meta-analysis pooling multiple population-based cohorts including the EchoGen consortium and the BWHHS, GRAPHIC, and WHII cohorts. Of note, the BWHHS, GRAPHIC, and WHII defined LVH by ECG criteria.
Harper et al (2013) (70)	LV mass	Candidate gene association study	In a cohort of 255 families comprising 1,425 individuals ascertained via a hypertensive proband, 10 SNPs at the miR-22 gene were genotyped with evidence of association between rs7223247 and LV mass.	Rs7223247 lies in the 3' UTR of a gene of unknown function, *TLCD2*, which is downstream from miR-22 and this genotype was responsible for 1% of the population variability in LV mass. Of note, LV mass was approximated by Sokolow–Lyon voltage (ECG).
Fox et al (2013) (69)	LV structure and function	GWAS	In the Candidate Gene Association Resource (CARe) Consortium, 6,765 African Americans were genotyped using Affymetrix Human SNP Array 6.0 (with imputation) for 2.5 million SNPs; four genetic loci reached genome-wide significance: rs4552931 in *UBE2V2* (P = 1.43 × 10^{-7}) for LV mass, rs7213314 in *WIPI1* (P = 1.68 × 10^{-7}) for LVIDd, rs1571099 in *PPADC1A* (P = 2.57 × 10^{-8}) for SWT, and rs9530175 in *KLF5* (P = 4.02 × 10^{-7}) for EF.	The CARe consortium consists of nine population-based cohort studies sponsored by the NHLBI, with four cohorts that included African Americans (ARIC, JHS, CARDIA, and MESA). Replication was performed in three smaller cohorts of African ancestry (GENOA, HyperGEN, and CHS). However, none of the four loci was confirmed in the lookups in EchoGen, a consortium of European ancestry, suggesting these loci may be unique to individuals of African ancestry; however, additional studies are warranted.
Zhi et al (2012) (44)	LVH	WES combined with in vitro functional assessment and linkage analyses	WES was conducted in seven African American sibling trios with high average familial LV mass indexed to height using Illumina HiSeq technology. WES findings were functionally assessed in human induced pluripotent stem cell derived cardiomyocytes with RNA sequencing to determine gene expression differences. Finally, candidate genes were prioritized by regional linkage evidence in the larger HyperGEN cohort and publicly available gene- and variant-based annotations. A total of 295 variants in 265 genes were associated with LV mass with 44 differentially expressed in hypertrophied cells and 5 meeting supportive criteria: *HLA-B, HTT, MTSS1, SLC5A12, and THBS1*.	This study uniquely combined data from WES with cellular experiments to assess functional significance. In addition, replication was performed using existing linkage analysis results for LV mass from the HyperGEN cohort to provide supportive evidence for the candidate genes discovered by WES.

TABLE 15.2 Selected Studies Identifying Genetic Risk Factors for Indices of Cardiac Structure and Function (*continued*)

Author (Year)	Quantitative Phenotype	Study Design	Main Findings	Notes
Hong et al (2012) (71)	LVH	GWAS	8,432 controls and 398 cases in the community-based Korea Association Resource (KARE) study were analyzed by Affymetrix SNP array 5.0 with validation in hospital-based samples (596 controls and 207 cases). Fourteen SNPs in eight loci were associated with LVH (5q35.1, 6p22.3-22.1, 8q24.2, 11p15, 11q21-22.1, 14q12, 17q11.2, and 19q13.1) with $P < 1 \times 10^{-5}$. The most significant SNP in the replication cohort was identified within the 19q13.1 region, which contains the *RYR1* gene.	Mutations in *RYR1*, which encodes a major calcium channel in skeletal muscle, have been reported to be associated with cardiovascular disease and may be implicated in the pathophysiology of LVH. Of note, in this study LVH was diagnosed by ECG using the Minnesota Code Classification System.
Arnett et al (2011) (1)	LV structure phenotypes	GWAS	From the HyperGEN study, 1,258 black and 1,316 white participants were genotyped using Affymetrix Genome-Wide Human SNP 6.0 Array. Of the five SNPs identified in blacks ($P \leq 10^{-6}$), only one (rs1436109; *NCAM1* intron1) replicated in GENOA with the same phenotype of PWT, $P = .025$. This SNP was also replicated in HyperGEN white families.	NCAM is upregulated during the remodeling period during the transition from LVH to heart failure in Dahl salt-sensitive rats offering biological plausibility for its role in LV remodeling.
Hall et al (2011) (90)	LV mass	Candidate gene association study	255 families comprising 1,425 individuals ascertained via a hypertensive proband were included, and seven SNPs, which together tagged common genetic variation in the *CD36* gene, were genotyped using a SEQUENOM MALDI-TOF instrument. There was evidence of association for rs1761663 (Echo $P = .003$, ECG $P = .001$).	The genotype at rs1761663 had independent effects on body mass index in addition to LV mass supporting its pleiotropic effects. Limitations include the lack of a replication cohort.
Shah et al (2011) (91)	LVH	Candidate gene association study	The discovery cohort consisted of 10,256 individuals who had their DNA typed using a customized gene array (the Illumina HumanCVD BeadChip 50K array) with replication in 11,777 individuals. Four loci were associated with ECG-LVH indices: 3p22.2 (*SCN5A*, rs6797133, $P = 1.22 \times 10^{-7}$), 12q13.3 (*PTGES3*, rs2290893, $P = 3.74 \times 10^{-8}$), 15q25.2 (*NMB*, rs2292462, $P = 3.23 \times 10^{-9}$), and 15q26.3 (*IGF1R*, rs4966014, $P = 1.26 \times 10^{-7}$).	The discovery cohort comprised three population-based cohorts including BWHHS, GRAPHIC, and WHII. The replication cohorts comprised individuals from BRHS, BRIGHT, and the PREVEND study. Analyses in both discovery and replication cohorts were restricted to subjects of European White descent. The HumanCVD BeadChip 50K array contains only about 10% of genes in the human genome.

(continued)

TABLE 15.2   Selected Studies Identifying Genetic Risk Factors for Indices of Cardiac Structure and Function (*continued*)

Author (Year)	Quantitative Phenotype	Study Design	Main Findings	Notes
Wang et al (2010) (66)	Left atrial enlargement	Linkage analysis and peak-wide association analysis	100 Dominican families comprising 1,350 individuals were studied, and linkage analysis revealed suggestive evidence on chromosome 10p19 (D10S1423, MLOD = 2.00) and 17p10 (D17S974, MLOD = 2.05). Peak-wide association demonstrated associations in *NTN1*, *MYH10*, *COX10*, and *MYOCD* genes (P = .00005 to .005).	*MYOCD* has also been shown to serve as a key transducer of hypertrophic signals in cardiomyocytes.
Vasan et al (2009) (24)	LV structure and function phenotypes	GWAS	The EchoGen consortium consisting of five community-based cohorts comprised the discovery sample (n = 12,612) with replication in two other community-based samples (n = 4,094). In the discovery cohort, five genetic loci were associated with four echocardiographic traits (LV mass, wall thickness, internal dimensions, and systolic dysfunction). In the replication cohort, only one locus was replicated (6q22 with LV diastolic dimensions).	The EchoGen consortium includes seven cohort studies (CHS, Rotterdam, MONICA, Gutenberg, FHS, SHIP, and Austrian Stroke Prevention). Discovery was performed in CHS, Rotterdam, MONICA, Gutenberg, and FHS with replication in SHIP and the Austrian Stroke Prevention Study. All cohorts comprised individuals of European ancestry. While the findings are novel, the loci explained a very small proportion of the variance of the traits (0.2%–0.5%).
Tang et al (2009) (76)	LV mass, dimensions, and diastolic function	Genome-wide linkage scan	In 434 African American families (1,344 individuals) and 284 white families (1,119 individuals) from the HyperGEN study, genome-wide linkage scan was performed for LV mass, wall thickness, and diastolic filling. LV phenotypes and composite factor scores were analyzed in a multipoint variance components linkage model (SOLAR). In whites, suggestive linkage was observed on chromosomes 2 and 4 for LV mass, chromosomes 3, 5, 10, and 17 for LV atrial phase peak filling velocity, and chromosome 10 for LV diastolic filling fraction. In African Americans, suggestive linkage was observed on chromosome 12 for LV mass, chromosome 21 for SWT, and chromosome 3 for LVIDd.	Linkage analysis was also performed with measures of body size, given that it is an important correlate of LV mass and significant genetic correlation was found in African Americans (P < .05) and in whites (P < .05) after adjustment for age and gender. Limitations included the lack of a replication cohort to confirm these findings.
Arnett et al (2009) (26)	LV mass	GWAS	In 101 cases and 101 controls selected from the extreme tails of LV mass index distribution, GWAS in Caucasians was performed using the Affymetrix GeneChip Human 100K Set from the HyperGEN study. Eleven SNPs were identified and successfully genotyped in the validation study of 704 Caucasians and 1,467 African Americans; five SNPs on chromosomes 5, 12, and 20 were significantly associated with LV mass after correction for multiple testing.	One of the identified SNPs, rs756529, is intragenic within *KCNB1*, which is dephosphorylated by calcineurin and has been previously reported as a candidate gene for LVH. Different association effects were noted in the validation study between those of European and African descent, which is not unexpected as the discovery cohort did not include African Americans.

TABLE 15.2 Selected Studies Identifying Genetic Risk Factors for Indices of Cardiac Structure and Function (*continued*)

Author (Year)	Quantitative Phenotype	Study Design	Main Findings	Notes
Arnett et al (2009) (21)	LV mass and relative wall thickness	Linkage and association analysis	Race-specific linkage analyses were conducted in the HyperGEN study in 885 siblings from 382 sibships. A broad band of peaks in both white and blacks was observed on chromosome 4 and selected candidate genes from this region (*NPY1R, NPY2R, NPY5R, SFRP2, CPE, IL15,* and *EDNRA*). In blacks, SNPs in *IL15, NPY2R,* and *NPY5R* showed strong evidence for association ($P < .0005$). In whites, *NPY2R, NPY5R,* and *SFRP2* SNPs offered suggestive association with one or more traits ($P < .05$).	Limitations include the absence of haplotype analysis and a replication cohort to confirm findings.
Vasan et al (2007) (23)	LV structural measurements	GWAS	In the community-based Framingham Heart Study, generalized estimating equations, family-based association tests, and variance components linkage were used to relate multivariable-adjusted trait residuals to 70,987 SNPs (human 100K GeneChip Affymetrix) restricted to autosomal SNPs with minor allele frequency ≥ 0.10, genotype call rate ≥ 0.80, and Hardy–Weinberg equilibrium ≥ 0.1. The top SNPs associated with traits were: LVIDd (rs1379659; *SLIT2, P* = 1.17×10^{-7}), LVIDs (rs105045; *KCNB2, P* = 5.18×10^{-6}), LV mass (rs10498091, *P* = 5.68×10^{-6}), and left atrial size (rs1935881; *FAM5C, P* = 6.56×10^{-6}).	This analysis served as a hypothesis-generating GWAS in a moderate-sized community-based sample. Limitations include the lack of a replication cohort to confirm these findings as well as extension to include other ethnicities as this sample was white and of European descent only.

Studies are listed in reverse chronological order.

Abbreviations: ARIC, Atherosclerosis Risk in Communities; BMI, body mass index; BWHHS, British Women's Heart and Health Study; CARDIA, Coronary Artery Risk Development in Young Adults; ECG, electrocardiogram; EF, ejection fraction; GENOA, Genetic Epidemiology Network of Atherosclerosis; GRAPHIC, The Genetic Regulation of Arterial Pressure of Humans in the Community; GWAS, genome-wide association study; HyperGEN, Hypertension Genetic Epidemiology Network; JHS, Jackson Heart Study; LV, left ventricular; LVH, left ventricular hypertrophy; LVIDd, left ventricular internal dimension at end diastole; LVIDs, left ventricular internal dimension at end systole; MESA, Multi-Ethnic Study of Atherosclerosis; MLOD, maximal log of odds; NHLBI, National Heart Lung, and Blood Institute; PWT, posterior wall thickness; SOLAR, sequential oligogenic linkage analysis routines; SNP, single nucleotide polymorphism; SWT, septal wall thickness; WES, whole-exome sequencing; WHII, Whitehall II Study.

Heart Study (FHS); three GWASs of cardiac structure/function (EchoGen, HyperGEN, and the National Heart, Lung, and Blood Institute [NHLBI] Candidate Gene Association Resource [CARe]); and finally, a study in HyperGEN that combined whole-exome sequencing (WES) with functional studies in induced pluripotent stem cell (iPS)-derived cardiomyocytes. Pre-GWASs of cardiac structure/function, particularly LVH, have been summarized in detail previously (8,78).

Morita et al studied 1,862 unrelated participants from FHS who had previously undergone echocardiography and had provided DNA samples (79). Of the 1,862 participants, 50 (3%) had unexplained increase in LV wall thickness, defined as septal or posterior wall thickness greater than 13 mm. These 50 unrelated individuals were further studied by sequencing of 8 sarcomere protein genes, 3 storage disease cardiomyopathy-causing genes, and 27 mitochondrial genes. These investigators found that 9/50 (18%) of these individuals, not previously known to have a genetic basis of LVH, had sarcomere protein or lipid storage gene mutations. Eight mutations in nine individuals were detected: seven mutations in five sarcomere protein genes (*MYH7*, *MYBPC3*, *TNNT2*, *TNNI3*, and *MYL3*), and one *GLA* gene mutation. This study is important because it demonstrates that single-gene mutations may be responsible for a significant number of individuals with unexplained LVH in the general population.

The EchoGen consortium investigators published the first large-scale GWAS of cardiac structure and function that included replication (24). In this study, high-density genotyping of SNPs (with imputation to the HapMap CEU panel) and echocardiographic data from five community-based cohorts ($n = 12,612$, all of European descent) were used to perform a GWAS of multiple traits including LV mass, internal dimensions, wall thickness, systolic dysfunction, and left atrial size (Figure 15.3). SNPs that were significantly associated with indices of cardiac structure/function in the first stage of the study were replicated in two additional cohorts ($n = 4,094$). Despite the relatively large number of individuals in this study, only one locus achieved genome-wide significance and was successfully replicated (6q22 locus associated with LV dimensions), and it only explained less than 1% of the trait variance. The EchoGen study highlighted the challenges of estimating cardiac structure and function from echocardiography, and the likely need for much larger studies in order to successfully perform GWAS for these traits.

In the HyperGEN and Genetic Epidemiology Network of Arteriopathy (GENOA) studies, echocardiography was used to determine the basis of LVH in whites and African Americans. These investigators identified genetic loci contributing to LVH and related traits through traditional approaches (ie, linkage (21,40,67,76,80,81)) and GWASs (1,26). In HyperGEN, a GWAS for LVH and related phenotypes was first conducted in African American participants, and then replicated in African American participants of GENOA. A variant in intron 1 of the *NCAM1* gene (rs1436109), identified first in HyperGEN African Americans, was associated with posterior wall thickness and relative wall thickness in GENOA, and then successfully replicated in HyperGEN whites. Studies in Dahl salt-sensitive rats have demonstrated that NCAM1 is upregulated in myocytes during the transition from hypertrophy to heart failure, giving biological plausibility to the role of *NCAM1* in LVH. In addition, as shown in Figure 15.4, pathway analysis demonstrates that *NCAM1* is connected with several genes that are expressed in cardiac tissue and are known to have a biological role in heart disease.

In the NHLBI CARe study, African Americans from multiple NHLBI population-based cohort studies underwent genotyping with the Affymetrix 6.0 platform, after which SNPs (imputed to HapMap) were examined for associations with indices of cardiac structure/function measured by echocardiography and CMR (69). In this study, variants in four genetic loci reached genome-wide significance (defined a priori as $P < 4.0 \times 10^{-7}$): rs4552931 in *UBE2V2* ($P = 1.43 \times 10^{-7}$) for LV mass, rs7213314 in *WIPI1* ($P = 1.68 \times 10^{-7}$) for LV internal diastolic diameter, rs1571099 in *PPAPDC1A* ($P = 2.57 \times 10^{-8}$) for septal wall thickness, and rs9530176 in *KLF5* ($P = 4.02 \times 10^{-7}$) for LV ejection fraction. None of the four loci replicated in cohorts of African ancestry was found to be associated with LV structure/function in whites enrolled in EchoGen, therefore suggesting that these variants are specific to those with African ancestry.

Zhi and colleagues performed an analysis of African Americans in HyperGEN by combining WES with iPS-derived cardiomyocytes stimulated to induce hypertrophy (44). HyperGEN African American participants from seven sibling trios ascertained based on high average familial wall thickness underwent WES. A total of 295 of the identified mutations (in 265 genes) were missense or nonsense mutations. Of the 265 genes, 44 candidate genes were differentially expressed in hypertrophied cells (compared to control cells). These genes were further filtered based on additional association tests, regional linkage data from the entire HyperGEN cohort, and gene- and variant-based annotations from publicly available databases. Of the five identified genes, *THBS1*, which encodes an adhesive glycoprotein that promotes matrix preservation in LVH induced by pressure overload was the most promising finding (44). This study was novel

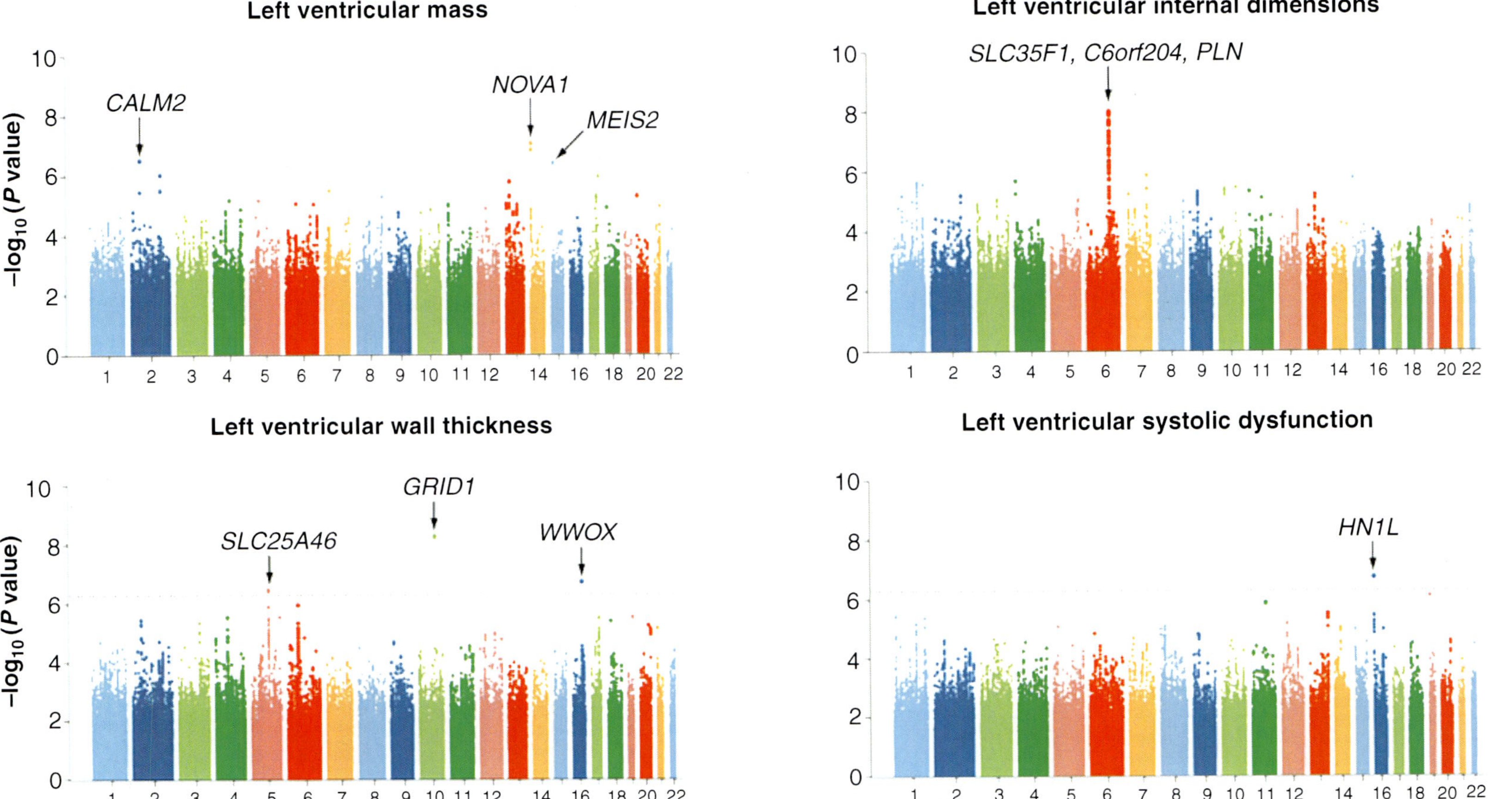

FIGURE 15.3 Genome-wide signal intensity plots for echocardiographic parameters in the EchoGen consortium. The graphs ("Manhattan plots") display the log P values (based on the fixed-effects meta-analysis) for single nucleotide polymorphisms against their genomic position for the following traits: left ventricular (LV) mass, LV internal dimension, LV wall thickness, and LV systolic dysfunction (ie, reduced LV ejection fraction). Within each chromosome, shown on the x-axis, the results are plotted from the p-terminal end. The horizontal dotted lines indicate the genome-wide significance threshold of $P = 5 \times 10^{-7}$.

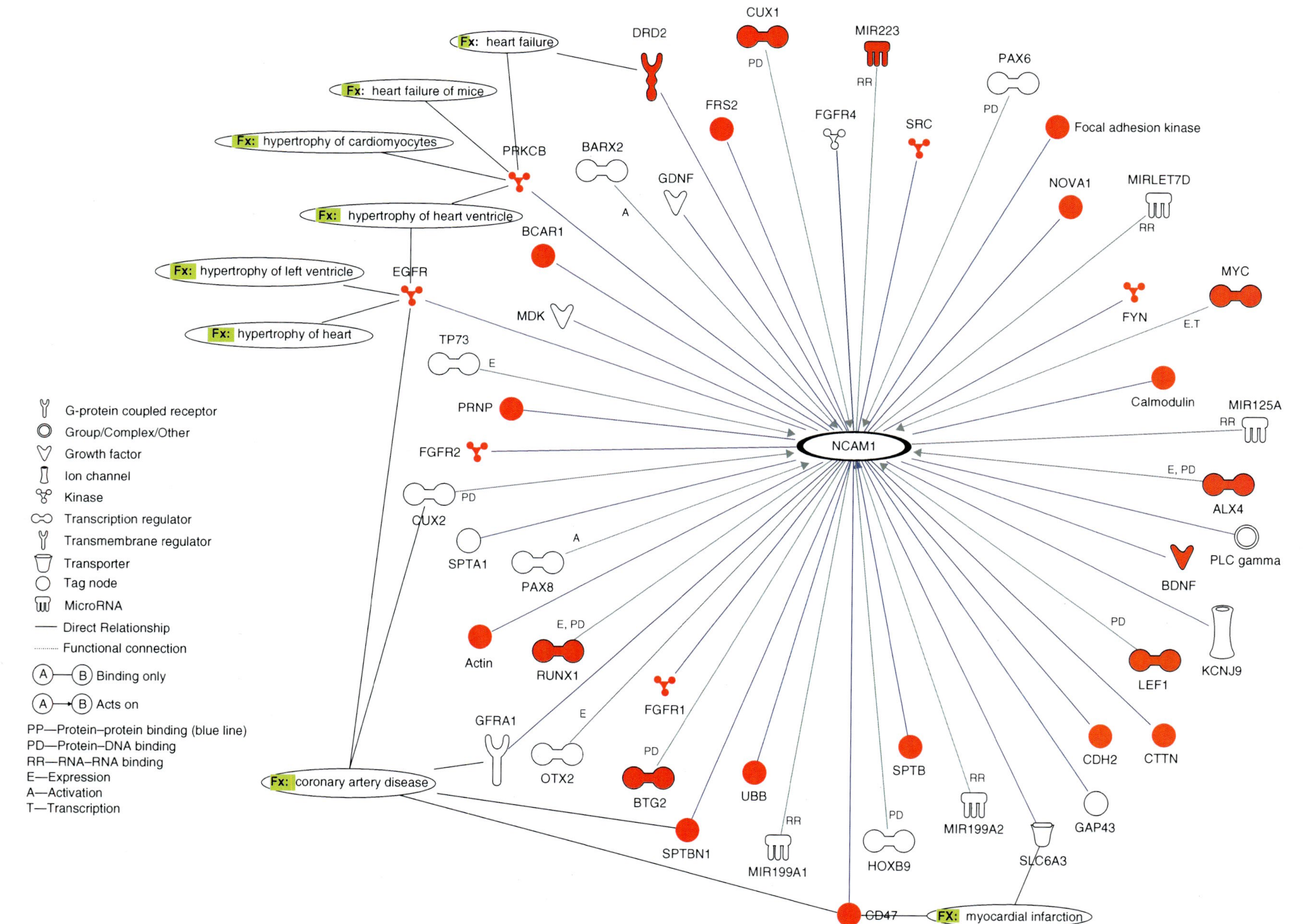

FIGURE 15.4 Interactions of *NCAM1* based on pathway analysis. Blue lines represent protein–protein interactions. Gray line interactions coded with letter. Red genes are expressed in cardiac tissue.

in its approach to LVH genetics, as it combined two cutting-edge technologies (WES and iPS cells) with a family-based design from a population-based study (HyperGEN).

FUTURE DIRECTIONS

There is the realization that, while GWASs have identified important loci, reliance on common variant-based studies is unlikely to solve the genetic puzzle for complex traits: LVH and other cardiac structural and functional traits are no exception. Thus, in order to enhance gene discovery for these traits, investigators will most likely need to search for rare and low-frequency variants with larger effect sizes through state-of-the-art exome-sequencing and analysis methods (82,83) using sampling from the extremes to select subjects for sequencing, a powerful method for the identification of rare genetic variants (84) (see also Chapter 8).

The notion that more than just GWAS is necessary for the genetic dissection of cardiac traits such as LVH is supported by the fact that a large amount of genetic variation remains unexplained. This is similar to most complex traits, and has been referred to as "missing heritability" (85). Missing heritability could result from low-frequency (minor allele frequency less than 5%) coding variants largely missed by the array-based genotyping of GWAS but more comprehensively assayed by WES, offering many advantages to the newer technology. For instance, the gene membership of exomic variants is typically clearly defined; however, GWAS variants often fall in intergenic regions. Second, annotation of exomic variants is often more informative, as our interpretation of protein-coding variants is far more detailed than noncoding variants. Third, exome sequencing provides a complete catalog of coding variants, and thus there is a good chance that the true causal variants, if protein-coding, are directly observed. This is in contrast to GWAS signals that "tag" the true underlying causal variant, and this tagging is less accurate for rare variants.

Beyond WES and other next-generation sequencing technologies, the use of iPS-derived cardiomyocytes from human blood samples, as done in the study by Zhi et al (44), may also provide novel insight into the molecular pathogenesis of LVH and other cardiac structure/function traits. These cardiomyocytes, derived from the DNA of patients involved in cohort studies such as HyperGEN, can be studied in several ways heretofore unimaginable without cardiac biopsy tissue. However, several questions remain: How do these cardiomyocytes compare to in situ cardiomyocytes within the hearts of study participants? Do exogenous, stimuli-induced changes in structure/function of iPS-derived cardiomyocytes mimic what happens in the human whole heart in response to increased afterload? What about the influence of epigenetic factors, which will not be present in the iPS-derived cardiomyocytes?

Finally, as described above, new approaches to phenotyping cardiac structure and function may allow for more successful genetic studies of these traits. In particular, speckle-tracking echocardiography and CMR look especially promising; combined with next-generation sequencing techniques, these imaging modalities will hopefully provide novel insight into the molecular pathogenesis of cardiac remodeling and dysfunction.

CONCLUSIONS

Abnormalities in cardiac structure and function are major risk factors for adverse cardiovascular events, and many of these abnormalities specifically mediate the transition from risk factors to overt, symptomatic heart failure. Given the considerable interindividual variation in cardiac remodeling and dysfunction in response to risk factors, combined with the heritability of many cardiac structure/function traits, it is likely that there are significant genetic risk factors for these traits. Thus far, LVH has been the most well-studied trait because of its importance as a prime risk factor for worse cardiovascular outcomes.

Despite knowledge that LVH imparts significant risk for adverse cardiovascular outcomes, limited progress has been made in identifying treatment strategies specifically targeted to its prevention or regression independent of the influence on BP. Treatment with antihypertensive medications results in only about one-third less incident LVH than in controls (86). In some antihypertensive (angiotensin-converting enzyme inhibitor, calcium channel blocker) trials, LV mass regression to the reference range occurred in only about half of those treated (87). No doubt this is due, at least in part, to the limited efficacy of antihypertensives to control BP. However, changes in LV mass during BP treatment are only weakly correlated to BP reduction; this suggests there is a nonhemodynamic component (88). These data identify an important "gap" in terms of LVH prevention and treatment that could be addressed by identifying specific genetic variants associated with LVH.

Unfortunately, common variant-based methods such as GWAS have become critical barriers. To fill the gap, we will likely need a combination of new approaches and techniques to determine the genetics of cardiac structure and function. Family-based cohorts, exome sequencing, studies using cardiomyocytes derived from iPS cells, epigenetics, gene–environment interaction studies, and advanced phenotyping techniques are all ways that will hopefully increase our ability to find genetic variants that influence cardiac structure/function. In particular, discovering rare variants (with large effects) that account for a significant amount of variation could (a) provide the basis for new strategies for screening individuals for more aggressive medical management of risk factors, and (b) stimulate development of novel treatments to prevent cardiac remodeling such as LVH and the progression from risk factors to heart failure.

REFERENCES

1. Arnett DK, Meyers KJ, Devereux RB, et al. Genetic variation in NCAM1 contributes to left ventricular wall thickness in hypertensive families. *Circ Res.* 2011;108(3):279–283.
2. Bikkina M, Levy D, Evans JC, et al. Left ventricular mass and risk of stroke in an elderly cohort. The Framingham Heart Study. *JAMA.* 1994;272(1):33–36.
3. Hawkins NM, Wang D, McMurray JJ, et al. Prevalence and prognostic implications of electrocardiographic left ventricular hypertrophy in heart failure: evidence from the CHARM programme. *Heart.* 2007;93(1):59–64.
4. Koren MJ, Devereux RB, Casale PN, Savage DD, Laragh JH. Relation of left ventricular mass and geometry to morbidity and mortality in uncomplicated essential hypertension. *Ann Intern Med.* 1991;114(5):345–352.
5. Levy D, Garrison RJ, Savage DD, Kannel WB, Castelli WP. Prognostic implications of echocardiographically determined left ventricular mass in the Framingham Heart Study. *N Engl J Med.* 1990;322(22):1561–1566.
6. Lonn E, Mathew J, Pogue J, et al. Relationship of electrocardiographic left ventricular hypertrophy to mortality and cardiovascular morbidity in high-risk patients. *Eur J Cardiovasc Prev Rehabil.* 2003;10(6):420–428.
7. Vasan RS, Larson MG, Levy D, Evans JC, Benjamin EJ. Distribution and categorization of echocardiographic measurements in relation to reference limits: the Framingham Heart Study: formulation of a height- and sex-specific classification and its prospective validation. *Circulation.* 1997;96(6):1863–1873.
8. Arnett DK, de las Fuentes L, Broeckel U. Genes for left ventricular hypertrophy. *Curr Hypertens Rep.* 2004;6(1):36–41.
9. Assimes TL, Narasimhan B, Seto TB, et al. Heritability of left ventricular mass in Japanese families living in Hawaii: the SAPPHIRe Study. *J Hypertens.* 2007;25(5):985–992.
10. Bella JN, MacCluer JW, Roman MJ, et al. Heritability of left ventricular dimensions and mass in American Indians: the Strong Heart Study. *J Hypertens.* 2004;22(2):281–286.
11. Chien KL, Hsu HC, Su TC, Chen MF, Lee YT. Heritability and major gene effects on left ventricular mass in the Chinese population: a family study. *BMC Cardiovasc Disord.* 2006;6:37.
12. de Simone G, Tang W, Devereux RB, et al. Assessment of the interaction of heritability of volume load and left ventricular mass: the HyperGEN offspring study. *J Hypertens.* 2007;25(7):1397–1402.
13. Juo SH, Di Tullio MR, Lin HF, et al. Heritability of left ventricular mass and other morphologic variables in Caribbean Hispanic subjects: the Northern Manhattan Family Study. *J Am Coll Cardiol.* 2005;46(4):735–737.
14. Mayosi BM, Keavney B, Kardos A, et al. Electrocardiographic measures of left ventricular hypertrophy show greater heritability than echocardiographic left ventricular mass. *Eur Heart J.* 2002;23(24):1963–1971.
15. Post WS, Larson MG, Myers RH, Galderisi M, Levy D. Heritability of left ventricular mass: the Framingham Heart Study. *Hypertension.* 1997;30(5):1025–1028.
16. Sharma P, Middelberg RP, Andrew T, Johnson MR, Christley H, Brown MJ. Heritability of left ventricular mass in a large cohort of twins. *J Hypertens.* 2006;24(2):321–324.
17. Kannel WB, Gordon T, Offutt D. Left ventricular hypertrophy by electrocardiogram. Prevalence, incidence, and mortality in the Framingham study. *Ann Intern Med.* 1969;71(1):89–105.
18. Gupta S, Berry JD, Ayers CR, et al. Left ventricular hypertrophy, aortic wall thickness, and lifetime predicted risk of cardiovascular disease: the Dallas Heart Study. *JACC Cardiovasc Imaging.* 2010;3(6):605–613.
19. Bluemke DA, Kronmal RA, Lima JA, et al. The relationship of left ventricular mass and geometry to incident cardiovascular events: the MESA (Multi-Ethnic Study of Atherosclerosis) study. *J Am Coll Cardiol.* 2008;52(25):2148–2155.
20. Mayosi BM, Avery PJ, Farrall M, Keavney B, Watkins H. Genome-wide linkage analysis of electrocardiographic and echocardiographic left ventricular hypertrophy in families with hypertension. *Eur Heart J.* 2008;29(4):525–530.
21. Arnett DK, Devereux RB, Rao DC, et al. Novel genetic variants contributing to left ventricular hypertrophy: the HyperGEN study. *J Hypertens.* 2009;27(8):1585–1593.
22. Tang W, Arnett DK, Devereux RB, et al. Identification of a novel 5-base pair deletion in calcineurin B (PPP3R1) promoter region and its association with left ventricular hypertrophy. *Am Heart J.* 2005;150(4):845–851.
23. Vasan RS, Larson MG, Aragam J, et al. Genome-wide association of echocardiographic dimensions, brachial artery endothelial function and treadmill exercise responses in the Framingham Heart Study. *BMC Med Genet.* 2007;8(Suppl 1):S2.
24. Vasan RS, Glazer NL, Felix JF, et al. Genetic variants associated with cardiac structure and function: a meta-analysis and replication of genome-wide association data. *JAMA.* 2009;302(2):168–178.
25. Parry HM, Donnelly LA, Van Zuydam N, et al. Genetic variants predicting left ventricular hypertrophy in a diabetic population: a Go-DARTS study including meta-analysis. *Cardiovasc Diabetol.* 2013;12:109.

26. Arnett DK, Li N, Tang W, et al. Genome-wide association study identifies single-nucleotide polymorphism in KCNB1 associated with left ventricular mass in humans: the HyperGEN Study. *BMC Med Genet.* 2009;10:43.

27. Gardin JM, Wagenknecht LE, Anton-Culver H, et al. Relationship of cardiovascular risk factors to echocardiographic left ventricular mass in healthy young black and white adult men and women. The CARDIA study. Coronary Artery Risk Development in Young Adults. *Circulation.* 1995;92(3):380–387.

28. Liebson PR, Grandits G, Prineas R, et al. Echocardiographic correlates of left ventricular structure among 844 mildly hypertensive men and women in the Treatment of Mild Hypertension Study (TOMHS). *Circulation.* 1993;87(2):476–486.

29. Cuspidi C, Negri F, Muiesan ML, et al. Prevalence and severity of echocardiographic left ventricular hypertrophy in hypertensive patients in clinical practice. *Blood Press.* 2010;20(1):3–9.

30. Bacchus R, Singh K, Ogeer I, Mungrue K. The occurrence of left ventricular hypertrophy in normotensive individuals in a community setting in North-East Trinidad. *Vasc Health Risk Manag.* 2011;7:327–332.

31. Schillaci G, Verdecchia P, Porcellati C, Cuccurullo O, Cosco C, Perticone F. Continuous relation between left ventricular mass and cardiovascular risk in essential hypertension. *Hypertension.* 2000;35(2):580–586.

32. Benjamin EJ, Levy D. Why is left ventricular hypertrophy so predictive of morbidity and mortality? *Am J Med Sci.* 1999;317(3):168–175.

33. Vakili BA, Okin PM, Devereux RB. Prognostic implications of left ventricular hypertrophy. *Am Heart J.* 2001;141(3):334–341.

34. Liao Y, Cooper RS, McGee DL, Mensah GA, Ghali JK. The relative effects of left ventricular hypertrophy, coronary artery disease, and ventricular dysfunction on survival among black adults. *JAMA.* 1995;273(20):1592–1597.

35. Levy D, Lauer MS. Left ventricular hypertrophy: echocardiographic assessment and prognostic implications. In: Sutton MSJ, ed. *Textbook of adult and pediatric echocardiography and Doppler.* Boston, MA: Blackwell Scientific Publications; 1997:342–352.

36. Post WS, Larson MG, Levy D. Impact of left ventricular structure on the incidence of hypertension. The Framingham Heart Study. *Circulation.* 1994;90(1):179–185.

37. de Simone G, Devereux RB, Chinali M, et al. Left ventricular mass and incident hypertension in individuals with initial optimal blood pressure: the Strong Heart Study. *J Hypertens.* 2008;26(9):1868–1874.

38. Shimbo D, Muntner P, Mann D, et al. Association of left ventricular hypertrophy with incident hypertension: the multi-ethnic study of atherosclerosis. *Am J Epidemiol.* 2011;173(8):898–905.

39. Jin Y, Kuznetsova T, Bochud M, et al. Heritability of left ventricular structure and function in Caucasian families. *Eur J Echocardiogr.* 2011;12(4):326–332.

40. Tang W, Arnett DK, Devereux RB, Atwood LD, Kitzman DW, Rao DC. Linkage of left ventricular early diastolic peak filling velocity to chromosome 5 in hypertensive African Americans: the HyperGEN echocardiography study. *Am J Hypertens.* 2002;15(7, Pt 1):621–627.

41. Fox ER, Klos KL, Penman AD, et al. Heritability and genetic linkage of left ventricular mass, systolic and diastolic function in hypertensive African Americans (From the GENOA Study). *Am J Hypertens.* 2010;23(8):870–875.

42. Della-Morte D, Beecham A, Rundek T, et al. A follow-up study for left ventricular mass on chromosome 12p11 identifies potential candidate genes. *BMC Med Genet.* 2011;12:100.

43. Lynch AI, Tang W, Shi G, Devereux RB, Eckfeldt JH, Arnett DK. Epistatic effects of ACE I/D and AGT gene variants on left ventricular mass in hypertensive patients: the HyperGEN study. *J Hum Hypertens.* 2012;26(2):133–140.

44. Zhi D, Irvin MR, Gu CC, et al. Whole-exome sequencing and an iPSC-derived cardiomyocyte model provides a powerful platform for gene discovery in left ventricular hypertrophy. *Front Genet.* 2012;3:92.

45. Newton-Cheh C. What can genetic studies of left ventricular mass tell us? *Circ Cardiovasc Genet.* 2011;4(6):581–584.

46. Devereux RB, Reichek N. Echocardiographic determination of left ventricular mass in man. Anatomic validation of the method. *Circulation.* 1977;55(4):613–618.

47. Lang RM, Bierig M, Devereux RB, et al. Recommendations for chamber quantification. *Eur J Echocardiogr.* 2006;7(2):79–108.

48. Pennell DJ. Cardiovascular magnetic resonance. *Circulation.* 2010;121(5):692–705.

49. Grothues F, Smith GC, Moon JC, et al. Comparison of interstudy reproducibility of cardiovascular magnetic resonance with two-dimensional echocardiography in normal subjects and in patients with heart failure or left ventricular hypertrophy. *Am J Cardiol.* 2002;90(1):29–34.

50. Busjahn CA, Schulz-Menger J, Abdel-Aty H, et al. Heritability of left ventricular and papillary muscle heart size: a twin study with cardiac magnetic resonance imaging. *Eur Heart J.* 2009;30(13):1643–1647.

51. Mor-Avi V, Lang RM, Badano LP, et al. Current and evolving echocardiographic techniques for the quantitative evaluation of cardiac mechanics: ASE/EAE consensus statement on methodology and indications endorsed by the Japanese Society of Echocardiography. *Eur J Echocardiogr.* 2011;12(3):167–205.

52. Geyer H, Caracciolo G, Abe H, et al. Assessment of myocardial mechanics using speckle tracking echocardiography: fundamentals and clinical applications. *J Am Soc Echocardiogr.* 2010;23(4):351–369; quiz 453–355.

53. Shah SJ, Aistrup GL, Gupta DK, et al. Ultrastructural and cellular basis for the development of abnormal myocardial mechanics during the transition from hypertension to heart failure. *Am J Physiol Heart Circ Physiol.* 2014;306(1):H88–H100.

54. Burkhoff D, Mirsky I, Suga H. Assessment of systolic and diastolic ventricular properties via pressure-volume analysis: a guide for clinical, translational, and basic researchers. *Am J Physiol Heart Circ Physiol.* 2005;289(2):H501–H512.

55. Klotz S, Dickstein ML, Burkhoff D. A computational method of prediction of the end-diastolic pressure-volume relationship by single beat. *Nat Protoc.* 2007;2(9):2152–2158.

56. Klotz S, Hay I, Dickstein ML, et al. Single-beat estimation of end-diastolic pressure-volume relationship: a novel method with potential for noninvasive application. *Am J Physiol Heart Circ Physiol.* 2006;291(1):H403–H412.

57. Chen CH, Fetics B, Nevo E, et al. Noninvasive single-beat determination of left ventricular end-systolic elastance in humans. *J Am Coll Cardiol.* 2001;38(7):2028–2034.

58. Bielen E, Fagard R, Amery A. The inheritance of left ventricular structure and function assessed by imaging and Doppler echocardiography. *Am Heart J.* 1991;121(6, Pt 1):1743–1749.

59. Garner C, Lecomte E, Visvikis S, Abergel E, Lathrop M, Soubrier F. Genetic and environmental influences on left ventricular mass. A family study. *Hypertension.* 2000;36(5):740–746.

60. Swan L, Birnie DH, Padmanabhan S, Inglis G, Connell JM, Hillis WS. The genetic determination of left ventricular mass in healthy adults. *Eur Heart J.* 2003;24(6):577–582.

61. Tang W, Arnett DK, Devereux RB, et al. Sibling resemblance for left ventricular structure, contractility, and diastolic filling. *Hypertension.* 2002;40(3):233–238.

62. Wang L, Beecham A, Di Tullio MR, et al. Novel quantitative trait locus is mapped to chromosome 12p11 for left ventricular mass in Dominican families: the Family Study of Stroke Risk and Carotid Atherosclerosis. *BMC Med Genet.* 2009;10:74.

63. Schunkert H, Brockel U, Hengstenberg C, et al. Familial predisposition of left ventricular hypertrophy. *J Am Coll Cardiol.* 1999;33(6):1685–1691.

64. Adams TD, Yanowitz FG, Fisher AG, et al. Heritability of cardiac size: an echocardiographic and electrocardiographic study of monozygotic and dizygotic twins. *Circulation.* 1985;71(1):39–44.

65. Palatini P, Amerena J, Nesbitt S, et al. Heritability of left atrial size in the Tecumseh population. *Eur J Clin Invest.* 2002;32(7):467–471.

66. Wang L, Di Tullio MR, Beecham A, et al. A comprehensive genetic study on left atrium size in Caribbean Hispanics identifies potential candidate genes in 17p10. *Circ Cardiovasc Genet.* 2010;3(4):386–392.

67. Arnett DK, Devereux RB, Kitzman D, et al. Linkage of left ventricular contractility to chromosome 11 in humans: the HyperGEN Study. *Hypertension.* 2001;38(4):767–772.

68. Dries DL. Natriuretic peptides and the genomics of left-ventricular hypertrophy. *Heart Fail Clin.* 2010;6(1):55–64.

69. Fox ER, Musani SK, Barbalic M, et al. Genome-wide association study of cardiac structure and systolic function in African Americans: the Candidate Gene Association Resource (CARe) study. *Circ Cardiovasc Genet.* 2013;6(1):37–46.

70. Harper AR, Mayosi BM, Rodriguez A, et al. Common variation neighbouring micro-RNA 22 is associated with increased left ventricular mass. *PLoS One.* 2013;8(1):e55061.

71. Hong KW, Shin DJ, Lee SH, et al. Common variants in RYR1 are associated with left ventricular hypertrophy assessed by electrocardiogram. *Eur Heart J.* 2012;33(10):1250–1256.

72. Kraja AT, Huang P, Tang W, et al. QTLs of factors of the metabolic syndrome and echocardiographic phenotypes: the hypertension genetic epidemiology network study. *BMC Med Genet.* 2008;9:103.

73. Rahman TJ, Mayosi BM, Hall D, et al. Common variation at the 11-beta hydroxysteroid dehydrogenase type 1 gene is associated with left ventricular mass. *Circ Cardiovasc Genet* 2011;4(2):156–162.

74. Rasmussen-Torvik LJ, North KE, Gu CC, et al. A population association study of angiotensinogen polymorphisms and haplotypes with left ventricular phenotypes. *Hypertension.* 2005;46(6):1294–1299.

75. Tang W, Devereux RB, Kitzman DW, et al. The Arg-16Gly polymorphism of the beta2-adrenergic receptor and left ventricular systolic function. *Am J Hypertens.* 2003;16(11, Pt 1):945–951.

76. Tang W, Devereux RB, Li N, et al. Identification of a pleiotropic locus on chromosome 7q for a composite left ventricular wall thickness factor and body mass index: the HyperGEN Study. *BMC Med Genet.* 2009;10:40.

77. Tang W, Devereux RB, Rao DC, et al. Associations between angiotensinogen gene variants and left ventricular mass and function in the HyperGEN study. *Am Heart J.* 2002;143(5):854–860.

78. Arnett DK. Genetic contributions to left ventricular hypertrophy. *Curr Hypertens Rep.* 2000;2(1):50–55.

79. Morita H, Larson MG, Barr SC, et al. Single-gene mutations and increased left ventricular wall thickness in the community: the Framingham Heart Study. *Circulation.* 2006;113(23):2697–2705.

80. Lynch AI, Arnett DK, Atwood LD, et al. A genome scan for linkage with aortic root diameter in hypertensive African Americans and whites in the Hypertension Genetic Epidemiology Network (HyperGEN) study. *Am J Hypertens.* 2005;18(5, Pt 1):627–632.

81. Sherva R, Miller MB, Lynch AI, et al. A Whole genome scan for pulse pressure/stroke volume ratio in African Americans: the HyperGEN Study. *Am J Hypertens.* 2007;20(4):398–402.

82. Kiezun A, Garimella K, Do R, et al. Exome sequencing and the genetic basis of complex traits. *Nat Genet.* 2012;44(6):623–630.

83. Kazma R, Bailey JN. Population-based and family-based designs to analyze rare variants in complex diseases. *Genet Epidemiol.* 2011;35(Suppl 1):S41–S47.

84. Liu DJ, Leal SM. A unified framework for detecting rare variant quantitative trait associations in pedigree and unrelated individuals via sequence data. *Hum Hered.* 2012;73(2):105–122.

85. Manolio TA, Collins FS, Cox NJ, et al. Finding the missing heritability of complex diseases. *Nature.* 2009;461(7265):747–753.

86. Moser M, Hebert PR. Prevention of disease progression, left ventricular hypertrophy and congestive heart failure in hypertension treatment trials. *J Am Coll Cardiol.* 1996;27(5):1214–1218.

87. Devereux RB, Palmieri V, Sharpe N, et al. Effects of once-daily angiotensin-converting enzyme inhibition and calcium channel blockade-based antihypertensive treatment regimens on left ventricular hypertrophy and diastolic filling in hypertension: the prospective randomized enalapril study evaluating regression of ventricular enlargement (preserve) trial. *Circulation.* 2001;104(11):1248–1254.

88. Gottdiener JS, Reda DJ, Massie BM, Materson BJ, Williams DW, Anderson RJ. Effect of single-drug therapy on reduction of left ventricular mass in mild to moderate hypertension: comparison of six antihypertensive agents. The Department of Veterans Affairs Cooperative Study Group on Antihypertensive Agents. *Circulation.* 1997;95(8):2007–2014.

89. Bella JN, Palmieri V, Kitzman DW, et al. Gender difference in diastolic function in hypertension (the HyperGEN study). *Am J Cardiol.* 2002;89(9):1052–1056.

90. Hall D, Mayosi BM, Rahman TJ, Avery PJ, Watkins HC, Keavney B. Common variation in the CD36 (fatty acid translocase) gene is associated with left-ventricular mass. *J Hypertens.* 2011;29(4):690–695.

91. Shah S, Nelson CP, Gaunt TR, et al. Four genetic loci influencing electrocardiographic indices of left ventricular hypertrophy. *Circ Cardiovasc Genet.* 2011;4(6):626–635.

16

C H A P T E R

Genetics of Heart Failure

Wolfgang Lieb and Ramachandran S. Vasan

TAKE HOME POINTS

1. The genetic basis of heart failure is complex and multifactorial, ranging from monogenic to polygenic forms of the disease, coupled with several comorbidities and environmental risk factors that contribute to the heart failure syndrome; with respect to the polygenic basis of heart failure in the community, familial clustering of both overt heart failure and left ventricular geometric patterns has been demonstrated.

2. Recent genome-wide approaches and comprehensive candidate gene analyses have identified several genetic loci associated with heart failure and sporadic dilated cardiomyopathy (DCM) in clinical and population-based settings. To date, the strongest evidence for association exists for a locus on chromosome 1p36, which has been associated with heart failure and sporadic DCM in various independent samples. However, compared to other complex cardiovascular disease (CVD) traits, the genetic architecture of heart failure is less well understood.

3. In the future, genetic testing for rare variants with technologies such as exome and whole-genome sequencing and additional molecular investigations, including gene expression, microRNA (miRNA), and epigenetic analyses, as well as metabolomic and proteomic profiling may be promising tools to deepen our understanding of the pathophysiology of heart failure.

CASE STUDY

A 75-year-old African American woman with a history of hypertension presented to the hospital with shortness of breath, exercise intolerance, and leg swelling. On initial evaluation in the emergency department, her blood pressure was 98/50 mmHg, heart rate 102 bpm, respiratory rate 18 breaths/minute, and oxygen saturation 94% while breathing room air. Her physical exam was remarkable for an elevated jugular venous pressure of 14 cm H_2O with a Kussmaul's sign (inspiratory increase in jugular venous pressure), diffuse point of maximal impulse, tachycardic but regular rhythm, an S_3 heart sound, and no murmurs. She had 2+ lower extremity edema and mild hepatomegaly. Laboratory testing revealed an elevated B-type natriuretic peptide (739 pg/mL, normal range less than 100 pg/mL), and a mildly elevated troponin-I of 0.39 ng/mL (normal range less than 0.03 ng/mL). Her electrocardiogram was unremarkable except for some nonspecific ST-T changes. A bedside echocardiogram revealed a left ventricular ejection fraction of 15% to 20%.

The patient was admitted to the hospital and underwent diuresis with intravenous loop diuretics and had symptomatic improvement. Despite persistent mild elevation of troponin enzymes, coronary angiography revealed no significant coronary artery disease. Her outpatient antihypertensive medications were initially held. As she improved clinically, low-dose angiotensin converting enzyme (ACE)-inhibitor therapy was started. However, the patient became acutely hypotensive with ACE-inhibitor therapy. Eventually she was transitioned to oral bumetanide and spironolactone and was discharged. During the first 3 months after her hospitalization, the patient was tried serially on ACE-inhibitors, angiotensin receptor blockers, and beta-blockers but could not tolerate any of these medications due to hypotension, lightheadedness, and dizziness. After 3 months, her ejection fraction remained 15% to 20%, so an implantable cardioverter-defibrillator was placed.

During her initial workup, the patient noted that two of her siblings and her father all had congestive heart failure as well. Therefore, the patient underwent screening for genetic causes of DCM. However, no mutations were detected. The cause of her familial heart failure syndrome was thought to be due to either a rare variant or a more typical polygenic form of heart failure.

Several months later, the patient presented to a new cardiologist for a second opinion. An echocardiogram (Figures 16.1 and 16.2, Video 16.1 [http://www.demosmedical.com/media/videos/Shah_Video_16_1.mp4]) was performed. It demonstrated increased left ventricular wall thickness, severe left ventricular systolic dysfunction (ejection fraction 20%), and severely reduced tissue Doppler e' and s' velocities. Her troponin was still mildly elevated, even in the outpatient setting. Given these findings, an endomyocardial biopsy was performed; it demonstrated diffuse, extensive amyloid deposition. Mass spectrometry of the amyloid fibrils revealed a mutated form of transthyretin, and genotyping of the *TTR* gene in the patient demonstrated a V122I mutation.

This case demonstrates the complex nature of heart failure, and the challenge of diagnosing the underlying genetic cause of a patient's heart failure syndrome, even when there is a clear familial aggregation of heart failure. This patient appeared to have an idiopathic DCM. However, several clues pointed to an atypical presentation for heart failure with reduced ejection fraction: a strong family history of heart failure, Kussmaul's sign on physical examination, persistently elevated low-level troponin, inability to tolerate neurohormonal therapy, and severely reduced tissue Doppler e' and s' velocities. Whether this particular patient had an idiopathic DCM with superimposed transthyretin cardiac amyloidosis is difficult to tease apart. Approximately 3% to 4% of African Americans are heterozygous for the V122I *TTR* mutation, however, and therefore should be considered in atypical heart failure cases such as this one, especially when there is a familial aggregation of heart failure. Future studies using whole-exome sequencing and other novel omics technologies in patients with heart failure may provide added insight into rare genetic variation and risk of heart failure.

Heart failure is a clinical syndrome that affects approximately 20 million individuals worldwide (1,2) and about 5.1 million adults (age greater than 20 years) in the United States, and the prevalence is expected to increase by 25% by the year 2030 in the United States (3). Its etiology is complex and multifactorial but ultimately converges to the inability of the heart to adequately supply the body with blood commensurate with its metabolic needs. The risk of developing heart failure over the life course is about 1 in 5 for men and women (4) and the prognosis of patients with heart failure is relatively poor. About 50% of patients with heart failure die within 5 years after diagnosis (3,5). In general, genes are thought to contribute to disease susceptibility, but the *impact* of genes on disease risk and the genetic architecture varies considerably in heart failure due to its etiological heterogeneity. Conceptually, one can differentiate between inherited cardiomyopathies as *monogenic traits* at one end of the distribution—diseases affecting

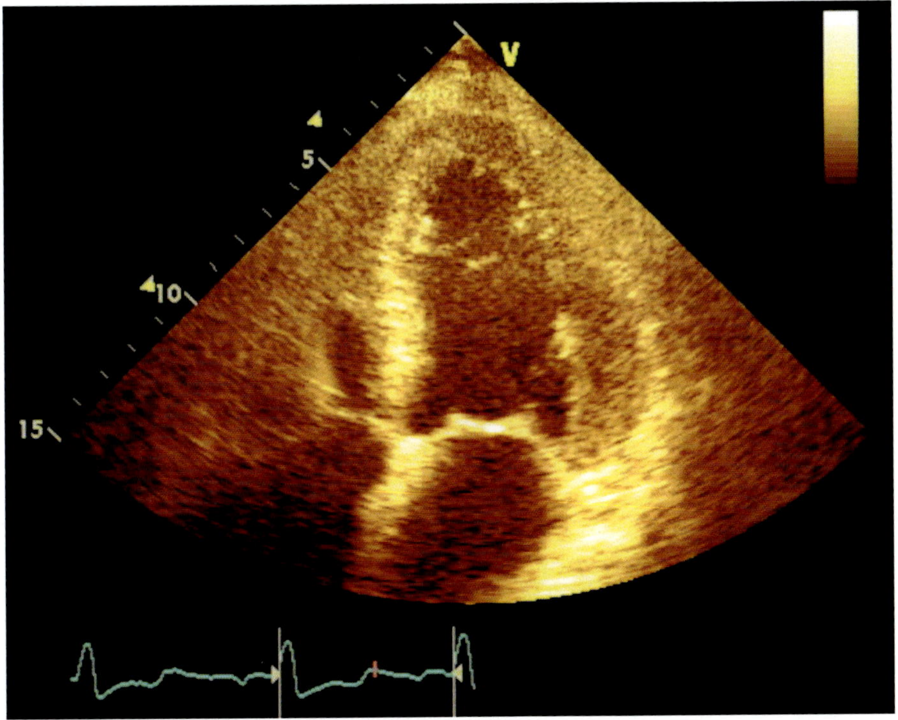

FIGURE 16.1 Apical 4-chamber view of the patient described in the case study. The left ventricle appears thickened, and the echocardiographic texture is consistent with an infiltrative process.

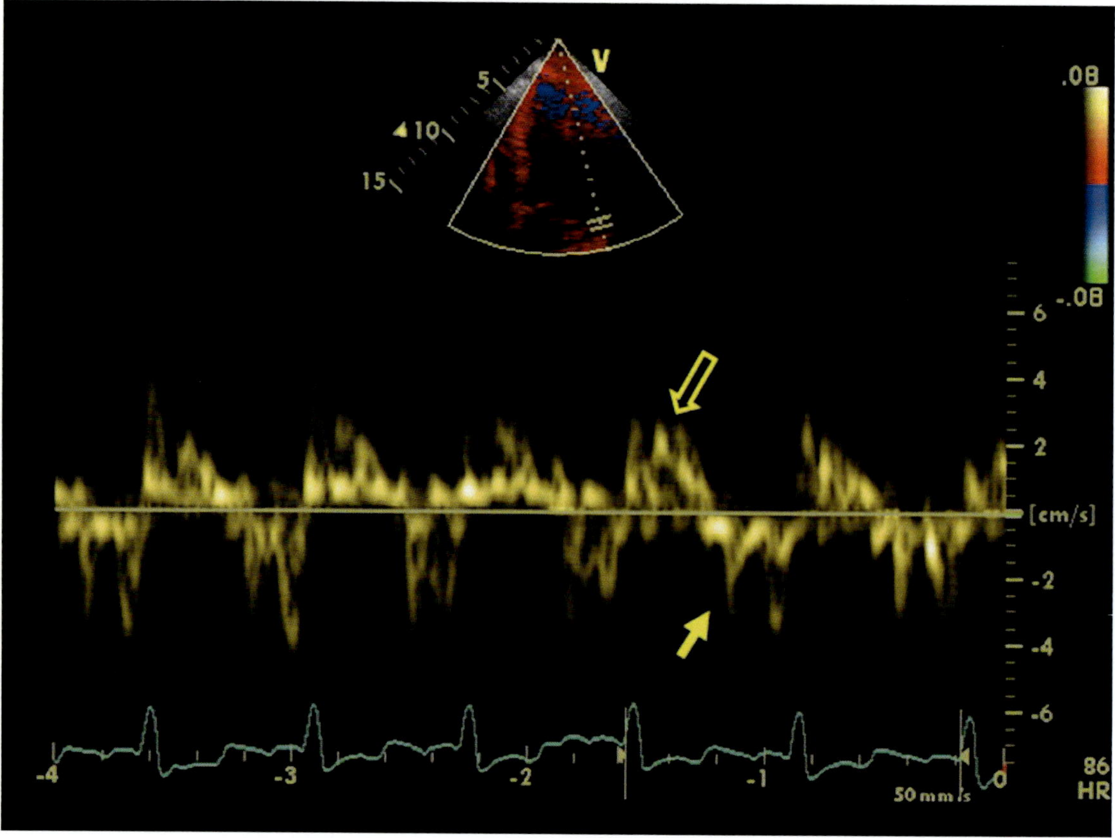

FIGURE 16.2 Tissue Doppler imaging of the lateral mitral annulus in the patient described in the case study. Tissue Doppler e' (solid arrow) and s' (open arrow) velocities are severely reduced, a common finding in cardiac amyloidosis (i.e., left ventricular radial function is typically preserved but longitudinal function is severely reduced).

the heart muscle (6), which are driven by a single, rare, mutation with strong effects on disease risk—and heart failure as a *polygenic* complex trait at the other end, where multiple genetic loci (each with a small-to-modest effect size) interact with several environmental factors over the life course to modify disease susceptibility (7).

GENETIC BASIS OF CARDIOMYOPATHIES

Based on morphological criteria, the most common inherited cardiomyopathies can be distinguished broadly into dilated cardiomyopathy (DCM) and hypertrophic cardiomyopathy (HCM) (6). Inherited cardiomyopathies are covered extensively in Chapter 17. Briefly, inherited cardiomyopathies are generally regarded as monogenic disease entities, caused by single mutations in defined genes. It is acknowledged, though, that the degree of clinical symptoms and the clinical course can vary substantially between carriers of cardiomyopathy-causing mutations (even between carriers of identical mutations) and that this variability is likely explained by the modifying influence of additional genetic and environmental determinants in patients with HCM or DCM (7,8).

The Genetics of Hypertrophic Cardiomyopathy

HCM is characterized by an increased left ventricular wall thickness that cannot be explained in its extent by "any other cardiac or systematic disease" (9). Echocardiographic data from the population-based Coronary Artery Risk Development in Young Adults (CARDIA) study suggest that about 0.2% (1 in 500) (3,10) of young adults are affected by HCM in the general population (3). There is substantial variability in the extent to which mutation carriers develop clinical symptoms (even within families; this phenomenon is referred to as *phenotypic heterogeneity* (7)). Genetic analyses revealed that HCM is caused by mutations in about 11 genes encoding sarcomeric or sarcomere-associated proteins (9). While the number of genes currently known to cause HCM is relatively small, substantial heterogeneity exists with respect to the *mutations* identified in these genes (referred to as *allelic heterogeneity* (11)). In total, more than 1,400 mutations in HCM genes have been reported (9) and most mutations are found in only one or few families (referred to as *private* mutations), which hampers the prediction of the clinical course based on individual mutations. HCM mutations segregate within families in an autosomal dominant fashion (9).

Recent progress in sequencing technologies (12) provided an opportunity to estimate the prevalence of rare and potentially harmful variants in cardiomyopathy-associated genes in the general population. Data from the 1000 Genome project (13–15) and from the community-based Framingham and Jackson Heart Studies (16) suggest that rare mutations in sarcomeric proteins, including the genes *MYH7* (beta-myosin heavy chain), *MYBPC3* (myosin-binding protein C), *TNNT2* (Troponin T), *ACTC1* (alpha-actin), *MYL2* (myosin regulatory light chain), *MYL3* (myosin essential light chain), *TNNI3* (troponin I), and *TPM1* (alpha-tropomyosin) are more common in the general population (15) than would be expected based on the prevalence of clinically diagnosed HCM (14–16). Rare (defined based on a minor allele frequency [MAF] less than 1%) nonsynonymous variants in genes encoding sarcomeric proteins were observed in 11.2% of individuals, and 0.6% of participants carried mutations of likely pathogenicity (16). On average, individuals with such variants had higher left ventricular wall thickness as compared to individuals without such variants or traditional risk factors (16). However, only four out of 22 individuals with rare and likely pathogenic sarcomeric variants had clinical evidence of HCM (16). Potential explanations for this discrepancy include a reduced penetrance of the gene or a possible lack of causality of the rare variants discovered, or potential underdiagnosis of HCM upon clinical and echocardiographic assessment (14).

The Genetics of Dilated Cardiomyopathy

DCM represents the third most common cause of heart failure in some series (17). A genetic cause of DCM is likely if no nongenetic cause can be identified, and if several members in a given family are affected (11,14). About 20% to 35% of patients with DCM have evidence of affected family members on systematic clinical evaluation (6,11,14). The exact prevalence of DCM in the community is not well known. Former estimates were in the magnitude of 1:2500 (6,17), but recent prevalence estimates are substantially higher, between 1 in 250 and 1 in 400 individuals (14). As noted, the frequency of potentially protein-altering rare variants in cardiomyopathy genes in the general population is likely higher than the actual prevalence estimates of clinical disease (14–16). This observation suggests that not all rare variants in cardiomyopathy genes are actually disease-causing, which presents a challenge for the genetic counseling of DCM and HCM families.

Locus heterogeneity is much more pronounced in familial idiopathic DCM. For DCM, mutations in more than 30 genes have been reported but causal mutations are only found in about 30% to 35% (11) of DCM families. The known DCM genes contribute to various biochemical networks and pathways, encoding, for example, components of the sarcomere, the nuclear envelope, the cytoskeleton, ion channels, and desmosomes (11,14,17).

While there is a clinical distinction of HCM and DCM based on morphological characteristics that is helpful for the clinical management of patients (7), there is substantial overlap in the genetic basis of both disease conditions (7,14,18). Different mutations in the same genes can manifest as either HCM or DCM. For example, mutations in the genes *TNNT2* (cardiac troponin T), *MYH6* (alpha-myosin heavy chain), *MYH7* (beta-myosin heavy chain), and *MYBPC3* (myosin-binding protein C) have been reported in patients with DCM and in patients with HCM (17,19–22). Often, the DCM- and HCM-causing mutations within the same gene have opposing functional consequences (18).

While patients with DCM and HCM are at increased risk of heart failure, these cardiomyopathies account only for a relatively small proportion of the overall heart failure burden at the population level, particularly in older adults (7). The following section, therefore, describes recent progress in the understanding of heart failure as a complex trait in the community and its polygenic basis.

THE GENETIC BASIS OF IDIOPATHIC DCM AND HEART FAILURE AT THE POPULATION LEVEL (HEART FAILURE AS POLYGENIC TRAIT)

In unselected samples from the community, **a familial aggregation of congestive heart failure** has been reported that is not fully explained by familial clustering of traditional risk factors (23). In general, the risk of developing heart failure in the community is substantially driven by the cumulative exposure to traditional risk factors over the life course, including, for example, blood pressure and body mass index (BMI) (24), both of which are heritable traits with a demonstrable genetic basis (25,26). Thus, part of the genetic predisposition to heart failure is likely due to heritable genetic variants predisposing to an unfavorable risk profile. However, even after accounting for major cardiac risk factors, including age, sex, systolic blood pressure, BMI, total/high-density lipoprotein (HDL) cholesterol, antihypertensive treatment, and diabetes, the clinical diagnosis of congestive heart failure in at least one parent was associated with an increased relative risk (odds ratio 1.70; 95% CI, 1.11 to 2.60) of developing heart failure in the offspring prospectively, and also conferred an increased risk for adverse cardiac remodeling (an important precursor of heart failure) in cross-sectional analyses (23). On a similar note, familial aggregation of different abnormal geometric patterns of the left ventricle has been reported (27,28). This observation supports the concept that genetic variability contributes to heart failure susceptibility in the community, beyond affecting traditional risk factors.

Initially, analyses on the genetics of polygenic heart failure in the population focused on common genetic variants (usually with an MAF above 5%) in candidate genes—genes in pathways that are thought to be relevant in cardiovascular physiology—including, for example, the renin–angiotensin system and the beta-adrenergic system (29–31). However, most candidate gene analyses for cardiovascular traits, including heart failure, have revealed rather inconsistent results (30–32).

Recent advances in genotyping technologies allowed the interrogation of the entire genome with respect to disease-associated genetic variants (genome-wide analyses). The advantage of such unbiased approaches is the potential of discovering completely new disease-associated genes and pathways, and thus new potential drug targets or clinical predictors. On a similar note, the ability to measure genetic variability in thousands of individuals at declining costs also brought hypothesis-driven research to a new level, enabling the simultaneous assessment of ten thousands of variants in thousands of candidate genes (33), scattered across the entire genome. As a consequence, several genome-wide association analyses and comprehensive candidate gene approaches have been conducted for incident and prevalent heart failure (Table 16.1) (34–36) in the past few years; and also unbiased analyses for important echocardiographic precursors of heart failure (37), and for mortality in patients with heart failure (38) were performed.

Within the Cohorts for Heart and Aging Research in Genomic Epidemiology (CHARGE) consortium (39), genome-wide association study (GWAS) data from four population-based cohorts were meta-analyzed and two genome-wide significant loci for incident heart failure, one locus on chromosome 15q22 in participants of European descent, and one locus on chromosome 12q14 in individuals of African ancestry (Table 16.1) (36) were reported. In parallel, the genome has been interrogated systematically for genetic predictors of mortality in patients with established heart failure. Such analyses identified a region on chromosome 3p22 that was associated with reduced survival in patients with heart failure (38).

Furthermore, comprehensive genetic approaches have been conducted in patients with nonfamilial forms of DCM (40). Such analyses revealed several loci associated with idiopathic DCM, for example, on chromosomes 1 (40), 3 (41), 6 (35), and 10 (40) (Table 16.1). Two genes in the critical region on chromosome 1p36 (*HSPB7* and *CLCNKA*) have been related to heart failure risk in several additional studies (34,41–43), making this the most consistently associated heart failure/idiopathic DCM locus to date. Experimental data suggest that variation in the nearby *CLCNKA* gene (rs10927887), encoding a renal chloride channel, ultimately explains the association signal (31,43). Furthermore, a locus on 6p21.3 within the major histocompatibility complex (MHC) was found to be significantly associated with DCM in several case–control cohorts (35); and based on a comprehensive analysis of approximately 30,000 single nucleotide polymorphisms (SNPs) in 2000 candidate genes with a putative role in cardiovascular physiology, also SNP rs6787362 in *FRMD4B* (chromosome 3p14.1) displayed evidence for association with advanced heart failure (41). Overall, these findings suggest that common genetic variation contributes to heart failure disease susceptibility, but like for many complex polygenic cardiovascular traits (44), common SNPs account for only a small fraction of the heritability of heart failure. In fact, compared to other CVD risk factors (eg, blood pressure (26)) or CVD endpoints (eg, myocardial infarction (45)), genome-wide approaches for heart failure have identified a much smaller number of consistently associated SNPs (Table 16.1). Ongoing studies are currently investigating the contribution of rare and low-frequency variants on the overall heart failure risk in the population and their potential impact as modifiers of disease severity in monogenic forms of the disease (46). Furthermore, the interaction between multiple genes and between genes and environmental factors with respect to heart failure risk needs to be addressed in more detail in the future. Novel molecular techniques that might contribute to an improved understanding of the pathophysiological basis of heart failure risk are summarized in the next section.

ADDITIONAL MOLECULAR INVESTIGATIONS TO GAIN INSIGHT INTO THE BIOLOGICAL BASIS OF HEART FAILURE RISK

Heart failure risk is determined by complex biological networks, where DNA sequence variation is just one contributing factor. Therefore, additional research activities aim to unravel the contribution of gene expression, epigenetic influences, and metabolomic markers on sporadic DCM and heart failure risk. In contrast to DNA sequence variation, which

TABLE 16.1 Association of Common Genetic Variation With Heart Failure and Dilated Cardiomyopathy (DCM)

SNP ID	Locus	Candidate Gene(s)	Risk Allele	RAF	OR	95% CI	P Value	Phenotype	Reference
rs1739843	1p36.13	HSPB7, CLCNKA	C	0.61	1.39	1.28, 1.54	5.28×10^{-13}	Idiopathic DCM	(34,40,41)
rs6787362	3p14.1	FRMD4B	A	0.90	1.49	1.25, 1.75	6.09×10^{-6}	Advanced heart failure	(41)
rs12638540	3p22	CMTM7	G	0.04	1.53	1.01, 2.31	3.21×10^{-7}	Mortality in HF patients	(38)
rs9262636	6p21	HCG22, HLA-C	G	0.2	1.2	1.11, 1.28	4.9×10^{-9}	Idiopathic DCM	(35)
rs2234962	10q26	BAG3	T	0.80	1.52	1.22,1.89	4.0×10^{-12}	Sporadic DCM	(40)
rs11172782	12q14	LRIG3	G	0.29	1.46	1.03, 2.09	6.7×10^{-8}	Incident heart failure	(36)
rs10519210	15q22	USP3, CA12	G	0.03	1.53	1.05, 2.24	1.4×10^{-8}	Incident heart failure	(36)

Abbreviations: BAG3, BCL2-associated athanogene 3; CA12, carbonic anhydrase XII; CLCNKA, chloride channel; CMTM7, CKLF-like MARVEL transmembrane domain containing 7; DCM, dilated cardiomyopathy; FRMD4B, FERM domain-containing protein 4B; HCG22, HLA-complex group 22; HF, heart failure; HLA, human leukocyte antigen; HSPB7, heat shock 27 kDa protein family, member 7; Ka, voltage-sensitive; LRIG3, leucine-rich repeats and immunoglobulin-like domain 3; OR, odds ratio; RAF, risk allele frequency; USP 3, ubiquitin-specific peptidase 3.

is by nature "static," the aforementioned processes may explain the dynamic nature of the heart failure disease process.

MicroRNAs are short noncoding RNA sequences with important regulatory function, which are increasingly evaluated with respect to CVD, including heart failure (47,48). Several miRNA signatures have been identified in various biological samples (myocardial tissue, plasma, whole peripheral blood, peripheral blood mononuclear cells) of patients with heart failure, which were significantly different from signatures in controls, making them potential candidates for diagnostic subclassification or risk stratification (49–52). Some biomarker profiles also correlate with the disease stage (52). On a parallel note, genome-wide approaches reported important differences in histone methylation pattern (53) and mRNA splicing (54) between myocardial biopsies of DCM patients and controls, pointing in part toward known (eg, sarcomeric genes), but also new genes that might be relevant to heart failure risk. Furthermore, serum metabolic profiling revealed a distinct biomarker pattern, able to discriminate between patients with heart failure and healthy controls in one investigation (55). Thus, more and more methodological approaches are available to assess genome-wide molecular and biochemical information related to heart failure. It will be a key challenge for future research to merge and combine these different lines of evidence to elucidate the entire biological system governing normal cardiac function, cardiac dysfunction, and heart failure development and progression, to understand its regulation, and ultimately to use such information for improved clinical prediction, prevention, and therapy of heart failure.

SUMMARY

The genetic basis of heart failure is complex and multifactorial, ranging from monogenic to polygenic forms of the disease. Familial forms of DCM and particularly HCM are classic examples of inherited cardiomyopathies, caused (mainly) by single, rare, mutations with strong impact on disease risk although there is substantial allelic heterogeneity for both disease forms. In most cases, HCM and DCM segregate in an autosomal dominant fashion within families, implying a 50% chance of transmitting the disease-causing mutation to the offspring. For DCM, other transmission patterns have been reported as well. There is substantial variability in the clinical presentation, even between carriers of identical mutations, warranting further research

to explore genetic and clinical modifiers of disease severity and clinical course in DCM and HCM. In all cardiomyopathy patients, a comprehensive family history with thorough assessment of potential CVD manifestations in first-degree relatives is recommended (21).

With respect to the polygenic basis of heart failure in the community, familial clustering of both overt heart failure and left ventricular geometric patterns has been demonstrated. Recent genome-wide approaches and comprehensive candidate gene analyses have identified several genetic loci associated with heart failure and sporadic DCM in clinical and population-based settings. To date, the strongest evidence for association exists for a locus on chromosome 1p36, which has been associated with heart failure and sporadic DCM in various independent samples (34,40,41). However, compared to other complex CVD traits, the genetic architecture of heart failure is less well understood. Additional molecular investigations, including gene expression, miRNA and epigenetic analyses, as well as metabolomic and proteomic profiling might be promising tools to deepen our understanding of the pathophysiology of heart failure. In this context, it will be particularly important to integrate multiple layers of molecular and clinical information with the ultimate goal of improved risk prediction, prevention, and patient management.

REFERENCES

1. Joseph SM, Cedars AM, Ewald GA, et al. Acute decompensated heart failure: Contemporary medical management. *Tex Heart Inst J.* 2009;36(6):510–520.
2. Roger VL. Epidemiology of heart failure. *Circ Res.* 2013;113(6):646–659.
3. Go AS, Mozaffarian D, Roger VL, et al. Executive summary: heart disease and stroke statistics—2013 update: a report from the American Heart Association. *Circulation.* 2013;127(1):143–152.
4. Lloyd-Jones DM, Larson MG, Leip EP, et al. Lifetime risk for developing congestive heart failure: the Framingham Heart Study. *Circulation.* 2002;106(24):3068–3072.
5. Levy D, Kenchaiah S, Larson MG, et al. Long-term trends in the incidence of and survival with heart failure. *N Engl J Med.* 2002;347(18):1397–1402.
6. Towbin JA, Bowles NE. The failing heart. *Nature.* 2002;415(6868):227–233.
7. Cahill TJ, Ashrafian H, Watkins H. Genetic cardiomyopathies causing heart failure. *Circ Res.* 2013;113(6):660–675.
8. Kelly M, Semsarian C. Multiple mutations in genetic cardiovascular disease: a marker of disease severity? *Circ Cardiovasc Genet.* 2009;2(2):182–190.
9. Maron BJ, Maron MS. Hypertrophic cardiomyopathy. *Lancet.* 2013;381(9862):242–255.

10. Maron BJ, Gardin JM, Flack JM, et al. Prevalence of hypertrophic cardiomyopathy in a general population of young adults. Echocardiographic analysis of 4111 subjects in the cardia study. Coronary artery risk development in (young) adults. *Circulation*. 1995;92(4):785–789.

11. Hershberger RE, Siegfried JD. Update 2011: clinical and genetic issues in familial dilated cardiomyopathy. *J Am Coll Cardiol*. 2011;57(16):1641–1649.

12. Cirulli ET, Goldstein DB. Uncovering the roles of rare variants in common disease through whole-genome sequencing. *Nat Rev Genet*. 2010;11(6):415–425.

13. Abecasis GR, Altshuler D, Auton A, et al. A map of human genome variation from population-scale sequencing. *Nature*. 2010;467(7319):1061–1073.

14. Hershberger RE, Hedges DJ, Morales A. Dilated cardiomyopathy: the complexity of a diverse genetic architecture. *Nat Rev Cardiol*. 2013;10(9):531–547.

15. Golbus JR, Puckelwartz MJ, Fahrenbach JP, et al. Population-based variation in cardiomyopathy genes. *Circ Cardiovasc Genet*. 2012;5(4):391–399.

16. Bick AG, Flannick J, Ito K, et al. Burden of rare sarcomere gene variants in the Framingham and Jackson Heart Study cohorts. *Am J Hum Genet*. 2012;91(3):513–519.

17. Maron BJ, Towbin JA, Thiene G, et al. Contemporary definitions and classification of the cardiomyopathies: An American Heart Association Scientific Statement from the Council on Clinical Cardiology, Heart Failure and Transplantation Vommittee; Quality of Care and Outcomes Research and Functional Genomics and Translational Biology Interdisciplinary Working Groups; and Council On Epidemiology And Prevention. *Circulation*. 2006;113(14):1807–1816.

18. Watkins H, Ashrafian H, Redwood C. Inherited cardiomyopathies. *N Engl J Med*. 2011;364(17):1643–1656.

19. Hershberger RE, Pinto JR, Parks SB, et al. Clinical and functional characterization of tnnt2 mutations identified in patients with dilated cardiomyopathy. *Circ Cardiovasc Genet*. 2009;2(4):306–313.

20. Watkins H, McKenna WJ, Thierfelder L, et al. Mutations in the genes for cardiac troponin T and alpha-tropomyosin in hypertrophic cardiomyopathy. *N Engl J Med*. 1995;332(16):1058–1064.

21. Hershberger RE, Lindenfeld J, Mestroni L, et al. Genetic evaluation of cardiomyopathy—a Heart Failure Society of America Practice Guideline. *J Card Fail*. 2009;15(2):83–97.

22. Hershberger RE, Norton N, Morales A, et al. Coding sequence rare variants identified in mybpc3, myh6, tpm1, tnnc1, and tnni3 from 312 patients with familial or idiopathic dilated cardiomyopathy. *Circ Cardiovasc Genet*. 2010;3(2):155–161.

23. Lee DS, Pencina MJ, Benjamin EJ, et al. Association of parental heart failure with risk of heart failure in offspring. *N Engl J Med*. 2006;355(2):138–147.

24. Lee DS, Massaro JM, Wang TJ, et al. Antecedent blood pressure, body mass index, and the risk of incident heart failure in later life. *Hypertension*. 2007;50(5):869–876.

25. Speliotes EK, Willer CJ, Berndt SI, et al. Association analyses of 249,796 individuals reveal 18 new loci associated with body mass index. *Nat Genet*. 2010;42(11):937–948.

26. Padmanabhan S, Newton-Cheh C, Dominiczak AF. Genetic basis of blood pressure and hypertension. *Trends Genet*. 2012;28(8):397–408.

27. Lam CS, Liu X, Yang Q, et al. Familial aggregation of left ventricular geometry and association with parental heart failure: the Framingham Heart Study. *Circ Cardiovasc Genet*. 2010;3(6):492–498.

28. Schunkert H, Brockel U, Hengstenberg C, et al. Familial predisposition of left ventricular hypertrophy. *J Am Coll Cardiol*. 1999;33(6):1685–1691.

29. Zakrzewski-Jakubiak M, de Denus S, Dube MP, et al. Ten renin-angiotensin system-related gene polymorphisms in maximally treated canadian caucasian patients with heart failure. *Br J Clin Pharmacol*. 2008;65(5):742–751.

30. Kitsios G, Zintzaras E. Genetic variation associated with ischemic heart failure: a huge review and meta-analysis. *Am J Epidemiol*. 2007;166(6):619–633.

31. Dorn GW, 2nd. The genomic architecture of sporadic heart failure. *Circ Res*. 2011;108(10):1270–1283.

32. Morgan TM, Krumholz HM, Lifton RP, Spertus JA. Nonvalidation of reported genetic risk factors for acute coronary syndrome in a large-scale replication study. *JAMA*. 2007;297(14):1551–1561.

33. Keating BJ, Tischfield S, Murray SS, et al. Concept, design and implementation of a cardiovascular gene-centric 50 k SNP array for large-scale genomic association studies. *PLoS One*. 2008;3(10):e3583.

34. Stark K, Esslinger UB, Reinhard W, et al. Genetic association study identifies hspb7 as a risk gene for idiopathic dilated cardiomyopathy. *PLoS Genet*. 2010;6(10):e1001167.

35. Meder B, Ruhle F, Weis T, et al. A genome-wide association study identifies 6p21 as novel risk locus for dilated cardiomyopathy. *Eur Heart J*. 2014;35(16):1069–1077.

36. Smith NL, Felix JF, Morrison AC, et al. Association of genome-wide variation with the risk of incident heart failure in adults of european and african ancestry: a prospective meta-analysis from the cohorts for heart and aging research in genomic epidemiology (CHARGE) consortium. *Circ Cardiovasc Genet*. 2010;3(3):256–266.

37. Vasan RS, Glazer NL, Felix JF, et al. Genetic variants associated with cardiac structure and function: a meta-analysis and replication of genome-wide association data. *JAMA*. 2009;302(2):168–178.

38. Morrison AC, Felix JF, Cupples LA, et al. Genomic variation associated with mortality among adults of european and african ancestry with heart failure: The cohorts for heart and aging research in genomic epidemiology consortium. *Circ Cardiovasc Genet*. 2010;3(3):248–255.

39. Psaty BM, O'Donnell CJ, Gudnason V, et al. Cohorts for heart and aging research in genomic epidemiology (charge) consortium: design of prospective meta-analyses of genome-wide association studies from 5 cohorts. *Circ Cardiovasc Genet*. 2009;2(1):73–80.

40. Villard E, Perret C, Gary F, et al. A genome-wide association study identifies two loci associated with heart failure due to dilated cardiomyopathy. *Eur Heart J*. 2011;32(9):1065–1076.

41. Cappola TP, Li M, He J, et al. Common variants in hspb7 and frmd4b associated with advanced heart failure. *Circ Cardiovasc Genet*. 2010;3(2):147–154.

42. Matkovich SJ, Van Booven DJ, Hindes A, et al. Cardiac signaling genes exhibit unexpected sequence diversity in sporadic cardiomyopathy, revealing hspb7 polymorphisms associated with disease. *J Clin Invest.* 2010;120(1):280–289.

43. Cappola TP, Matkovich SJ, Wang W, et al. Loss-of-function DNA sequence variant in the clcnka chloride channel implicates the cardio-renal axis in interindividual heart failure risk variation. *Proc Natl Acad Sci USA.* 2011;108(6):2456–2461.

44. Manolio TA, Collins FS, Cox NJ, et al. Finding the missing heritability of complex diseases. *Nature.* 2009;461(7265):747–753.

45. Lieb W, Vasan RS. Genetics of coronary artery disease. *Circulation.* 2013;128(10):1131–1138.

46. Roncarati R, Viviani Anselmi C, Krawitz P, et al. Doubly heterozygous LMNA and TTN mutations revealed by exome sequencing in a severe form of dilated cardiomyopathy. *Eur J Hum Genet.* 2013;21(10):1105–1111.

47. Mo YY. MicroRNA regulatory networks and human disease. *Cell Mol Life Sci.* 2012;69(21):3529–3531.

48. Kumarswamy R, Thum T. Non-coding rnas in cardiac remodeling and heart failure. *Circ Res.* 2013;113(6):676–689.

49. Naga Prasad SV, Duan ZH, Gupta MK, et al. Unique microrna profile in end-stage heart failure indicates alterations in specific cardiovascular signaling networks. *J Biol Chem.* 2009;284(40):27487–27499.

50. Gupta MK, Halley C, Duan ZH, et al. Mirna-548c: a specific signature in circulating pbmcs from dilated cardiomyopathy patients. *J Mol Cell Cardiol.* 2013;62131–62141.

51. Tijsen AJ, Creemers EE, Moerland PD, et al. Mir423-5p as a circulating biomarker for heart failure. *Circ Res.* 2010;106(6):1035–1039.

52. Vogel B, Keller A, Frese KS, et al. Multivariate mirna signatures as biomarkers for non-ischaemic systolic heart failure. *Eur Heart J.* 2013;34(36):2812–2823.

53. Haas J, Frese KS, Park YJ, et al. Alterations in cardiac DNA methylation in human dilated cardiomyopathy. *EMBO Mol Med.* 2013;5(3):413–429.

54. Kong SW, Hu YW, Ho JW, et al. Heart failure-associated changes in rna splicing of sarcomere genes. *Circ Cardiovasc Genet.* 2010;3(2):138–146.

55. Tenori L, Hu X, Pantaleo P, et al. Metabolomic fingerprint of heart failure in humans: a nuclear magnetic resonance spectroscopy analysis. *Int J Cardiol.* 2013;168(4):e113–e115.

Inherited Cardiomyopathies

Michael A. Burke

TAKE HOME POINTS

1. Many previously idiopathic cardiomyopathies are now known to be genetic diseases, including hypertrophic cardiomyopathy, some cases of idiopathic dilated cardiomyopathy, and arrhythmogenic right ventricular cardiomyopathy. However, the genetic basis for these conditions remains incomplete.
2. Strict criteria are used to define a variant as pathogenic. While genetic associations abound in the literature, definitive evidence proving pathogenicity is lacking for many genes that have been found to be associated with cardiomyopathies.
3. Genotype–phenotype correlations are variable, even in individuals with the same mutation. As such, the primary clinical role for genetic testing at present is for family screening and not for clinical management of affected individuals.
4. The genetic landscape of inherited cardiomyopathies is rapidly evolving. New technologies such as next-generation sequencing (NGS), comparative genomic hybridization, and digital droplet polymerase chain reaction (PCR) promise to further refine our understanding of the genetic basis of these diseases.

CASE STUDY

A 25-year-old male experienced sudden cardiac arrest due to ventricular fibrillation while playing basketball; he was successfully resuscitated. On initial evaluation, he reported a 3- to 4-year history of intermittent lightheadedness after physical exertion but no syncope. His father died suddenly at age 45. A paternal aunt had a long-standing history of effort intolerance of unclear etiology. He reported having four healthy siblings and one healthy child. Physical examination revealed a 2/6 harsh, late-peaking systolic murmur that increased after a squat-to-stand maneuver. Based on his history and physical examination, a familial cardiomyopathy was suspected. His electrocardiogram was abnormal, demonstrating generous voltage and inferior and lateral ST-depression and T-wave inversion (Figure 17.1). Echocardiography demonstrated asymmetric septal hypertrophy with systolic anterior motion of the mitral valve (Figure 17.2, Videos 17.1 and 17.2 [http://www.demosmedical.com/media/videos/Shah_Video_17_1.mp4; http://www.demosmedical.com/media/videos/Shah_Video_17_2.mp4]), confirming a diagnosis of hypertrophic cardiomyopathy (HCM). The patient was started on a beta-blocker and received an implantable cardioverter defibrillator (ICD).

In a case like the one presented here, clinical screening should be pursued in all adolescent and adult first-degree relatives. The siblings of the patient's father (and their children) should also be screened for HCM since the father died of sudden death at a young age. All clinically affected relatives should receive appropriate treatment as indicated. The patient should be referred to a genetic counselor and genetic testing recommended. If a putative disease-causing mutation is found, all at-risk relatives can be screened for this mutation. Genotype-negative relatives need no further clinical testing while genotype-positive, phenotype-negative relatives require continued clinical screening to monitor for development of the disease.

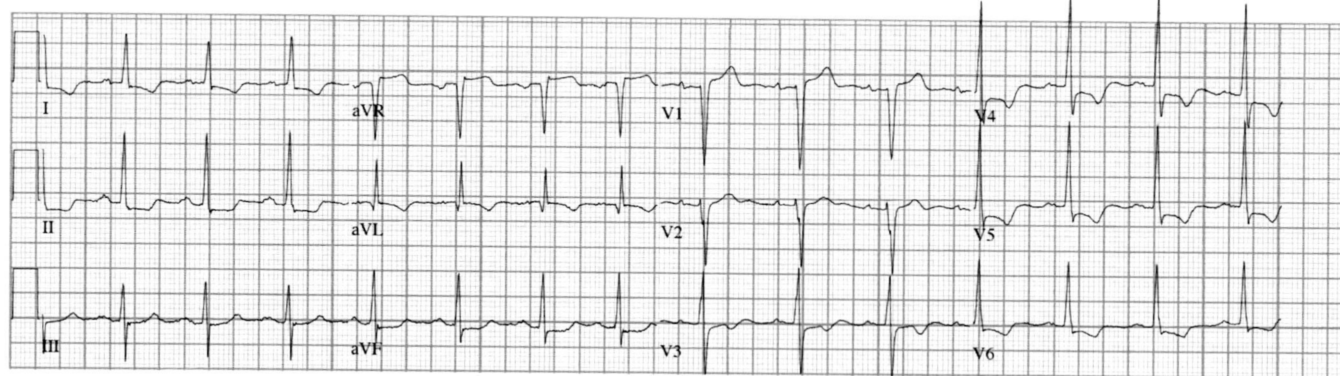

FIGURE 17.1 Electrocardiogram of the patient described in the case study. Note the increased voltage and left ventricular hypertrophy with strain pattern in the lateral precordial leads, nonspecific findings that can be seen in hypertrophic cardiomyopathy.

Nearly 300 years ago, the Italian physician Giovanni Maria Lancisi (1654–1720) published what is believed to be the first account of a familial cardiomyopathy (CM), describing a four-generation family with sudden death and right ventricular aneurysm (1). This description predated Gregor Mendel's landmark genetic studies by approximately 130 years, the discovery of DNA as the genetic material by 225 years, and the identification of the first CM disease-causing mutation by 262 years. The remarkable progress in genetic and genomic technologies over the past 25 years has brought to light the considerable role of genetics in cardiovascular disease. Hundreds of mutations in a myriad of genes have been discovered to cause a variety of cardiac pathologies. These encompass several forms of CM previously denoted as idiopathic, including hypertrophic cardiomyopathy (HCM), nonischemic dilated cardiomyopathy (DCM), arrhythmogenic right ventricular cardiomyopathy (ARVC), left ventricular noncompaction (LVNC), as well as a host of rare syndromes with cardiac manifestations. These diseases, which occur in families and as sporadic cases, are now recognized to be single-gene disorders.

Genetic mutations associated with CM are present at conception and have a high penetrance, yet the emergence of a clinically detectable phenotype often takes years to decades to manifest. Disease severity also varies, ranging from clinically insignificant phenotypic abnormalities to life-threatening arrhythmias and fulminant heart failure (HF). This is true even in family members with the same mutation. Further, different mutations in the same gene, called allelic variants, can cause different forms of CM or seemingly unrelated cardiac abnormalities, yet a specific mutation typically only produces one phenotype.

The study of genetics has provided remarkable new insight into these diseases and ongoing research seeks to understand the molecular mechanisms by which these mutations cause CM, with the hope of revealing novel therapies. These new discoveries are being rapidly incorporated into clinical practice, affecting how physicians manage individuals and families affected by these conditions. In this chapter, we explore the genetic basis of and clinical role for genetic investigation in CM.

HYPERTROPHIC CARDIOMYOPATHY

Clinical and anatomic descriptions of HCM date back hundreds of years in the medical literature, culminating in the late 1950s with the characterization of the pathologic entity we recognize today (2,3).

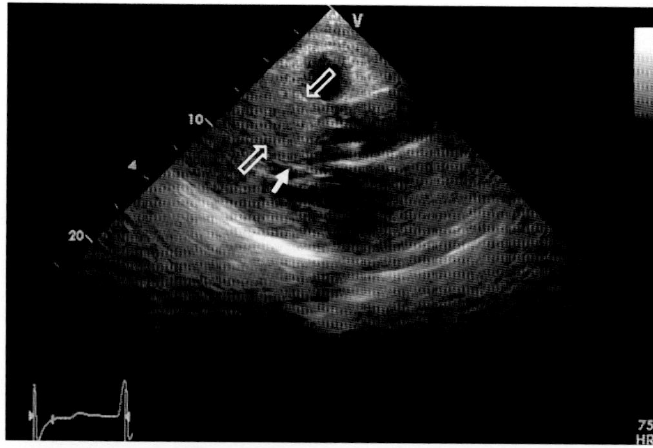

FIGURE 17.2 Echocardiographic parasternal long-axis view of the patient described in the case study. Note the systolic anterior motion of the mitral valve (solid arrow) along with the severe asymmetric septal hypertrophy (open arrows), measuring 2.6 cm in diameter at end-diastole.

Over the years, more than 80 names have been used to describe HCM, including commonly used terms such as idiopathic hypertrophic subaortic stenosis, muscular subaortic stenosis, or hypertrophic obstructive cardiomyopathy, which were based on the pathophysiology of the more extreme cases characterized at the time. In current use, HCM indicates a primary disorder of the heart that is unaccompanied by pathology in another organ, is characterized by unexplained left ventricular hypertrophy (LVH) and arises from a dominant mutation in a sarcomere gene that can typically be identified by robust genetic testing, though the etiology remains unknown in some sporadic cases (4).

Clinical and Pathophysiologic Characteristics

The pathologic hallmark of HCM is LVH in the absence of an obvious cause. HCM is common, being present in approximately 1 in 500 individuals (5). Diagnosis requires exclusion of secondary causes of LVH such as hypertension and aortic stenosis, or physiologic hypertrophy as seen in highly trained athletes. At the microscopic level, HCM is characterized by myocyte hypertrophy, myofibrillar disarray, thickening and narrowing of muscular arterioles, inadequate capillary density, and interstitial fibrosis (6).

Classically, HCM results in asymmetric septal hypertrophy, though almost any pattern of LVH can be associated with the disease (7,8). The defining feature of HCM is left ventricular outflow tract (LVOT) obstruction whereby systolic anterior motion of the mitral valve causes temporary and variable mechanical obstruction to blood flow. This functional obstruction results in the classic symptoms of lightheadedness and dyspnea on exertion, and is present in approximately 70% of those coming to clinical attention (9).

The natural history of HCM is broad, presenting from infancy to old age. Symptoms range from asymptomatic to severe, while changes in ventricular morphology range from mild LVH to severe LVH associated with LVOT obstruction, mitral valve disease, and extensive interstitial fibrosis. Early studies were typically performed at tertiary referral centers on those with more extreme disease who demonstrated poor prognosis. However, unselected patient cohort studies have shown that the majority of patients with HCM have a more benign course and most have normal life expectancy (10–12). HCM-related mortality is approximately 1.0% to 1.3% per year and is higher in children (12,13). In those diagnosed after age 50, HCM does not increase mortality beyond that expected in the general population (12).

Complications of HCM associated with higher morbidity and mortality are present in a minority of patients and generally fall into one of three potentially overlapping clinical syndromes: (a) atrial fibrillation, (b) HF, and (c) sudden cardiac death (SCD) (14). Atrial fibrillation is common, occurring in 20% to 25% of unselected patients with HCM, and is associated with a 6% prevalence of stroke (15,16). Development of advanced HF symptoms (functional class III or IV) may also occur in approximately 20% (11). However, progression to systolic dysfunction and end stage disease is rare, occurring in less than 5% (12,17,18). SCD is the most dreaded complication and the most frequent cause of death in those with HCM (19). SCD associated with HCM often occurs in young, seemingly healthy individuals, is the most common cause of death in young athletes, and unfortunately may be the first manifestation of the disease (20). Identification of at-risk individuals serves as a central reason for genetic screening of first-degree relatives in HCM.

Genetic Basis of Disease

Groundbreaking genetic research over the last 25 years has identified HCM as a monogenic, primary myocardial disorder caused by mutations in genes that encode the protein components of the cardiac sarcomere (Figure 17.3). Inheritance is autosomal dominant, and as such, first-degree relatives of HCM patients have a 50% risk of developing the disease. The lifelong penetrance of HCM is nearly 100% but because of latency in disease onset, longitudinal clinical evaluation is required in at-risk individuals unless genetic testing has excluded a known disease-causing familial mutation.

Disease Genes

Early genetic studies used linkage analysis in large families with well-characterized disease to elucidate the genetic basis of HCM. Detailed study of a large family with HCM and SCD linked the disease to chromosome 14 (21). Targeted DNA sequencing revealed a point mutation that converted a highly conserved arginine residue to a glutamine (Arg403Glu) in the beta-myosin heavy chain gene (*MYH7*) (22), thus identifying the first disease-causing gene in HCM. Additional families were found to have other mutations in *MYH7* while others were not linked to this locus, marking HCM as genetically heterogeneous (23,24). Subsequent analyses revealed disease-causing mutations in the genes for alpha-tropomyosin (*TPM1*) and cardiac troponin T (*TNNT2*), thus defining HCM as a disease of the sarcomere (25–27). This concept has been confirmed by the discovery of additional disease-causing mutations in the genes for cardiac myosin binding protein-C (*MYBPC3*) (28,29), myosin regulatory light chain (*MYL2*), myosin

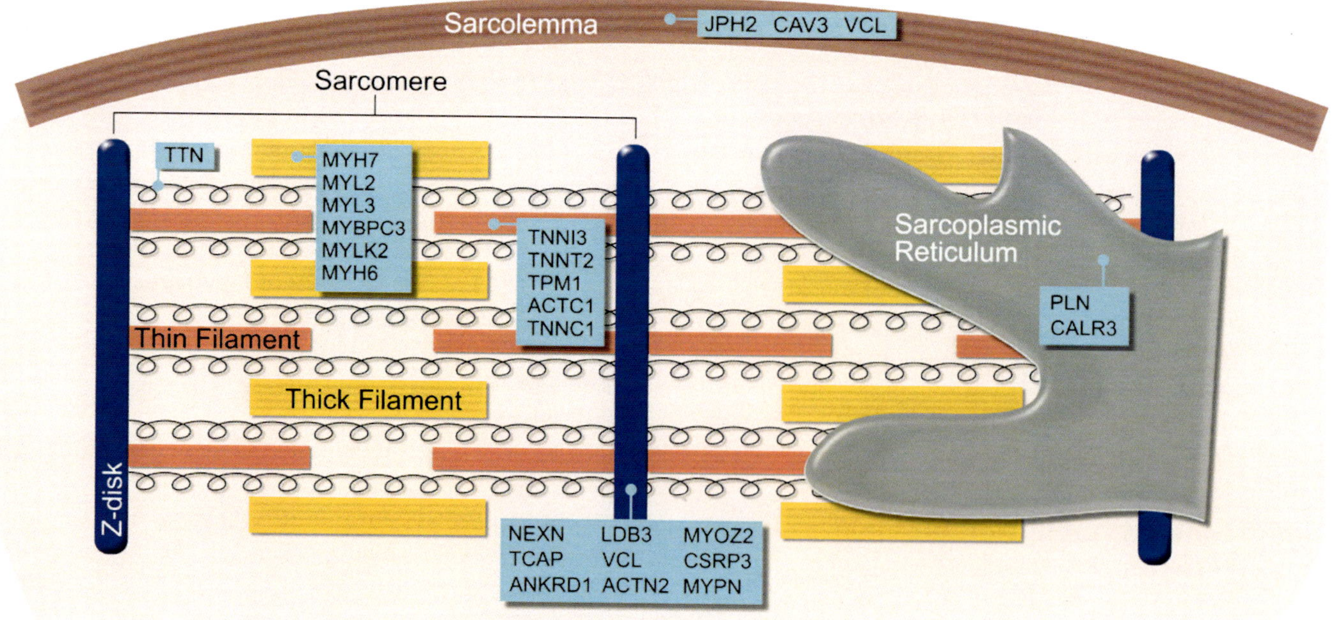

FIGURE 17.3 Definitive and possible hypertrophic cardiomyopathy genes. The image shows putative disease-causing hypertrophic cardiomyopathy genes and the subcellular locations of the gene products (ie, proteins).
Source: Illustration by Steven Claas.

essential light chain (*MYL3*) (30), cardiac troponin I (*TNNI3*) (31), and cardiac alpha-actin (*ACTC1*) (32,33), all myofilament components of the cardiac sarcomere (Table 17.1).

Many mutations identified in these eight genes meet strict criteria for defining a variant as pathogenic. These are nonsynonymous variants that alter an amino acid residue that is highly conserved across species, are absent from large numbers of healthy control individuals, show statistically significant cosegregation with affected individuals within a family, and are identified via an unbiased genetic approach (ie, genomic linkage analysis, genome-wide association study (GWAS), or whole exome/genome sequencing). Further, the same or other novel pathogenic mutations can be identified in afflicted individuals who are unrelated. Additional definitive evidence comes from animal models in which some disease-causing human mutations have been demonstrated to recapitulate the disease phenotype (Table 17.2).

The yield of genetic testing for these eight pathogenic HCM mutations is variably reported between 38% and 63% (34–36), suggesting additional genes are yet to be discovered. Indeed, many additional disease susceptibility genes have been found to be associated with HCM but lack the degree of rigor used to identify the eight confirmed pathogenic mutations. Of these additional genes, only myozenin 2 (37) and alpha-actinin 2 (38) were identified in unbiased

analyses. Another probable pathogenic gene identified by a candidate-gene approach is *CSRP3*, encoding the cardiac LIM protein, which was shown to cosegregate in a large family (39,40). Identification and characterization of disease-associated mutations in additional families would support their causality in HCM and would expand the confirmed pathogenic mutations to encompass the Z-disc of the cardiac sarcomere. A host of other genes have been implicated in families or individuals but further validation is required to determine their pathogenic significance (Table 17.1). The giant protein titin, encoded by the TTN gene, had previously been proposed as a putative HCM disease gene (41), but multiple studies taking a candidate-gene approach have not found further proof of pathogenic mutations (36,42,43), and the causative role of this gene in HCM remains uncertain.

Genetic Heterogeneity

HCM displays marked genetic heterogeneity with over 1,000 different disease-associated mutations characterized. Most mutations are specific to a given family (so called "private" mutations). There are genetic hot spots but haplotype analysis of identical mutations in separate families has confirmed that most are recent, independent mutational events (44). The majority of known mutations are point mutations, whereby a single base pair is altered, changing the underlying amino acid sequence (missense mutation) or resulting

TABLE 17.1 Gene Mutations Associated With Hypertrophic Cardiomyopathy

Gene Symbol	Encoded Protein	Cellular Location
Definite disease-causing genes		
MYBPC3	Cardiac myosin binding protein C	Sarcomere, thick filament
MYH7	Beta-myosin heavy chain	Sarcomere, thick filament
MYL2	Regulatory myosin light chain	Sarcomere, thick filament
MYL3	Essential myosin light chain	Sarcomere, thick filament
TNNI3	Cardiac troponin I	Sarcomere, thin filament
TNNT2	Cardiac troponin T	Sarcomere, thin filament
TPM1	Alpha-tropomyosin	Sarcomere, thin filament
ACTC1	Cardiac alpha-actin	Sarcomere, thin filament
Probable disease-causing genes		
ACTN2	Alpha-actinin 2	Sarcomere, Z-disc
MYOZ2	Myozenin 2	Sarcomere, Z-disc
CSRP3	Cardiac LIM protein	Sarcomere, Z-disc
MYPN	Myopalladin	Sarcomere, Z-disc
HCM-associated genes of uncertain pathogenic significance		
TNNC1	Cardiac troponin C	Sarcomere, thin filament
MYH6	Alpha-myosin heavy chain	Sarcomere, thick filament
MYLK2	Myosin light chain kinase 2	Sarcomere, thick filament
NEXN	Nexilin	Sarcomere, Z-disc
TCAP	Telethonin	Sarcomere, Z-disc
ANKRD1	Cardiac ankyrin repeat protein	Sarcomere, Z-disc
LDB3	LIM-binding domain 3 protein*	Sarcomere, Z-disc
VCL	Vinculin	Sarcomere, Z-disc, intercalated disc and sarcolemma
JPH2	Junctophilin 2	Sarcolemma
CAV3	Caveolin-3	Sarcolemma
PLN	Phospholamban	Sarcoplasmic reticulum
CALR3	Calreticulin 3	Sarcoplasmic reticulum
PLN	Phospholamban	Sarcoplasmic reticulum

*LIM-binding domain 3 protein is also commonly known as Cypher/ZASP.

in a premature stop codon leading to the production of a truncated protein (nonsense mutation). Missense mutations predominate in nearly all genes except *MYBPC3*, where many types of mutations have been discovered.

The most commonly implicated gene is *MYBPC3*, while the second most frequently affected gene is *MYH7* (Table 17.3). Combined, *MYBPC3* and *MYH7* account for at least 75% of identified mutations and up to 50% of all clinically recognized cases, while other pathogenic HCM disease genes account for less than 10% of cases. This demographic spectrum holds true across different populations around the world (34,36,45–49). The substantial number of mutation-negative patients observed in clinical studies again confirms that the genetic landscape of HCM remains incomplete.

Functional Consequences of Gene Mutations

With few exceptions, HCM mutations produce dominant-negative proteins that are incorporated into the myofilaments, resulting in defective sarcomeres. The balance of available data suggests that incorporation of mutant proteins produces enhanced

TABLE 17.2   Criteria for Defining a Genetic Variant as Pathogenic

Criterion	Description and Significance
Nonsynonymous genetic variant	A genetic variant that alters the amino acid sequence
Conservation	Conservation of the affected amino acid in different species through evolution suggests a critical role of that particular residue
Control population	Lack of identification in a large number of unaffected controls demonstrates the mutation is unlikely to be an unrelated polymorphism
Cosegregation	Demonstrates that the mutation is found in those with disease only (incomplete penetrance is common); identification of the disease phenotype in those *without* the mutation proves nonpathogenicity
Unbiased genetic approach	Genome-wide analysis (linkage analysis, GWAS, whole exome/genome sequencing) reduces the possibility of identifying a coincidental and unrelated mutation or a disease-modifying mutation
Multiplicity of events	Demonstration that a gene/mutation may be causal in 2 or more probands greatly increases confidence in pathogenicity
Animal models of disease	Recapitulation of the disease phenotype by the mutation strongly supports causality and can confirm possible pathogenicity suggested by *in silico* analysis

Abbreviation: GWAS, genome-wide association study.

TABLE 17.3   Frequency of Pathogenic Mutations in HCM

Gene	Frequency of HCM Probands (%)	Frequency of Mutation Positive Probands (%)
MYBPC3	11–35	42–66
MYH7	10–25	22–40
TNNI3	3–4	6–7
TNNT2	2–4	2–7
TPM1	<2	6
MYL2	2–3	4–6
MYL3	~1	1–3
ACTC1	~1	<3

contraction but impaired relaxation (50–54). Different mutations trigger distinct signaling pathways, which in turn may lead to divergent mechanisms of ventricular remodeling (55).

Thick Filament Proteins.   Mutations in *MYH7* are common in the head and neck regions; this is the force-generating region and contains binding sites for ATP as well as the proteins essential light chain and regulatory light chain (56). Cumulative data gathered from in vitro studies, animal models, and analysis of individuals with *MYH7* mutations have shown that such mutations appear to cause a gain-of-function effect that enhances cardiac contractility (57). Those with *MYH7* mutations present earlier in life and tend to have substantial hypertrophy. Specific mutations increase the risk of SCD (Arg719Trp, Arg403Gln) and end-stage HF (Arg453Cys), while many others are associated with a more benign prognosis (58,59).

Different mutations in *MYL3* and *MYL2* can produce an atypical pattern of LVH with mid LV chamber and papillary muscle involvement or apical hypertrophy in addition to the more classic asymmetric septal hypertrophy. A wide clinical spectrum has also been reported with different mutations in these genes (60).

Cardiac myosin binding protein C (MyBPC) is a structural support protein that binds to myosin and titin. One MyBPC protein interacts with several myosin crowns, such that the protein has unequal distribution along the length of the sarcomere (61). Mutations in *MYBPC3* are unique among disease-causing HCM mutations in that the majority of disease-causing mutations are nonsense mutations, splice-site variants, and insertion/deletion mutations, which produce truncated proteins (60). The means by which abnormal MyBPC produces disease are also unique. While *MYBPC3* missense mutations may still produce dominant-negative proteins that are incorporated into the sarcomere, mutations that truncate the MyBPC transcript and protein can be cleared by cell surveillance mechanisms before incorporation, resulting in haploinsufficiency (disease produced by an inadequate amount of wild-type protein) (62,63).

HCM caused by *MYBPC3* can present at any age but generally presents later in life and displays less morbidity and lower penetrance than other genes (64,65). Supporting this is the fact that founder mutations causing HCM are most commonly found in *MYBPC3* (57). One such mutation found in 4% of individuals with Indian ancestry, arose approximately 33 ± 23 thousand years ago and is associated with a sevenfold increased risk of HF late in life (66). Maintenance of such a defect through thousands of years would only occur via neutral selection, with no impact during childbearing years; it is only recently, with increased life expectancy, that the disease has become apparent.

Thin Filament Proteins. Mutations in *TNNI3* cause a wide range of clinical phenotypes, with striking variability even within the same family (67). A substantial portion of those with *TNNI3* mutations progress to systolic dysfunction and end-stage HF (68). *TNNT2* mutations were initially associated with high rates of SCD despite less LVH (27,69), but other studies have shown more variable phenotypes (70). Confirming this heterogeneity, *TPM1* mutations have been associated with varying onset and degrees of LVH (71) and at least some mutations are associated with high rates of SCD (72).

Genetics in Clinical Practice

Genotype–Phenotype Correlation

At the outset of the medical genetic era, much optimism existed that genetics would lead to a paradigm shift in the ability to predict outcomes and affect management. Furthering this notion, early studies of single mutations in large families seemingly linked specific mutations to prognosis (56,58,64). Many clinical correlates between genotype and phenotype exist: individuals with an identified pathogenic mutation are more likely to be younger, have a family history of HCM, and are more likely to have significant LVH (35). Ventricular morphology also predicts the ability to identify a mutation on genetic testing, with classic asymmetric septal hypertrophy being associated with an identifiable mutation in one of the eight pathogenic HCM genes in approximately 80% of cases (73).

However, with increased knowledge of the genetic basis of HCM has come the realization that genotype–phenotype correlations are limited by the marked phenotypic complexity of this disease. Large studies have revealed that an individual mutation in unrelated individuals cannot reliably predict outcomes (74). Reasons for this include the presence of compound or digenic variants (mutations in the same gene or other sarcomere genes that may accentuate or compensate for the identified pathogenic mutation) (75), or the effect of modifier genes and environmental factors (76). Consequently, while genotype can provide additional information, it cannot supplant conventional clinical risk predictors in HCM (4).

Genetic Testing for HCM

Almost as soon as the genetic basis for HCM was established, attempts were made to identify the disease in unaffected family members (77). However, clinical application of genetic testing lagged far behind research until automated DNA sequencing technology became readily available in the first decade of the 21st century. Since that time, genetic testing has become an integral part in the management of patients with HCM.

Clinical Applications of Genetic Testing. Genetic testing is used for two main purposes clinically. First, the identification of pathogenic mutations confirms the diagnosis of myofilament-associated HCM with virtually 100% certainty and eliminates any possible ambiguity in diagnosis. Identification of a pathogenic mutation also allows for screening of first-degree family members, which is the predominant role for genetic testing at the present time.

Current guidelines recommend frequent clinical screening for family members of affected individuals (Table 17.4). This causes both an economic and psychological burden. Genetic testing identifies first-degree family members who are at risk for developing overt diseases in the future, allowing for continued surveillance, while negating the need for further screening in those who are genotype negative (4).

Commercial Genetic Testing. Presently, genetic testing for HCM is available through multiple companies. Commercially available genetic testing for many disorders can be found at www.genetests.org, a resource funded by the National Institutes of Health; specific tests offered by different laboratories can be found by following the "panel directory" link. Clinical application was initially hampered by high costs and lack of insurance coverage. However, as DNA sequencing technology has evolved, costs have come down substantially, resulting in better coverage of genetic testing by insurance providers.

Typically, the exons plus or minus flanking splice sites of all eight pathogenic genes as well as a variable number of associated genes are sequenced to assess for variants. Screening panels are typically in the $1500 to $5000 range depending on the number of genes sequenced. Once a mutation is identified,

TABLE 17.4 Recommended Clinical Screening Guidelines for HCM in the Absence of Genetic Testing or Identified Pathologic Mutation

Age Range	Recommendation
12 to 18–21 years*	Echocardiography and 12-lead electrocardiogram (ECG) every 12–18 months
Greater than 18–21 years	Echocardiography and 12-lead ECG: 1. At symptom onset 2. Every 5 years 3. More frequently as clinically indicated (ie, change in symptoms, malignant familial history, late-onset familial disease)

*Screening is optional under 12 years of age unless there is a family history of advanced complications or sudden cardiac death early in life, the patient is a competitive athlete, has developed symptoms, or other clinical information has raised suspicion of early left ventricular hypertrophy.

targeted analysis at significantly reduced costs (often only a few hundred dollars) can be performed in first-degree family members.

Genetic testing generally produces one of five results: (a) Identification of a known pathogenic mutation. This is the most clinically useful outcome as it confirms the diagnosis, establishes etiology, and gives a target for familial screening. (b) Identification of a probable pathogenic mutation. This is potentially useful but supportive evidence, such as confirmation of cosegregation within the family or identification in unrelated individuals, may also be necessary. (c) Identification of a variant of unknown significance. These may be novel causative variants or could represent rare background polymorphisms; they require further testing. (d) Identification of benign variants. These are polymorphisms that represent genetic variation in the general population such as previously described single nucleotide polymorphisms or synonymous variants, and are not related to disease. (e) No mutation is identified. Failure to identify a pathogenic mutation (scenarios c–e) is a common outcome as not all causative genes and mutations are known. In these instances, continued clinical surveillance rather than genetic testing of first-degree relatives is necessary (4).

The Genotype-Positive/Phenotype-Negative Individual. Genetic testing has given rise to a new patient subset: the genotype-positive/phenotype-negative individual. This group has raised several new clinical questions. First, which if any should receive prophylactic ICD therapy? Family history of SCD is a strong risk factor in HCM, and ICDs mitigate this risk (78). Additionally, which individuals should be excluded from participation in competitive sports? Guideline recommendations clearly state that those with overt HCM should be excluded from intense competitive sports due to a known increased risk of SCD (4).

However, the risk to those without phenotypic disease is unknown. Other questions surround appropriate screening, clinical outcomes, and treatment for preclinical disease.

This uncertainty stems from two key facts: (a) a complete lack of outcome data in such patients, and (b) a paucity of knowledge about the molecular mechanisms that link gene mutation and pathologic disease. Nevertheless, this new subset will allow for the unique chance to study the natural history of HCM longitudinally including a more accurate analysis of penetrance and the study of pharmacologic intervention before HCM becomes clinically evident. Further, in concert with animal models, the study of such patients should yield a better understanding of the molecular basis for the development of LVH in HCM.

METABOLIC CARDIOMYOPATHIES

Mutations in several genes that encode metabolic proteins can also produce substantial LVH (Table 17.5). Findings on echocardiography or cardiac MRI frequently mimic HCM, leading to misdiagnosis. Endomyocardial biopsy is typically needed to establish the correct diagnosis. The myofibrillar disarray characteristic of HCM is absent. Instead, there is diffuse myocyte hypertrophy associated with numerous cytoplasmic vacuoles containing lipid (*GLA* mutations), lysosomal remnants (*LAMP2* mutations), and glycogen (*PRKAG2* and *GAA* mutations). Inheritance patterns are also varied, including autosomal dominant (*PRKAG2* mutations), X-linked (*GLA* and *LAMP2* mutations), and autosomal recessive patterns (*GAA* mutations). Frequently called HCM phenocopies, these metabolic cardiomyopathies represent a distinct subset of inherited CM.

TABLE 17.5 Gene Mutations Associated With Metabolic Cardiomyopathies

Gene	Encoded Protein	Disease	Inheritance
GLA	Alpha-galactosidase A	Fabry disease	*X*-linked
LAMP2	Lysosome-associated membrane protein 2	Danon disease	*X*-linked
PRKAG2	AMP-activated protein kinase, gamma$_2$-subunit (noncatalytic)	PRKAG2 cardiomyopathy	Autosomal dominant
GAA	Alpha-glucosidase	Pompe disease	Autosomal recessive

Fabry Disease

Mutations of the *GLA* gene encoding alpha-galactosidase A lead to deficient enzyme activity and accumulation of neutral glycosphingolipids that have terminal alpha-galactosyl moieties, causing Fabry disease (79). It is a heterogeneous disorder affecting the heart, kidneys, skin, eyes, and nervous system, and is rare, with an estimated incidence of 1 in 50,000. Fabry disease is inherited in an X-lined recessive fashion, meaning males are more commonly affected, and females are typically only affected when two defective alleles are inherited.

Cardiac manifestations can be found in conjunction with or in isolation of systemic manifestations (80–82). Lipid accumulation occurs in the myocardium, vascular endothelium, and valvular tissue, resulting variably in LVH, aberrant conduction, ventricular dysfunction, myocardial infarction, and valvular disease (83,84). Concentric LVH is the most common cardiac manifestation and without treatment, LV function may progressively worsen (85,86).

Fabry disease accounts for at least 1% of cases of unexplained LVH coming to clinical attention, and may account for approximately 10% of cases presenting after the age of 40 (87–89). When suspected, the diagnosis can often be made noninvasively by assessment of alpha-galactosidase enzymatic activity, though false-negative rates may be high, particularly in women (90). Definitive diagnosis can be made by endomyocardial biopsy. Interestingly, not all individuals with clinically confirmed Fabry cardiomyopathy have an identified gene mutation in *GLA*. This could be due to the older techniques used in published studies or could be due to the presence of other unidentified mutations in the *GLA* gene, regulatory elements, or in other genes that result in the disease phenotype. Genetic testing for *GLA* is commercially available and is frequently included in testing panels for HCM.

Danon Disease

Mutations of the *LAMP2* gene, encoding lysosomal associated membrane protein 2, results in Danon disease (91,92). LAMP2 is a lysosomal membrane protein that plays a role in autophagy. Deficiency of LAMP2 protein leads to accumulation of vacuoles filled with glycogen and cellular proteins, which in turn leads to cellular dysfunction (93). Like Fabry disease, Danon disease is *X*-linked and therefore, males are more frequently affected. Classically, Danon disease is a multisystem disorder affecting the nervous system, skeletal muscle, liver, and heart, resulting in mental retardation, skeletal myopathy, and HF (91). However, more recent studies have shown that Danon CM can exist in isolation (94).

The heart in Danon disease shows myocyte hypertrophy and diffuse vacuolization, which lead variably to massive LVH, DCM, and ventricular pre-excitation (94,95). Danon disease is typically diagnosed in childhood or adolescence and represents an underdiagnosed cause of LVH in this population (94,96). Prognosis is grim with death frequently occurring before the third decade of life in affected males and by the fourth to fifth decade in females (95). Genetic testing is commercially available, typically in conjunction with HCM panels.

Glycogen Storage Disorders

Pompe disease (also called glycogen storage disease type II or acid-maltase deficiency) is an autosomal recessive disorder caused by mutations in *GAA*, encoding alpha-glucosidase. It is rare with an incidence of not greater than 1 in 40,000 (97). The absolute level of enzymatic activity determines disease onset, which can present at any age. The most severe form of disease occurs with presentation during infancy, which is also the only time when significant cardiac involvement is observed. When present in infancy, severe cardiac hypertrophy is present and the disease is often fatal without treatment. Specific mutations are not associated with the development of CM but the autosomal recessive nature of the disease (meaning two diseased alleles must be inherited) and the scarcity of index cases (the disease is 100 times less common than HCM) limit the confidence with which any one particular mutation can be associated with a specific disease feature. Multiple animal models have confirmed that disruption of the *GAA* gene replicates human disease, including CM.

An additional glycogen storage CM occurs with mutations in *PRKAG2*, encoding the gamma$_2$-subunit of adenosine monophosphate-activated protein kinase (AMPK) (98–100). AMPK is a master regulator of cellular metabolism; its activity is regulated via the gamma-subunit. Mutations in *PRKAG2* result in constitutive activation of AMPK, leading to excessive accumulation of glycogen in intramyocardial vacuoles (101). This leads to both LVH and conduction abnormalities, in particular ventricular pre-excitation due to disruption of the annulus fibrosis by glycogen-filled myocytes (102). Unlike other metabolic cardiomyopathies, *PRKAG2* cardiomyopathy is inherited in an autosomal dominant fashion. Genetic testing is typically offered via commercially available HCM genetic testing panels.

RESTRICTIVE CARDIOMYOPATHY

Some evidence suggests that there is also a genetic basis for the rare condition of restrictive cardiomyopathy (RCM). Linkage analysis defined a mutation in *TNNI3* as the cause of RCM in a large family (103). However, multiple family members met clinical criteria for HCM and histologic analysis in some of those with RCM showed myocyte hypertrophy, fibrosis, and myofibrillar disarray consistent with HCM, a finding confirmed in other patients with clinically apparent RCM (103,104). Additional mutations in *ACTC1* (104), *TNNT2* (105), and an unidentified gene at locus 10q23.3 (106,107) have been associated with RCM but phenotypic heterogeneity, including HCM, DCM, and desmin-associated myopathy existed in affected first-degree relatives. In aggregate, these data suggest that RCM simply reflects restrictive physiology in the context of a different genetic cardiomyopathy, rather than a distinct genetic entity.

DILATED CARDIOMYOPATHY

LV dilatation is the end stage of a variety of cardiac maladies. It is the consequence of myocardial infarction and chronic ischemic heart disease in over 60% of cases (108). Historically, nonischemic DCM has been categorized based on the presence or absence of identifiable, secondary causes. These include infectious and noninfectious myocarditis, toxins such as cocaine and alcohol, endocrine disease, autoimmune disorders, peripartum cardiomyopathy, infiltrative, inflammatory, and storage disorders, stress-induced LV dysfunction, endomyocardial disorders, and sustained tachyarrhythmias. Despite this long list of secondary causes, up to 50% of cases of nonischemic DCM have no identifiable etiology (known as idiopathic DCM). Investigation over the past 20 years has uncovered a possible genetic explanation for many cases of idiopathic DCM.

Clinical and Pathophysiologic Characteristics

Nonischemic DCM is a primary myocardial disorder characterized by ventricular dilatation and contractile dysfunction with a prevalence estimated at 1:2,500 (109). Microscopically, DCM is characterized by myocyte hypertrophy and fibrosis but lacks the myofibrillar disarray characteristic of HCM. LV dilatation leads to systolic dysfunction, progressive symptomatic HF, arrhythmias, conduction system disease, thromboembolism, and death. The natural history covers the entire human life span with overt disease manifesting between infancy and old age. The spectrum of disease is broad with some having no clinical findings in association with the dilated phenotype (asymptomatic LV dysfunction) and others presenting with fulminant HF requiring mechanical circulatory support or heart transplantation.

DCM is a progressive disease with an annual incidence for overt HF of 10% in asymptomatic individuals (110,111). Individuals with asymptomatic DCM of idiopathic etiology are at a 6.5-fold increased risk of HF (111). In contrast to HCM, prognosis with DCM is significantly worse than the general population, though outcomes are improving with the implementation of guideline-based treatment (112).

Genetic Basis of Disease

Idiopathic DCM has an incidence of approximately 4–8/100,000 per year depending on the population studied (113,114). Idiopathic DCM appears to be familial in at least 25% to 50% of cases, though estimates vary widely depending on the study (115). Inheritance is generally autosomal dominant, but there are autosomal recessive, *X*-linked, and matrilineal modes of inheritance as well. Penetrance is variably age-dependent and can be incomplete, even in family members carrying the same genetic mutation. Genetic forms of DCM are also phenotypically diverse, frequently being associated with other cardiac and noncardiac pathologies.

Disease Genes and Putative Functional Consequences

Whereas HCM is clearly a disease of the cardiac sarcomere, mutations that cause DCM are much more diverse, affecting proteins found throughout the cell (Figure 17.4). Over 50 genes have been associated with DCM to date (Table 17.6). In contrast to HCM, robust genetic evidence supporting definitive pathogenicity is lacking for most DCM genes. Sequence data in the general population indicate that rare nonsynonymous variants are actually present in frequencies higher than the estimated population prevalence of DCM (116). This highlights the important point that the presence of a rare variant alone does not imply pathogenicity; additional evidence is needed to confirm pathogenicity, just as with HCM (Table 17.2).

Mutations in Sarcomere and Z-Disc Genes. The most prevalent genetic cause of idiopathic DCM is mutation of the *TTN* gene encoding the giant protein titin, accounting for approximately 20% of all inherited DCM cases (43,117). Causative mutations in *TTN* include nonsense, frameshift, or splicing mutations, as well as copy number variations, all of which produce truncated proteins. The molecular mechanisms by which truncated titin causes DCM remain unknown.

Definitive pathogenic mutations have been shown in other sarcomere genes as well, including *MYH7* (118,119), *ACTC1* (120), *TPM1* (121), and *TNNI3* (122,123), while possible disease-causing mutations or variants of uncertain significance have been identified in *MYBPC3* (124,125), *TNNC1* (126,127), and *MYH6* (128).

Defects in the Z-disc of the sarcomere can also be associated with DCM. The best evidence for involvement of the cardiac Z-disc is mutation in the gene encoding Bcl2-associated athanogene 3 (*BAG3*), a chaperone protein localized in the Z-disc that is believed to play a role in transduction of mechanical signals from the sarcomere (129–131). Many other Z-disc genes have been implicated in DCM, albeit with varying degrees of confidence (Table 17.6).

Mutations Affecting Electrolyte Homeostasis. Pathogenic DCM mutations extend beyond the sarcomere to encompass a variety of proteins that regulate myocyte electrolyte homeostasis. Phospholamban, encoded by *PLN*, regulates calcium

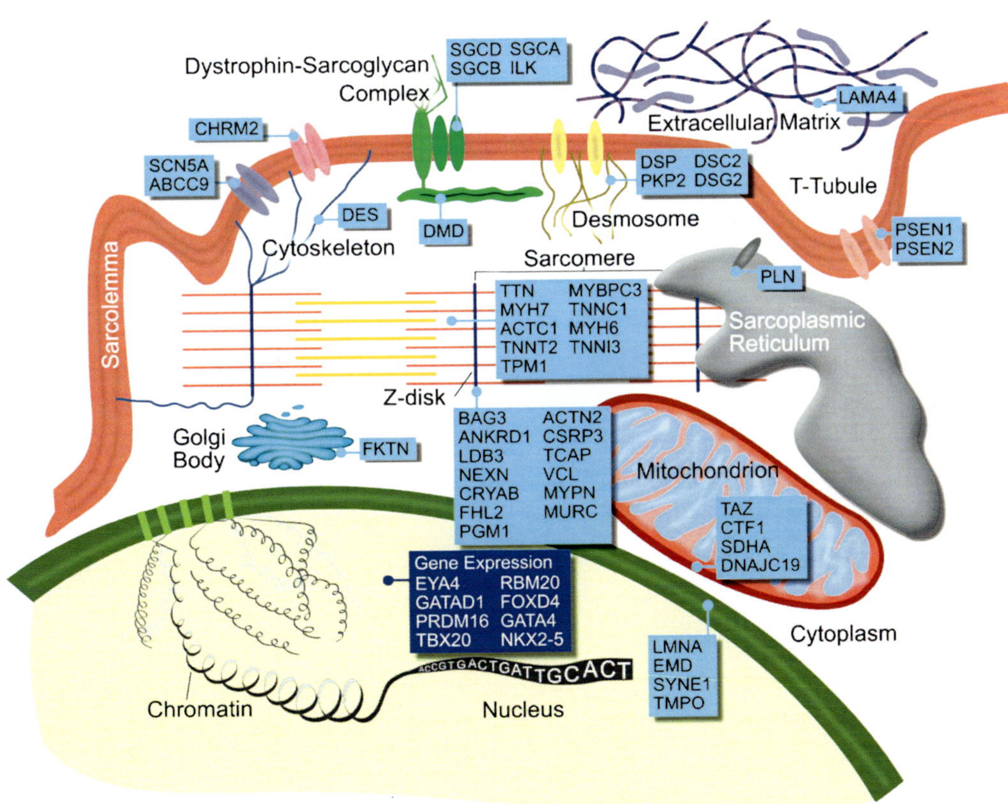

FIGURE 17.4 Definitive and possible dilated cardiomyopathy genes. The image shows putative disease-causing dilated cardiomyopathy genes and the subcellular locations of the gene products (ie, proteins).

Source: Illustration by Steven Claas.

TABLE 17.6 Gene Mutations Associated With Dilated Cardiomyopathy

Gene	Encoded Protein	Possible Coexistent Phenotypes
Definitive disease-causing genes		
TTN	Titin	None
MYH7	Beta-myosin heavy chain	None
ACTC1	Cardiac alpha-actin	None
TNNT2	Cardiac troponin T	None
TPM1	Alpha-tropomyosin	None
TNNI3	Cardiac troponin I	None
BAG3	Bcl2-associated athanogene 3	None
DES	Desmin	Skeletal myopathy, conduction system disease
DMD	Dystrophin	Duchenne muscular dystrophy, Becker muscular dystrophy
DNAJC19	DNAJ (Hsp40) homolog	DCMA
EMD	Emerin	Emery–Dreifuss muscular dystrophy type 1
LMNA	Lamin A/C	Conduction system disease, Emery–Dreifuss muscular dystrophy type 2
PLN	Phospholamban	None
RBM20	RNA-binding protein, 20	None
SCN5A	Voltage-gated sodium channel type V, alpha-subunit	Supraventricular and ventricular arrhythmias
TAZ	Tafazzin	Barth syndrome
Possible disease-causing genes or those of uncertain pathogenic significance		
MYBPC3	Cardiac myosin binding protein C	None
TNNC1	Cardiac troponin C	None
ANKRD1	Cardiac ankyrin repeat protein	None
LDB3	LIM-binding domain 3 protein	Myofibrillar myopathy
NEXN	Nexilin	None
SGCD	Delta-sarcoglycan	Limb-girdle muscular dystrophy, type 2F
EYA4	Eyes absent homolog 4	Sensorineural hearing loss
PSEN1	Presenilin-1	None*
MYPN	Myopalladin	None
GATAD1	GATA zinc finger domain containing 1 protein	None
GATA4	GATA-binding protein 4	None
CHRM2	Muscarinic cholinergic receptor	Supraventricular and ventricular arrhythmias
MYH6	Alpha-myosin heavy chain	None
ACTN2	Alpha-actinin 2	None
CSRP3	Cardiac LIM protein	Limb-girdle muscular dystrophy
TCAP	Telethonin	None
ABCC9	Sulfonylurea receptor 2	Cantú syndrome
TBX20	T-box 20	CHD
NKX2-5	NK2 homeobox 5	CHD
CTF1	Cardiotrophin-1	None
VCL	Vinculin	None
SGCB	Beta-sarcoglycan	Limb-girdle muscular dystrophy, type 2E
SGCA	Alpha-sarcoglycan	Limb-girdle muscular dystrophy, type 2D
FHL2	Four and a half limb domains 2	None

(continued)

TABLE 17.6 Gene Mutations Associated With Dilated Cardiomyopathy (*continued*)

Gene	Encoded Protein	Possible Coexistent Phenotypes
FKTN	Fukutin	Fukuyama-type muscular dystrophy
FOXD4	Forkhead box protein D4	Obsessive compulsive disorder and suicidality
LAMA4	Laminin, alpha4	None
ILK	Integrin-linked kinase	None
PSEN2	Presenilin-2	None*
SDHA	Succinate dehydrogenase subunit A	LVNC
CRYAB	Alpha-B crystallin	Skeletal myopathy
MURC	Muscle-related coiled-coil protein	None
SYNE1	Nesprin-1	Emery–Dreifuss muscular dystrophy type 4
PRDM16	PR domain containing 16	LVNC
PGM1	Phosphoglucomutase 1	Glycogenosis type XIV
TMPO	Thymopoietin	None
DSC2	Desmocollin 2	None
DSG2	Desmoglein 2	None
DSP	Desmoplakin	Palmoplantar keratoderma
PKP2	Plakophilin 2	None
Genetic loci linked to DCM without a known disease-associated mutation		
9q13-q22	MIM CMD1B	None
2q14-q22	MIM CMD1H	Conduction system disease
6q12-q16	MIM CMD1K	None
7q22-q31	MIM CMD1Q	None
9q22-q31	Unknown	None
1p36.13	Unknown	None

*Allelic variants for Alzheimer's dementia.
Abbreviations: CHD, congenital heart disease; DCMA, dilated cardiomyopathy with ataxia; LVNC, left ventricular noncompaction; MIM, Mendelian inheritance in man.

uptake by the sarcoplasmic/endoplasmic reticulum calcium transporting ATPase (SERCA2a), thereby functioning as a molecular brake on sarcoplasmic reticulum calcium cycling. Multiple distinct mutations have been described, which can result in either superinhibition or a total loss of inhibition of SERCA2a while causing DCM, a phenotype reproduced in several animal models (132–134).

Extensive data implicate the cardiac voltage-gated sodium channel, encoded by *SCN5A*, in the development of DCM (135–137). Screening has identified variants in *SCN5A* in 1.7% to 2.6% of unrelated individuals with idiopathic DCM (138,139). Sodium channels are sarcolemmal transmembrane proteins that play a critical role in the development of the cardiac action potential. Not surprisingly, families with such mutations also have a high burden of arrhythmias leading some to call the associated syndrome "arrhythmic dilated cardiomyopathy" (139). There are many allelic variants in *SCN5A* including those causing Brugada syndrome (140), idiopathic

ventricular fibrillation (141), and familial atrial fibrillation (142). Interestingly, different mutations in *SCN5A* may produce DCM via different mechanisms. Animal models of the common D1275N mutation demonstrate reduced channel current and suggest electromechanical dyssynchrony as the underlying cause of DCM (143). In contrast, the R222Q mutation leads to increased channel current and affected individuals may develop DCM secondary to an excessively high burden of ventricular ectopy (144,145).

Mutations Altering Structural Proteins.   DCM can also be caused by mutations in a variety of structural proteins, from the nucleus to the cytoskeleton to the sarcolemma. Interestingly, mutations in most of these genes are also associated with skeletal myopathy, presumably due to the importance of such structural proteins in all muscle cells.

Mutations in several genes encoding nuclear envelope structural proteins have been associated with DCM. Definitive genetic evidence links mutations in *LMNA* (146) and *EMD* (147,148) to DCM, while limited evidence also suggests a possible role for mutations in *SYNE1* (149,150). Each is associated with different forms of Emery–Dreifuss muscular dystrophy (EDMD1—*EMD*, EDMD2—*LMNA*, EDMD4—*SYNE1*) and can present with or without skeletal myopathy.

Lamin A/C, encoded by the *LMNA* gene, is a ubiquitously expressed inner nuclear membrane protein that plays a role in the maintenance of proper nuclear structure. Mutations in *LMNA* lead to DCM with conduction system abnormalities (146). Inheritance is autosomal dominant. Skeletal myopathy is also common, but DCM can occur in isolation. Over 200 distinct *LMNA* mutations have been identified, accounting for approximately 10% of idiopathic DCM cases and up to a third of those who have familial DCM with progressive conduction disease and arrhythmias (115). *LMNA* mutations are associated with poor prognosis (151). Some data suggest that deficiency of LMNA results in an inappropriate response to mechanical strain, but the exact mechanism by which mutations lead to DCM remains unclear.

Mutations in the cytoskeletal intermediate filament desmin, encoded by the *DES* gene, produce a heterogeneous group of conditions known as desminopathies, which include DCM that is frequently associated with skeletal myopathy and conduction disease (152,153). Some mutations produce an isolated cardiac phenotype, though this is rare (154). *DES* mutations are relatively uncommon, being identified in approximately 1% to 2% of unselected cases of DCM (155). Despite this, many mutations have been found and inheritance patterns vary. Early evidence suggested that disruption of the intermediate filament network was a central pathologic component to desminopathies (153,156,157), but more recent evidence suggests that DCM associated with certain *DES* mutations still occurs in the presence of apparently preserved intermediate filament networks (155). Thus the mechanism by which *DES* mutations cause DCM remains uncertain.

Disruption of the sarcolemmal dystrophin–sarcoglycan complex is also associated with DCM. Duchenne muscular dystrophy (DMD) and Becker muscular dystrophy (BMD) are both caused by mutations in the dystrophin gene, *DMD*. Cardiac abnormalities are very common in these patients, with DCM being the most frequent. The disease is *X*-linked and mortality is high in males, particularly when DCM is present, with death typically occurring in the third decade. Interestingly, mutations in specific exons may predict the development or absence of DCM, suggesting changes to different regions of the large dystrophin protein have differing effects on myocardial function

(158). Mutations in *DMD* that affect the N-terminus cause isolated *X*-linked DCM without apparent skeletal myopathy (159–162). This rare allelic condition is also associated with progressive HF and early mortality (163). A common feature of DMD, BMD, and *X*-linked DCM is an absolute reduction or absence of dystrophin. This is believed to alter the response to mechanical stress, though the precise mechanisms remain elusive.

Mutations Affecting Gene Expression. DCM can also be caused by mutations in genes that produce proteins regulating gene expression, including numerous signaling molecules, transcription factors, transcriptional coactivators, RNA-binding proteins, and chromatin modifiers. *RBM20* encodes RNA-binding motif 20 protein, a chaperone protein that binds messenger RNA and regulates splicing. Multiple groups have identified mutation of a very specific hot spot in exon 9 of *RBM20* as a cause of highly penetrant DCM (164–167). RBM20 regulates splicing of the different isoforms of titin. Mutations in this hot spot cause a complete loss of the splicing function in RBM20; hence these mutations indirectly cause DCM secondary to improper splicing of titin by RBM20 (168).

A dominant loss-of-function mutation in *EYA4*, an evolutionarily conserved transcriptional coactivator, has been identified as a probable cause of DCM associated with sensorineural hearing loss (169,170). Mutations in the cardiac transcription factor *NKX2-5* typically cause congenital heart disease (171); however, some individuals without congenital heart disease or with hemodynamically insignificant lesions develop DCM (172). This data, along with the association of several other cardiac transcription factors with DCM (Table 17.6), suggests that these genes, which typically govern cardiac development, may also be essential for maintenance of cardiac regulatory networks throughout life.

Genetics in Clinical Practice

Genotype–Phenotype Correlation

DCM is frequently associated with a variety of noncardiac phenotypes (Table 17.6). Prior to the advent of genetic testing, these associations were not well recognized. In this way, genetic testing has helped to better characterize the phenotype of these individuals, permitting discovery and treatment of disease features (including DCM in those with other phenotypes) that may have gone undiagnosed in affected individuals.

As with HCM, early speculation held that genetics could lead to a major transformation in DCM management. However, the marked phenotypic and genetic diversity of idiopathic DCM relative to HCM, the diverse possible etiology of apparently sporadic cases, and the lower population prevalence of DCM generally blunt the ability to use genetic information to guide management (173).

One notable exception is in individuals with *LMNA* mutations, who have a high incidence of progressive conduction disease and SCD (146,174,175). Participation in competitive sports even before onset of phenotypic disease is associated with a higher incidence of SCD later in life, making avoidance of such activities an important recommendation in affected individuals (174). Two studies have shown that prophylactic ICD placement successfully treats a high percentage of malignant ventricular arrhythmias independent of ejection fraction in *LMNA+* patients receiving a pacemaker (176,177). While the presence of an *LMNA* mutation independent of standard criteria is not an indication for ICD implantation, a low threshold for prophylactic ICD placement should be considered in those receiving a pacemaker. Likewise, consideration should be given to pacemaker implantation in *LMNA+* individuals receiving an ICD who do not otherwise meet criteria for dual chamber pacemaker implantation.

Genetic Testing for DCM

Clinical Applications of Genetic Testing. One of the main roles of genetic testing in DCM is in screening first-degree family members of affected individuals with a confirmed disease-causing mutation. Genetic testing for DCM is more appropriately used to establish the etiology of DCM and to possibly define associated phenotypes, rather than to confirm the diagnosis, which is usually not in doubt phenotypically as can be the case with HCM. Unfortunately, even when a mutation is identified, one may be faced with ambiguity as the prevalence of definitively pathogenic DCM mutations varies considerably with the causal gene, age of the mutation carrier, and other unknown factors.

Commercial Genetic Testing. Genetic testing for DCM is commercially available but test sensitivity is generally lower than for HCM, with a higher detection rate in familial disease than sporadic cases (178). This is partly due to the higher degree of genetic and etiologic diversity of DCM but is also due to substantial variability in the genes tested by commercially available genetic testing panels. The recent identification of *TTN* mutations as a cause of up to 20% of cases of idiopathic and familial DCM should increase the yield of genetic testing for DCM

substantially as this gene becomes incorporated into standard panels. Comprehensive genetic testing of all possible associated gene mutations using modern techniques such as next-generation sequencing should produce yields approaching or exceeding 50%, which is similar to that seen with HCM.

The Genotype-Positive/Phenotype-Negative Individual. Research directed to the genotype-positive/phenotype-negative DCM patient holds the promise of providing a better understanding of the natural history of genetic DCM and the potential that treatments could alter the natural history of the disease. As these important endeavors remain works in progress, the clinical utility of genetics in the management of DCM is expected to undergo substantial changes in the coming decades.

LEFT VENTRICULAR NONCOMPACTION

LVNC is a rare CM characterized by a thick endocardium with prominent trabeculations and deep endocardial recesses (179). As with other forms of CM, there is heterogeneity in clinical symptoms and age of onset, but many patients will develop DCM with progressive HF. LVNC usually coexists with other congenital heart defects and is frequently associated with neuromuscular disorders. Rarely, LVNC can be coexistent with HCM (180). Up to half of cases are familial and the disease is genetically heterogeneous, with mutations in sarcomere genes, structural genes, mitochondrial genes, and signaling molecules (Table 17.7). Genetic screening identifies a possible disease-causing variant in only approximately 40% of cases (181). As the prevalence of LVNC is low, data supporting definitive pathogenesis for most mutations are generally lacking.

This unique phenotype is believed to be the result of incomplete compaction of the LV endocardium during embryogenesis. However, this has been debated for two reasons: first, there is no evidence demonstrating incomplete embryonic myocardial compaction in humans, and, second, there is evidence suggesting noncompaction can develop in the adult after such acute insults as myocarditis. Further, there is significant phenotypic overlap in families with LVNC with some members having typical LVNC, while others carrying the same disease-associated mutation are affected with HCM (often the apical variant) or DCM. As a result, some have suggested that LVNC could be a stage of disease rather than a distinct clinical entity. Genetic analysis has settled this controversy, confirming that arrested embryologic development does occur.

TABLE 17.7 Gene Mutations Associated With Left Ventricular Noncompaction Cardiomyopathy

Gene	Encoded Protein	Associated Findings or Syndromes*
Definitive disease genes		
TAZ	Tafazzin	DCM, Barth syndrome
MIB1	Mindbomb homolog 1	None
MYH7	Myosin heavy chain	None
LVNC-associated genes of uncertain pathogenic significance		
LDB3	LIM-binding domain 3 protein	DCM, myofibrillar myopathy
DTNA	Alpha-dystrobrevin	Barth syndrome
ACTC1	Cardiac alpha-actin	Apical HCM
TNNT2	Cardiac troponin T	None
NKX2.5	NK2 homeobox 5	None
TPM1	Alpha-tropomyosin	None
MYBPC3	Cardiac myosin binding protein C	None
PRDM16	PR domain containing 16	1p36 deletion syndrome
Genetic loci linked to LVNC without a known disease-associated mutation		
11p15	MIM LVNC2	None

*All LVNC-associated genes are also associated with a variety of other congenital heart defects, which are not listed here.
Abbreviation: MIM, Mendelian inheritance in man.

Mutations in *MIB1* were implicated as the cause of LVNC in two unrelated families and genetic manipulation in animal models recapitulates this phenotype in embryos (182). This confirms that, in at least a subset of cases of LVNC, the disease is due to embryonic malformation of the myocardium.

Genetic testing for LVNC is commercially available either as a stand-alone test targeting LVNC-associated genes or in combination with other cardiomyopathy panels. However, as this disease is only beginning to emerge genetically, the yield of testing is relatively low and should only be pursued for the purpose of family screening where indicated.

ARRHYTHMOGENIC RIGHT VENTRICULAR DYSPLASIA/CARDIOMYOPATHY

ARVC is a relatively uncommon CM with a prevalence of approximately 1 in 5,000 (183). It is characterized by fibro-fatty replacement of the myocardium with resultant ventricular dysfunction and arrhythmias. The disease predominantly affects the RV but the LV can be affected to varying degrees. Mortality is quite high with this disorder, owing to a high rate of SCD secondary to malignant ventricular arrhythmias. ARVC accounts for up to 20% of sudden deaths in young adults, second only to HCM, making screening

and early identification of paramount importance (184,185).

Genetic analyses over the past 20 years have shown that ARVC is an inherited disorder that primarily affects the desmosome. These cell-surface structures mediate cell-to-cell adhesion via a complex of structural and transmembrane proteins. Desmosomes play an important role in structural support of tissues exposed to mechanical stress and have been highly conserved through vertebrate evolution. The first mutation identified was in the *JUP* gene, encoding junction plakoglobin, in patients with Naxos disease, a syndromic disorder characterized by ARVC, keratoderma, and wooly hair due to abnormalities in desmosomes in cardiac and epidermal tissue (186,187). A similar syndromic disorder, Carvajal syndrome, is due to mutations in another desmosomal gene, *DSP*, encoding desmoplakin (188).

ARVC is genetically heterogeneous with mutations in several genes, including all major components of the cardiac desmosome (Table 17.8). Syndromic forms are autosomal recessive while the majority of isolated ARVC cases are autosomal dominant. Penetrance is very variable and is both mutation and age-dependent. In addition, the incidence of digenic and compound mutations is between 7% and 42% when all five desmosomal genes are sequenced (189–194). This suggests that at least some of the published mutations are probably not pathogenic

TABLE 17.8 Gene Mutations Associated With Arrhythmogenic Right Ventricular Dysplasia/Cardiomyopathy

Gene	Encoded Protein	Associated Syndromes
Definitive disease genes		
JUP	Plakoglobin	Naxos syndrome
DSP	Desmoplakin	Carvajal syndrome
PKP2	Plakophilin 2	None
DSG2	Desmoglein 2	None
DSC2	Desmocollin 2	None
TMEM43	Transmembrane protein 43	None
ARVC-associated genes of uncertain pathogenic significance		
TGFB3	Transforming growth factor beta3	None
TTN	Titin	None
Genetic loci linked to ARVC without a known disease-associated mutation		
14q12-q22	MIM ARVD3	None
2q32.1-q32.3	MIM ARVD4	None
10p14-p12	MIM ARVD6	None

Abbreviation: MIM, Mendelian inheritance in man.

in isolation and that the role of modifier mutations may be quite significant. Supporting this is the finding that putative pathogenic mutations can be identified in a substantial number of apparently healthy controls (195).

Commercial genetic testing is available and the yield can exceed 50% (196). Again, the primary role is for family screening. However, the combination of incomplete penetrance, frequent digenic or compound mutations, and occasional mutations in apparently healthy individuals complicates the interpretation of genetic testing results in a given patient. Incorporation of family history and clinical screening should be used in conjunction with genetic testing for diagnosis. Further, genetic testing cannot be used to guide management in isolation of family history (173).

THE FUTURE OF GENETICS IN CARDIOMYOPATHY RESEARCH AND CLINICAL PRACTICE

Tremendous progress has been made in defining the genetic etiology of CM over the last quarter century. However, approximately 50% of individuals with CM do not have an identified gene mutation. Many genes are yet to be revealed as causative in CM. Further, the role played by

modifier genes and epigenetic changes in CM has yet to be explored. The landscape of cardiovascular genetics is currently undergoing a sea-shift with the rise of high-throughput, whole genome analysis. GWAS are being used to identify new disease susceptibility loci (131,197). NGS techniques (see Chapter 8), specifically whole exome sequencing (WES) and whole genome sequencing (WGS), have the potential to markedly expand the list of disease-causing genes (167,198). Importantly, these techniques are unbiased approaches to gene discovery, permitting assessment of all genes in the genome, including those that are not immediately apparent as candidate disease genes. In addition, WGS allows for the study of variants outside of the immediate vicinity of the coding regions of a gene. This has the potential to reveal disease-associated mutations in noncoding regulatory elements such as promoter, enhancer/silencer, expression quantitative trait loci, microRNA, and intronic sequences, revealing both novel modes of pathogenesis and new biology.

A novel application of NGS technology is comprehensive sequencing of the RNA transcriptome (RNA-seq), which identifies all expressed genes in a tissue or cell of interest. This permits the study of disease-associated expression patterns, splicing variants, and network/pathway analysis, thus providing a uniquely comprehensive view of global gene expression. A relatively unexplored area of genetic variation in CM research is the study of large genomic structural variations. NGS techniques such as mate-pair sequencing and split-read mapping, as well as other techniques such as comparative genomic hybridization and digital droplet PCR, are now being applied to study structural variation in cardiac disease (130).

The application of genetics in clinical cardiology is transforming from one of slow and cumbersome discovery to one of immense data generation via high-throughput and exquisitely sensitive techniques. These new approaches hold the promise of revealing a better understanding of the biology underlying disease-causing mutations, which in turn could translate into novel disease management strategies and therapeutic interventions.

REFERENCES

1. Lancisi G. *De motu cordis et aneurysmatibus*. Rome: Giovanni Maria Salvioni; 1728.
2. Coats CJ, Hollman A. Hypertrophic cardiomyopathy: Lessons from history. *Heart*. 2008;94:1258–1263.
3. Maron BJ, Maron MS, Wigle ED, Braunwald E. The 50-year history, controversy, and clinical implications of left ventricular outflow tract obstruction in hypertrophic cardiomyopathy from idiopathic hypertrophic subaortic stenosis to hypertrophic cardiomyopathy: from idiopathic hypertrophic subaortic stenosis to hypertrophic cardiomyopathy. *J Am Coll Cardiol*. 2009;54:191–200.
4. Gersh BJ, Maron BJ, Bonow RO, et al. 2011 ACCF/AHA guideline for the diagnosis and treatment of hypertrophic cardiomyopathy: a report of the American College of Cardiology Foundation/American Heart Association Task Force on Practice Guidelines. *Circulation*. 2011;124:e783–e831.
5. Maron BJ, Gardin JM, Flack JM, et al. Prevalence of hypertrophic cardiomyopathy in a general population of young adults. Echocardiographic analysis of 4111 subjects in the CARDIA Study. Coronary Artery Risk Development in (Young) Adults. *Circulation*. 1995;92:785–789.
6. Maron BJ, Bonow RO, Cannon RO, et al. Hypertrophic cardiomyopathy. Interrelations of clinical manifestations, pathophysiology, and therapy (1). *N Engl J Med*. 1987;316:780–789.
7. Wigle ED, Rakowski H, Kimball BP, Williams WG. Hypertrophic cardiomyopathy. Clinical spectrum and treatment. *Circulation*. 1995;92:1680–1692.
8. Nagueh SF, Mahmarian JJ. Noninvasive cardiac imaging in patients with hypertrophic cardiomyopathy. *J Am Coll Cardiol*. 2006;48:2410–2422.
9. Maron MS, Olivotto I, Zenovich AG, et al. Hypertrophic cardiomyopathy is predominantly a disease of left ventricular outflow tract obstruction. *Circulation*. 2006;114:2232–2239.
10. Spirito P, Chiarella F, Carratino L, et al. Clinical course and prognosis of hypertrophic cardiomyopathy in an outpatient population. *N Engl J Med*. 1989;320:749–755.
11. Cannan CR, Reeder GS, Bailey KR, et al. Natural history of hypertrophic cardiomyopathy. A population-based study, 1976 through 1990. *Circulation*. 1995;92:2488–2495.
12. Maron BJ, Casey SA, Poliac LC, et al. Clinical course of hypertrophic cardiomyopathy in a regional United States cohort. *JAMA*. 1999;281:650–655.
13. Kofflard MJ, Waldstein DJ, Vos J, ten Cate FJ. Prognosis in hypertrophic cardiomyopathy observed in a large clinic population. *Am J Cardiol*. 1993;72:939–943.
14. Maron BJ. Hypertrophic cardiomyopathy: a systematic review. *JAMA*. 2002;287:1308–1320.
15. Olivotto I, Cecchi F, Casey SA, et al. Impact of atrial fibrillation on the clinical course of hypertrophic cardiomyopathy. *Circulation*. 2001;104:2517–2524.
16. Maron BJ, Olivotto I, Bellone P, et al. Clinical profile of stroke in 900 patients with hypertrophic cardiomyopathy. *J Am Coll Cardiol*. 2002;39:301–307.
17. Thaman R, Gimeno JR, Murphy RT, et al. Prevalence and clinical significance of systolic impairment in hypertrophic cardiomyopathy. *Heart*. 2005;91:920–925.
18. Harris KM, Spirito P, Maron MS, et al. Prevalence, clinical profile, and significance of left ventricular remodeling in the end-stage phase of hypertrophic cardiomyopathy. *Circulation*. 2006;114:216–225.
19. Maron BJ, Olivotto I, Spirito P, et al. Epidemiology of hypertrophic cardiomyopathy-related death: revisited in a large non-referral-based patient population. *Circulation*. 2000;102:858–864.
20. Maron BJ, Epstein SE, Roberts WC. Causes of sudden death in competitive athletes. *J Am Coll Cardiol*. 1986;7:204–214.
21. Jarcho JA, McKenna W, Pare JA, et al. Mapping a gene for familial hypertrophic cardiomyopathy to chromosome 14q1. *N Engl J Med*. 1989;321:1372–1378.
22. Geisterfer-Lowrance AA, Kass S, Tanigawa G, et al. A molecular basis for familial hypertrophic cardiomyopathy: a beta cardiac myosin heavy chain gene missense mutation. *Cell*. 1990;62:999–1006.
23. Tanigawa G, Jarcho JA, Kass S, et al. A molecular basis for familial hypertrophic cardiomyopathy: an alpha/beta cardiac myosin heavy chain hybrid gene. *Cell*. 1990;62:991–998.
24. Solomon SD, Jarcho JA, McKenna W, et al. Familial hypertrophic cardiomyopathy is a genetically heterogeneous disease. *J Clin Invest*. 1990;86:993–999.
25. Watkins H, MacRae C, Thierfelder L, et al. A disease locus for familial hypertrophic cardiomyopathy maps to chromosome 1q3. *Nat Genet* 1993;3:333–337.
26. Thierfelder L, Watkins H, MacRae C, et al. Alpha-tropomyosin and cardiac troponin T mutations cause familial hypertrophic cardiomyopathy: a disease of the sarcomere. *Cell*. 1994;77:701–712.
27. Watkins H, McKenna WJ, Thierfelder L, et al. Mutations in the genes for cardiac troponin T and alpha-tropomyosin in hypertrophic cardiomyopathy. *N Engl J Med*. 1995;332:1058–1064.
28. Watkins H, Conner D, Thierfelder L, et al. Mutations in the cardiac myosin binding protein-C gene on chromosome 11 cause familial hypertrophic cardiomyopathy. *Nat Genet*. 1995;11:434–437.

29. Bonne G, Carrier L, Bercovici J, et al. Cardiac myosin binding protein-C gene splice acceptor site mutation is associated with familial hypertrophic cardiomyopathy. *Nat Genet.* 1995;11:438–440.

30. Poetter K, Jiang H, Hassanzadeh S, et al. Mutations in either the essential or regulatory light chains of myosin are associated with a rare myopathy in human heart and skeletal muscle. *Nat Genet.* 1996;13:63–69.

31. Kimura A, Harada H, Park JE, et al. Mutations in the cardiac troponin I gene associated with hypertrophic cardiomyopathy. *Nat Genet.* 1997;16:379–382.

32. Mogensen J, Klausen IC, Pedersen AK, et al. Alpha-cardiac actin is a novel disease gene in familial hypertrophic cardiomyopathy. *J Clin Invest.* 1999;103:R39–R43.

33. Olson TM, Doan TP, Kishimoto NY, et al. Inherited and de novo mutations in the cardiac actin gene cause hypertrophic cardiomyopathy. *J Mol Cell Cardiol.* 2000;32:1687–1694.

34. Richard P, Charron P, Carrier L, et al. Hypertrophic cardiomyopathy: distribution of disease genes, spectrum of mutations, and implications for a molecular diagnosis strategy. *Circulation.* 2003;107:2227–2232.

35. Van Driest SL, Ommen SR, Tajik AJ, et al. Yield of genetic testing in hypertrophic cardiomyopathy. *Mayo Clin Proc.* 2005;80:739–744.

36. Andersen PS, Havndrup O, Hougs L, et al. Diagnostic yield, interpretation, and clinical utility of mutation screening of sarcomere encoding genes in Danish hypertrophic cardiomyopathy patients and relatives. *Hum Mutat.* 2009;30:363–370.

37. Osio A, Tan L, Chen SN, et al. Myozenin 2 is a novel gene for human hypertrophic cardiomyopathy. *Circ Res.* 2007;100:766–768.

38. Chiu C, Bagnall RD, Ingles J, et al. Mutations in alpha-actinin-2 cause hypertrophic cardiomyopathy: a genome-wide analysis. *J Am Coll Cardiol.* 2010;55:1127–1135.

39. Geier C, Perrot A, Ozcelik C, et al. Mutations in the human muscle LIM protein gene in families with hypertrophic cardiomyopathy. *Circulation.* 2003;107:1390–1395.

40. Geier C, Gehmlich K, Ehler E, et al. Beyond the sarcomere: CSRP3 mutations cause hypertrophic cardiomyopathy. *Hum Mol Genet.* 2008;17:2753–2765.

41. Satoh M, Takahashi M, Sakamoto T, et al. Structural analysis of the titin gene in hypertrophic cardiomyopathy: identification of a novel disease gene. *Biochem Biophys Res Commun.* 1999;262:411–417.

42. Bos JM, Poley RN, Ny M, et al. Genotype-phenotype relationships involving hypertrophic cardiomyopathy-associated mutations in titin, muscle LIM protein, and telethonin. *Mol Genet Metab.* 2006;88:78–85.

43. Herman DS, Lam L, Taylor MRG, et al. Truncations of titin causing dilated cardiomyopathy. *N Engl J Med.* 2012;366:619–628.

44. Watkins H, Thierfelder L, Anan R, et al. Independent origin of identical beta cardiac myosin heavy-chain mutations in hypertrophic cardiomyopathy. *Am J Hum Genet.* 1993;53:1180–1185.

45. Erdmann J, Daehmlow S, Wischke S, et al. Mutation spectrum in a large cohort of unrelated consecutive patients with hypertrophic cardiomyopathy. *Clin Genet.* 2003;64:339–349.

46. Millat G, Bouvagnet P, Chevalier P, et al. Prevalence and spectrum of mutations in a cohort of 192 unrelated patients with hypertrophic cardiomyopathy. *Eur J Med Genet.* 2010;53:261–267.

47. Curila K, Benesova L, Penicka M, et al. Spectrum and clinical manifestations of mutations in genes responsible for hypertrophic cardiomyopathy. *Acta Cardiol.* 2012;67:23–29.

48. Brito D, Miltenberger-Miltenyi G, Vale Pereira S, et al. Sarcomeric hypertrophic cardiomyopathy: genetic profile in a Portuguese population. *Rev Port Cardiol.* 2012;31:577–587.

49. Liu W, Liu W, Hu D, et al. Mutation spectrum in a large cohort of unrelated chinese patients with hypertrophic cardiomyopathy. *Am J Cardiol.* 2013;112:585–589.

50. Tardiff JC, Hewett TE, Palmer BM, et al. Cardiac troponin T mutations result in allele-specific phenotypes in a mouse model for hypertrophic cardiomyopathy. *J Clin Invest.* 1999;104:469–481.

51. Tyska MJ, Hayes E, Giewat M, et al. Single-molecule mechanics of R403Q cardiac myosin isolated from the mouse model of familial hypertrophic cardiomyopathy. *Circ Res.* 2000;86:737–744.

52. Sanbe A, Nelson D, Gulick J, et al. In vivo analysis of an essential myosin light chain mutation linked to familial hypertrophic cardiomyopathy. *Circ Res.* 2000;87:296–302.

53. James J, Zhang Y, Osinska H, et al. Transgenic modeling of a cardiac troponin I mutation linked to familial hypertrophic cardiomyopathy. *Circ Res.* 2000;87:805–811.

54. Debold EP, Schmitt JP, Patlak JB, et al. Hypertrophic and dilated cardiomyopathy mutations differentially affect the molecular force generation of mouse alpha-cardiac myosin in the laser trap assay. *Am J Physiol Heart Circ Physiol.* 2007;293:H284–H291.

55. Ertz-Berger BR, He H, Dowell C, et al. Changes in the chemical and dynamic properties of cardiac troponin T cause discrete cardiomyopathies in transgenic mice. *Proc Natl Acad Sci USA.* 2005;102:18219–18224.

56. Watkins H, Rosenzweig A, Hwang DS, et al. Characteristics and prognostic implications of myosin missense mutations in familial hypertrophic cardiomyopathy. *N Engl J Med.* 1992;326:1108–1114.

57. Seidman CE, Seidman JG. Identifying sarcomere gene mutations in hypertrophic cardiomyopathy: a personal history. *Circ Res.* 2011;108:743–750.

58. Anan R, Greve G, Thierfelder L, et al. Prognostic implications of novel beta cardiac myosin heavy chain gene mutations that cause familial hypertrophic cardiomyopathy. *J Clin Invest.* 1994;93:280–285.

59. Ko YL, Chen JJ, Tang TK, et al. Malignant familial hypertrophic cardiomyopathy in a family with a 453Arg-->Cys mutation in the beta-myosin heavy chain gene: coexistence of sudden death and end-stage heart failure. *Hum Genet.* 1996;97:585–590.

60. Harris SP, Lyons RG, Bezold KL. In the thick of it: HCM-causing mutations in myosin binding proteins of the thick filament. *Circ Res.* 2011;108:751–764.

61. Zoghbi ME, Woodhead JL, Moss RL, Craig R. Three-dimensional structure of vertebrate cardiac muscle myosin filaments. *Proc Natl Acad Sci USA.* 2008;105:2386–2390.

62. Marston S, Copeland ON, Jacques A, et al. Evidence from human myectomy samples that MYBPC3 mutations cause hypertrophic cardiomyopathy through haploinsufficiency. *Circ Res*. 2009;105:219–222.

63. van Dijk SJ, Dooijes D, dos Remedios C, et al. Cardiac myosin-binding protein C mutations and hypertrophic cardiomyopathy: haploinsufficiency, deranged phosphorylation, and cardiomyocyte dysfunction. *Circulation*. 2009;119:1473–1483.

64. Niimura H, Bachinski LL, Sangwatanaroj S, et al. Mutations in the gene for cardiac myosin-binding protein C and late-onset familial hypertrophic cardiomyopathy. *N Engl J Med*. 1998;338:1248–1257.

65. Page SP, Kounas S, Syrris P, et al. Cardiac myosin binding protein-C mutations in families with hypertrophic cardiomyopathy: disease expression in relation to age, gender, and long term outcome. *Circ Cardiovasc Genet*. 2012;5:156–166.

66. Dhandapany PS, Sadayappan S, Xue Y, et al. A common MYBPC3 (cardiac myosin binding protein C) variant associated with cardiomyopathies in South Asia. *Nat Genet*. 2009;41:187–191.

67. Mogensen J, Murphy RT, Kubo T, et al. Frequency and clinical expression of cardiac troponin I mutations in 748 consecutive families with hypertrophic cardiomyopathy. *J Am Coll Cardiol*. 2004;44:2315–2325.

68. Kokado H, Shimizu M, Yoshio H, et al. Clinical features of hypertrophic cardiomyopathy caused by a Lys183 deletion mutation in the cardiac troponin I gene. *Circulation*. 2000;102:663–669.

69. Moolman JC, Corfield VA, Posen B, et al. Sudden death due to troponin T mutations. *J Am Coll Cardiol*. 1997;29:549–555.

70. Tardiff JC. Thin filament mutations: developing an integrative approach to a complex disorder. *Circ Res*. 2011;108:765–782.

71. Coviello DA, Maron BJ, Spirito P, et al. Clinical features of hypertrophic cardiomyopathy caused by mutation of a "hot spot" in the alpha-tropomyosin gene. *J Am Coll Cardiol*. 1997;29:635–640.

72. Karibe A, Tobacman LS, Strand J, et al. Hypertrophic cardiomyopathy caused by a novel alpha-tropomyosin mutation (V95A) is associated with mild cardiac phenotype, abnormal calcium binding to troponin, abnormal myosin cycling, and poor prognosis. *Circulation*. 2001;103:65–71.

73. Binder J, Ommen SR, Gersh BJ, et al. Echocardiography-guided genetic testing in hypertrophic cardiomyopathy: septal morphological features predict the presence of myofilament mutations. *Mayo Clin Proc*. 2006;81:459–467.

74. Maron BJ, Maron MS, Semsarian C. Genetics of hypertrophic cardiomyopathy after 20 years: clinical perspectives. *J Am Coll Cardiol*. 2012;60:705–715.

75. Ingles J, Doolan A, Chiu C, et al. Compound and double mutations in patients with hypertrophic cardiomyopathy: implications for genetic testing and counselling. *J Med Genet*. 2005;42:e59.

76. Marian AJ. Modifier genes for hypertrophic cardiomyopathy. *Curr Opin Cardiol*. 2002;17:242–252.

77. Rosenzweig A, Watkins H, Hwang DS, et al. Preclinical diagnosis of familial hypertrophic cardiomyopathy by genetic analysis of blood lymphocytes. *N Engl J Med*. 1991;325:1753–1760.

78. Bos JM, Maron BJ, Ackerman MJ, et al. Role of family history of sudden death in risk stratification and prevention of sudden death with implantable defibrillators in hypertrophic cardiomyopathy. *Am J Cardiol*. 2010;106:1481–1486.

79. Bernstein HS, Bishop DF, Astrin KH, et al. Fabry disease: six gene rearrangements and an exonic point mutation in the alpha-galactosidase gene. *J Clin Invest*. 1989;83:1390–1399.

80. Colucci WS, Lorell BH, Schoen FJ, et al. Hypertrophic obstructive cardiomyopathy due to Fabry's disease. *N Engl J Med*. 1982;307:926–928.

81. von Scheidt W, Eng CM, Fitzmaurice TF, et al. An atypical variant of Fabry's disease with manifestations confined to the myocardium. *N Engl J Med*. 1991;324:395–399.

82. Nakao S, Takenaka T, Maeda M, et al. An atypical variant of Fabry's disease in men with left ventricular hypertrophy. *N Engl J Med*. 1995;333:288–293.

83. Ferrans VJ, Hibbs RG, Burda CD. The heart in Fabry's disease. A histochemical and electron microscopic study. *Am J Cardiol*. 1969;24:95–110.

84. Becker AE, Schoorl R, Balk AG, van der Heide RM. Cardiac manifestations of Fabry's disease. Report of a case with mitral insufficiency and electrocardiographic evidence of myocardial infarction. *Am J Cardiol*. 1975;36:829–835.

85. Linhart A, Palecek T, Bultas J, et al. New insights in cardiac structural changes in patients with Fabry's disease. *Am Heart J*. 2000;139:1101–1108.

86. Shah JS, Lee P, Hughes D, et al. The natural history of left ventricular systolic function in Anderson-Fabry disease. *Heart*. 2005;91:533–534.

87. Sachdev B, Takenaka T, Teraguchi H, et al. Prevalence of Anderson-Fabry disease in male patients with late onset hypertrophic cardiomyopathy. *Circulation*. 2002;105:1407–1411.

88. Chimenti C, Pieroni M, Morgante E, et al. Prevalence of Fabry disease in female patients with late-onset hypertrophic cardiomyopathy. *Circulation*. 2004;110:1047–1053.

89. Monserrat L, Gimeno-Blanes JR, Marín F, et al. Prevalence of fabry disease in a cohort of 508 unrelated patients with hypertrophic cardiomyopathy. *J Am Coll Cardiol*. 2007;50:2399–2403.

90. Linthorst GE, Poorthuis BJHM, Hollak CEM. Enzyme activity for determination of presence of Fabry disease in women results in 40% false-negative results. *J Am Coll Cardiol*. 2008;51:2082; author reply 2082–2083.

91. Danon MJ, Oh SJ, DiMauro S, et al. Lysosomal glycogen storage disease with normal acid maltase. *Neurology*. 1981;31:51–57.

92. Nishino I, Fu J, Tanji K, et al. Primary LAMP-2 deficiency causes X-linked vacuolar cardiomyopathy and myopathy (Danon disease). *Nature*. 2000;406:906–910.

93. Tanaka Y, Guhde G, Suter A, et al. Accumulation of autophagic vacuoles and cardiomyopathy in LAMP-2-deficient mice. *Nature*. 2000;406:902–906.

94. Arad M, Maron BJ, Gorham JM, et al. Glycogen storage diseases presenting as hypertrophic cardiomyopathy. *N Engl J Med.* 2005;352:362–372.

95. Maron BJ, Roberts WC, Arad M, et al. Clinical outcome and phenotypic expression in LAMP2 cardiomyopathy. *JAMA.* 2009;301:1253–1259.

96. Yang Z, McMahon CJ, Smith LR, et al. Danon disease as an underrecognized cause of hypertrophic cardiomyopathy in children. *Circulation.* 2005;112:1612–1617.

97. van der Ploeg AT, Reuser AJJ. Pompe's disease. *Lancet.* 2008;372:1342–1353.

98. MacRae CA, Ghaisas N, Kass S, et al. Familial hypertrophic cardiomyopathy with Wolff-Parkinson-White syndrome maps to a locus on chromosome 7q3. *J Clin Invest.* 1995;96:1216–1220.

99. Gollob MH, Green MS, Tang AS, et al. Identification of a gene responsible for familial Wolff-Parkinson-White syndrome. *N Engl J Med.* 2001;344:1823–1831.

100. Blair E, Redwood C, Ashrafian H, et al. Mutations in the gamma(2) subunit of AMP-activated protein kinase cause familial hypertrophic cardiomyopathy: evidence for the central role of energy compromise in disease pathogenesis. *Hum Mol Genet.* 2001;10:1215–1220.

101. Arad M, Benson DW, Perez-Atayde AR, et al. Constitutively active AMP kinase mutations cause glycogen storage disease mimicking hypertrophic cardiomyopathy. *J Clin Invest.* 2002;109:357–362.

102. Arad M, Moskowitz IP, Patel VV, et al. Transgenic mice overexpressing mutant PRKAG2 define the cause of Wolff-Parkinson-White syndrome in glycogen storage cardiomyopathy. *Circulation.* 2003;107:2850–2856.

103. Mogensen J, Kubo T, Duque M, et al. Idiopathic restrictive cardiomyopathy is part of the clinical expression of cardiac troponin I mutations. *J Clin Invest.* 2003;111:209–216.

104. Kaski JP, Syrris P, Burch M, et al. Idiopathic restrictive cardiomyopathy in children is caused by mutations in cardiac sarcomere protein genes. *Heart.* 2008;94:1478–1484.

105. Menon SC, Michels VV, Pellikka PA, et al. Cardiac troponin T mutation in familial cardiomyopathy with variable remodeling and restrictive physiology. *Clin Genet.* 2008;74:445–454.

106. Zhang J, Kumar A, Kaplan L, et al. Genetic linkage of a novel autosomal dominant restrictive cardiomyopathy locus. *J Med Genet.* 2005;42:663–665.

107. Zhang J, Kumar A, Stalker HJ, et al. Clinical and molecular studies of a large family with desmin-associated restrictive cardiomyopathy. *Clin Genet.* 2001;59:248–256.

108. Gheorghiade M, Sopko G, De Luca L, et al. Navigating the crossroads of coronary artery disease and heart failure. *Circulation.* 2006;114:1202–1213.

109. Maron BJ, Towbin JA, Thiene G, et al. Contemporary definitions and classification of the cardiomyopathies: an American Heart Association Scientific Statement from the Council on Clinical Cardiology, Heart Failure and Transplantation Committee; Quality of Care and Outcomes Research and Functional Genomics and Translational Biology Interdisciplinary Working Groups; and Council on Epidemiology and Prevention. *Circulation.* 2006;113:1807–1816.

110. Investigators TS. Effect of enalapril on mortality and the development of heart failure in asymptomatic patients with reduced left ventricular ejection fractions. The SOLVD Investigators. *N Engl J Med.* 1992;327:685–691.

111. Wang TJ, Evans JC, Benjamin EJ, et al. Natural history of asymptomatic left ventricular systolic dysfunction in the community. *Circulation.* 2003;108:977–982.

112. Castelli G, Fornaro A, Ciaccheri M, et al. Improving survival rates of patients with idiopathic dilated cardiomyopathy in tuscany over 3 decades: impact of evidence-based management. *Circ Heart Fail.* 2013;6:913–921.

113. Codd MB, Sugrue DD, Gersh BJ, Melton LJ. Epidemiology of idiopathic dilated and hypertrophic cardiomyopathy. A population-based study in Olmsted County, Minnesota, 1975–1984. *Circulation.* 1989;80:564–572.

114. Miura K, Nakagawa H, Morikawa Y, et al. Epidemiology of idiopathic cardiomyopathy in Japan: results from a nationwide survey. *Heart.* 2002;87:126–130.

115. McNally EM, Golbus JR, Puckelwartz MJ. Genetic mutations and mechanisms in dilated cardiomyopathy. *J Clin Invest.* 2013;123:19–26.

116. Golbus JR, Puckelwartz MJ, Fahrenbach JP, et al. Population-based variation in cardiomyopathy genes. *Circ Cardiovasc Genet.* 2012;5:391–399.

117. Gerull B, Gramlich M, Atherton J, et al. Mutations of TTN, encoding the giant muscle filament titin, cause familial dilated cardiomyopathy. *Nat Genet.* 2002;30:201–204.

118. Kamisago M, Sharma SD, DePalma SR, et al. Mutations in sarcomere protein genes as a cause of dilated cardiomyopathy. *N Engl J Med.* 2000;343:1688–1696.

119. Villard E, Duboscq-Bidot L, Charron P, et al. Mutation screening in dilated cardiomyopathy: prominent role of the beta myosin heavy chain gene. *Eur Heart J.* 2005;26:794–803.

120. Olson TM, Michels VV, Thibodeau SN, et al. Actin mutations in dilated cardiomyopathy, a heritable form of heart failure. *Science.* 1998;280:750–752.

121. Lakdawala NK, Dellefave L, Redwood CS, et al. Familial dilated cardiomyopathy caused by an alpha-tropomyosin mutation: the distinctive natural history of sarcomeric dilated cardiomyopathy. *J Am Coll Cardiol.* 2010;55:320–329.

122. Murphy RT, Mogensen J, Shaw A, et al. Novel mutation in cardiac troponin I in recessive idiopathic dilated cardiomyopathy. *Lancet.* 2004;363:371–372.

123. Carballo S, Robinson P, Otway R, et al. Identification and functional characterization of cardiac troponin I as a novel disease gene in autosomal dominant dilated cardiomyopathy. *Circ Res.* 2009;105:375–382.

124. Daehmlow S, Erdmann J, Knueppel T, et al. Novel mutations in sarcomeric protein genes in dilated cardiomyopathy. *Biochem Biophys Res Commun.* 2002;298:116–120.

125. Møller DV, Andersen PS, Hedley P, et al. The role of sarcomere gene mutations in patients with idiopathic dilated cardiomyopathy. *Eur J Hum Genet.* 2009;17:1241–1249.

126. Mogensen J, Murphy RT, Shaw T, et al. Severe disease expression of cardiac troponin C and T mutations in patients with idiopathic dilated cardiomyopathy. *J Am Coll Cardiol.* 2004;44:2033–2040.

127. Hershberger RE, Norton N, Morales A, et al. Coding sequence rare variants identified in MYBPC3, MYH6, TPM1, TNNC1, and TNNI3 from 312 patients with familial or idiopathic dilated cardiomyopathy. *Circ Cardiovasc Genet.* 2010;3:155–161.

128. Carniel E, Taylor MRG, Sinagra G, et al. Alpha-myosin heavy chain: a sarcomeric gene associated with dilated and hypertrophic phenotypes of cardiomyopathy. *Circulation.* 2005;112:54–59.

129. Arimura T, Ishikawa T, Nunoda S, et al. Dilated cardiomyopathy-associated BAG3 mutations impair Z-disc assembly and enhance sensitivity to apoptosis in cardiomyocytes. *Hum Mutat.* 2011;32:1481–1491.

130. Norton N, Li D, Rieder MJ, et al. Genome-wide studies of copy number variation and exome sequencing identify rare variants in BAG3 as a cause of dilated cardiomyopathy. *Am J Hum Genet.* 2011;88:273–282.

131. Villard E, Perret C, Gary F, et al. A genome-wide association study identifies two loci associated with heart failure due to dilated cardiomyopathy. *Eur Heart J.* 2011;32:1065–1076.

132. Schmitt JP, Kamisago M, Asahi M, et al. Dilated cardiomyopathy and heart failure caused by a mutation in phospholamban. *Science.* 2003;299:1410–1413.

133. Haghighi K, Kolokathis F, Pater L, et al. Human phospholamban null results in lethal dilated cardiomyopathy revealing a critical difference between mouse and human. *J Clin Invest.* 2003;111:869–876.

134. Haghighi K, Kolokathis F, Gramolini AO, et al. A mutation in the human phospholamban gene, deleting arginine 14, results in lethal, hereditary cardiomyopathy. *Proc Natl Acad Sci USA.* 2006;103:1388–1393.

135. Olson TM, Keating MT. Mapping a cardiomyopathy locus to chromosome 3p22–p25. *J Clin Invest.* 1996;97:528–532.

136. McNair WP, Ku L, Taylor MRG, et al. SCN5A mutation associated with dilated cardiomyopathy, conduction disorder, and arrhythmia. *Circulation.* 2004;110:2163–2167.

137. Olson TM, Michels VV, Ballew JD, et al. Sodium channel mutations and susceptibility to heart failure and atrial fibrillation. *JAMA.* 2005;293:447–454.

138. Hershberger RE, Parks SB, Kushner JD, et al. Coding sequence mutations identified in MYH7, TNNT2, SCN5A, CSRP3, LBD3, and TCAP from 313 patients with familial or idiopathic dilated cardiomyopathy. *Clin Transl Sci.* 2008;1:21–26.

139. McNair WP, Sinagra G, Taylor MRG, et al. SCN5A mutations associate with arrhythmic dilated cardiomyopathy and commonly localize to the voltage-sensing mechanism. *J Am Coll Cardiol.* 2011;57:2160–2168.

140. Chen Q, Kirsch GE, Zhang D, et al. Genetic basis and molecular mechanism for idiopathic ventricular fibrillation. *Nature.* 1998;392:293–296.

141. Watanabe H, Nogami A, Ohkubo K, et al. Electrocardiographic characteristics and SCN5A mutations in idiopathic ventricular fibrillation associated with early repolarization. *Circ Arrhythm Electrophysiol.* 2011;4:874–881.

142. Ellinor PT, Nam EG, Shea MA, et al. Cardiac sodium channel mutation in atrial fibrillation. *Heart Rhythm.* 2008;5:99–105.

143. Watanabe H, Yang T, Stroud DM, et al. Striking in vivo phenotype of a disease-associated human SCN5A mutation producing minimal changes in vitro. *Circulation.* 2011;124:1001–1011.

144. Laurent G, Saal S, Amarouch MY, et al. Multifocal ectopic Purkinje-related premature contractions: a new SCN5A-related cardiac channelopathy. *J Am Coll Cardiol.* 2012;60:144–156.

145. Mann SA, Castro ML, Ohanian M, et al. R222Q SCN5A mutation is associated with reversible ventricular ectopy and dilated cardiomyopathy. *J Am Coll Cardiol.* 2012;60:1566–1573.

146. Fatkin D, MacRae C, Sasaki T, et al. Missense mutations in the rod domain of the lamin A/C gene as causes of dilated cardiomyopathy and conduction-system disease. *N Engl J Med.* 1999;341:1715–1724.

147. Bione S, Maestrini E, Rivella S, et al. Identification of a novel X-linked gene responsible for Emery-Dreifuss muscular dystrophy. *Nat Genet.* 1994;8:323–327.

148. Bione S, Small K, Aksmanovic VM, et al. Identification of new mutations in the Emery-Dreifuss muscular dystrophy gene and evidence for genetic heterogeneity of the disease. *Hum Mol Genet.* 1995;4:1859–1863.

149. Zhang Q, Bethmann C, Worth NF, et al. Nesprin-1 and -2 are involved in the pathogenesis of Emery Dreifuss muscular dystrophy and are critical for nuclear envelope integrity. *Hum Mol Genet.* 2007;16:2816–2833.

150. Puckelwartz MJ, Kessler EJ, Kim G, et al. Nesprin-1 mutations in human and murine cardiomyopathy. *J Mol Cell Cardiol.* 2010;48:600–608.

151. Taylor MRG, Fain PR, Sinagra G, et al. Natural history of dilated cardiomyopathy due to lamin A/C gene mutations. *J Am Coll Cardiol.* 2003;41:771–780.

152. Goldfarb LG, Park KY, Cervenáková L, et al. Missense mutations in desmin associated with familial cardiac and skeletal myopathy. *Nat Genet.* 1998;19:402–403.

153. Dalakas MC, Park KY, Semino-Mora C, et al. Desmin myopathy, a skeletal myopathy with cardiomyopathy caused by mutations in the desmin gene. *N Engl J Med.* 2000;342:770–780.

154. Li D, Tapscoft T, Gonzalez O, et al. Desmin mutation responsible for idiopathic dilated cardiomyopathy. *Circulation.* 1999;100:461–464.

155. Taylor MRG, Slavov D, Ku L, et al. Prevalence of desmin mutations in dilated cardiomyopathy. *Circulation.* 2007;115:1244–1251.

156. Muñoz-Mármol AM, Strasser G, Isamat M, et al. A dysfunctional desmin mutation in a patient with severe generalized myopathy. *Proc Natl Acad Sci USA.* 1998;95:11312–11317.

157. Sjöberg G, Saavedra-Matiz CA, Rosen DR, et al. A missense mutation in the desmin rod domain is associated with autosomal dominant distal myopathy, and exerts a dominant negative effect on filament formation. *Hum Mol Genet.* 1999;8:2191–2198.

158. Jefferies JL, Eidem BW, Belmont JW, et al. Genetic predictors and remodeling of dilated cardiomyopathy in muscular dystrophy. *Circulation.* 2005;112:2799–2804.

159. Muntoni F, Cau M, Ganau A, et al. Brief report: deletion of the dystrophin muscle-promoter region associated with X-linked dilated cardiomyopathy. *N Engl J Med*. 1993;329:921–925.

160. Muntoni F, Wilson L, Marrosu G, et al. A mutation in the dystrophin gene selectively affecting dystrophin expression in the heart. *J Clin Invest*. 1995;96:693–699.

161. Milasin J, Muntoni F, Severini GM, et al. A point mutation in the 5' splice site of the dystrophin gene first intron responsible for X-linked dilated cardiomyopathy. *Hum Mol Genet*. 1996;5:73–79.

162. Ortiz-Lopez R, Li H, Su J, et al. Evidence for a dystrophin missense mutation as a cause of X-linked dilated cardiomyopathy. *Circulation*. 1997;95:2434–2440.

163. Berko BA, Swift M. X-linked dilated cardiomyopathy. *N Engl J Med*. 1987;316:1186–1191.

164. Brauch KM, Karst ML, Herron KJ, et al. Mutations in ribonucleic acid binding protein gene cause familial dilated cardiomyopathy. *J Am Coll Cardiol*. 2009;54:930–941.

165. Li D, Morales A, Gonzalez-Quintana J, et al. Identification of novel mutations in RBM20 in patients with dilated cardiomyopathy. *Clin Transl Sci*. 2010;3:90–97.

166. Millat G, Bouvagnet P, Chevalier P, et al. Clinical and mutational spectrum in a cohort of 105 unrelated patients with dilated cardiomyopathy. *Eur J Med Genet*. 2011;54:e570–e575.

167. Wells QS, Becker JR, Su YR, et al. Whole exome sequencing identifies a causal RBM20 mutation in a large pedigree with familial dilated cardiomyopathy. *Circ Cardiovasc Genet*. 2013;6:317–326.

168. Guo W, Schafer S, Greaser ML, et al. RBM20, a gene for hereditary cardiomyopathy, regulates titin splicing. *Nat Med*. 2012;18:766–773.

169. Schönberger J, Levy H, Grünig E, et al. Dilated cardiomyopathy and sensorineural hearing loss: a heritable syndrome that maps to 6q23-24. *Circulation*. 2000;101:1812–1818.

170. Schönberger J, Wang L, Shin JT, et al. Mutation in the transcriptional coactivator EYA4 causes dilated cardiomyopathy and sensorineural hearing loss. *Nat Genet*. 2005;37:418–422.

171. Schott JJ, Benson DW, Basson CT, et al. Congenital heart disease caused by mutations in the transcription factor NKX2-5. *Science*. 1998;281:108–11.

172. Costa MW, Guo G, Wolstein O, et al. Functional characterization of a novel mutation in NKX2-5 associated with congenital heart disease and adult-onset cardiomyopathy. *Circ Cardiovasc Genet*. 2013;6:238–247.

173. Hershberger RE, Lindenfeld J, Mestroni L, et al. Genetic evaluation of cardiomyopathy—a Heart Failure Society of America practice guideline. *J Card Fail*. 2009;15:83–97.

174. Pasotti M, Klersy C, Pilotto A, et al. Long-term outcome and risk stratification in dilated cardiolaminopathies. *J Am Coll Cardiol*. 2008;52:1250–1260.

175. van Rijsingen IAW, Arbustini E, Elliott PM, et al. Risk factors for malignant ventricular arrhythmias in lamin A/C mutation carriers a European cohort study. *J Am Coll Cardiol*. 2012;59:493–500.

176. Meune C, Van Berlo JH, Anselme F, et al. Primary prevention of sudden death in patients with lamin A/C gene mutations. *N Engl J Med*. 2006;354:209–210.

177. Anselme F, Moubarak G, Savouré A, et al. Implantable cardioverter-defibrillators in lamin A/C mutation carriers with cardiac conduction disorders. *Heart Rhythm*. 2013;10:1492–1498.

178. van Spaendonck-Zwarts KY, van Rijsingen IAW, Van den Berg MP, et al. Genetic analysis in 418 index patients with idiopathic dilated cardiomyopathy: overview of 10 years' experience. *Eur J Heart Fail*. 2013;15:628–636.

179. Chin TK, Perloff JK, Williams RG, et al. Isolated noncompaction of left ventricular myocardium. A study of eight cases. *Circulation*. 1990;82:507–513.

180. Kelley-Hedgepeth A, Towbin JA, Maron MS. Images in cardiovascular medicine. Overlapping phenotypes: left ventricular noncompaction and hypertrophic cardiomyopathy. *Circulation*. 2009;119:e588–e589.

181. Hoedemaekers YM, Caliskan K, Michels M, et al. The importance of genetic counseling, DNA diagnostics, and cardiologic family screening in left ventricular noncompaction cardiomyopathy. *Circ Cardiovasc Genet*. 2010;3:232–239.

182. Luxán G, Casanova JC, Martínez-Poveda B, et al. Mutations in the NOTCH pathway regulator MIB1 cause left ventricular noncompaction cardiomyopathy. *Nat Med*. 2013;19:193–201.

183. Basso C, Bauce B, Corrado D, Thiene G. Pathophysiology of arrhythmogenic cardiomyopathy. *Nat Rev Cardiol*. 2012;9:223–233.

184. Dalal D, Nasir K, Bomma C, et al. Arrhythmogenic right ventricular dysplasia: a United States experience. *Circulation*. 2005;112:3823–3832.

185. Nattel S, Schott JJ. Arrhythmogenic right ventricular dysplasia type 1 and mutations in transforming growth factor beta3 gene regulatory regions: a breakthrough? *Cardiovasc Res*. 2005;65:302–304.

186. Coonar AS, Protonotarios N, Tsatsopoulou A, et al. Gene for arrhythmogenic right ventricular cardiomyopathy with diffuse nonepidermolytic palmoplantar keratoderma and woolly hair (Naxos disease) maps to 17q21. *Circulation*. 1998;97:2049–2058.

187. McKoy G, Protonotarios N, Crosby A, et al. Identification of a deletion in plakoglobin in arrhythmogenic right ventricular cardiomyopathy with palmoplantar keratoderma and woolly hair (Naxos disease). *Lancet*. 2000;355:2119–2124.

188. Norgett EE, Hatsell SJ, Carvajal-Huerta L, et al. Recessive mutation in desmoplakin disrupts desmoplakin-intermediate filament interactions and causes dilated cardiomyopathy, woolly hair and keratoderma. *Hum Mol Genet*. 2000;9:2761–2766.

189. den Haan AD, Tan BY, Zikusoka MN, et al. Comprehensive desmosome mutation analysis in north americans with arrhythmogenic right ventricular dysplasia/cardiomyopathy. *Circ Cardiovasc Genet*. 2009;2:428–435.

190. Christensen AH, Benn M, Bundgaard H, et al. Wide spectrum of desmosomal mutations in Danish patients with arrhythmogenic right ventricular cardiomyopathy. *J Med Genet*. 2010;47:736–744.

191. Bauce B, Nava A, Beffagna G, et al. Multiple mutations in desmosomal proteins encoding genes in arrhythmogenic right ventricular cardiomyopathy/dysplasia. *Heart Rhythm*. 2010;7:22–29.

192. Fressart V, Duthoit G, Donal E, et al. Desmosomal gene analysis in arrhythmogenic right ventricular dysplasia/cardiomyopathy: spectrum of mutations and clinical impact in practice. *Europace*. 2010;12:861–868.

193. Xu T, Yang Z, Vatta M, et al. Compound and digenic heterozygosity contributes to arrhythmogenic right ventricular cardiomyopathy. *J Am Coll Cardiol*. 2010;55:587–597.

194. Barahona-Dussault C, Benito B, Campuzano O, et al. Role of genetic testing in arrhythmogenic right ventricular cardiomyopathy/dysplasia. *Clin Genet*. 2010;77:37–48.

195. Kapplinger JD, Landstrom AP, Salisbury BA, et al. Distinguishing arrhythmogenic right ventricular cardiomyopathy/dysplasia-associated mutations from background genetic noise. *J Am Coll Cardiol*. 2011;57:2317–2327.

196. Cox MGPJ, van der Zwaag PA, van der Werf C, et al. Arrhythmogenic right ventricular dysplasia/cardiomyopathy: pathogenic desmosome mutations in index-patients predict outcome of family screening: Dutch arrhythmogenic right ventricular dysplasia/cardiomyopathy genotype-phenotype follow-up study. *Circulation*. 2011;123:2690–2700.

197. Meder B, Rühle F, Weis T, et al. A genome-wide association study identifies 6p21 as novel risk locus for dilated cardiomyopathy. *Eur Heart J*. 2014;35:1069–1077.

198. Meder B, Haas J, Keller A, et al. Targeted next-generation sequencing for the molecular genetic diagnostics of cardiomyopathies. *Circ Cardiovasc Genet*. 2011;4:110–122.

Genetics of Pulmonary Hypertension

Ankit A. Desai, Julio D. Duarte, Sunit Singla, and Roberto F. Machado

TAKE HOME POINTS

1. In the heritable form of World Health Organization (WHO) Group I pulmonary arterial hypertension (PAH), approximately 70% of cases are associated with mutations in either *BMPR2* (most common) or *ALK1/ENG*, leaving a smaller subset of patients with recently documented caveolin-1 (*CAV1*) mutations and the remainder still without a known genetic mechanism.
2. Based on animal studies, reduced function of *BMPR2* appears neither necessary nor sufficient to cause PAH; therefore, disease development likely requires a "double hit" from genetic mutations and environmental factors.
3. In patients with less clearly heritable forms of pulmonary hypertension (PH), including those with secondary forms of PH, detailed mechanistic investigations have revealed a complex and intricate web of genetic pathways that appear to modulate the complex PH phenotype.

CASE STUDY

A 39-year-old woman presented to her physician with complaints of slowly progressive shortness of breath, exercise intolerance, and chest pain for the past 6 months. She was previously healthy and had no medical problems, and she played sports as a child and adolescent without any problems. At first she thought that her symptoms were due to weight gain and a sedentary lifestyle; however, her symptoms had progressed to the point that she began having worsening shortness of breath, lightheadedness, and dizziness during exertion. In addition, she noted that 2 weeks prior to presentation, she had an episode of syncope while climbing stairs. She denied taking any medications currently, but did admit to anorexigen use in the past. She was a nonsmoker, rarely consumed alcohol, and denied illicit drug use.

On physical examination, her vital signs were as follows: blood pressure 94/70 mmHg; heart rate 92 bpm; respiratory rate 16 breaths/minute; and oxygen saturation 94% while breathing room air. She had multiple telangiectasias visible on her face and mucosal surfaces, and throughout her body. Her jugular venous pressure was elevated at 14 cm H_2O with prominent A and V waves; her lungs were clear to auscultation; and her cardiac exam revealed a prominent right ventricular lift, regular rate and rhythm, a loud P2, a right-sided S4, and a 2/6 holosystolic murmur at the left sternal border, which increased with inspiration. Her abdominal exam was benign, and she had 1+ lower extremity edema bilaterally.

Based on the clinical presentation and physical exam, a diagnosis of PAH was suspected, and further testing was performed. An electrocardiogram revealed a normal sinus rhythm and evidence of right atrial enlargement and right ventricular hypertrophy. Chest radiography demonstrated clear lungs with prominent central pulmonary arteries. An echocardiogram demonstrated the following: small left ventricle with normal left ventricular systolic function; severely enlarged right ventricle with significant right ventricular systolic dysfunction; interventricular septal flattening in systole and diastole, consistent with right ventricular pressure and volume

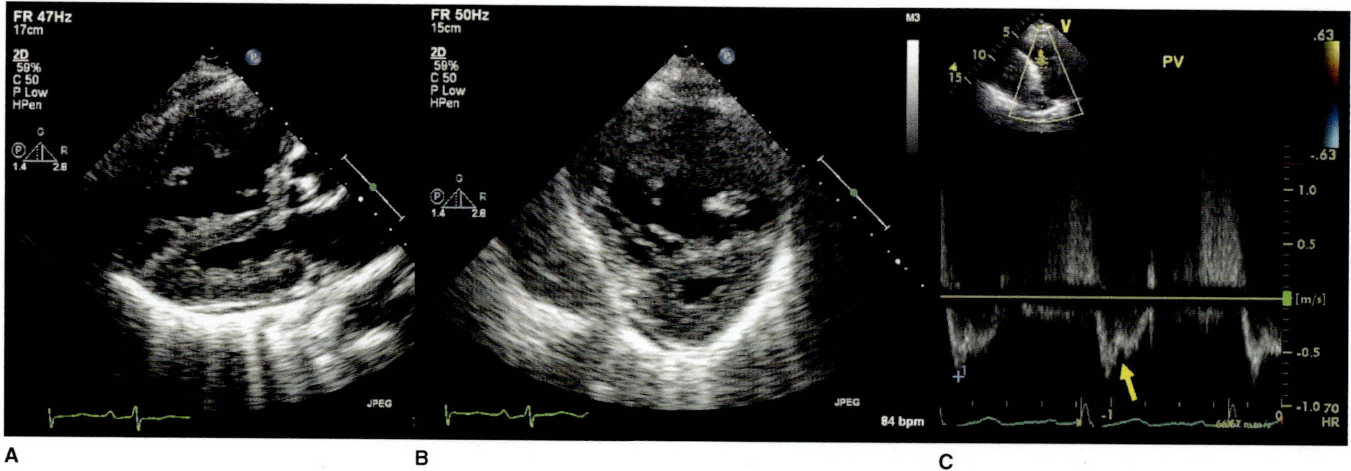

FIGURE 18.1 Echocardiographic images from the case study. Panel A (parasternal long-axis view) and Panel B (parasternal short-axis view) both demonstrate a severely dilated right ventricle, small left ventricle, and interventricular septal flattening. Panel C demonstrates mid-systolic notching of the right ventricular outflow pulse-wave Doppler profile, indicative of severely decreased compliance of the pulmonary vasculature (consistent with severe pulmonary arterial hypertension).

overload (Figures 18.1A–B, Videos 18.1A–B [http://www.demosmedical.com/media/videos/Shah_Video_18_1A.mp4; http://www.demosmedical.com/media/videos/Shah_Video_18_1B.mp4]); moderate tricuspid regurgitation; elevated pulmonary artery systolic pressure (PASP) of 85 to 90 mmHg, based on an estimated right atrial pressure (RAP) of 15 to 20 mmHg; and mid-systolic notching of the right ventricular outflow pulse-wave Doppler profile (Figure 18.1C).

The patient was treated with loop diuretics. She had improvement of symptoms and her weight decreased by 15 lbs. Invasive hemodynamic testing confirmed the diagnosis of PAH (Figure 18.2). Subsequent workup included a low-probability ventilation/perfusion (V/Q) scan on the lungs, no evidence of parenchymal lung disease on high-resolution chest computed tomography, and reduced diffusing capacity of carbon monoxide (DL_{CO}) on pulmonary function testing. Further laboratory testing ruled out human immunodeficiency virus infection, connective tissue disease, or liver disease. There was also no evidence of congenital heart disease. Based on these findings, the patient was diagnosed with anorexigen-induced PAH and referred to a PH specialist.

Upon referral to a PH specialist, further details of the family history were obtained. The patient's mother and father died in a car accident when she was a child, so she was raised by her grandparents. The patient's grandfather (Figure 18.3) and his siblings all had multiple telangiectasias, and her grandfather carried a diagnosis of congestive heart failure with frequent hospitalizations. Based on these findings, a diagnosis of hereditary hemorrhagic telangiectasia (HHT) was entertained, and the patient and her grandfather underwent genetic testing, which revealed a missense mutation in exon 10 of the *ALK1* gene known to be associated with HHT and PAH. In addition, her grandfather underwent echocardiography and invasive hemodynamic testing, both of which revealed moderate PAH.

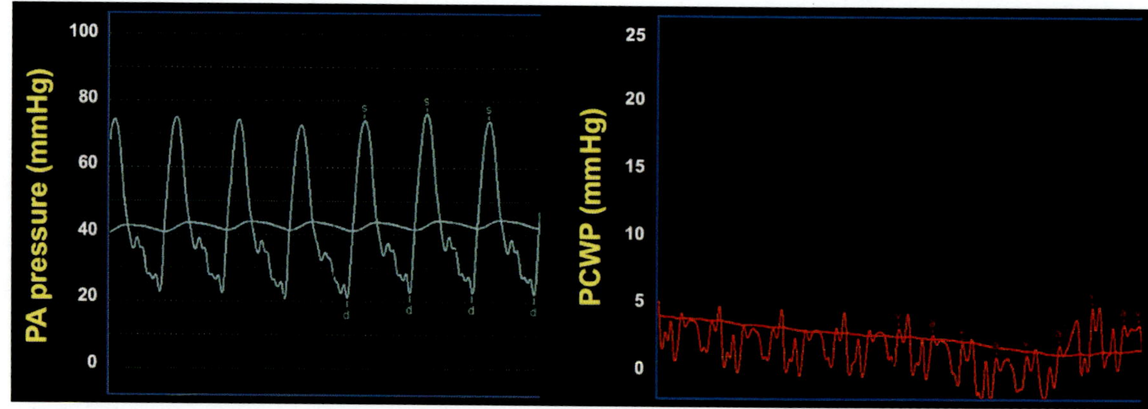

FIGURE 18.2 Invasive hemodynamic tracings from the case study. The pulmonary artery pressure was 76/22 (mean 43) mmHg and the pulmonary capillary wedge pressure was 3 mmHg at end expiration. The cardiac output was 4 L/min, and the pulmonary vascular resistance was severely elevated at 10 Wood units.

Abbreviations: PA, pulmonary artery; PCWP, pulmonary capillary wedge pressure.

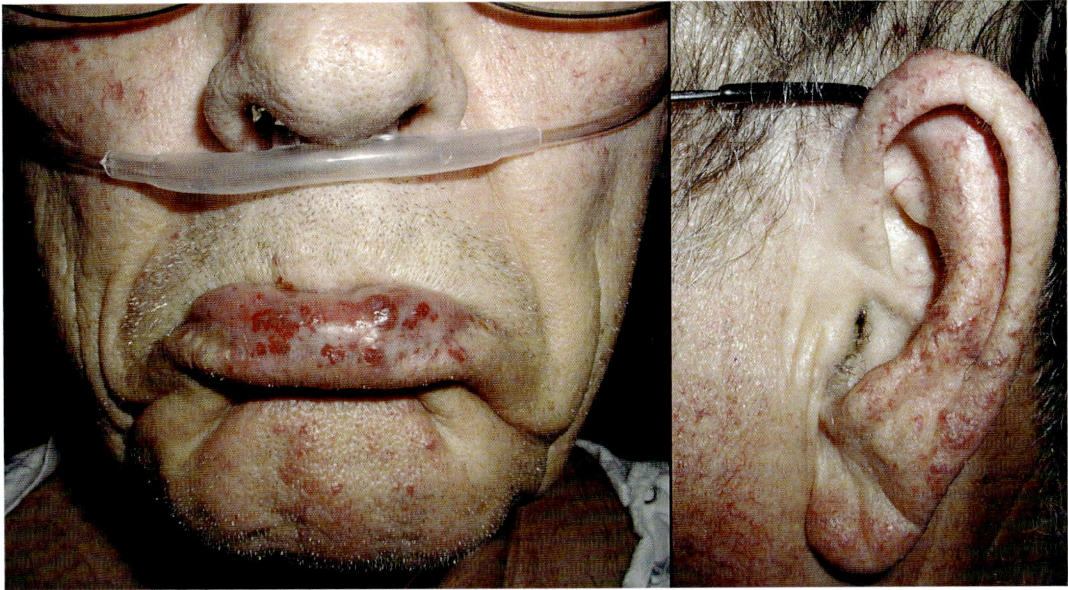

FIGURE 18.3 Representative telangectasias from the grandfather in the case study. The left panel demonstrates multiple oral telangectasias on the lower lip; the right panel demonstrates telangectasias on the ear.

PH is a group of disorders defined hemodynamically by increased mean pulmonary artery pressure (mPAP) greater than or equal to 25 mmHg on right heart catheterization. PH, particularly PAH, is typically considered a life-threatening disease of the pulmonary vasculature associated with progressive right ventricular failure and high morbidity and mortality. However, within the last four decades, significant progress has been made to better understand and categorize groups of diseases that can lead to or are associated with PH. In this chapter, we provide: (a) a brief review of the clinical aspects of PH; (b) an overview of the genetic aspects of PH; (c) detailed analysis of specific genes and molecular pathways that have been found to contribute to PH; (d) review novel omics technologies such as gene expression profiling and epigenetics, which should advance our understanding of PH pathogenesis; and (e) review the available clinical genetic testing for patients and families with known or suspected heritable forms of PAH.

OVERVIEW OF PULMONARY HYPERTENSION

The current clinical classification divides PH into five major groups according to pathophysiological and therapeutic features: (a) PAH; (b) PH due to left heart disease; (c) PH due to lung diseases and/or hypoxia; (d) chronic thromboembolic PH (CTEPH); and (e) PH due to miscellaneous mechanisms (1). Although causes that lead to PH development differ between the groups, several cellular and histologic pathways are common across the various groups including evidence of vasoconstriction, endothelial dysfunction, abnormal cell proliferation, inflammation, and thrombosis (2).

Doppler echocardiography is usually the screening test of choice for PH in a patient typically presenting with dyspnea. With this technique, PASP can be approximated by Doppler ultrasound measurement of maximum tricuspid regurgitant jet velocity (TRV) together with an estimation of RAP based on size and collapsibility of the inferior vena cava (3). PASP is then calculated by utilizing the relationship between these variables described in the equation: PASP = $(4 \times TRV^2)$ + RAP. When PASP determined by this method is greater than 36 mmHg, PH is considered a possible diagnosis, and in particular, when it is greater than 50 mmHg, PH is likely (4). Confirmation requires right heart catheterization documentation of mPAP greater than or equal to 25 mmHg at rest, a pulmonary capillary wedge pressure less than or equal to 15 mmHg, and increased pulmonary vascular resistance greater than 3 Wood units (2,5). Subsequent diagnostic workup is aimed at identifying and mitigating factors known to contribute to PH pathogenesis (1,6).

There are several registries that have helped to define the epidemiology of PH including the first registry from the National Institutes of Health (NIH) in 1981 (7). More recent registries include French National registry, Pulmonary Hypertension Connection registry, the Mayo registry, and the Registry to Evaluate

Early and Long-Term PAH Disease Management (REVEAL) (7). The latter four registries vary in patient population, numbers of newly and previously diagnosed patients, duration of observation, survival monitoring, and other factors. Recently, one of the largest registries from the United States, REVEAL, provided a detailed landscape of the incidence, prevalence, prognosis, and other clinical characteristics (7,8). An estimated incidence of approximately two cases per million per year was observed with a mean age at presentation in adults who range from 36 to 50 years, although individuals of any age can be affected (7,8). After the advent of modern therapies, survival rates at 1, 3, and 5 years for adults with idiopathic and familial pulmonary arterial hypertension (IPAH) are now 91 ± 2%, 74 ± 2%, 65 ± 3%, and 59 ± 3% compared with previous estimated survival rates of 68%, 47%, 36%, and 32%, respectively, using the NIH risk equation (9).

Despite progress in the treatment and management of PAH, a cure is not yet available. Recent investigations in humans, preclinical models, and in vitro work have shed greater understanding of disease progression, leading to increased awareness of the vast molecular complexity and network of underlying mechanisms. These studies have revealed overlapping pathways that are common in the various subtypes of PH while also providing novel concepts of therapeutic approaches that may further improve survival and quality of life. Although progressive right ventricular dysfunction contributes to worse functional capacity and prognosis, most of the therapeutic strategies have focused on the inciting pathogenic processes, which involve the pulmonary vasculature.

PAH appears to affect the majority of precapillary pulmonary vessels but predominantly involves small to medium size pulmonary arteries (10). Vascular changes include intimal hyperplasia, medial hypertrophy, adventitial proliferation, thrombosis in situ, and varying degrees of inflammation (10). Furthermore, a hallmark of severe, end-stage disease is the formation of a vessel "neointima," characterized by increased deposition of extracellular matrix and myofibroblasts (10). Plexiform lesions can predominate, characterized by hyperproliferation of endothelial-like cells leading to vaso-obstructive effects (11). PAH is also characterized by decreased serotonin in platelets and elevation of plasma serotonin (12). The dysfunctional endothelium produces increased mitogenic and vasoconstrictor compounds, such as endothelin and thromboxane, compared to a decrease in vasodilators, such as prostacyclin and nitric oxide (NO) (12). A procoagulant state features elevated levels of fibrinopeptide A and plasminogen activator inhibitor-1 and reduced levels of tissue plasminogen activator (12).

OVERVIEW OF PULMONARY HYPERTENSION GENETICS

Much of what is currently known about the possible underlying genetic predisposition to PH of any cause has been informed by the study of those patients for whom no clear associated condition is found, and who, therefore, were deemed to have idiopathic PAH (formerly primary PH). Within this group a familial, heritable form of PAH (HPAH) has long been recognized (13). Based on pedigree analysis of 14 cases of HPAH described between 1954 and 1984, the inheritance pattern in HPAH has been characterized as autosomal dominant with a variable and markedly reduced penetrance of 10% to 20% (14,15). This incomplete penetrance strongly suggests the presence of additional genetic and/or environmental disease-modifying factors. The only well-established modifying factor thus far is female gender, possibly related to hormonal factors, which impart a 1.7-fold greater risk for penetrance in HPAH (16). Reduced penetrance also suggests that a heritable component of disease may exist in any number of cases of PH otherwise thought to be non-familial on the basis of limited family histories and small pedigrees.

The concept of genetic anticipation in HPAH has been observed in pedigree analyses as well. Family members with PH in successive generations have had life expectancies approximately 10 years less than the previous generation (17). Possible molecular mechanisms explaining this phenomenon have been observed in other genetic disorders in the form of trinucleotide repeat expansions and progressive telomere shortening, but have not been found in HPAH, raising the question of an ascertainment bias underlying anticipation (18).

With these observations of the genetic behavior of HPAH, two independent teams of investigators at Vanderbilt University and Columbia University compiled HPAH pedigrees, documenting affected and unaffected individuals, and collecting DNA for genome-wide linkage analysis. In the late 1990s, both teams successfully identified a PH gene locus on chromosome 2q31-32 later found to carry mutations in a member of the transforming growth factor-beta (TGF-beta) superfamily of receptors, bone morphogenetic protein receptor type II (*BMRPR2*) (19–22). Mutations in other TGF-beta family members, activin-receptor-like kinase 1 (*ACVRL1* or ALK1) or endoglin (*ENG*), had already been shown to be involved in the pathogenesis of HHT syndrome and associated with the development of PH in this vascular disorder (23).

BMPR2 AND TGF-BETA

The TGF-beta superfamily comprises many cytokine growth factors that regulate a variety of cellular functions including proliferation, migration, differentiation, apoptosis, and the extracellular matrix (24). TGF-beta members are highly evolutionarily conserved across species and are classified into subfamilies such as TGF-beta ligands, receptors, accessory molecules, activins, and bone morphogenetic proteins (BMPs). A member of the TGF-beta superfamily of growth factor receptors, BMPs are cytokines that regulate growth, differentiation, and apoptosis of multiple cell types, as well as contributing to maintenance and repair of tissues (24). For ligand binding, *BMPR2* forms a tetrameric complex with other receptors such as *BMPR1A*, *BMPR1B*, and *ALK1*. Once activated, this receptor complex stimulates the R-Smad family of signal transduction proteins (*Smad-1, 2, 3, 5,* or *8*) (25). These signal proteins recruit the nuclear chaperone Smad-4 for travel into the nucleus and interaction with nuclear cofactors to affect gene expression (Figure 18.4) (25).

The *BMPR2* gene contains 13 exons that code for four functional domains of the type II TGF-beta receptor: an extracellular ligand binding domain, a transmembrane region, a serine/threonine kinase domain, and a cytoplasmic tail. A total of 298 PH-associated mutations in *BMPR2* have been identified, spanning all four functional domains of the protein product, and including virtually every type of mutation: frameshift, missense, nonsense, splice site mutations at intron/exon boundaries, and partial gene rearrangements (24,26). Ancestral mutations do not appear to exist, possibly because of the mutation's effect on reproductive fitness (24). With the exception of missense mutations, the majority (approximately 70%) of mutations predict premature truncation of the protein (24). These result in the activation of nonsense-mediated decay (NMD) surveillance machinery (27), making haploinsufficiency (27,28) the most common mechanism by which *BMPR2* mutations lead to PH predisposition. Missense mutations (approximately 30%), on the other hand, occur commonly in the highly conserved extracellular and kinase domains (exons 2, 3, 6–9, 11, and 12) where they exert a dominant negative effect on

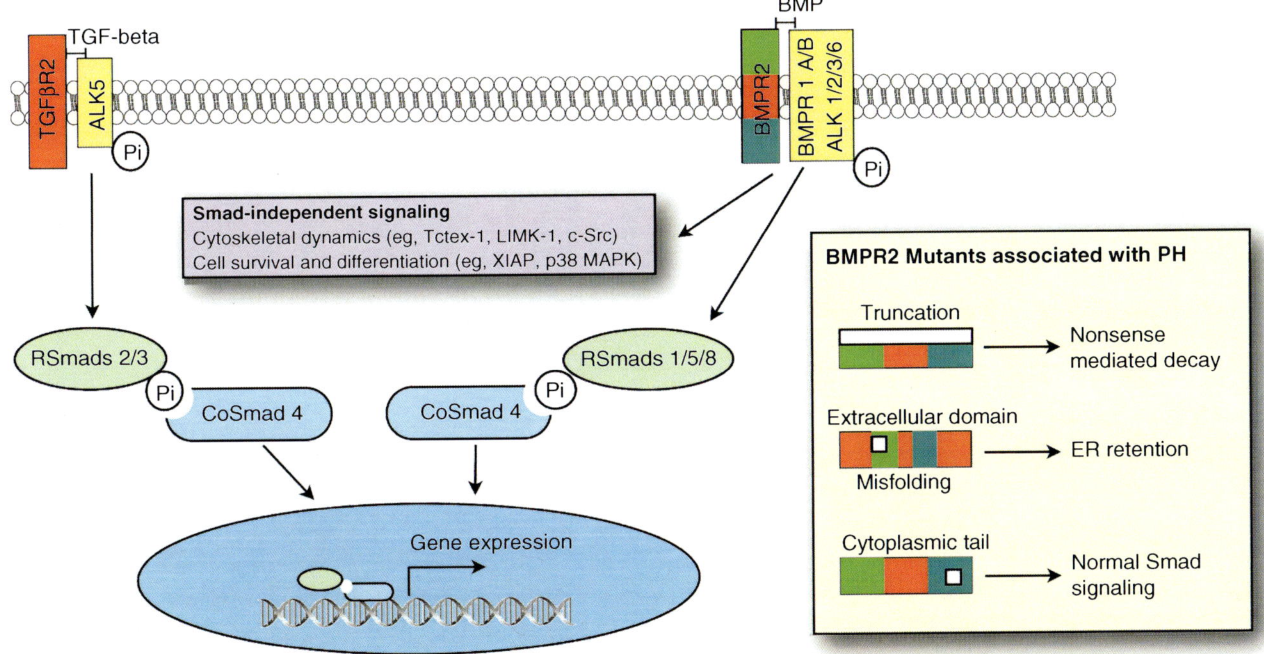

FIGURE 18.4 TGF-β pathway, *BMPR2*, and PAH. *BMPR2* signaling in vascular smooth muscle and endothelium during the development of PH is regulated by both the canonical TGF-β superfamily Smad-mediated pathway and Smad-independent elements. Upon binding the BMP ligand, the *BMPR2* receptor forms a complex with either a *BMPR1* or ALK coreceptor at the cell surface. Subsequent phosphorylation events trigger recruitment of RSmads, which complex with CoSmads, translocate to the nucleus, and regulate gene transcriptional changes underlying cellular processes critical to PH pathogenesis such as endothelial apoptosis and smooth muscle proliferation. Alternatively, the *BMPR2* receptor complex may transduce signals in a Smad-independent manner (purple box) through modulation of a number of other mediators regulating cytoskeletal changes and cell survival/differentiation. (Inset) *BMPR2* mutants associated with PH may affect *BMPR2* expression, cell surface recruitment, or signal transduction. *BMPR2* mutations result in protein truncation most commonly leading to haploinsufficiency, misfolding and resulting in defective trafficking to the cell surface, or occur in the cytoplasmic tail, which do not appear to have significant effect on Smad signaling and may mediate PH pathogenesis through Smad-independent pathways.

BMPR2 function by disruption of normal trafficking of both mutant and wild-type receptors to the cell surface (29,30). These mutations can cause retention within the endoplasmic reticulum, perhaps due to abnormal protein folding (29). Patients with missense mutations also appear to have worse survival (31). Amino acid substitutions in the cytoplasmic tail of *BMPR2* (7 of 298, 2%), in contrast to truncating mutations, account for approximately 15% of PAH cases and are associated with normal levels of Smad activation (24).

Homozygous inactivating mutation of *BMPR2* is lethal in the embryonic stage (32), thus segregating in an autosomal dominant pattern. Loss of *BMPR2* signaling in pulmonary artery smooth muscle cells (PASMCs) results in increased proliferation (33), a well-known component of the PH phenotype. *BMPR2* is additionally implicated by the observation that targeted delivery of the gene to the pulmonary vascular endothelium improves hemodynamics and pulmonary artery remodeling in two rodent models of PAH (34). Most patients with HPAH have reduced *BMPR2* protein expression as, to a lesser extent, do patients with other forms of PAH (35). Nonetheless, since reduced function of *BMPR2* appears neither necessary nor sufficient to cause PAH, disease development likely requires a "double hit" from genetic mutations and environmental factors. Further supporting this latter theory, genome-wide analyses in patients with PAH detected somatic deletions (including one patient with germline *BMPR2* mutation and somatic deletion of *SMAD8*—a double hit), compared with no deletions detected in controls (36).

Several series of studies have examined the impact of *BMPR2* mutation on the clinical course of PH. The largest of these is a multicenter French registry of PH cases during a 1-year period (37). Retrospective analysis of patient outcomes revealed that *BMPR2* mutation carriers compared to a noncarrier PH cohort were younger at diagnosis, exhibited worse hemodynamic parameters at diagnosis, underwent transplant or suffered mortality earlier after diagnosis, and died at younger ages (37). Findings from a Chinese cohort indicate that males with *BMPR2* mutations have a higher mortality risk, which was not observed in females (38). Two other studies have additionally shown a markedly reduced prevalence of vasoreactivity at right heart catheterization in *BMPR2* mutation carriers (39,40).

Approximately 70% of HPAH cases are associated with mutations in either *BMPR2* or *ALK1/ENG*, leaving a smaller subset of patients with recently documented *CAV1* mutations and the remainder, still without a known genetic mechanism (11). *BMPR2* mutations have also been found in small subsets of patients with fenfluramine exposure or with congenital heart disease (37,41), while none have been

identified in PH associated with scleroderma or HIV disease (42,43). In contrast to approximately 30% of HPAH patients, nearly all PAH patients with fenfluramine exposure studies to date have missense *BMPR2* mutations. Furthermore, the mutations do not appear to be functional based on in silico and in vitro data, further emphasizing the role of modifier genes or the environment.

While *BMPR2* has been identified as a key player in the development of PAH, the exact mechanism of its effect on PAH is not well understood, and genetic or nongenetic modifiers of PAH development are outlined in the following and in Figure 18.5. The roles of these pathways are less clear at this time. Mutations in other genes involved with *BMPR2* signaling such as *SMAD* genes have also been associated with PAH. However, these mutations have been reported as single cases only, and the impact of these mutations on PAH as a whole is not yet known (44).

Mutations in *ALK1* and *ENG* are associated with HHT, which can also lead to the development of PAH (45–47). HHT is an autosomal dominant disorder characterized by cutaneous telangiectasias and arteriovenous malformations. Patients with HHT can sometimes develop PAH that is clinically indistinguishable from HPAH. Of the 83 known mutations associated with HHT, 16 (approximately 20%) are also associated with the development of PH (48). Along with *BMPR2*, endoglin and *ALK1* are membrane-bound receptors in the TGF-beta signaling superfamily, which controls diverse cellular processes including cell differentiation, proliferation, and apoptosis. A majority of mutations in *ALK1* are missense mutations in functional domains of the receptor, which lead to defective trafficking and cytoplasmic retention of the mutant (47). Rarely, *ALK1* mutations may cause PAH without clinical signs of HHT (47,49). In particular, mutations in *ALK1* exon 10 were more frequently associated with development of PAH than other mutations in *ALK1* (50). Finally, in an analysis of the impact of both *BMPR2* and *ALK1* mutations on the course of childhood HPAH or IPAH (age less than 16 years), investigators found that a mutation in either gene was associated with significantly worse cumulative survival compared to a noncarrier cohort with 5-year survival rates of 54%, 64%, and 90% in *BMPR2* or *ALK1* mutation carriers and nonmutation carriers, respectively (51).

CAVEOLIN-1

As noted, only 70% of HPAH cases and roughly 30% of idiopathic PH are associated with *BMPR2* or *ALK1* mutations. Recently, in an effort to

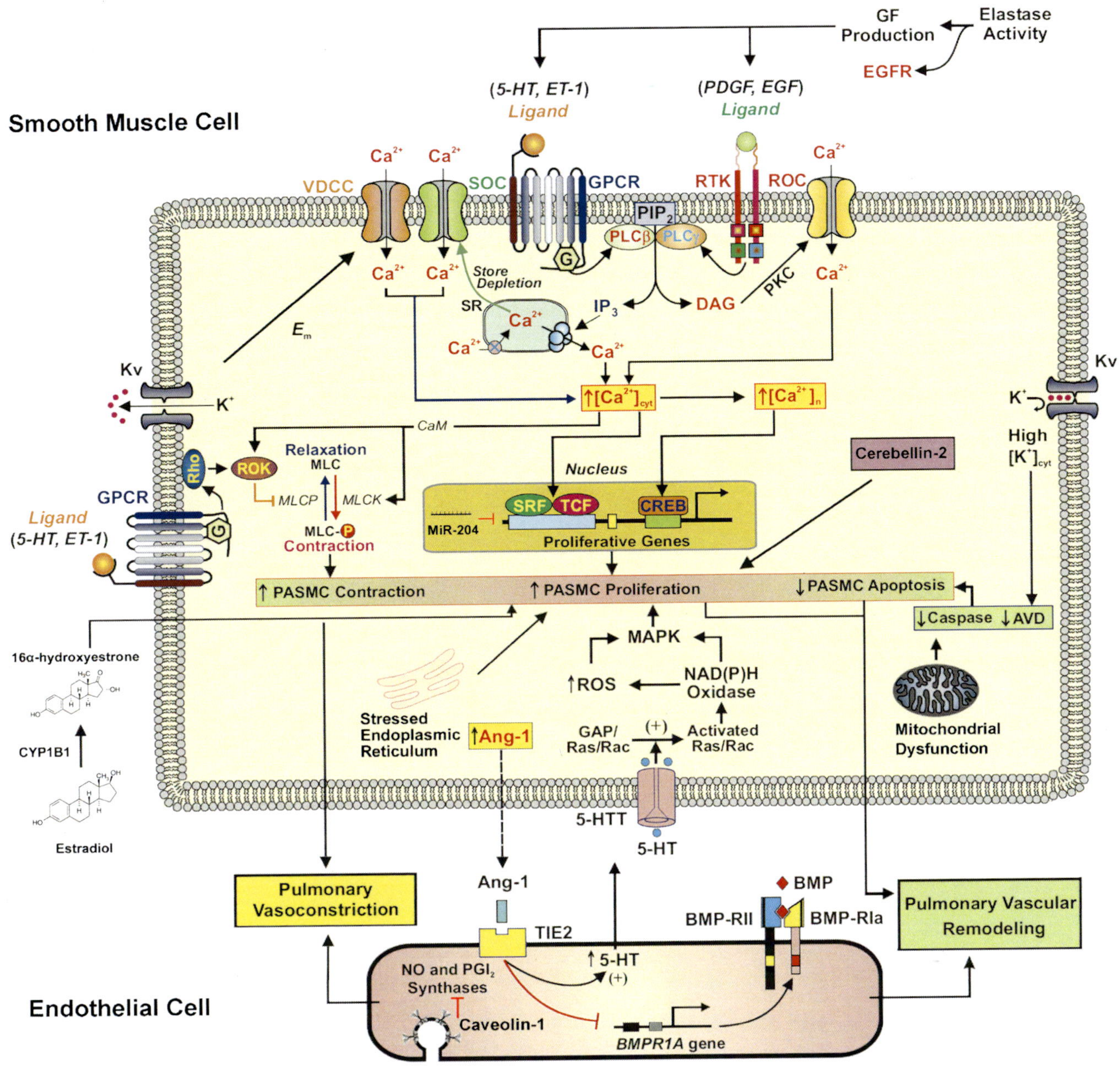

FIGURE 18.5 Modifier genes and PAH. A complex network of modifier genes and their signaling cascades has been investigated in PAH with significant overlap and common pathways and interactions between the endothelium, smooth muscle cells, and the adventitia highlighted.

Source: Adapted with permission from Ref. (136). Morrell NW, Adnot S, Archer SL, et al. Cellular and molecular basis of pulmonary arterial hypertension. *J Am Coll Cardiol*. 2009;54:S20–S31.

elucidate novel candidate genes in PH, Austin et al (52) used whole-exome sequencing to study a three-generation family with HPAH not associated with *BMPR2/ALK1* mutation. Interestingly, a PH-associated frameshift mutation was discovered in caveolin-1, the structural protein of caveolae predominantly found in endothelial cells and not associated with TGF-beta signaling. The same mutation was also discovered in several non-PH members of the family, demonstrating incomplete penetrance similar to what had been observed with *BMPR2*.

Sequencing of the gene that encodes *CAV1* in a validation cohort of unrelated HPAH and IPAH cases identified additional novel mutations associated with disease. All *CAV1* gene variants were associated with decreased protein levels measured by both Western blotting of lung tissue extracts and by immunostaining of lung histologic specimens. The mechanism by which caveolin deficiency contributes to PH pathogenesis remains unknown. Caveolin-1, a protein involved in cellular membrane invagination, is highly expressed in the lung. *CAV1*

TABLE 18.1 Candidate Gene Studies in PAH. Based on Search Terms "Pulmonary Hypertension" and "GWAS," "SNP," "Polymorphism," or "Gene Association" on PubMed, a Comprehensive Table of Significant Nucleotide Polymorphisms Is Listed, Which Are Associated With PH Based on Nearly all Candidate Gene Studies in PAH Reported to Date Beyond Previously Discussed Studies on *BMPR2, ALK1, CAV1,* and Cerebellin-2

Gene	Other Genes Screened	Pathway	Type of PH (N)	Phenotype	Association	Reference
EDN1 (ET-1)	*NOS3*	NO and endothelin-1	IPAH (77)	RHC-PAH	EDN1- rs10478694 and rs5370 associated	(65)
HLA-DRB1/ HLA-DQB1	None	Immune/inflammatory	IPAH (19)	RHC-PAH	HLA-DRB1*0406 and DQB1*0302 associated	(94)
TRPC6 (TRP6)	None	TRPC	IPAH (268)	RHC-PAH	254(C->G) SNP associated	(62)
KCNA5 (Kv1.5)	None	Kv channel	IPAH (188)	RHC-PAH	17 SNPs associated	(85)
MAOA	None	Serotonin	IPAH (77)	RHC-PAH	Both alleles not associated	(113)
TGF-β1	None	TGF-beta	HPAH (81)	RHC-PAH	-509 C/T and codon 10 CC genotypes associated with age of diagnosis	(114)
3q22	Genome-wide	NA	HPAH (164)	RHC-PAH	3q22 locus associated with HPAH in *BMPR2*-mutation carriers	(115)
KCNA5 (Kv1.5)	None	Kv channel	SSc (179)	RHC-PAH	Rs10744676 not associated	(86)
KCNA5 (Kv1.5)	None	Kv channel	SSc (119)	RHC-PAH	Rs10744676 associated	(87)
SLC6A4 (SERT1)	None	Serotonin	SSc (90)	RHC-PAH	All SNPs not associated	(80)
TNFAIP3	None	Ubiquitin-modifying enzyme	SSc (unknown)	PAH (unknown)	rs5029939 associated	(116)
IL23R	None	Immune/inflammatory	SSc (133)	RHC-PAH	rs11209026 and rs11465804 associated	(96)
CXCL12 (SDF-1)	None	Immune/inflammatory	SSc (150)	PAH (unknown)	SDF1-3'GA/AA genotype associated	(95)
ACVRL1 (ALK1)	None	TGF-beta signaling	SSc (29)	RHC-PAH	6bINS associated	(117)
BMPR2	*ALK1, ENG, TGFBR2*	TGF-beta signaling	SSc (54 + 219)	RHC-PAH	All SNPs not associated	(118)
TLR2	*TLR4, 7–9*	Immune/inflammatory	SSc (40)	RHC-PAH	rs5743704 associated	(119)
PLAUR (uPAR)	None	Signal transduction	SSc (92)	RHC-PAH	rs344781 associated	(120)
NOS2	None	NO	SSc (20)	RHC-PAH	-1026 and -277 promoter SNPs associated	(66)
CX3CR1	None	Immune/inflammatory	SSc (50)	TTE-defined RVSP	249II and 280MM associated	(121)
ACE	None	RAS	NPPH (44)	TTE-defined NPPH	DD II, DI genotypes not associated	(122)

TABLE 18.1 Candidate Gene Studies in PAH. Based on Search Terms "Pulmonary Hypertension" and "GWAS," "SNP," "Polymorphism," or "Gene Association" on PubMed, a Comprehensive Table of Significant Nucleotide Polymorphisms Is Listed, Which Are Associated With PH Based on Nearly all Candidate Gene Studies in PAH Reported to Date Beyond Previously Discussed Studies on *BMPR2, ALK1, CAV1*, and Cerebellin-2 (*continued*)

Gene	Other Genes Screened	Pathway	Type of PH (N)	Phenotype	Association	Reference
CRHR1	12 genes	HPA pathway	NPPH (88)	TTE-defined NPPH	rs4458044 associated	(123)
NF1	None	N/A	NF1 (8)	PH (not defined)	Not applicable	(124)
SLC6A4 (SERT1)	None	Serotonin	Heart failure (90)	TTE-defined PH	Promoter LL/SS/IS genotype associated	(90)
SLC6A4 (SERT1)	*HTR2A, NOS3*	Serotonin	COPD (26)	RHC-PAH	Promoter LL/SS/IS genotype associated	(67)
SFTPB	*EPHX1, SERPINE2, GSTP1, TGFB1*	Surfactant	COPD (389)	RHC- and TTE-defined PH	rs2118177 associated	(125)
NOS3	*ACE, EDN1, PTGIS, SLC6A4, VEGFA*	Multiple	COPD cases (278)	TTE-defined RVSP	NOS3-VNTR genotype associated	(126)
SLC6A4 (SERT1)	None	Serotonin	COPD (67)	RHC-PAH	Promoter LL/SS/IS genotype associated	(77)
IL6	None	Immune/inflammatory	COPD (50)	RHC-PAH	G/G SNP (−174G/C) was associated	(92)
IL6	None	Immune/inflammatory	COPD (148)	RHC-PAH	G/G SNP (−174G/C) was associated	(93)
GALNT13	10 genes	Glycosylation	Sickle cell (112)	TTE-defined RVSP	rs799813 associated	(88)
ADRB1 and *TGF-beta*1	49 genes	TGF-beta and adrenergic signaling	Sickle cell (120)	TTE-defined RVSP	5 SNPs associated	(127)
SLC6A4 (SERT1)	None	Serotonin	Congenital (287)	TTE-defined PH	5-HTTLPR and 10/10 homozygote of 5-HTTVNTR associated	(81)
15 genes	31 genes	Multiple	Portopulmonary (31)	RHC-PAH	29 SNPs associated	(128)
SLC6A4 (SERT1)	None	Serotonin	Portopulmonary (30)	RHC-PAH	All genotypes not associated	(82)
FGA	6 Genes	Coagulation	CTEPH (214)	RHC-PAH	Thr312Ala genotype associated	(129)
FGA	None	Coagulation	CTEPH (190)	RHC-PAH	Ins of 28bp in 3' UTR associated	(130)
PTGIS (PGI2)	None	Prostacyclin	CTEPH (90)	RHC-PAH	Promoter-VNTR not associated	(56)
ACE	None	RAS	CTEPH (95)	RHC-PAH	ID/DD alleles not associated	(89)
ACE	None	RAS	High-altitude PH (27)	RHC-PAH	LL allele associated	(91)

(*continued*)

TABLE 18.1 Candidate Gene Studies in PAH. Based on Search Terms "Pulmonary Hypertension" and "GWAS," "SNP," "Polymorphism," or "Gene Association" on PubMed, a Comprehensive Table of Significant Nucleotide Polymorphisms Is Listed, Which Are Associated With PH Based on Nearly all Candidate Gene Studies in PAH Reported to Date Beyond Previously Discussed Studies on *BMPR2, ALK1, CAV-1*, and Cerebellin-2 (*continued*)

Gene	Other Genes Screened	Pathway	Type of PH (N)	Phenotype	Association	Reference
SLC6A4 (SERT1)	None	Serotonin	Mixed (74 IPAH, 35 CTEPH)	RHC-PAH	Promoter LL/SS/IS genotype not associated	(131)
TSP1	None	Glycoprotein	Mixed (60 HPAH, 70 PAH)	PAH (undefined)	Asp362Asn and IVS8+ 255 G/A associated	(132)
SLC6A4 (SERT1)	None	Serotonin	IPAH and HPAH (83 + 166)	RHC-PAH	Promoter LL/SS/IS genotype associated with age of diagnosis	(79)
SLC6A4 (SERT1)	None	Serotonin	Mixed (528 total; 259 IPAH, 136 APAH, 82 HPAH)	RHC-PAH	Promoter LL/SS/IS and VNTR polymorphisms not associated	(78)
SLC6A4 (SERT1)	eNOS, 5-HTR-2A, BMPR2	Serotonin	Mixed (16 IPAH, 16 CTEPH)	RHC-PAH	Promoter LL/SS/IS genotype associated	(133)
SLC6A4 (SERT1)	None	Serotonin	Mixed	RHC-PAH	Promoter LL/SS/IS genotype not associated	(134)
EDNRA (ET-A)	None	Endothelin-1	Mixed	RHC-PAH	His323His TT allele associated	(69)
AGTR1 (AT1R)	*AGT, CMA1, CYP11B2, ACE*	RAS	Mixed	RHC-PAH	1166C SNP associated with age of diagnosis	(135)

Abbreviations: ACE, angiotensin-converting enzyme; ADRB1, adrenergic beta-1 receptor 1; AGT, angiotensinogen; AGTR1, angiotensin II type 1 receptor; CMA1, cardiac chymase A; CRH, corticotropin-releasing hormone; CRHBP, CRH-binding protein; CRHR1, CRH receptor 1; CTEPH, chronic thromboembolic pulmonary hypertension; CX3CR1, CX3C chemokine receptor 1; CYP11B2, aldosterone synthase; DD/ID, deletion–insertion; EDN1, endothelin-1; GALNT13, UDP-N-acetyl-a-D-galactosamine-polypeptide N-acetylgalactosaminyl-transferase 13; HPA pathway, hypothalamic–pituitary–adrenal pathway; MAO, monoamine oxidases; NF1, neurofibromatosis type I; NPPH, persistent pulmonary hypertension of the newborn; RAS, renin–aldosterone–angiotensin axis; RHC-PAH, right heart catheterization-defined PAH; RVSP, right ventricular systolic pressure; SDF-1, stromal-cell-derived factor 1; SLC6A4, serotonin transporter; Thr312Ala, fibrinogen Aalpha substitution of threonine with alanine at position 312; TLR, toll-like receptor; TNFAIP3, tumor necrosis factor/alpha-induced protein 3; TRPC6, transient receptor potential cation channel, subfamily C, member 6; TTE, transthoracic echocardiography; uPAR, urokinase-type plasminogen activator receptor; VEGF, vascular endothelial growth factor.

knockout mice exhibit decreased alveolar spaces, increased wall thickening, fibrosis, and eventually develop PH (53,54). Whether these mutations in *CAV1* are common enough to be generalizable to clinical practice is not yet known.

Beyond investigations of heritability, a variety of modifier genes and their respective signaling pathways have been elucidated in PAH. Decades of studies have been performed as listed in Table 18.1 evaluating genetic susceptibility to different types and cohorts of PAH in molecules residing outside of the TGF-beta pathway. The table lists nearly all published studies to date that provide evidence for both the presence and absence of association for polymorphisms/alleles arising from candidate genes surveyed across PH groups. Organized by type of PH and particular candidate genes of focus, no study to date has reported a validated single nucleotide polymorphism (SNP) in PAH across independent large PAH cohorts outside of mutations in the TGF-beta pathway described earlier.

CEREBELLIN-2

The first recently published genome-wide association study (GWAS) in patients with familial or idiopathic PAH (55) did not find any SNPs that reached genome-wide significance. However, based on trial design between two separate cohorts—discovery and validation—the authors reported an association between PAH (with no known *BMPR2* mutations) and two polymorphisms near the gene that encodes cerebellin-2 (*CBLN2*) (55). They also reported higher expression of *CBLN2* in the pulmonary arterial endothelial cells (PAECs) of patients with PAH compared to those without PAH (55). Further, when *CBLN2* protein was added to PASMCs, decreased proliferation was observed (55). However, the mechanism by which *CBLN2* influences PAH development has yet to be determined.

PROSTACYCLIN

Although the prostacyclin (PGI_2) pathway has an established pathophysiological role in PAH (24), no GWAS or linkage studies to date have found evidence of association with any particular genotype of PGI_2 synthase (the enzyme that forms the endogenous vasodilator prostacyclin) and the PH phenotype. In fact, a targeted study investigated the potential association of a promoter tandem repeat polymorphism with patients who have CTEPH and found no association (56). Nonetheless, PGI_2 signaling cascade initiates with prostacyclin receptor activation, leading to increased cyclic adenosine monophosphate and resulting in vasodilatory, antiproliferative, and decreased platelet aggregation effects. PAH patients have reduced levels of endogenous PGI_2 and reduced expression of PGI_2 synthase in the lung (57). In addition, prostacyclin receptor knockout mice develop severe hypoxia-induced PH, and mice that overexpress prostacyclin develop much less hypoxia-induced PH, compared to wild type (58,59). Several successful PAH-specific therapies have been developed that target the PGI_2 pathway (60).

CALCIUM SIGNALING

An increase in cytosolic Ca^{2+} concentration in PASMCs is a major trigger for pulmonary vasoconstriction modulating cell proliferation and migration, two major causes for pulmonary vascular remodeling. Both nuclear and cytosolic Ca^{2+} promote proliferation by activating Ca^{2+}-dependent kinases (eg, CaMK), immediate early genes, and other transcription factors (eg, *c-Fos*, nuclear factor of activated T-cells [*NFAT*], cAMP response element binding protein [CREB]), which are necessary for cell growth (5,11,43,102), and interact with protein kinase C (PKC) and calmodulin (CaM) or activate the proteins involved in the cell cycle (cyclins and cyclin-dependent kinases). Increases in cytosolic Ca^{2+}, which occur via decreased function of voltage-dependent potassium (Kv) channels or activating voltage-dependent calcium channels, and CaM binding activate myosin light chain kinase. The kinase phosphorylates the regulatory light chain of myosin and activation of myosin ATPase. Adenosine triphosphate (ATP) is then hydrolyzed to power cross-bridging cycles between myosin and actin filaments, leading to cellular contraction (106,107) and subsequently, pulmonary vasoconstriction. One of these "store-operated" calcium channels (ie, capacitative Ca^{2+} entry; those channels which contribute to the influx of Ca^{2+} entry via the plasma membrane and are activated by the emptying of intracellular Ca^{2+} stores) is the transient receptor potential (TRP). Upregulation of TRP channels is associated with calcium ion influx, vasoconstriction, and vascular smooth muscle proliferation, and have been implicated in PH (61). In fact, a -254($C\rightarrow G$) SNP in the *TRPC6* promoter was shown to be associated with IPAH and linked to abnormal *TRPC6* transcription as well as nuclear factor-kappaB, an inflammatory transcription factor, activity (62).

NITRIC OXIDE AND REACTIVE OXYGEN SPECIES

The endothelium plays a critical role in maintaining the balance between vasoconstrictor (eg, endothelin-1, thromboxane A_2) and vasodilator (eg, prostaglandin I_2, NO) states. PAH-related vasoconstriction therefore is in part related to endothelial dysfunction, as evidenced by a lack of vasodilator response to acetylcholine, bradykinin, or calcium ionophore (63). NO is partly produced from the metabolism of L-arginine by oxide NOS enzymes. There are three isoforms of the enzyme: inducible NO synthase (iNOS), neuronal NO synthase (nNOS), and predominantly, endothelial NO synthase (eNOS). NO signaling in PAH is related to multiple potential mechanisms, including uncoupling and dysregulation of eNOS, effects on NO-cGMP signaling, and catabolism of NO by reactive oxygen species (ROS) or hemoglobin (64). Alterations in ROS, reactive nitrogen species (RNS), and NO signaling pathways all have been described in PH, with decreased NO bioavailability and increased ROS and RNS production (63). Increased expression of nicotinamide adenine dinucleotide phosphate (NADPH) oxidases and xanthine oxidoreductase, uncoupling of *eNOS*, and reduction in mitochondrial number, as well as impaired mitochondrial function all have been described in PAH (63). Supported by these functional studies, several candidate gene studies have been reported describing the association of PH with various genes involved in the NO pathway including *NOS2* and *NOS3* (eNOS) (65–67). Based on the wealth of knowledge of NO and ROS pathways, several therapies have been suggested for PH including inhaled NO, nitroglycerin, nebulized sodium nitrite, soluble guanylate cyclase, all of which augment NO pathways, and L-NAME (decreases ROS), dichloroacetate (restores normal oxidative metabolism in rats), and trimetazidine (inhibitor of fatty acid metabolism) (63). Selective inhibitors of the cGMP-specific phosphodiesterase type 5 (PDE-5) enzyme, such as sildenafil and tadalafil, have shown clear efficacy in the treatment of PAH. The blockade of the PDE-5 prevents the normal hydrolysis of cGMP, increasing NO-derived pulmonary vasodilation and inhibition of smooth muscle cell growth (63).

ENDOTHELIN-1

Endothelin (ET)-1 is a 21 amino acid polypeptide that is mitogenic and may induce vasoconstriction as well as vasodilation, depending on signaling from at least two different endothelin G-protein-coupled receptors, ET_A and ET_B (68). During PAH, an increase of endothelin-1 has been demonstrated in both plasma and lung tissue in human patient samples, as well as in animal models of PAH (68). ET-1 exerts contrasting vasoactive effects under normal physiological conditions based on the expression and distribution of the receptor types. ET_A receptors are located predominantly on smooth muscle cells and mediate vasoconstriction, proliferation, hypertrophy, cell migration, and fibrosis (68). ET_B receptors are found predominantly on vascular endothelial cells and promote vasodilation by increasing NO synthesis, prostacyclin production, and antiproliferative properties and prevent apoptosis (68). ET_B in smooth muscle generates a vasoconstrictive signal, similar to ET_A (68). As PAH progresses, the cellular distribution of the endothelin-1 receptors changes, with decreased expression of vasodilatory endothelial ET_B and increased expression of both constrictive ET_A and ET_B in the vascular smooth muscle cells. Both the selective ET_A receptor blocking agent ambrisentan, and the dual ET_A and ET_B receptor blocker bosentan have been approved for therapy for PAH. Given these findings, several candidate gene studies have been performed with ET-1 in populations of PAH. One study reported a significant association of two ET-1 SNPs, rs10478694 (odds ratio [OR]—3.485; $P = .013$) and rs5370 (OR—3.378, $P = .03$) with IPAH (65). Another study in a mixed PAH cohort reported that patients with the His323His TT genotype of ET-1 had a lower cardiac index (2.0 ± 0.6 vs 2.3 ± 0.6; $P = .05$) and a higher indexed pulmonary vascular resistance (18.8 ± 9.6 vs 14.2 ± 6.9; $P = .01$) compared to those without the TT genotype (69).

SEX HORMONES

IPAH occurs predominantly in women, with three times the incidence compared to men, suggesting a pathophysiological role for sex hormones in PH. Both estrogen and activation of the estrogen receptor beta are protective against the development of PH (70). Decreased expression of the estrogen metabolizing enzyme CYP1B1 correlates closely with the development of PAH in female *BMPR2* mutation carriers, but not in males (71). Similarly, the *CYP1B1* Asn453Ser polymorphism was associated with four-fold higher penetrance of PAH in females with *BMPR2* mutations (72). The mechanism for this interaction appears to involve CYP1B1-mediated metabolism of estrogen to 16alpha-hydroxyestrone, which increases PASMC proliferation (73). Testosterone, via calcium channel blockade of the vascular smooth muscle, also has been shown to be a coronary and pulmonary vasodilator.

SEROTONIN

The increased incidence of PAH in patients taking the anorexigenic drugs aminorex and dexfenfluramine, both of which are serotonin transporter (SERT1) substrates and indirect serotinergic agonists, supports now significant evidence indicating that serotonin is also centrally involved in the development of PH (74). Elevated serotonin levels have themselves also been implicated in the pathology of PAH (75). Serotonin is synthesized in pulmonary artery endothelial cells by the enzyme tryptophan hydroxylase 1 (TPH1). Serotonin can constrict smooth muscle cells via 5-HT1B, 2A, and 2B receptors, or cause constriction via its mitogenic effects after entering the cell via SERT1 (59,76). Clinically, a gain-of-function polymorphism in *SLC6A4* (SERT1) was associated with increased severity of PH in chronic obstructive pulmonary disease (COPD), but this association was not replicated in PAH populations, with or without *BMPR2* mutations (77,78). Several other studies have evaluated association of polymorphisms within *SLC6A4*, 5-HT receptors, or related signaling molecules with different subgroups of PAH (including IPAH, scleroderma-, portopulmonary-, congenital, mixed, COPD-associated PH) and reported mixed results with no genotypes to date, which have been validated (67,77–82). Terguride is an oral, potent antagonist of serotonin (5-HT$_{2A}$ and 5-HT$_{2B}$) receptors that has been suggested as a target for PAH (60).

Kv CHANNELS

Potassium channels are tetrameric, membrane-spanning proteins that selectively conduct K+ to maintain the membrane potential (EM) and hence, vascular tone. K+ channels also participate in vascular remodeling by regulating cell proliferation and apoptosis (83,84). PASMCs express voltage-gated (Kv), inward rectifier (Kir), calcium-sensitive (KCa), and two-pore (K2P) channels. Kv channels have also been associated with hyperpolarized mitochondria and activation of hypoxia-induced factor 1 (HIF-1alpha), despite normoxic conditions (83,84). This metabolic shift is known as the Warburg effect, which is often noted in cancers. Loss of PASMC Kv1.5 and Kv2.1 contributes to the pathogenesis of PAH by causing a sustained depolarization, which increases intracellular calcium and K+, thereby stimulating cell proliferation and inhibiting apoptosis, respectively (83,84). Three studies to date have examined polymorphisms within the *KCNA5* gene, which encodes Kv1.5. One study sequenced *KCNA5* from IPAH patients and identified 17 SNPs; two of these SNPs correlated with the NO-mediated decrease in pulmonary arterial pressure (85). Allele frequency of two other SNPs in patients with a history of fenfluramine and phentermine use was significantly different from patients who had never taken the anorexigens. Another set of studies reported the inability to validate the rs10744676 SNP from *KCNA5* in two separate cohorts in scleroderma-associated PAH (86,87). Restoring Kv expression (via *KCNA5* gene therapy, dichloroacetate, or anti-survivin therapy) reduces experimental PAH. In conclusion, K+ channels regulate pulmonary vascular tone and remodeling and constitute potential therapeutic targets in the regression of PAH.

NONGROUP I PH AND OTHER PATHWAYS

Table 18.1 highlights several additional pathways including genes related to immunity, glycosylation, renin–angiotensin system, coagulation, and other cell signaling. Both novel and overlapping pathways as well as genetic modifiers have been reported for various subgroups of PH. New findings are highlighted by the discovery of SNPs in UDP-N-acetyl-a-D-galactosamine-polypeptide N-acetylgalactosaminyl-transferase 13 (GALNT13) in sickle cell disease associated PH, which underscores potential disease specificity (88). Each of these pathways has also been well studied in PAH development and genetic data appear to provide further support of their potential pathophysiological role across PH groups. This is best illustrated by angiotensin converting enzyme (ACE). SNPs in ACE have demonstrated association in CTEPH (89), heart-failure-associated PH (90), and high-altitude PH (91). Importantly, some of these genes and pathways appear to be important for and reported only in nongroup I PH. For example, IL-6 appears to be related to COPD-associated PH and has been validated in two small cohorts (92,93). Nonetheless, immune-mediated pathways are important across PH group genes such as *HLA-DRB1* (94), *CXCL12* (95), and *IL23R* (96) in IPAH and scleroderma-associated PAH. The studies also underscore the importance of validation in larger sample sizes, a challenging task in a rare disease. For example, several candidate gene studies have been performed in scleroderma-associated PAH, but none have validated a previously established SNP association; rather, some have shown lack of association for the same gene SNP in other PAH groups (78).

ENCODE ERA

As evidenced by the studies in Table 18.1, many reported SNPs fall into noncoding regions with unknown and indecipherable functions. Recently, with the emergence and growth of sequencing technology (ie, 1000 Genomes Project at www.1000genomes .org), there is a resurgence of more detailed mapping of human variation in normal and diseased states. Still, nearly all current approaches focus on direct changes in protein-coding genes (nonsynonymous mutations). However, most polymorphisms occur in intragenic or intronic regions as evidenced from most GWAS (55). Identification of potential regulatory changes that perturb these sites may help to better detect truly functional variants and interpret their impact. Based on this goal, in part, the National Human Genome Research Institute (NHGRI) launched a public research consortium named the Encyclopedia of DNA Elements (ENCODE), in September 2003, to carry out a project to identify all functional elements in the human genome sequence (97). A series of publications resulting from ENCODE revealed that approximately 20% of noncoding DNA in the human genome is involved in the regulation of the expression of coding genes while an additional 60% is transcribed with no known function (97). Furthermore, the expression of each coding gene is controlled by multiple regulatory sites located both near and distant from the gene (97) underscoring the complexity of gene regulation. With recently developed bioinformatic tools such as the RegulomeDB (98), interpretation of regulatory variants in the human genome via high-throughput, experimental datasets from ENCODE and computational predictions has become available and can identify putative regulatory potential and functional variants. These novel approaches, resources, and bioinformatics have not yet been fully leveraged and reported in PAH, but have the capacity to potently unravel the intricate network of modifier genes and genetic regulation leading to the disease.

MRNA EXPRESSION PROFILING

Several studies have been published using genome-wide (or large gene screen) expression profiling platforms to assess pathways that are associated with PH (68). Lung and circulating blood specimens from preclinical models and humans with PH have been utilized to extract genomic information and novel bioinformatic approaches have been applied to detect novel candidate genes/biomarkers for PH (88,99). Potential new therapeutic targets, such as receptor tyrosine kinases, have been extensively studied using array-based platforms to elucidate pharmacogenomic responses (100,101). Limitations for transcriptional profiling in PAH have traditionally been sample size, phenotype quality, and standardization of techniques for the various studies, which has led to the lack of reproducibility (68). Using more narrow and highly defined phenotypes and to date, the largest transcriptomic patient sample set, lung tissue microarray profiling in patients with pulmonary fibrosis recently demonstrated a specific gene expression signal that detects the presence of associated PAH validated in two independent cohorts (102). Other studies have integrated various high-throughput genome-wide genomic and genetic approaches in an effort to increase the power of prediction and highlight candidate genes and expression quantitative trait loci (QTLs) that are novel and established in PH (88). The ultimate goal for genomic approaches is the development of novel systems-based analysis approaches of pulmonary vascular disease, which may provide unique insights.

MicroRNA EXPRESSION PROFILING

MicroRNAs (miRNAs) are small noncoding sequences of ribonucleic acid (RNA) that can regulate many genes and networks (see Chapter 6). Given their upstream position in genetic pathways, their potential for pathogenic role has been well described in cardiovascular diseases and cancer. Investigations of their role in PAH have recently emerged. One study (103) observed correlation of miR-204 downregulation with PAH severity. The study linked a *STAT3*-miR-204-*SHP2*-Src kinase-*NFAT* regulatory axis revolving around PASMC proliferation and antiapoptotic phenotypes. The study also elucidated a potential novel PH therapeutic agent (synthetic miR-204 to the lungs). Aberrant miRNA signaling has also been observed in hypoxic PH and pulmonary remodeling (104–106). Additionally, upregulation of miR-145 was associated with the development of PH, which may stem from *BMPR2* mutations in PAH patients (107). More recently, miR-150 levels were associated as an independent predictor of survival in PAH (hazard ratio, 0.53; $P = .010$) across three separate cohorts with trends of expression levels consistent in plasma, circulating microvesicles from patients with PAH, and lung tissue from rodent models of PH (108).

EPIGENETICS

An epigenetic trait is defined as a stably heritable phenotype resulting from changes in a chromosome without alterations in DNA sequence and is considered essential to cell reprogramming (see Chapter 5). Recently, epigenetic regulation of gene expression has been suggested to play a role in the pathogenesis of PAH (109). Promoter and enhancer DNA methylation was associated with regulation of SOD-2, which is known to be reduced in PAH (109). Histone deacetylases (HDACs) catalyze removal of acetyl groups from lysine residues in a variety of proteins. HDACs have mainly been studied in the context of chromatin, where they regulate gene transcription by deacetylating nucleosomal histones (110). The 18 mammalian HDACs are grouped into four classes. Dysregulation of HDACs is associated with a variety of pathophysiological processes, including cancer and inflammatory diseases. Increased expression of Class I HDACs (HDAC1) is present in pulmonary arteries from patients with PH and in preclinical models of PH (110). Based on these findings, HDAC inhibitors have been evaluated in treating PH and right ventricular remodeling (110).

GENETIC TESTING

The goals of genetic testing in the diagnostic workup of PH are to identify family members at risk of developing PH who would need clinical surveillance for PH and to guide family planning. Clinical testing for mutations in *BMPR2*, *ALK1*, and *ENG* is currently available from laboratories in both North America and Europe. *ALK1* and *ENG* are only examined for mutations if there is a personal or family history of clinical symptoms consistent with HHT such as mucocutaneous telangiectasias, recurrent epistaxis, or arteriovenous malformations. In the remaining majority of cases, *BMPR2* is evaluated by complete sequencing of the entire gene, which can cost anywhere from $1000 to $3000. Once a mutation has been identified in the first individual, subsequent specific screening of any family members for the identified mutation costs approximately $300 to $500. Based on the prevalence statistics noted previously, the greatest yield for genetic testing is in patients in whom pedigree analysis confirms HPAH across multiple generations (70%) followed by those with seemingly sporadic, idiopathic PH (20%).

The role of a geneticist experienced in PH in all cases where genetic testing is being considered is invaluable as the psychosocial impact of both a positive and negative result can be significantly complex for the family dynamic as well as potentially devastating for any young children who may be affected (see also Chapter 10). In these circumstances, a geneticist may provide detailed genetic counseling based on a complete pedigree obtained by contacting all known family members.

Since there is a 50% chance that any *BMPR2* mutation has been transmitted from a carrier to a child, and given that reduced penetrance dictates that such a mutation would cause disease in only 20% of cases, the pretest probability for an asymptomatic child of a *BMPR2* mutation carrier to develop PH is approximately 10% (18). If the child is found not to have acquired the mutation, his or her risk for developing PH then drops to 1 in 500,000, which is the risk observed in the general population. On the other hand, if the mutation has been transmitted, posttest probability only becomes the rate of penetrance of *BMPR2* mutations (20%) (18). Therefore genetic testing is of greatest utility in eliminating related individuals from undergoing serial clinical evaluation for PH.

There are currently no evidence-based guidelines as to the optimal approach to monitoring *BMPR2* mutation carriers for the development of PH. The latest recommendations based on expert consensus advise asymptomatic mutation carriers to undergo clinical and echocardiographic screening for PH every 3 to 5 years (111). The rationale for periodic screening is that individuals who may be diagnosed early in the course of the disease may benefit from early treatment. However, there are currently no data supporting this rationale, although studies examining it are currently in progress. The situation is further complicated by an absence of effective therapies targeting specific gene-mutation relevant mechanisms.

Prenatal testing is also available for *BMPR2* mutations already identified in a prospective parent. Chorionic villus sampling can be performed in the tenth week of pregnancy to determine whether the fetus has acquired the mutation, but is rarely useful in decision making given the markedly reduced penetrance of such mutations (24). However, women who are *BMPR2* mutation carriers and who have been advised not to become pregnant may utilize preimplantation genetic diagnosis when pursuing in vitro fertilization for surrogacy. Genetic testing can be performed on the embryo prior to transfer and only embryos without the familial mutation may be implanted. This is an expensive procedure, which may not be covered by insurance in the United States, and may not be available in other countries (24). Additionally it may be complicated by an increased susceptibility for spontaneous, de novo genetic disorders. Nevertheless, it has been successfully performed in at least one case of *BMPR2*-mutation-related familial HPAH reported in the literature (112).

CONCLUSIONS

Significant progress has been made in the field of PH. Clinically, further understanding of the classification and causes of PH has helped to screen the patients at highest risk and provide a standardized framework, resulting in increased early detection and awareness of PH, all of which has fueled improvement in the quality of care provided to patients. Simultaneously with the growth in the clinical understanding, there has been immense expansion of the cellular and molecular knowledge base that underlies the pathophysiology of PH. These mechanistic investigations have revealed a much more complex and intricate web of genetic pathways that appear to modulate this complex phenotype. The integration of molecular, genomic, and clinical medicine in the postgenome era provides the promise of novel information on genetic variation and pathophysiological cascades. The current challenge is to translate these discoveries rapidly into viable biomarkers that identify susceptible populations and into the development of personally targeted therapies.

REFERENCES

1. Badesch DB, Champion HC, Sanchez MA, et al. Diagnosis and assessment of pulmonary arterial hypertension. *J Am Coll Cardiol.* 2009;54:S55–S66.
2. McLaughlin VV, Archer SL, Badesch DB, et al. ACCF/AHA 2009 expert consensus document on pulmonary hypertension a report of the American College of Cardiology Foundation Task Force on Expert Consensus Documents and the American Heart Association developed in collaboration with the American College of Chest Physicians; American Thoracic Society, Inc.; and the Pulmonary Hypertension Association. *J Am Coll Cardiol.* 2009;53:1573–1619.
3. Bossone E, Bodini BD, Mazza A, Allegra L. Pulmonary arterial hypertension: the key role of echocardiography. *Chest.* 2005;127:1836–1843.
4. Galie N, Hoeper MM, Humbert M, et al. Guidelines for the diagnosis and treatment of pulmonary hypertension. *Eur Respir J.* 2009;34:1219–1263.
5. Badesch DB, Abman SH, Ahearn GS, et al. Medical therapy for pulmonary arterial hypertension: ACCP evidence-based clinical practice guidelines. *Chest.* 2004;126:35S–62S.
6. Desai AA, Machado RF. Diagnostic and therapeutic algorithm for pulmonary arterial hypertension. *Pulm Circ.* 2011;1:122–124.
7. Benza RL, Gomberg-Maitland M, Frost AE, et al. Development of prognostic tools in pulmonary arterial hypertension: lessons from modern day registries. *Thromb Haemost.* 2012;108:1049–1060.
8. Benza RL, Miller DP, Gomberg-Maitland M, et al. Predicting survival in pulmonary arterial hypertension: insights from the Registry to Evaluate Early and Long-Term Pulmonary Arterial Hypertension Disease Management (REVEAL). *Circulation.* 2010;122:164–172.
9. Benza RL, Miller DP, Barst RJ, Bet al. An evaluation of long-term survival from time of diagnosis in pulmonary arterial hypertension from the REVEAL Registry. *Chest.* 2012;142:448–456.
10. Chan SY, Loscalzo J. Pathogenic mechanisms of pulmonary arterial hypertension. *J Mol Cell Cardiol.* 2008;44:14–30.
11. Cogan JD, Pauciulo MW, Batchman AP, et al. High frequency of BMPR2 exonic deletions/duplications in familial pulmonary arterial hypertension. *Am J Respir Crit Care Med.* 2006;174:590–598.
12. Wilkins MR. Pulmonary hypertension: the science behind the disease spectrum. *Eur Respir Rev.* 2012;21:19–26.
13. Dresdale DT, Michtom RJ, Schultz M. Recent studies in primary pulmonary hypertension, including pharmacodynamic observations on pulmonary vascular resistance. *Bull NY Acad Med.* 1954;30:195–207.
14. Loyd JE, Primm RK, Newman JH. Familial primary pulmonary hypertension: clinical patterns. *Am Rev Respir Dis.* 1984;129:194–197.
15. Sztrymf B, Yaici A, Girerd B, Humbert M. Genes and pulmonary arterial hypertension. *Respiration.* 2007;74:123–132.
16. Rich S, Dantzker DR, Ayres SM, et al. Primary pulmonary hypertension. A national prospective study. *Ann Intern Med.* 1987;107:216–223.
17. Loyd JE, Butler MG, Foroud TM, et al. Genetic anticipation and abnormal gender ratio at birth in familial primary pulmonary hypertension. *Am J Respir Crit Care Med.* 1995;152:93–97.
18. Fessel JP, Loyd JE, Austin ED. The genetics of pulmonary arterial hypertension in the post-BMPR2 era. *Pulm Circ.* 2011;1:305–319.
19. Deng Z, Morse JH, Slager SL, et al. Familial primary pulmonary hypertension (gene PPH1) is caused by mutations in the bone morphogenetic protein receptor-II gene. *Am J Hum Genet.* 2000;67:737–744.
20. Lane KB, Machado RD, Pauciulo MW, et al. Heterozygous germline mutations in BMPR2, encoding a TGF-beta receptor, cause familial primary pulmonary hypertension. *Nat Genet.* 2000;26:81–84.
21. Morse JH, Jones AC, Barst RJ, et al. Mapping of familial primary pulmonary hypertension locus (PPH1) to chromosome 2q31-q32. *Circulation.* 1997;95:2603–2606.
22. Nichols WC, Koller DL, Slovis B, et al. Localization of the gene for familial primary pulmonary hypertension to chromosome 2q31-32. *Nat Genet.* 1997;15:277–280.
23. Marchuk DA. Genetic abnormalities in hereditary hemorrhagic telangiectasia. *Curr Opin Hematol.* 1998;5:332–338.
24. Machado RD, Eickelberg O, Elliott CG, et al. Genetics and genomics of pulmonary arterial hypertension. *J Am Coll Cardiol.* 2009;54:S32–S42.
25. Shi Y, Massague J. Mechanisms of TGF-beta signaling from cell membrane to the nucleus. *Cell.* 2003;113:685–700.
26. Machado RD, Aldred MA, James V, et al. Mutations of the TGF-beta type II receptor BMPR2 in pulmonary arterial hypertension. *Hum Mutat.* 2006;27:121–132.
27. Nasim MT, Ghouri A, Patel B, et al. Stoichiometric imbalance in the receptor complex contributes to dysfunctional BMPR-II mediated signalling in pulmonary arterial hypertension. *Hum Mol Genet.* 2008;17:1683–1694.

28. Machado RD, Pauciulo MW, Thomson JR, et al. BMPR2 haploinsufficiency as the inherited molecular mechanism for primary pulmonary hypertension. *Am J Hum Genet.* 2001;68:92–102.

29. Li W, Dunmore BJ, Morrell NW. Bone morphogenetic protein type II receptor mutations causing protein misfolding in heritable pulmonary arterial hypertension. *Proc Am Thorac Soc.* 2010;7:395–398.

30. Rudarakanchana N, Flanagan JA, Chen H, et al. Functional analysis of bone morphogenetic protein type II receptor mutations underlying primary pulmonary hypertension. *Hum Mol Genet.* 2002;11:1517–1525.

31. Austin ED, Phillips JA, Cogan JD, et al. Truncating and missense BMPR2 mutations differentially affect the severity of heritable pulmonary arterial hypertension. *Respir Res.* 2009;10:87.

32. Beppu H, Kawabata M, Hamamoto T, et al. BMP type II receptor is required for gastrulation and early development of mouse embryos. *Dev Biol.* 2000; 221:249–258.

33. Jurasz P, Courtman D, Babaie S, Stewart DJ. Role of apoptosis in pulmonary hypertension: from experimental models to clinical trials. *Pharmacol Ther.* 2010;126:1–8.

34. Reynolds AM, Holmes MD, Danilov SM, Reynolds PN. Targeted gene delivery of BMPR2 attenuates pulmonary hypertension. Eur Respir J. 2012;39:329–343.

35. Atkinson C, Stewart S, Upton PD, et al. Primary pulmonary hypertension is associated with reduced pulmonary vascular expression of type II bone morphogenetic protein receptor. *Circulation.* 2002; 105:1672–1678.

36. Aldred MA, Comhair SA, Varella-Garcia M, et al. Somatic chromosome abnormalities in the lungs of patients with pulmonary arterial hypertension. *Am J Respir Crit Care Med.* 2010;182:1153–1160.

37. Humbert M, Sitbon O, Chaouat A, et al. Pulmonary arterial hypertension in France: results from a national registry. *Am J Respir Crit Care Med.* 2006;173:1023–1030.

38. Liu D, Wu WH, Mao YM, et al. BMPR2 mutations influence phenotype more obviously in male patients with pulmonary arterial hypertension. *Circ Cardiovasc Genet.* 2012;5:511–518.

39. Elliott CG, Glissmeyer EW, Havlena GT, et al. Relationship of BMPR2 mutations to vasoreactivity in pulmonary arterial hypertension. *Circulation.* 2006; 113:2509–2515.

40. Rosenzweig EB, Morse JH, Knowles JA, et al. Clinical implications of determining BMPR2 mutation status in a large cohort of children and adults with pulmonary arterial hypertension. *J Heart Lung Transplant.* 2008;27:668–674.

41. Roberts KE, McElroy JJ, Wong WP, et al. BMPR2 mutations in pulmonary arterial hypertension with congenital heart disease. *Eur Respir J.* 2004;24:371–374.

42. Nunes H, Humbert M, Sitbon O, et al. Prognostic factors for survival in human immunodeficiency virus-associated pulmonary arterial hypertension. *Am J Respir Crit Care Med.* 2003;167:1433–1439.

43. Tew MB, Arnett FC, Reveille JD, Tan FK. Mutations of bone morphogenetic protein receptor type II are not found in patients with pulmonary hypertension and underlying connective tissue diseases. *Arthritis Rheum.* 2002;46:2829–2830.

44. Shintani M, Yagi H, Nakayama T, et al. A new nonsense mutation of SMAD8 associated with pulmonary arterial hypertension. *J Med Genet.* 2009;46:331–337.

45. Trembath RC, Thomson JR, Machado RD, et al. Clinical and molecular genetic features of pulmonary hypertension in patients with hereditary hemorrhagic telangiectasia. *N Engl J Med.* 2001; 345:325–334.

46. Chaouat A, Coulet F, Favre C, et al. Endoglin germline mutation in a patient with hereditary haemorrhagic telangiectasia and dexfenfluramine associated pulmonary arterial hypertension. *Thorax.* 2004; 59:446–448.

47. Harrison RE, Flanagan JA, Sankelo M, et al. Molecular and functional analysis identifies ALK-1 as the predominant cause of pulmonary hypertension related to hereditary haemorrhagic telangiectasia. *J Med Genet.* 2003;40:865–871.

48. Prigoda NL, Savas S, Abdalla SA, et al. Hereditary haemorrhagic telangiectasia: mutation detection, test sensitivity and novel mutations. *J Med Genet.* 2006; 43:722–728.

49. Fujiwara M, Yagi H, Matsuoka R, et al. Implications of mutations of activin receptor-like kinase 1 gene (ALK1) in addition to bone morphogenetic protein receptor II gene (BMPR2) in children with pulmonary arterial hypertension. *Circ J.* 2008;72:127–133.

50. Girerd B, Montani D, Coulet F, et al. Clinical outcomes of pulmonary arterial hypertension in patients carrying an ACVRL1 (ALK1) mutation. *Am J Respir Crit Care Med.* 2010;181:851–861.

51. Chida A, Shintani M, Yagi H, et al. Outcomes of childhood pulmonary arterial hypertension in BMPR2 and ALK1 mutation carriers. *Am J Cardiol.* 2012; 110:586–593.

52. Austin ED, Ma L, LeDuc C, et al. Whole exome sequencing to identify a novel gene (caveolin-1) associated with human pulmonary arterial hypertension. *Circ Cardiovasc Genet.* 2012;5:336–343.

53. Drab M, Verkade P, Elger M, et al. Loss of caveolae, vascular dysfunction, and pulmonary defects in caveolin-1 gene-disrupted mice. *Science.* 2001;293: 2449–2452.

54. Zhao YY, Liu Y, Stan RV, et al. Defects in caveolin-1 cause dilated cardiomyopathy and pulmonary hypertension in knockout mice. *Proc Natl Acad Sci USA.* 2002;99:11375–11380.

55. Germain M, Eyries M, Montani D, et al. Genomewide association analysis identifies a susceptibility locus for pulmonary arterial hypertension. *Nat Genet.* 2013;45:518–521.

56. Amano S, Tatsumi K, Tanabe N, et al. Polymorphism of the promoter region of prostacyclin synthase gene in chronic thromboembolic pulmonary hypertension. *Respirology.* 2004;9:184–189.

57. Tuder RM, Cool CD, Geraci MW, et al. Prostacyclin synthase expression is decreased in lungs from patients with severe pulmonary hypertension. *Am J Respir Crit Care Med.* 1999;159:1925–1932.

58. Hoshikawa Y, Voelkel NF, Gesell TL, et al. Prostacyclin receptor-dependent modulation of pulmonary vascular remodeling. *Am J Respir Crit Care Med.* 2001; 164:314–318.

59. Geraci MW, Gao B, Shepherd DC, et al. Pulmonary prostacyclin synthase overexpression in transgenic mice protects against development of hypoxic pulmonary hypertension. *J Clin Invest.* 1999;103: 1509–1515.

60. Sitbon O, Morrell N. Pathways in pulmonary arterial hypertension: the future is here. *Eur Respir Rev.* 2012;21:321–327.

61. Firth AL, Remillard CV, Yuan JX. TRP channels in hypertension. *Biochim Biophys Acta.* 2007;1772: 895–906.

62. Yu Y, Keller SH, Remillard CV, et al. A functional single-nucleotide polymorphism in the TRPC6 gene promoter associated with idiopathic pulmonary arterial hypertension. *Circulation.* 2009;119:2313–2322.

63. Tabima DM, Frizzell S, Gladwin MT. Reactive oxygen and nitrogen species in pulmonary hypertension. *Free Radic Biol Med.* 2012;52:1970–1986.

64. Bowers R, Cool C, Murphy RC, et al. Oxidative stress in severe pulmonary hypertension. *Am J Respir Crit Care Med.* 2004;169:764–769.

65. Vadapalli S, Rani HS, Sastry B, Nallari P. Endothelin-1 and endothelial nitric oxide polymorphisms in idiopathic pulmonary arterial hypertension. *Int J Mol Epidemiol Genet.* 2010;1:208–213.

66. Kawaguchi Y, Tochimoto A, Hara M, et al. NOS2 polymorphisms associated with the susceptibility to pulmonary arterial hypertension with systemic sclerosis: contribution to the transcriptional activity. *Arthritis Res Ther.* 2006;8:R104.

67. Ulrich S, Hersberger M, Fischler M, et al. Genetic polymorphisms of the serotonin transporter, but not the 2a receptor or nitric oxide synthetase, are associated with pulmonary hypertension in chronic obstructive pulmonary disease. *Respiration.* 2010;79:288–295.

68. Geraci MW. Integrating molecular genetics and systems approaches to pulmonary vascular diseases. *Pulm Circ.* 2013;3:171–175.

69. Calabro P, Limongelli G, Maddaloni V, et al. Analysis of endothelin-1 and endothelin-1 receptor A gene polymorphisms in patients with pulmonary arterial hypertension. *Intern Emerg Med.* 2012;7:425–430.

70. Umar S, Iorga A, Matori H, et al. Estrogen rescues preexisting severe pulmonary hypertension in rats. *Am J Respir Crit Care Med.* 2011;184:715–723.

71. West J, Cogan J, Geraci M, et al. Gene expression in BMPR2 mutation carriers with and without evidence of pulmonary arterial hypertension suggests pathways relevant to disease penetrance. *BMC Med Genomics.* 2008;1:45.

72. Austin ED, Cogan JD, West JD, et al. Alterations in oestrogen metabolism: implications for higher penetrance of familial pulmonary arterial hypertension in females. *Eur Respir J.* 2009;34:1093–1099.

73. White K, Johansen AK, Nilsen M, et al. Activity of the estrogen-metabolizing enzyme cytochrome P450 1B1 influences the development of pulmonary arterial hypertension. *Circulation.* 2012;126:1087–1098.

74. Maclean MR, Dempsie Y. The serotonin hypothesis of pulmonary hypertension revisited. *Adv Exp Med Biol.* 2010;661:309–322.

75. Herve P, Launay JM, Scrobohaci ML, et al. Increased plasma serotonin in primary pulmonary hypertension. *Am J Med.* 1995;99:249–254.

76. Lee SL, Wang WW, Moore BJ, Fanburg BL. Dual effect of serotonin on growth of bovine pulmonary artery smooth muscle cells in culture. *Circ Res.* 1991;68:1362–1368.

77. Eddahibi S, Chaouat A, Morrell N, et al. Polymorphism of the serotonin transporter gene and pulmonary hypertension in chronic obstructive pulmonary disease. *Circulation.* 2003;108:1839–1844.

78. Machado RD, Koehler R, Glissmeyer E, et al. Genetic association of the serotonin transporter in pulmonary arterial hypertension. *Am J Respir Crit Care Med.* 2006;173:793–797.

79. Willers ED, Newman JH, Loyd JE, et al. Serotonin transporter polymorphisms in familial and idiopathic pulmonary arterial hypertension. *Am J Respir Crit Care Med.* 2006;173:798–802.

80. Wipff J, Bonnet P, Ruiz B, et al. Association study of serotonin transporter gene (SLC6A4) in systemic sclerosis in European Caucasian populations. *J Rheumatol.* 2010;37:1164–1167.

81. Cao H, Gu H, Qiu W, et al. Association study of serotonin transporter gene polymorphisms and ventricular septal defects related possible pulmonary arterial hypertension in Chinese population. *Clin Exp Hypertens.* 2009;31:605–614.

82. Roberts KE, Fallon MB, Krowka MJ, et al. Serotonin transporter polymorphisms in patients with portopulmonary hypertension. *Chest.* 2009;135:1470–1475.

83. Moudgil R, Michelakis ED, Archer SL. The role of K+ channels in determining pulmonary vascular tone, oxygen sensing, cell proliferation, and apoptosis: implications in hypoxic pulmonary vasoconstriction and pulmonary arterial hypertension. *Microcirculation.* 2006;13:615–632.

84. Archer SL, Michelakis ED, Thebaud B, et al. A central role for oxygen-sensitive K+ channels and mitochondria in the specialized oxygen-sensing system. *Novartis Found Symp.* 2006;272:157–171; Discussion 71–75, 214–217.

85. Remillard CV, Tigno DD, Platoshyn O, et al. Function of Kv1.5 channels and genetic variations of KCNA5 in patients with idiopathic pulmonary arterial hypertension. *Am J Physiol Cell Physiol.* 2007;292: C1837–C1853.

86. Bossini-Castillo L, Simeon CP, Beretta L, et al. KCNA5 gene is not confirmed as a systemic sclerosis-related pulmonary arterial hypertension genetic susceptibility factor. *Arthritis Res Ther.* 2012;14: R273.

87. Wipff J, Dieude P, Guedj M, et al. Association of a KCNA5 gene polymorphism with systemic sclerosis-associated pulmonary arterial hypertension in the European Caucasian population. *Arthritis Rheum.* 2010;62:3093–3100.

88. Desai AA, Zhou T, Ahmad H, et al. A novel molecular signature for elevated tricuspid regurgitation velocity in sickle cell disease. *Am J Respir Crit Care Med.* 2012;186(4):359–368.

89. Tanabe N, Amano S, Tatsumi K, et al. Angiotensin-converting enzyme gene polymorphisms and prognosis in chronic thromboembolic pulmonary hypertension. *Circ J.* 2006;70:1174–1179.

90. Olson TP, Snyder EM, Frantz RP, Turner ST, Johnson BD. Repeat length polymorphism of the serotonin transporter gene influences pulmonary artery pressure in heart failure. *Am Heart J.* 2007;153: 426–432.

91. Aldashev AA, Sarybaev AS, Sydykov AS, et al. Characterization of high-altitude pulmonary hypertension in the Kyrgyz: association with angiotensin-converting enzyme genotype. *Am J Respir Crit Care Med.* 2002;166:1396–1402.

92. Eddahibi S, Chaouat A, Tu L, et al. Interleukin-6 gene polymorphism confers susceptibility to pulmonary hypertension in chronic obstructive pulmonary disease. *Proc Am Thorac Soc.* 2006;3:475–476.

93. Chaouat A, Savale L, Chouaid C, et al. Role for interleukin-6 in COPD-related pulmonary hypertension. *Chest.* 2009;136:678–687.

94. Yoon SH, Oh HB, Kim HK, et al. Association of HLA class II genes with idiopathic pulmonary arterial hypertension in Koreans. *Lung.* 2007;185:145–149.

95. Manetti M, Liakouli V, Fatini C, et al. Association between a stromal cell-derived factor 1 (SDF-1/CXCL12) gene polymorphism and microvascular disease in systemic sclerosis. *Ann Rheum Dis.* 2009;68:408–411.

96. Agarwal SK, Gourh P, Shete S, et al. Association of interleukin 23 receptor polymorphisms with anti-topoisomerase-I positivity and pulmonary hypertension in systemic sclerosis. *J Rheumatol.* 2009;36:2715–2723.

97. Ecker JR, Bickmore WA, Barroso I, et al. Genomics: ENCODE explained. *Nature.* 2012;489:52–55.

98. Boyle AP, Hong EL, Hariharan M, et al. Annotation of functional variation in personal genomes using RegulomeDB. *Genome Res.* 2012;22:1790–1797.

99. Desai AA, Hysi P, Garcia JG. Integrating genomic and clinical medicine: searching for susceptibility genes in complex lung diseases. *Transl Res.* 2008;151:181–193.

100. Moreno-Vinasco L, Garcia JG. Receptor tyrosine kinase inhibitors in rodent pulmonary hypertension. *Adv Exp Med Biol.* 2010;661:419–434.

101. Moreno-Vinasco L, Gomberg-Maitland M, Maitland ML, et al. Genomic assessment of a multikinase inhibitor, sorafenib, in a rodent model of pulmonary hypertension. *Physiol Genomics.* 2008;33:278–291.

102. Rajkumar R, Konishi K, Richards TJ, et al. Genomewide RNA expression profiling in lung identifies distinct signatures in idiopathic pulmonary arterial hypertension and secondary pulmonary hypertension. *Am J Physiol Heart Circ Physiol.* 2010;298:H1235–H1248.

103. Courboulin A, Paulin R, Giguere NJ, et al. Role for miR-204 in human pulmonary arterial hypertension. *J Exp Med.* 2011;208:535–548.

104. Guo L, Qiu Z, Wei L, et al. The microRNA-328 regulates hypoxic pulmonary hypertension by targeting at insulin growth factor 1 receptor and L-type calcium channel-alpha1C. *Hypertension.* 2012;59:1006–1013.

105. Gou D, Ramchandran R, Peng X, et al. miR-210 has an antiapoptotic effect in pulmonary artery smooth muscle cells during hypoxia. *Am J Physiol Lung Cell Mol Physiol.* 2012;303:L682–L691.

106. Yang S, Banerjee S, Freitas A, et al. miR-21 regulates chronic hypoxia-induced pulmonary vascular remodeling. *Am J Physiol Lung Cell Mol Physiol.* 2012;302:L521–L529.

107. Caruso P, Dempsie Y, Stevens HC, et al. A role for miR-145 in pulmonary arterial hypertension: evidence from mouse models and patient samples. *Circ Res.* 2012;111:290–300.

108. Rhodes CJ, Wharton J, Boon RA, et al. Reduced microRNA-150 is associated with poor survival in pulmonary arterial hypertension. *Am J Respir Crit Care Med.* 2013;187:294–302.

109. Archer SL, Marsboom G, Kim GH, et al. Epigenetic attenuation of mitochondrial superoxide dismutase 2 in pulmonary arterial hypertension: a basis for excessive cell proliferation and a new therapeutic target. *Circulation.* 2010;121:2661–2671.

110. Zhao L, Chen CN, Hajji N, et al. Histone deacetylation inhibition in pulmonary hypertension: therapeutic potential of valproic acid and suberoylanilide hydroxamic acid. *Circulation.* 2012;126:455–467.

111. McGoon M, Gutterman D, Steen V, et al. Screening, early detection, and diagnosis of pulmonary arterial hypertension: ACCP evidence-based clinical practice guidelines. *Chest.* 2004;126:14S–34S.

112. Frydman N, Steffann J, Girerd B, et al. Pre-implantation genetic diagnosis in pulmonary arterial hypertension due to BMPR2 mutation. *Eur Respir J.* 2012;39:1534–1535.

113. Vadapalli S, Katta S, Sastry BK, Nallari P. MAO-A promoter polymorphism and idiopathic pulmonary arterial hypertension. *J Genet.* 2010;89:e43–e45.

114. Phillips JA, 3rd, Poling JS, Phillips CA, et al. Synergistic heterozygosity for TGFbeta1 SNPs and BMPR2 mutations modulates the age at diagnosis and penetrance of familial pulmonary arterial hypertension. *Genet Med.* 2008;10:359–365.

115. Rodriguez-Murillo L, Subaran R, Stewart WC, et al. Novel loci interacting epistatically with bone morphogenetic protein receptor 2 cause familial pulmonary arterial hypertension. *J Heart Lung Transplant.* 2010;29:174–180.

116. Dieude P, Guedj M, Wipff J, et al. Association of the TNFAIP3 rs5029939 variant with systemic sclerosis in the European Caucasian population. *Ann Rheum Dis.* 2010;69:1958–1964.

117. Wipff J, Kahan A, Hachulla E, et al. Association between an endoglin gene polymorphism and systemic sclerosis-related pulmonary arterial hypertension. *Rheumatology.* 2007;46:622–625.

118. Koumakis E, Wipff J, Dieude P, et al. TGFbeta receptor gene variants in systemic sclerosis-related pulmonary arterial hypertension: results from a multicentre EUSTAR study of European Caucasian patients. *Ann Rheum Dis.* 2012;71:1900–1903.

119. Broen JC, Bossini-Castillo L, van Bon L, et al. A rare polymorphism in the gene for Toll-like receptor 2 is associated with systemic sclerosis phenotype and increases the production of inflammatory mediators. *Arthritis Rheum.* 2012;64:264–271.

120. Manetti M, Allanore Y, Revillod L, et al. A genetic variation located in the promoter region of the UPAR (CD87) gene is associated with the vascular complications of systemic sclerosis. *Arthritis Rheum.* 2011;63:247–256.

121. Marasini B, Cossutta R, Selmi C, et al. Polymorphism of the fractalkine receptor CX3CR1 and systemic sclerosis-associated pulmonary arterial hypertension. *Clin Dev Immunol.* 2005;12:275–279.

122. De Jesus LC, Kazzi SN, Dahmer MK, et al. Role of angiotensin-converting enzyme gene polymorphism in persistent pulmonary hypertension of the newborn. *Acta paediatr.* 2011;100:1326–1330.

123. Byers HM, Dagle JM, Klein JM, et al. Variations in CRHR1 are associated with persistent pulmonary hypertension of the newborn. *Pediatr Res.* 2012;71:162–167.

124. Montani D, Coulet F, Girerd B, et al. Pulmonary hypertension in patients with neurofibromatosis type I. *Medicine*. 2011;90:201–211.

125. Castaldi PJ, Hersh CP, Reilly JJ, Silverman EK. Genetic associations with hypoxemia and pulmonary arterial pressure in COPD. *Chest*. 2009;135:737–744.

126. Shaw JG, Dent AG, Passmore LH, et al. Genetic influences on right ventricular systolic pressure (RVSP) in chronic obstructive pulmonary disease (COPD). *BMC Pulm Med*. 2012;12:25.

127. Ashley-Koch AE, Elliott L, Kail ME, et al. Identification of genetic polymorphisms associated with risk for pulmonary hypertension in sickle cell disease. *Blood*. 2008;111:5721–5726.

128. Roberts KE, Fallon MB, Krowka MJ, et al. Genetic risk factors for portopulmonary hypertension in patients with advanced liver disease. *Am J Respir Crit Care Med*. 2009;179:835–842.

129. Suntharalingam J, Goldsmith K, van Marion V, et al. Fibrinogen Aalpha Thr312Ala polymorphism is associated with chronic thromboembolic pulmonary hypertension. *Eur Respir J*. 2008;31:736–741.

130. Chen Z, Nakajima T, Tanabe N, et al. Susceptibility to chronic thromboembolic pulmonary hypertension may be conferred by miR-759 via its targeted interaction with polymorphic fibrinogen alpha gene. *Hum Genet*. 2010;128:443–452.

131. Koehler R, Olschewski H, Hoeper M, et al. Serotonin transporter gene polymorphism in a cohort of German patients with idiopathic pulmonary arterial hypertension or chronic thromboembolic pulmonary hypertension. *Chest*. 2005;128:619S.

132. Maloney JP, Stearman RS, Bull TM, et al. Loss-of-function thrombospondin-1 mutations in familial pulmonary hypertension. *Am J Physiol Lung Cell Mol Physiol*. 2012;302:L541–L554.

133. Ulrich S, Szamalek-Hoegel J, Hersberger M, et al. Sequence variants in BMPR2 and genes involved in the serotonin and nitric oxide pathways in idiopathic pulmonary arterial hypertension and chronic thromboembolic pulmonary hypertension: relation to clinical parameters and comparison with left heart disease. *Respiration*. 2010;79:279–287.

134. Baloira A, Nunez M, Cifrian J, et al. Polymorphisms in the serotonin transporter protein (SERT) gene in patients with pulmonary arterial hypertension. *Arch Bronconeumol*. 2012;48:77–80.

135. Chung WK, Deng L, Carroll JS, et al. Polymorphism in the angiotensin II type 1 receptor (AGTR1) is associated with age at diagnosis in pulmonary arterial hypertension. *J Heart Lung Transplant*. 2009;28:373–379.

136. Morrell NW, Adnot S, Archer SL, et al. Cellular and molecular basis of pulmonary arterial hypertension. *J Am Coll Cardiol*. 2009;54:S20–S31.

19

C H A P T E R

Genetics of Blood Lipids, Lipoproteins, and Related Phenotypes

Steven A. Claas and Donna K. Arnett

TAKE HOME POINTS

1. Genetic dissection of triglyceride (TG) variability has strengthened our understanding of lipid metabolism and the effects of diet and drugs, and the large number of known and novel genes that have been implicated as being involved in TG metabolism has expanded the range of operative pathways and potential therapeutic targets.

2. Mendelian and genome-wide association studies (GWAS) of the genetics of low-density lipoprotein cholesterol (LDL-C) have been complementary and have shown that any of the pathways influencing circulating LDL-C represent potential therapeutic targets that could prove clinically significant with respect to cardiovascular development and outcomes; the current enthusiasm for PCSK9 inhibitors for LDL-C lowering provides an illustrative example.

3. Genetic studies of high-density lipoprotein cholesterol (HDL-C) have been less consistent compared to TGs and LDL-C in establishing a clear link between HDL-C and cardiovascular disease (CVD) risk, a finding that suggests that not all mechanisms that lead to HDL-C variability are equal; consequently, not all HDL-C pathways may be viable therapeutic targets or predictors of drug response.

4. Genetic studies of blood lipids, lipoproteins, and related phenotypes have not only advanced our knowledge of lipid metabolism and its role in CVD and related therapies; they have been instrumental in advancing our more fundamental understanding of human genetics.

CASE STUDY

A 32-year-old previously healthy accountant presented to the emergency department with severe chest discomfort, shortness of breath, and diaphoresis. Vital signs revealed a blood pressure of 162/98 mmHg, heart rate 112 bpm, respiratory rate 24 breaths/minute, and oxygen saturation 86% on room air. The patient was in visible distress, tachypneic, and unable to speak in full sentences. Physical exam revealed elevated jugular venous pressure; bilateral crackles on lung exam; a normal point of maximal impulse, tachycardic but regular rhythm, an S4 heart sound, and no murmurs or rubs; and multiple tendon xanthomas. She was given aspirin, placed on oxygen by face mask, and an electrocardiogram (ECG) was immediately performed (Figure 19.1) showing inferior ST elevations in leads II, III, and aVF, consistent with an acute ST-elevation myocardial infarction (MI).

Based on the ECG findings, the patient was intubated and taken urgently to the cardiac catheterization laboratory where coronary angiography was performed. There was a 99% stenosis of the proximal right coronary artery (Figure 19.2) with diffuse disease distally. There were also a 70% proximal left anterior descending and 80% proximal left circumflex coronary artery stenoses. The patient was stabilized medically, underwent coronary artery bypass grafting surgery, and discharged from the hospital 1 week later.

During her hospitalization, the patient's husband revealed that she had recently undergone laboratory testing as part of a health screening fair at her workplace. At the time, she was concerned about the results because she had a family history of premature coronary artery

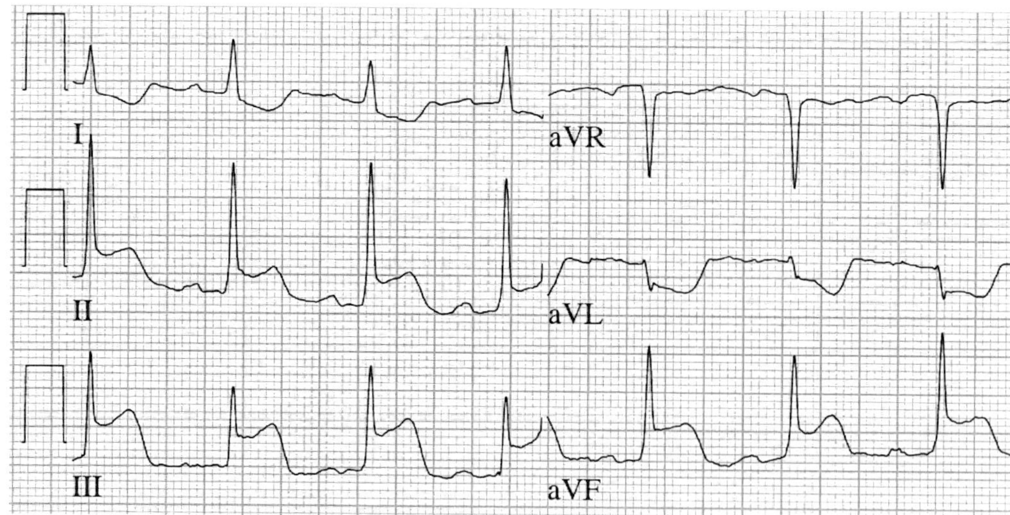

FIGURE 19.1   Electrocardiogram of the patient described in the case study.

Source: R. Kannan Mutharasan, MD.

disease (CAD). Her father had an MI when he was 41 years old, and he underwent coronary artery bypass grafting surgery 7 years later. Of her father's siblings, three out of five had premature CAD, as did the patient's grandfather. The results of the patient's evaluation at the health fair were as follows: blood pressure 132/72 mmHg, body mass index 28 kg/m², total cholesterol 426 mg/dl, TGs 200 mg/dL, HDL-C 38 mg/dL, and LDL-C 348 mg/dL. She was a nonsmoker, consumed alcohol occasionally, and exercised three times a week.

Because of the patient's premature CAD, a diagnosis of autosomal dominant (familial) hypercholesterolemia was considered likely. However, genetic testing revealed no mutations in the *LDLR* or *APOB* genes. She was treated aggressively with rosuvastatin and multiple other lipid lowering drugs, and ultimately underwent LDL apheresis. Further genetic testing demonstrated a heterozygous D374Y gain-of-function mutation in the *PCSK9* gene, known to be associated with an unpredictably severe clinical phenotype. Her two children were screened and were both found to have the same *PCSK9* mutation; both also had elevated cholesterol levels. The patient entered a clinical trial for a *PCSK9* inhibitor, and her children began treatment for hypercholesterolemia.

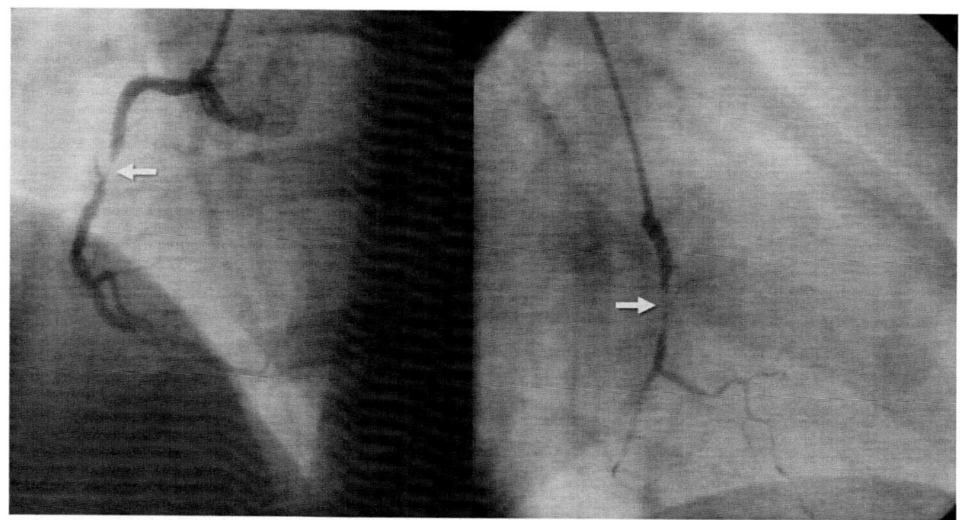

FIGURE 19.2   Angiographic appearance of the right coronary artery in the patient described in the case study. The arrows point to the severe, 99% proximal right coronary artery stenosis.

Source: R. Kannan Mutharasan, MD.

Blood lipid concentrations are a significant contributor to CVD (1). Hypertriglyceridemia (ie, high levels of TGs) and mixed dyslipidemia (ie, high levels of LDL-C and TG combined with decreased levels of HDL-C) are important in the development and progression of atherosclerosis (2). The atherogenic effects of elevated TG may be explained, in part, by postprandial lipoprotein remnants (3,4), which can enter the arterial intima and contribute to the formation of foam cells and the atherosclerotic plaque (5). Lowering TG levels can reduce CVD risk (6). LDL-C is necessary for cellular function, but high concentrations of oxidized LDL-C can also attract and promote migration of macrophages and can lead to endothelial dysfunction and atheroma formation (7). HDL's antiatherogenic effects are believed to stem, at least partially, from its ability to promote cholesterol efflux from macrophages in the arterial wall (8). Some HDL particles also possess anti-inflammatory, antithrombotic, and antioxidant properties (9,10).

Although blood lipids can be significantly influenced by environmental factors, lipid phenotypes appear to be genetically complex: multiple sequence variations and epigenetic factors may interact with the environment to influence blood lipids. Heritability estimates for TG range from 35% to 38%; for LDL-C, from 40% to 50%; and for HDL-C, from 40% to 60% (11). Lipid phenotypes used in genetic studies have often comprised the typical clinical fasting measures but have also included postprandial phenotypes and response phenotypes (eg, to dietary or pharmacological interventions). Lipid assays not normally used in the clinic, for example those that fractionate species by particle size (12) or detailed molecular assessment (ie, "lipidomic" species) (13), have also been scrutinized for genetic associations. Whereas the standard fasting phenotypes offer the benefit of having historic clinical relevance (and are often conveniently available in many study populations), it is becoming increasingly evident that the study of lipid metabolism should include postprandial parameters (14): large population-based studies have shown that a delayed elimination of postprandial TG-rich lipoproteins is associated with increased CVD risk (15–17). Additionally, the more granular lipidomic traits may, in some cases, be more strongly associated with genetic variants than standard clinical phenotypes and may provide deeper mechanistic or therapeutic insights. Ultimately, dissecting the genetic contributions to hypercholesterolemia and other lipid abnormalities may one day inform the development of novel treatments and diagnostics and thereby improve clinical practice.

SINGLE-GENE LIPID DISEASES

A number of single-gene diseases (sometimes referred to as Mendelian or simple diseases, in contrast to complex, multigenic disorders) involving lipid metabolism have long been documented. Studies of these diseases have been important in establishing the connection between hypercholesterolemia and atherosclerosis as well as being instrumental in the development of drugs used to treat elevated cholesterol (18). As a group, these monogenic diseases are referred to as autosomal dominant hypercholesterolemia (ADH). The most common and potentially severe form of ADH is familial hypercholesterolemia (FH), which causes elevated serum LDL-C concentrations. Normally, LDL-C binds to LDL receptors and the LDL-C/LDL-receptor complex is degraded in the liver, thereby removing circulating LDL-C. FH results from any of over 1,000 known mutations in the low-density lipoprotein receptor (*LDLR*) gene (19,20). Individuals homozygous for *LDLR* mutations are often unresponsive to dietary and pharmacological intervention and often develop clinically significant CVD before reaching age 30 (21). Heterozygote FH individuals also have somewhat less elevated LDL-C and are more responsive to environmental and drug intervention (22).

ADH syndromes involving other genes have been documented (22). For example, some mutations in the proprotein convertase, subtilisin/kexin-type, 9 (*PCSK9*) gene produce clinical symptoms similar to that of heterozygous FH (23–25), a condition variously referred to as ADH3, HCHOLA3, or FH3. The PCSK9 enzyme binds to and enhances the breakdown of LDL receptors. Gain-of-function mutations in *PCSK9* result in fewer LDL receptors and slower removal of LDL-C from the circulation (23). Conversely, loss-of-function mutations can lower LDL-C and be cardioprotective (26). A number of methods are being tested to therapeutically inhibit PCSK9 to reduce LDL-C (27–29). Structural changes in apolipoprotein B 100 (the major lipoprotein constituent of LDL-C) caused by mutations in the apolipoprotein B gene reduce the affinity of the LDL-C complex for the LDL receptor, thereby retarding removal of LDL-C from circulation. The hypercholesterolemia resulting from *APOB* mutations is referred to as familial defective apo B-100 (FDB) and typically results in serum LDL-C concentrations that are 2.6 mmol/L (100 mg/dL) higher than that of individuals without mutations (30). Population-level studies estimate that 5% to 53% of those with clinical ADH phenotypes do not have mutations in any of these three known ADH genes (31). This diagnostic gap suggests other, as yet undiscovered, monogenic forms of ADH may exist.

COMPLEX DISEASES AND PHENOTYPES

Mendelian genetic disorders are typically caused by relatively rare variants in single genes that have a large phenotypic effect in individuals and high penetrance but low frequency in populations. In contrast, complex genetic diseases, such as atherosclerosis, may be influenced by multiple (and interacting) sequence variants, epigenetic processes, and gene–environment interactions (see also Chapters 5 and 9). The genetic component of complex diseases is thought to be the aggregate of multiple small effects; the relatively small phenotypic variation (compared to monogenic traits) of intermediate traits (such as blood lipids) disposes individuals and populations toward disease development. As outlined in Chapter 2, the very nature of complex diseases—genetic heterogeneity, gene–gene interactions, gene–environment interactions, incomplete penetrance, lack of phenotypic specificity, and the natural history of disease progression (32)— makes genetic scrutiny of disease endpoints difficult. By studying intermediate phenotypes, such as blood lipids, with known contributions to common diseases, such as atherosclerosis and other CVDs, investigators hope to dissect the genetic basis of complex diseases. That said, and as noted above, traits such as fasting LDL-C, HDL-C, TGs, and the response of these phenotypes to dietary and pharmacological interventions are themselves complex phenotypes, with single-gene disorders accounting for a small percentage of the total population variance in blood lipid levels. As is the case with other complex disease phenotypes, our understanding of the genetics of blood lipids has evolved slowly and often as a function of advances in molecular genetics (eg, genotyping and sequencing technologies) as well as shifts in the very paradigms that undergird our assumptions about complex diseases.

Candidate Gene Associations

For more than 20 years, the genes known or suspected (in some cases from study of monogenic diseases) to be involved in lipoprotein synthesis, structure, transport, and transformation have been studied for association with phenotypic variation in plasma lipids as well as CVD endpoints. Early studies used restriction fragment length polymorphism (RFLP)/electrophoresis techniques to genotype these candidate genes. These hypothesis-testing experiments were among the first to explore the nascent "common disease/ common variant" paradigm (33,34) of complex disease phenotypes; these studies also demonstrated the difficulty of making easy generalizations about genotype–phenotype associations across populations. For

example, a common variant (-75 G/A, rs670) in a promoter region of *APOA1*, the gene that encodes apolipoprotein A1 (apo A-1), the primary apolipoprotein constituent in HDL-C and an important signaling protein in reverse cholesterol transport from peripheral tissues to the liver for its excretion (35), was associated with 20% higher plasma apo A-1 in Chinese male never-smokers (but not associated in male ever-smokers or females) (36); significant associations between this variant and apo A-1 and HDL-C were also reported for European females, with smoking status also modulating (although not erasing) the strength of the association (37). However, no significant associations between this variant and HDL-C were found in Hispanic (38), Canadian (39), Japanese (40), or Finnish populations (41). Although the approaches employed by genetic epidemiologists have changed, many common variants in these candidate genes, including rs670 in *APOA1*, continue to prove to be significantly associated with lipid phenotypes in some populations (42), but uncertainties often remain about their routes of influence, possible interacting factors, and the genetic heterogeneity of the phenotypes among populations.

Microsatellite Linkage Findings

Whereas studies of individual candidate genes test hypotheses about genes and variants that are known or suspected to play a role in dyslipidemias, scans of the entire genome are agnostic in their assumptions. Such studies are capable of discovering novel associations, that is, they can identify loci not previously known or suspected to be involved in lipid synthesis, structure, transport, or transformation. Early whole-genome scans relied upon linkage analyses that use genetic, pedigree, and phenotype data collected from families. Linkage scans make use of the decreased randomization during meiotic recombination of chromosomal loci physically nearer each other. Therefore, phenotypes that follow inheritance patterns similar to typed polymorphic markers (such as microsatellites) are likely to be influenced by genetic variation in the chromosomal region of the marker. Such studies are hypothesis-generating: the regions implicated during whole-genome linkage scans must be further scrutinized (by genetic sequencing, for example) to positively identify possible causal variants. A review of the literature suggests that some 50 whole-genome linkage studies of lipid traits have been published for various populations. If judged by the number of statistically significant or suggestive findings these studies produced, linkage approaches have been hugely successful. A review published in 2004 that tabulated the results of 32 lipid-related whole-genome linkage studies conducted from 1998 to 2003 (which might be considered the height of microsatellite linkage study era) reported 152 loci with suggestive or significant (LOD score greater than or equal to 1.7 or *P* less than or equal to .0023) evidence of linkage. Suggestive or better results were found for at least one locus on every chromosome. Chromosomes 2, 13, 15, and 21 harbored more findings than would be expected by chance alone, suggesting that these chromosomes indeed held gene variants influencing lipoprotein and/or lipid levels (43). However, if judged by the number of replicated findings or linkage findings that had translational importance, the linkage approach yielded a meager harvest for complex lipid traits. Indeed, the authors of the 2004 review concluded that the "avalanche" of susceptibility loci for lipid phenotypes combined with the inability to pinpoint loci (a characteristic of microsatellite linkage studies) made any claims of replication tricky, at best: with hits throughout and covering large regions of the genome, the possibility of false positive claims of replication was significant (43). These factors, in conjunction with sometimes disparate study designs, dubious comparisons of diverse lipid phenotypes, and the intrinsic difficulties associated with complex lipid traits (genetic heterogeneity, epistasis, and so on), meant that the era of whole-genome microsatellite linkage studies of lipid traits largely ended with equal amounts of optimism and uncertainty (44).

Genome-Wide Association and Other Microarray Studies

The development of genotyping microarrays ("gene chips") capable of interrogating millions of single nucleotide polymorphisms (SNPs) throughout the genome (with the possibility of predicting millions more through statistical imputation) merged the association methods of candidate gene studies with the agnostic, hypothesis-generating methods of whole-genome linkage methods. The design of the gene chips used in GWAS was predicated upon the common disease/common variant hypothesis. The SNPs typed by GWAS arrays typically have population minor allele frequencies (MAFs) greater than or equal to 5%. As of early 2014, the U.S. National Human Genome Research Institute's Division of Genomic Medicine's GWAS catalog listed 75 studies of more than 20 lipid-related phenotypes with over 600 significant or suggestive (as deemed by each study's authors) genotype–phenotype associations spread across the genome. Figure 19.3 graphically depicts the approximate chromosomal positions of the significant lipid associations (with data from GWAS through May 2013). Table 19.1 lists genes containing variants (and genes with variants nearby) significantly associated with a subset of lipid phenotypes.

To those familiar with lipid metabolism and the history of lipid genetics, a quick scan of Table 19.1 will reveal many familiar characters. Genes known

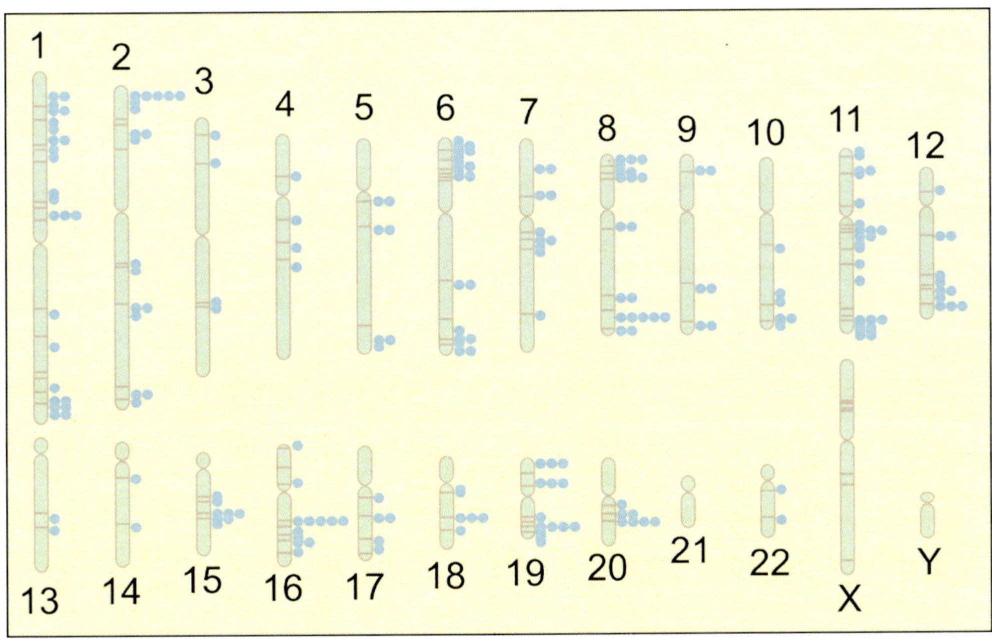

FIGURE 19.3 Approximate positions of significant (as judged by study authors) genotype–lipid phenotype associations from genome-wide association studies published before May 2013. Multiple dots at one locus indicate significant associations were reported in more than one lipid phenotype category (total cholesterol, low-density lipoprotein cholesterol, high-density lipoprotein cholesterol, triglyceride, and lipid measures).

Source: National Human Genome Research GWAS Diagram Browser (www.ebi.ac.uk/fgpt/gwas).

TABLE 19.1   Genes Containing Variants (and Genes With Variants Nearby) Significantly or Suggestively (as Specified by Study Authors) Associated With a Subset of Lipid Phenotypes

HDL cholesterol

ABCA1 (45–52), *ABCA8* (49), *ACAA2* (46), *ADM* (49), *AMPD3* (49), *ANGPTL4* (47,49), *APOA* (53), *APOA1* (47,49,50), *APOA4* (47,49,50), *APOA5* (47,49,50), *APOB* (48–50) *APOC* (53), *APOC1* (49), *APOC2* (49), *APOC3* (47,49,50), *APOE* (48,49,51), *ARL15* (49), *C12orf51* (51), *C6orf106* (49), *CCDC92* (49), *CD36* (53), *CETP* (45–57), *CITED2* (49), *CMIP* (49), *COBLL1* (49), *CTCF* (48), *DOCK6* (49,52), *FADS1* (47,49,50), *FADS2* (47–49), *FADS3* (47–49), *FOLH1* (48), *GALNT2* (45–47,49,50), *GFOD2* (50), *HERPUD1* (52,57), *HNF4A* (47,49), *IRS1* (49), *KANK2* (52), *KCTD10* (50), *KLF14* (49), *LACTB* (49), *LCAT* (45,47,49,50,54), *LILRA3* (49), *LILRB2* (49), *LIPC* (45–50,52,54), *LIPG* (45–50,56), *LOC55908* (49,52,53), *LPA* (49), *LPL* (45–51,53,56), *LRP1* (49), *LRP4* (49), *MACF1* (49), *MADD* (48), *MC4R* (49), *MLXIPL* (49), *MMAB* (45,47,49,50), *MVK* (45,47,49,50), *MYL2* (51), *MYO1H* (50), *NIPSNAP3A* (52), *NIPSNAP3B* (52), *NLRC5* (52), *NR1H3* (49,54), *NUP93* (57), *OAS3* (51), *PABPC4* (49), *PDE3A* (49), *PGS1* (49), *PLTP* (47,49), *PPP1R3B* (49,52,53), *PRMT8* (48), *PVRL2* (51), *RAB3D* (52), *RPS3A* (49), *SBNO1* (49), *SCARB1* (49), *SLC12A3* (52,57), *SLC39A8* (49), *SPC24* (52), *STARD3* (49), *TOMM40* (48,51), *TRIB1* (49), *TRPS1* (49), *TSPAN16* (52), *TTC39B* (47,49,50), *UBASH3B* (49), *UBE2L3* (49), *UBE3B* (50), *ZNF259* (50), *ZNF648* (49), *ZNF664* (49)

Hypertriglyceridemia

APOA1 (52), *APOA4* (52), *APOA5* (52,58), *APOB* (58), *APOC3* (52), *BAZ1B* (52), *BCL7B* (52), *BMP1* (52), *BUD13* (52), *C18orf45* (52), *C2orf16* (52), *CABLES1* (52), *FNDC4* (52), *GCKR* (52,58), *HAVCR1* (52), *HAVCR2* (52), *IFT172* (52), *INTS10* (52), *KIAA0999* (52), *KRTCAP3* (52), *LGI3* (52), *LPL* (52,58), *MLXIPL* (52,58), *MTMR10* (52), *MTMR15* (52), *NRBP1* (52), *PHYHIP* (52), *PIWIL2* (52), *POLR3D* (52), *PPM1G* (52), *REEP4* (52), *SFTPC* (52), *SLC18A1* (52), *TBL2* (52), *TIMD4* (52), *TRPM1* (52), *VPS37D* (52), *ZNF259* (52), *ZNF512* (52)

LDL cholesterol

ABCG5 (48,49), *ABCG8* (47,49,53), *ABO* (49), *ANGPTL3* (49), *APO cluster* (54), *APOA1* (48–50,59), *APOA4* (48–50), *APOA5* (48–50,59), *APOB* (45–50,53,54,60–62), *APOC1* (45–47,49,50,60,63), *APOC2* (46,47,49,50,63), *APOC3* (48–50,59), *APOC4* (45–47,50,63), *APOE* (45–50,53,63,64), *AR* (54), *B3GALT4* (45), *BRAP* (49), *BUD13* (50), *CBLN3* (49), *CELSR2* (45–50,54,59,60), *CETP* (49), *CILP2* (45–47,49,50), *CR1L* (54), *CSPG3* (49), *CYP7A1* (49), *DHX38* (49), *DNAH11* (48,49), *DOCK7* (48,49), *FADS1* (49,54), *FADS2* (48,49,54), *FADS3* (48,49), *FRK* (49), *GCKR* (59), *GMPR* (50), *GPAM* (49), *HAVCR1* (47,49), *HFE* (49), *HIST1H4C* (49), *HLA* (49), *HMGCR* (46–50,63), *HNF1A* (47,49), *HP* (49), *HPR* (49), *IDOL* (49), *IRF2BP2* (49), *KIAA0323* (49), *LDLR* (45–50,53,54), *LDLRAP1* (49), *LPA* (49), *MAFB* (47,49), *MOSC1* (49), *MYLIP* (50), *NCAN* (45,47,48), *NPC1L1* (49), *OSBPL7* (49), *PBX4* (46,47,49), *PCSK9* (45–47,49,50,53), *PLEC1* (49), *PPP1R3B* (49,50), *PSRC1* (45–47,49,54,59,60), *SF4* (50), *SORT1* (45–47,49,54), *ST3GAL4* (49), *TIMD4* (47,49), *TMEM57* (49), *TOMM20* (48,49), *TOP1* (49), *TRIB1* (48–50), *ZNF259* (50)

Lipid metabolism phenotypes

ABCA1 (65), *ABCG5* (66), *ABCG8* (66), *ALB* (65), *ANGPTL3* (65,66), *APOA1* (65), *APOA4* (65), *APOA5* (65), *APOB* (66), *APOC1* (65,66), *APOC2* (65), *APOC3* (65), *APOE* (65,66), *CELSR2* (66), *CETP* (65,66), *CPT1A* (65), *FADS1* (65), *FADS2* (65), *FADS3* (65), *FCGR2A* (65), *FCGR2B* (65), *GCKR* (66), *LDLR* (65,66), *LIPC* (65,66), *LIPG* (65), *LPL* (66), *LPL* (65), *MLXIPL* (65), *PCSK9* (65), *PDXC1* (65), *PLTP* (65,66), *PPP1R11* (65), *PSRC1* (66), *SPRT1* (66), *TRIB1* (66)

Triglycerides

AFF1 (49,50), *ALDH2* (67), *AMAC1L2* (47), *ANGPTL3* (45–47,49,50,68), *ANKRD55* (49), *APOA* (53), *APOA* (53), *APOA1* (45–51,68–70), *APOA4* (45–51,68,70), *APOA5* (45–51,67,68,70), *APOB* (46–50,71), *APOC* (53), *APOC1* (49,53), *APOC2* (49), *APOC3* (45–51,68,70), *APOE* (48,49), *APOE cluster* (71), *ATG4C* (46), *BAZ1B* (50), *BCL7B* (46,50), *BUD13* (46,50,51), *C5orf35* (50), *C8orf35* (69), *CAPN3* (49), *CCDC92* (49), *CETP* (49,71), *CILP2* (45–47,49,50), *COBLL1* (49), *CSPG3* (49), *CTF1* (49), *CYP26A1* (49), *DOCK7* (46,48–50), *DSCAML1* (70), *FADS1* (47,49,50), *FADS2* (47,49), *FADS3* (47,49), *FRMD5* (49), *GALNT2* (45,46,49), *GCKR* (45–51,53,68), *HAVCR1* (49), *HLA* (49), *IRS1* (49), *JMJD1C* (49), *KIAA0999* (69), *KLHL8* (49), *LIPC* (45,49), *LOC440069* (69), *LOC645044* (69), *LPL* (45–51,53,67–69,71), *LRP1* (49), *MAP3K1* (49), *MGC13125* (69), *MLXIPL* (45–51,68,69), *MSL2L1* (49), *NAT2* (49), *NCAN* (45,47,48), *PBX4* (46,47,49), *PINX1* (49), *PLA2G6* (49), *PLTP* (47,49), *SLC18A1* (69), *TBL2* (46,50,51), *TIMD4* (49), *TOMM40* (48), *TRIB1* (45–47,49,50,68), *TYW1B* (49), *XKR6* (47,49), *ZNF101* (50), *ZNF259* (46,50,51), *ZNF664* (49)

Source: Data from the U.S. National Human Genome Research Institute, Division of Genomic Medicine's Catalog of Published Genome-Wide Association Studies (www.genome.gov/Gwastudies).

to play roles in lipid synthesis, structure, transport, and transformation (for example, genes in the *APOA/APOC3* cluster on chromosome 11, the cholesterol ester transfer protein [*CETP*] gene), *LDLR*) have been well represented in GWAS findings. This is encouraging for a number of reasons. First,

identifying common variants associated with genes of known or suspected importance (something that has not been the case with some other phenotypes studied by GWAS) bolstered researchers' confidence in the ability of the GWAS approach to identify novel common variants associated with lipid traits. Second, although some gene products may be known to play a role in lipid metabolism, identifying the common variants associated with interindividual variation of a trait may provide insights into molecular mechanisms, disease etiology, and novel pharmacological interventions. Importantly, many replicated GWAS findings have also pointed to novel genes not previously suspected to play a role in lipid synthesis, structure, transport, or transformation. Although the accretion of new findings constantly changes the landscape of "novel" associations, early lipid trait GWAS (and meta-analyses of GWAS) reported that 37% to 62% of their significant associations with total cholesterol, LDL-C, HDL-C, or TG were novel (45,47,49).

The lipid and lipoprotein associations outlined in Table 19.1 are for baseline traits. A number of GWAS have also explored the response of plasma lipids and lipoproteins to pharmacological interventions. Fenofibrate (and the newer fenofibric acid formulation) is a peroxisome proliferator-activated receptor alpha activator that lowers TG levels with ancillary effects on lowering very low-density lipoproteins (VLDLs) and LDL and raising HDL. This drug is commonly prescribed to treat hypertriglyceridemia, although considerable interindividual variation of response to the drug has been observed (72). A study of the response of lipid particle size distribution to treatment with fenofibrate revealed that a variant in the adenosylhomocysteinase-like 2 (*AHCYL2*) gene was associated with VLDL particle size distribution response to the drug. The protein encoded by *AHCYL2* may be involved in the conversion of S-adenosylhomocysteine to L-homocysteine and adenosine. High levels of S-adenosylhomocysteine inhibit methylation of phospholipids, proteins, DNA, and ribonucleic acid (RNA); are associated with insulin resistance; and have been posited to be a sensitive indicator of CVD (73). As with fenofibrate, considerable interindividual variation has been observed in response to statins (HMG-CoA reductase inhibitors), indicated for elevated LDL-C and capable of lowering CVD risk (74). Response to statin treatment has been investigated in a number of studies. A meta-analysis of three statin GWAS reported an association between *CLMN* (the calmin gene) and total cholesterol reduction (75). The function of calmin and its role in mediating cholesterol response to statin remain unknown. Another GWAS in 3,895 individuals and replicated in 14,810 additional participants failed to find significant genetic associations with LDL-C and APOB response to simvastatin, suggesting that common variants may not influence lipid response to statin therapy at levels detectable in modestly sized discovery populations (76). These and other studies indicate that the research front for lipid pharmacogenetics still sits soundly in the discovery phase.

DNA microarrays can be used in applications other than standard, common variant, SNP association studies. For example, some GWAS chips can be used for "virtual karyotyping" to assess copy number variants (CNVs)–differences in the number of copies of particular DNA segments. Like SNPs, CNVs can be associated with diseases and disease-related traits. To date, only one small study has used this approach in the context of lipid phenotypes: these investigators identified seven CNV regions significantly associated with hyperlipidemia and MI (77). Specialized targeted genotyping platforms have been created that interrogate common and rare variants identified as potentially important in previous sequencing studies and GWAS. Some of these arrays focus on coding regions ("exome chips"), while others include variants implicated in a particular disease domain. A recent study used a CVD chip with follow-up with an exome chip to identify novel common and rare variants in *FTO* (the fat mass and obesity associated gene), *SERPINA12* (the serpin peptidase inhibitor, clade A [alpha-1 antiproteinase, antitrypsin], member 12 gene), and *ITGAL* (the integrin, alpha L [antigen CD11A (p180), lymphocyte-function-associated antigen 1; alpha polypeptide gene]) that were associated with PON1 (paraoxonase/arylesterase 1) activity. PON1 is a component of HDL and is responsible for some of the cardioprotective effects of HDL (78). A meta-analysis of 32 studies using a chip with about 50,000 SNPs in 2,000 candidate genes (previously implicated in various diseases and disease phenotypes) identified 24 novel SNPs associated with HDL-C, LDL-C, TG, and/or total cholesterol in genes known to play a role in lipid traits. This study also found associations in 23 genes previously unknown to play a role in lipid phenotypes (79). Studies that employ these targeted arrays stand midway between agnostic GWAS and hypothesis-driven candidate gene studies: they leverage a priori knowledge of disease and lipid metabolism in studies that can, in effect, test thousands of hypotheses simultaneously. As long as large-scale, whole-genome sequencing studies stay out of reach for budgetary reasons, targeted arrays will have a place in the arsenal of tools used by those exploring lipid genetics.

While several years of GWAS and other microarray studies of lipid-related phenotypes have produced a mind-numbing harvest of "hits," including many in intergenic regions and in genes seemingly unrelated to lipid and lipoprotein synthesis, structure, transport, and transformation, one question begs to be answered: So what? As with many other quests in complex disease genetics, the road from the discovery of genotype–lipid phenotype associations to clinical relevance appears neither short nor straight.

It is important to remember that, despite the often well-replicated nature of genetic associations with lipid phenotypes and the seemingly pinpoint localization of a single nucleotide, the findings of GWAS do not necessarily (or even usually) point to actual causal variants or even genes; rather they reveal variants that tag a region of linkage disequilibrium (LD) within which the causal variant(s) might lie. For example, in one lipid GWAS meta-analysis, of the 175 SNPs that were significantly associated with lipid traits, only 54 were nonsense or missense mutations. Identifying functional variants often requires additional follow-up. One approach to identifying functional genes involves determining whether a variant is associated with gene expression levels (an expression quantitative trait locus or eQTL) in a relevant tissue, for lipid phenotypes usually in the liver where lipoproteins, cholesterols, and phospholipids are synthesized and prepared for transport. Although a SNP implicated in an eQTL study may simply be in LD with the functional SNP, such a finding at least narrows down which nearby gene is operative. For example, SNP rs9987289 in an uncharacterized locus (LOC157273) on chromosome 8 was significantly associated with HDL-C via GWAS. The SNP was also associated with expression of the nearby protein phosphatase 1, regulatory subunit 3B (*PPP1R3B*) gene in the liver (49). Evidence such as this spurred subsequent fine mapping of the *PPP1R3B* locus (using a targeted array), which further refined the set of probable causal variants potentially regulating circulating HDL-C levels (80). Mouse experiments that knock down gene expression can also be used to determine which genes that have been inconclusively implicated in GWAS might be functional with respect to lipids and lipoproteins. For example, intergenic SNPs near *TRIB1* were found to be associated with LDL-C and HDL-C via GWAS (47). Knockdown studies of *Trib1* in mice confirmed the functional role of this gene in the context of lipids and CVD (81). Complete sequencing of candidate regions identified via these methods (and, eventually, whole-genome sequencing) can provide definitive evidence of causal variation. A more detailed understanding of exactly which variants contribute to the heritability of lipid and lipoprotein traits is expected to move us one step closer to a more satisfying answer to the "So what?" question.

A significant point in the recent history of genetic epidemiology came with the articulation of the problem of "missing heritability" (82). Put succinctly, most of the variants identified via GWAS as being associated with complex diseases and traits confer relatively small increments of risk and seem to explain only a small proportion of the total heritability. In the lipid domain, the effects of lipid-associated variants discovered via GWAS and other microarray studies explain about 10% to 12% of the total heritability of these traits (79,83). The issue of missing

heritability has been robustly discussed in the literature, with evidence from lipid and lipoprotein studies figuring importantly. Some investigators suggest that gene–gene interaction (84,85) and gene–environment interaction (86) could account for missing heritability. Others believe that, from a practical perspective, the concern over missing heritability misses the point of disease genomics. For example, one GWAS identified a common variant in the region of the 3-hydroxy-3-methylglutaryl-CoA reductase (*HMGCR*) gene as associated with LDL-C. The product of this gene is the rate-limiting enzyme in the production of cholesterol. The variant exerts a modest 2.5 mg/dL effect on LDL-C. However, statin drugs, which inhibit the HMGCR enzyme, have been shown to lower LDL-C by 14 to 70 mg/dL (49,87,88). This example from lipid genetics suggests that GWASs are well poised to identify potentially biologically important genes, but the effect size of any single variant in a gene must not be considered indicative of the ultimate clinical potential of GWAS findings. Others have claimed that the heritability is not missing, but rather has gone unmeasured in human studies (89), largely because of the incomplete picture of genetic associations detectable by GWAS arrays. One study of LDL-C associations compared the percentage of heritability explained by five common variants identified by GWAS to that explained by rare and common variants identified via fine mapping at previously implicated LDL-C loci. The GWAS-identified SNPs explained 3.1% of the total LDL-C heritability. Using the fine-map-identified variants discovered in apolipoprotein E (*APOE*), apolipoprotein C1 (*APOC1*), apolipoprotein C2 (*APOC2*), sortilin 1 (*SORT*), *LDLR*, *APOB*, and *PCSK9*, 6.5% of the total heritability was explained. By extrapolating the trend observed in the study of seven genes to all known LDL-C-associated loci, the authors predicted that the proportion of variance explained could double from 12% to 24%, exceeding half of the genetic variance for LDL-C (90).

SEQUENCING STUDIES AND THE ROLE OF RARE VARIANTS

Studies such as the one described have created something of a paradigm shift in the field of complex disease genomics. Although the common disease/common variant hypothesis has not been discarded, the field has expanded its gaze to include the study of rare (and low-frequency) variants. The search for rare variants influencing lipid traits has only just begun, and the degree to which rare mutations might account for missing heritability in this domain remains an open question (83,91). Whole-exome and whole-genome sequencing approaches have been used to discover rare genetic defects that cause Mendelian lipid disorders, such as a rare form of severe hypercholesterolemia found through whole-genome sequencing (92), and rare variants in the angiopoietin-like 3 (*ANGPTL3*) gene causing a form of familial combined hypolipidemia found using whole-exome analysis (93). Whole-exome sequencing has recently also been used to investigate the role of rare and low-frequency variants in normal population variation of LDL-C. About 2000 individuals were exome sequenced, including 544 selected from the upper and lower tails (greater than 98th, less than 2nd percentiles) of the LDL-C distribution. Additional participants were exome sequenced (N = 1300) or targeted genotyped (N = 52,200) in follow-up analyses. Significant associations were found with the burden of rare and low-frequency variants in *PNPLA5* (the patatin-like phospholipase domain containing 5 gene; this finding was replicated in an additional approximately 2000 individuals) as well as in previously unidentified variants in *PCSK9*, *LDLR* and *APOB*. "Burden tests" are a necessary statistical maneuver used in sequencing studies in lieu of traditional association tests. The authors of this study made a number of important observations: first, the functional variants identified varied widely in both their MAF and in their mode of action (for example, loss-of-function or missense variants). This is important because it suggests that a range of bioinformatic and statistical tests must be used concomitantly in sequencing studies of complex traits and diseases to thoroughly capture putatively functional variants. Second, the effect sizes for the four associated genes ranged from approximately 24 to 116 mg/dl LDL-C, substantially higher than effects observed for variants discovered by GWAS. The proportion of LDL-C variance explained by these four genes was approximately 5.4%; this is significant when compared to 10% to 12% explained by the hundreds of variants implicated in GWASs (94). This study not only discovered novel genetic associations with LDL-C; it also established an important methodological proof-of-concept and offered an optimistic picture of the next phase of lipid genomics.

EPIGENETICS

The field of epigenetics concerns the study of heritable changes in gene activity caused by external modifications to DNA, not by changes in the nucleotide sequence of DNA. DNA methylation and histone modification are examples of epigenetic changes, both of which can play a role in gene expression by affecting chromatin structure and modifying the availability of coding regions to transcription mechanisms.

Unlike DNA sequence variation, epigenetic variation is susceptible to both inherited and environmental factors. Like sequencing studies, epigenetic studies of lipid phenotypes are in their fledgling stages. As is the case with genetic sequence studies, epigenetics studies of lipids have tested hypotheses regarding particular candidate genes and regions and agnostically scanned the entire genome for novel epigenetic associations (in epigenome-wide association studies or EWAS). One candidate gene study found that the degree of methylation of *IGF2* (the insulin-like growth factor 2 [somatomedin A] gene) is cross-sectionally associated with TG/HDL-C ratio in obese children and adolescents (95). Another reported that hypermethylation of the proopiomelanocortin (*POMC*) gene was associated with TG in children (96). An EWAS in a sample of 21 men with FH with replication in an additional 276 discovered HDL-C associations with differentially methylated loci in the promoter region of *TNNT1* (the troponin T type 1 [skeletal, slow] gene) (97). A more detailed discussion of the mechanics and early fruits of lipid epigenetics studies is beyond the scope of this chapter. It is already clear, however, that epigenetics offers a new frontier for investigators of lipid-related traits.

REFINING LIPID PHENOTYPES

The bulk of this chapter has contextualized the accumulation of knowledge of lipid genetics primarily in light of evolving genotyping and sequencing technologies. However, our understanding of lipid phenotypes (and our ability to measure them) has evolved as well. For example, some studies have demonstrated that measurement of both concentration and distribution of HDL, LDL, and VLDL subclasses based on lipoprotein particle size improves CVD risk assessment beyond that of the traditional clinical measurements (98,99); certain lipoprotein subclass profiles may be more atherogenic than others. One GWAS used lipid particle size phenotypes to determine if genetic variants that were associated with lipoprotein particle diameter measures in Caucasians were also associated with the same lipid subclass measures in other racial/ethnic groups. These investigators found that variants in the APOB region were associated with mean VLDL in Caucasian and Hispanic samples but not in African American or Chinese samples. Variants in *LIPC* (the lipase, hepatic gene) were suggestively associated with mean HDL diameter, but only in Caucasians. They concluded that differences across racial strata in lipoprotein subclass profiles are, in part, genetically determined (100).

Emerging data have shown that profiling small molecule lipid species—an area of inquiry known as lipidomics—can enhance our understanding of the

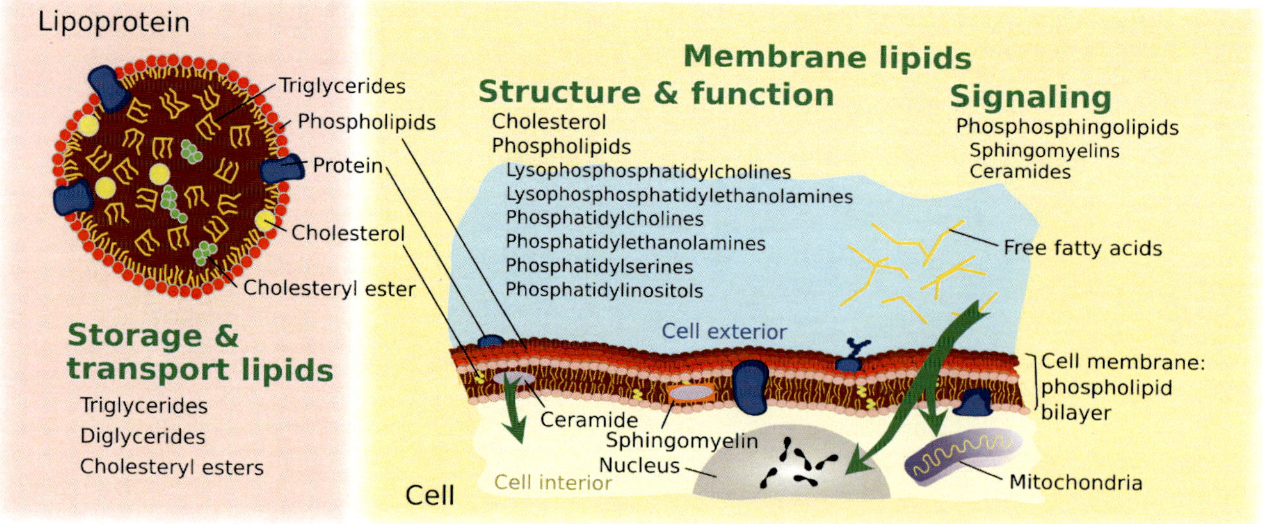

FIGURE 19.4 Classes of small molecule lipids typically assayed using lipidomic approaches.

complexity and dynamics of lipid functions and have driven discoveries in lipid biochemistry, physiology, and nutrition (101–103). The "lipidome" (a subset of the "metabolome") comprises the entire spectrum of lipid species in a biological system (104). Although wide application of lipidomics has been hampered by cost, progress in mass spectrometry (MS)-based techniques now permits cost-efficient profiling of large numbers of lipid species in biological samples (104). For example, one lipid research consortium recently identified and quantified more than 500 lipids representing the major species of the phosphosphingolipid, glycerolipid, phospholipid, fatty acid, sterol, and prenol classes in fasting human plasma (105). See Figure 19.4 for a summary of lipidomic phenotypes. The constituents of the lipidome are expected to be pivotal phenotypes necessary to understand complex diseases that have been, so far, recalcitrant to genetic dissection. Importantly, intermediate lipid species may be "closer to the gene" than lipids routinely assessed within the clinic. Previous studies have suggested lipidomic profiles are both heritable and under strong genetic control. One study used quantitative MS-based metabolomic profiling in 117 individuals in an investigation of premature CAD. They found very strong heritability for free fatty acids (48%–59%) and acylcarnitines (23%–79%) (106). Genetic variants highlighted in a metabolomic GWAS are in or near proteins linked to metabolism, relatively common (MAF greater than 20%), and have exceptionally high effect sizes (accounting for 10%–60% of the total variance in the metabolite) (107). A comprehensive understanding of genetic variation in small molecule lipid phenotypes has a potentially wide range of biomedical and pharmaceutical applications.

CONCLUSIONS

To date, the clinical significance of studies of Mendelian and complex genetic lipid disorders depends somewhat on the lipid species in question. Genetic dissection of TG variability has strengthened our understanding of lipid metabolism and the effects of diet and drugs (108), and the large number of known and novel genes that have been implicated as being involved in TG metabolism has expanded the range of operative pathways and potential therapeutic targets. In the realm of LDL-C, a positive diagnosis of ADH can trigger lifestyle and drug interventions that can significantly reduce the risk of CVD. Although the findings would not be considered clinically novel, evidence from studies of Mendelian disorders has reinforced the clear relationship between elevated LDL-C and increased CVD risk; GWAS have largely augmented this evidence. The implication is that elements in any of the pathways influencing circulating LDL-C all represent potential therapeutic targets that could prove clinically significant with respect to CVD development and outcomes (109). The current enthusiasm over PCSK9 inhibitors for LDL-C lowering provides an illustrative example (110). Genetic studies of HDL-C have been less consonant in establishing a link between HDL-C and CVD risk. This too, however, may prove to be illuminating as it suggests that not all mechanisms that lead to HDL-C variability are equal; consequently, not all HDL-C pathways may be viable therapeutic targets or predictors of drug response (109,111). It should be emphasized that genetic studies of blood lipids, lipoproteins, and related phenotypes have not only advanced our knowledge of lipid metabolism and its role in CVD and related therapies, but they have been instrumental in advancing (indeed, complicating) our more fundamental understanding of human genetics. Any feelings of discontent about the slow march toward novel clinical applications are more than offset by optimistic anticipation of discoveries that lie ahead.

REFERENCES

1. Kannel WB, Dawber TR, Kagan A, et al. Factors of risk in the development of coronary heart disease—six year follow-up experience. The Framingham Study. *Ann Intern Med.* 1961;55:33–50.
2. Sarwar N, Danesh J, Eiriksdottir G, et al. Triglycerides and the risk of coronary heart disease: 10,158 incident cases among 262,525 participants in 29 Western prospective studies. *Circulation.* 2007;115:450–458.
3. Zilversmit DB. Atherogenic nature of triglycerides, postprandial lipidemia, and triglyceride-rich remnant lipoproteins. *Clin Chem.* 1995;41:153–158.
4. Karpe F. Postprandial lipoprotein metabolism and atherosclerosis. *J Intern Med.* 1999;246:341–355.
5. Gianturco SH, Lin AH, Hwang SL, et al. Distinct murine macrophage receptor pathway for human triglyceride-rich lipoproteins. *J Clin Invest.* 1988;82:1633–1643.
6. Havel RJ. Remnant lipoproteins as therapeutic targets. *Curr Opin Lipidol.* 2000;11:615–620.
7. Mitra S, Deshmukh A, Sachdeva R, et al. Oxidized low-density lipoprotein and atherosclerosis implications in antioxidant therapy. *Am J Med Sci.* 2011;342:135–142.
8. Rosenson RS, Brewer HB, Jr, Davidson WS, et al. Cholesterol efflux and atheroprotection: advancing the concept of reverse cholesterol transport. *Circulation.* 2012;125:1905–1919.
9. Rosenson RS, Brewer HB, Jr, Chapman MJ, et al. HDL measures, particle heterogeneity, proposed nomenclature, and relation to atherosclerotic cardiovascular events. *Clin Chem.* 2011;57:392–410.
10. Camont L, Chapman MJ, Kontush A. Biological activities of HDL subpopulations and their relevance to cardiovascular disease. *Trends Mol Med.* 2011;17:594–603.
11. Weiss LA, Pan L, Abney M, Ober C. The sex-specific genetic architecture of quantitative traits in humans. *Nat Genet.* 2006;38:218–222.
12. Wojczynski MK, Glasser SP, Oberman A, et al. High-fat meal effect on LDL, HDL, and VLDL particle size and number in the Genetics of Lipid-Lowering Drugs and Diet Network (GOLDN): an interventional study. *Lipids Health Dis.* 2011;10:181.
13. Garcia-Rios A, Perez-Martinez P, Delgado-Lista J, et al. Nutrigenetics of the lipoprotein metabolism. *Mol Nutr Food Res.* 2012;56:171–183.
14. Kolovou GD, Mikhailidis DP, Kovar J, et al. Assessment and clinical relevance of non-fasting and postprandial triglycerides: an expert panel statement. *Curr Vasc Pharmacol.* 2011;9:258–270.
15. Mora S, Rifai N, Buring JE, Ridker PM. Fasting compared with nonfasting lipids and apolipoproteins for predicting incident cardiovascular events. *Circulation.* 2008;118:993–1001.
16. Stampfer MJ, Krauss RM, Ma J, et al. A prospective study of triglyceride level, low-density lipoprotein particle diameter, and risk of myocardial infarction. *JAMA.* 1996;276:882–888.
17. Nordestgaard BG, Benn M, Schnohr P, Tybjaerg-Hansen A. Nonfasting triglycerides and risk of myocardial infarction, ischemic heart disease, and death in men and women. *JAMA.* 2007;298:299–308.
18. Rader DJ, Cohen J, Hobbs HH. Monogenic hypercholesterolemia: new insights in pathogenesis and treatment. *J Clin Invest.* 2003;111:1795–1803.
19. Jensen HK. The molecular genetic basis and diagnosis of familial hypercholesterolemia in Denmark. *Dan Med Bull.* 2002;49:318–345.
20. De Castro-Oros I, Pocovi M, Civeira F. The genetic basis of familial hypercholesterolemia: inheritance, linkage, and mutations. *Appl Clin Genet.* 2010;3:53–64.
21. Goldstein J, Hobbs H, Brown M. Familial hypercholesterolemia. In: Beaudet A, Sly W, Valle D, eds. *The Metabolic and Molecular Bases of Inherited Disease.* New York, NY: McGraw-Hill; 2001:2863–2913.
22. Robinson JG. Management of familial hypercholesterolemia: a review of the recommendations from the National Lipid Association Expert Panel on Familial Hypercholesterolemia. *J Manag Care Pharm.* 2013;19:139–149.
23. Abifadel M, Varret M, Rabes JP, et al. Mutations in PCSK9 cause autosomal dominant hypercholesterolemia. *Nat Genet.* 2003;34:154–156.
24. Dubuc G, Chamberland A, Wassef H, et al. Statins upregulate PCSK9, the gene encoding the proprotein convertase neural apoptosis-regulated convertase-1 implicated in familial hypercholesterolemia. *Arterioscler Thromb Vasc Biol.* 2004;24:1454–1459.
25. Horton JD, Cohen JC, Hobbs HH. PCSK9: a convertase that coordinates LDL catabolism. *J Lipid Res.* 2009;50(Suppl):S172–S177.
26. Cohen J, Pertsemlidis A, Kotowski IK, et al. Low LDL cholesterol in individuals of African descent resulting from frequent nonsense mutations in PCSK9. *Nat Genet.* 2005;37:161–165.
27. Graham MJ, Lemonidis KM, Whipple CP, et al. Antisense inhibition of proprotein convertase subtilisin/kexin type 9 reduces serum LDL in hyperlipidemic mice. *J Lipid Res.* 2007;48:763–767.
28. Wierzbicki AS, Hardman TC, Viljoen A. Inhibition of pro-protein convertase subtilisin kexin 9 [corrected] (PCSK-9) as a treatment for hyperlipidaemia. *Expert Opin Investig Drugs.* 2012;21:667–676.
29. Sjouke B, Kusters DM, Kastelein JJ, Hovingh GK. Familial hypercholesterolemia: present and future management. *Curr Cardiol Rep.* 2011;13:527–536.
30. Innerarity TL, Weisgraber KH, Arnold KS, et al. Familial defective apolipoprotein B-100: low density lipoproteins with abnormal receptor binding. *Proc Natl Acad Sci USA.* 1987;84:6919–6923.
31. Ahmad Z, Adams-Huet B, Chen C, Garg A. Low prevalence of mutations in known loci for autosomal dominant hypercholesterolemia in a multiethnic patient cohort. *Circ Cardiovasc Genet.* 2012;5:666–675.

32. Lander ES, Schork NJ. Genetic dissection of complex traits. *Science.* 1994;265:2037–2048.

33. Lander ES. The new genomics: global views of biology. *Science.* 1996;274:536–539.

34. Collins FS, Guyer MS, Charkravarti A. Variations on a theme: cataloging human DNA sequence variation. *Science.* 1997;278:1580–1581.

35. Fielding CJ, Fielding PE. Molecular physiology of reverse cholesterol transport. *J Lipid Res.* 1995;36:211–228.

36. Saha N, Tay JS, Low PS, Humphries SE. Guanidine to adenine (G/A) substitution in the promoter region of the apolipoprotein AI gene is associated with elevated serum apolipoprotein AI levels in Chinese non-smokers. *Genet Epidemiol.* 1994;11:255–264.

37. Talmud PJ, Ye S, Humphries SE. Polymorphism in the promoter region of the apolipoprotein AI gene associated with differences in apolipoprotein AI levels: the European Atherosclerosis Research Study. *Genet Epidemiol.* 1994;11:265–280.

38. Barre DE, Guerra R, Verstraete R, et al. Genetic analysis of a polymorphism in the human apolipoprotein A-I gene promoter: effect on plasma HDL-cholesterol levels. *J Lipid Res.* 1994;35:1292–1296.

39. Minnich A, DeLangavant G, Lavigne J, et al. G-->A substitution at position -75 of the apolipoprotein A-I gene promoter. Evidence against a direct effect on HDL cholesterol levels. *Arterioscler Thromb Vasc Biol.* 1995;15:1740–1745.

40. Akita H, Chiba H, Tsuji M, et al. Evaluation of G-to-A substitution in the apolipoprotein A-I gene promoter as a determinant of high-density lipoprotein cholesterol level in subjects with and without cholesteryl ester transfer protein deficiency. *Hum Genet.* 1995;96:521–526.

41. Miettinen HE, Korpela K, Hamalainen L, Kontula K. Polymorphisms of the apolipoprotein and angiotensin converting enzyme genes in young North Karelian patients with coronary heart disease. *Hum Genet.* 1994;94:189–192.

42. Al-Bustan SA, Al-Serri AE, Annice BG, et al. Re-sequencing of the APOAI promoter region and the genetic association of the -75G > A polymorphism with increased cholesterol and low density lipoprotein levels among a sample of the Kuwaiti population. *BMC Med Genet.* 2013;14:90.

43. Bosse Y, Chagnon YC, Despres JP, et al. Compendium of genome-wide scans of lipid-related phenotypes: adding a new genome-wide search of apolipoprotein levels. *J Lipid Res.* 2004;45:2174–2184.

44. Altmuller J, Palmer LJ, Fischer G, et al. Genomewide scans of complex human diseases: true linkage is hard to find. *Am J Hum Genet.* 2001;69:936–950.

45. Willer CJ, Sanna S, Jackson AU, et al. Newly identified loci that influence lipid concentrations and risk of coronary artery disease. *Nat Genet.* 2008;40:161–169.

46. Kathiresan S, Melander O, Guiducci C, et al. Six new loci associated with blood low-density lipoprotein cholesterol, high-density lipoprotein cholesterol or triglycerides in humans. *Nat Genet.* 2008;40:189–197.

47. Kathiresan S, Willer CJ, Peloso GM, et al. Common variants at 30 loci contribute to polygenic dyslipidemia. *Nat Genet.* 2009;41:56–65.

48. Aulchenko YS, Ripatti S, Lindqvist I, et al. Loci influencing lipid levels and coronary heart disease risk in 16 European population cohorts. *Nat Genet.* 2009;41:47–55.

49. Teslovich TM, Musunuru K, Smith AV, et al. Biological, clinical and population relevance of 95 loci for blood lipids. *Nature.* 2010;466:707–713.

50. Waterworth DM, Ricketts SL, Song K, et al. Genetic variants influencing circulating lipid levels and risk of coronary artery disease. *Arterioscler Thromb Vasc Biol.* 2010;30:2264–2276.

51. Kim YJ, Go MJ, Hu C, et al. Large-scale genome-wide association studies in East Asians identify new genetic loci influencing metabolic traits. *Nat Genet.* 2011;43:990–995.

52. Weissglas-Volkov D, Aguilar-Salinas CA, Nikkola E, et al. Genomic study in Mexicans identifies a new locus for triglycerides and refines European lipid loci. *J Med Genet.* 2013;50:298–308.

53. Coram MA, Duan Q, Hoffmann TJ, et al. Genome-wide characterization of shared and distinct genetic components that influence blood lipid levels in ethnically diverse human populations. *Am J Hum Genet.* 2013;92:904–916.

54. Sabatti C, Service SK, Hartikainen AL, et al. Genome-wide association analysis of metabolic traits in a birth cohort from a founder population. *Nat Genet.* 2009;41:35–46.

55. Hiura Y, Shen CS, Kokubo Y, et al. Identification of genetic markers associated with high-density lipoprotein-cholesterol by genome-wide screening in a Japanese population: the Suita study. *Circ J.* 2009;73:1119–1126.

56. Heid IM, Boes E, Muller M, et al. Genome-wide association analysis of high-density lipoprotein cholesterol in the population-based KORA study sheds new light on intergenic regions. *Circ Cardiovasc Genet.* 2008;1:10–20.

57. Ridker PM, Pare G, Parker AN, et al. Polymorphism in the CETP gene region, HDL cholesterol, and risk of future myocardial infarction: Genomewide analysis among 18 245 initially healthy women from the Women's Genome Health Study. *Circ Cardiovasc Genet.* 2009;2:26–33.

58. Johansen CT, Wang J, Lanktree MB, et al. Excess of rare variants in genes identified by genome-wide association study of hypertriglyceridemia. *Nat Genet.* 2010;42:684–687.

59. Wallace C, Newhouse SJ, Braund P, et al. Genome-wide association study identifies genes for biomarkers of cardiovascular disease: serum urate and dyslipidemia. *Am J Hum Genet.* 2008;82:139–149.

60. Sandhu MS, Waterworth DM, Debenham SL, et al. LDL-cholesterol concentrations: a genome-wide association study. *Lancet.* 2008;371:483–491.

61. Shen H, Damcott CM, Rampersaud E, et al. Familial defective apolipoprotein B-100 and increased low-density lipoprotein cholesterol and coronary artery calcification in the old order amish. *Arch Intern Med.* 2010;170:1850–1855.

62. Makela KM, Seppala I, Hernesniemi JA, et al. Genome-wide association study pinpoints a new functional apolipoprotein B variant influencing oxidized low-density lipoprotein levels but not cardiovascular events: AtheroRemo Consortium. *Circ Cardiovasc Genet.* 2013;6:73–81.

63. Burkhardt R, Kenny EE, Lowe JK, et al. Common SNPs in HMGCR in micronesians and whites associated with LDL-cholesterol levels affect alternative splicing of exon13. *Arterioscler Thromb Vasc Biol.* 2008;28:2078–2084.

64. Rasmussen-Torvik LJ, Pacheco JA, Wilke RA, et al. High density GWAS for LDL cholesterol in African Americans using electronic medical records reveals a strong protective variant in APOE. *Clin Transl Sci.* 2012;5:394–399.

65. Kettunen J, Tukiainen T, Sarin AP, et al. Genome-wide association study identifies multiple loci influencing human serum metabolite levels. *Nat Genet.* 2012;44:269–276.

66. Chasman DI, Pare G, Mora S, et al. Forty-three loci associated with plasma lipoprotein size, concentration, and cholesterol content in genome-wide analysis. *PLoS Genet.* 2009;5:e1000730.

67. Tan A, Sun J, Xia N, et al. A genome-wide association and gene-environment interaction study for serum triglycerides levels in a healthy Chinese male population. *Hum Mol Genet.* 2012;21:1658–1664.

68. Kamatani Y, Matsuda K, Okada Y, et al. Genome-wide association study of hematological and biochemical traits in a Japanese population. *Nat Genet.* 2010; 42:210–215.

69. Kooner JS, Chambers JC, Aguilar-Salinas CA, et al. Genome-wide scan identifies variation in MLXIPL associated with plasma triglycerides. *Nat Genet.* 2008;40:149–151.

70. Pollin TI, Damcott CM, Shen H, et al. A null mutation in human APOC3 confers a favorable plasma lipid profile and apparent cardioprotection. *Science.* 2008;322:1702–1705.

71. Saxena R, Voight BF, Lyssenko V, et al. Genome-wide association analysis identifies loci for type 2 diabetes and triglyceride levels. *Science.* 2007;316:1331–1336.

72. Keating GM, Ormrod D. Micronised fenofibrate: an updated review of its clinical efficacy in the management of dyslipidaemia. *Drugs.* 2002;62:1909–1944.

73. Frazier-Wood AC, Aslibekyan S, Borecki IB, et al. Genome-wide association study indicates variants associated with insulin signaling and inflammation mediate lipoprotein responses to fenofibrate. *Pharmacogenet Genomics.* 2012;22:750–757.

74. Shepherd J, Blauw GJ, Murphy MB, et al. Pravastatin in elderly individuals at risk of vascular disease (PROSPER): a randomised controlled trial. *Lancet.* 2002;360:1623–1630.

75. Barber MJ, Mangravite LM, Hyde CL, et al. Genome-wide association of lipid-lowering response to statins in combined study populations. *PLoS One.* 2010;5:e9763.

76. Postmus I, Verschuren JJ, de Craen AJ, et al. Pharmacogenetics of statins: achievements, whole-genome analyses and future perspectives. *Pharmacogenomics.* 2012;13:831–840.

77. Yip VL, Pirmohamed M. Expanding role of pharmacogenomics in the management of cardiovascular disorders. *Am J Cardiovasc Drugs.* 2013;13:151–162.

78. Kim DS, Burt AA, Crosslin DR, et al. Novel common and rare genetic determinants of paraoxonase activity: FTO, SERPINA12, and ITGAL. *J Lipid Res.* 2013;54:552–560.

79. Asselbergs FW, Guo Y, van Iperen EP, et al. Large-scale gene-centric meta-analysis across 32 studies identifies multiple lipid loci. *Am J Hum Genet.* 2012; 91:823–838.

80. Wu Y, Waite LL, Jackson AU, et al. Trans-ethnic fine-mapping of lipid loci identifies population-specific signals and allelic heterogeneity that increases the trait variance explained. *PLoS Genet.* 2013;9: e1003379.

81. Burkhardt R, Toh SA, Lagor WR, et al. Trib1 is a lipid- and myocardial infarction-associated gene that regulates hepatic lipogenesis and VLDL production in mice. *J Clin Invest.* 2010;120:4410–4414.

82. Manolio TA, Collins FS, Cox NJ, et al. Finding the missing heritability of complex diseases. *Nature.* 2009;461:747–753.

83. Talmud PJ, Yiannakouris N, Humphries SE. Lipoprotein association studies: taking stock and moving forward. *Curr Opin Lipidol.* 2011;22:106–112.

84. Ma L, Yang J, Runesha HB, et al. Genome-wide association analysis of total cholesterol and high-density lipoprotein cholesterol levels using the Framingham heart study data. *BMC Med Genet.* 2010;11:55.

85. Zuk O, Hechter E, Sunyaev SR, Lander ES. The mystery of missing heritability: Genetic interactions create phantom heritability. *Proc Natl Acad Sci USA.* 2012;109:1193–1198.

86. Corella D, Portoles O, Arriola L, et al. Saturated fat intake and alcohol consumption modulate the association between the APOE polymorphism and risk of future coronary heart disease: a nested case-control study in the Spanish EPIC cohort. *J Nutr Biochem.* 2011;22:487–494.

87. Willer CJ, Mohlke KL. Finding genes and variants for lipid levels after genome-wide association analysis. *Curr Opin Lipidol.* 2012;23:98–103.

88. Baigent C, Keech A, Kearney PM, et al. Efficacy and safety of cholesterol-lowering treatment: prospective meta-analysis of data from 90,056 participants in 14 randomised trials of statins. *Lancet.* 2005;366:1267–1278.

89. Bloom JS, Ehrenreich IM, Loo WT, et al. Finding the sources of missing heritability in a yeast cross. *Nature.* 2013;494:234–237.

90. Sanna S, Li B, Mulas A, Sidore C, et al. Fine mapping of five loci associated with low-density lipoprotein cholesterol detects variants that double the explained heritability. *PLoS Genet.* 2011;7:e1002198.

91. Singleton AB. Exome sequencing: a transformative technology. *Lancet Neurol.* 2011;10:942–946.

92. Rios J, Stein E, Shendure J, et al. Identification by whole-genome resequencing of gene defect responsible for severe hypercholesterolemia. *Hum Mol Genet.* 2010; 19:4313–4318.

93. Musunuru K, Pirruccello JP, Do R, et al. Exome sequencing, ANGPTL3 mutations, and familial combined hypolipidemia. *N Engl J Med.* 2010;363:2220–2227.

94. Lange LA, Hu Y, Zhang H, et al. Whole-exome sequencing identifies rare and low-frequency coding variants associated with LDL cholesterol. *Am J Hum Genet.* 2014;94:233–245.

95. Deodati A, Inzaghi E, Liguori A, et al. IGF2 methylation is associated with lipid profile in obese children. *Horm Res Paediatr.* 2013;79:361–367.

96. Yoo JY, Lee S, Lee HA, et al. Can proopiomelanocortin methylation be used as an early predictor of metabolic syndrome? *Diabetes Care.* 2014;37(3):734–739.

97. Guay SP, Voisin G, Brisson D, et al. Epigenome-wide analysis in familial hypercholesterolemia identified new loci associated with high-density lipoprotein cholesterol concentration. *Epigenomics.* 2012;4: 623–639.

98. Colhoun HM, Otvos JD, Rubens MB, et al. Lipoprotein subclasses and particle sizes and their relationship with coronary artery calcification in men and women with and without type 1 diabetes. *Diabetes.* 2002;51:1949–1956.

99. Otvos JD, Collins D, Freedman DS, et al. Low-density lipoprotein and high-density lipoprotein particle subclasses predict coronary events and are favorably changed by gemfibrozil therapy in the Veterans Affairs High-Density Lipoprotein Intervention Trial. *Circulation.* 2006;113:1556–1563.

100. Frazier-Wood AC, Manichaikul A, Aslibekyan S, et al. Genetic variants associated with VLDL, LDL and HDL particle size differ with race/ethnicity. *Hum Genet.* 2013;132:405–413.

101. Bou Khalil M, Hou W, Zhou H, et al. Lipidomics era: accomplishments and challenges. *Mass Spectrom Rev.* 2010;29:877–929.

102. German JB, Gillies LA, Smilowitz JT, et al. Lipidomics and lipid profiling in metabolomics. *Curr Opin Lipidol.* 2007;18:66–71.

103. Fernandis AZ, Wenk MR. Membrane lipids as signaling molecules. *Curr Opin Lipidol.* 2007;18: 121–128.

104. Stock J. The emerging role of lipidomics. *Atherosclerosis.* 2012;221:38–40.

105. Quehenberger O, Armando AM, Brown AH, et al. Lipidomics reveals a remarkable diversity of lipids in human plasma. *J Lipid Res.* 2010;51:3299–3305.

106. Shah SH, Hauser ER, Bain JR, et al. High heritability of metabolomic profiles in families burdened with premature cardiovascular disease. *Mol Syst Biol.* 2009; 5:258.

107. Suhre K, Shin SY, Petersen AK, et al. Human metabolic individuality in biomedical and pharmaceutical research. *Nature.* 2011;477:54–60.

108. Tai ES, Ordovas JM. Clinical significance of apolipoprotein A5. *Curr Opin Lipidol.* 2008;19:349–354.

109. Strong A, Rader DJ. Clinical implications of lipid genetics for cardiovascular disease. *Curr Cardiovasc Risk Rep.* 2010;4:461–468.

110. Weinreich M, Frishman WH. Antihyperlipidemic therapies targeting PCSK9. *Cardiol Rev.* 2014;22(3): 140–146.

111. Aslibekyan S, Straka RJ, Irvin MR, et al. Pharmacogenomics of high-density lipoprotein-cholesterol-raising therapies. *Expert Rev Cardiovasc Ther.* 2013; 11:355–364.

Genetic Applications in Coronary Artery Disease

Stephen Pan, Juyong Brian Kim, Nehal N. Mehta, and Joshua W. Knowles

TAKE HOME POINTS

1. A tier-based system is a useful framework for evaluating the utility of genetic testing for coronary artery disease (CAD): examples include genetic testing for familial hypercholesterolemia (FH) in at-risk individuals (Tier 1—recommended for clinical use); pharmacogenomics testing for *CYP2C19* variants to predict clopidogrel response (Tier 2—holds future promise for clinical utility, but not ready for widespread use); and genetic testing of common variants (eg, 9p21) for the prediction of CAD risk (Tier 3—no current clinical utility/validity).
2. In genome-wide association studies (GWAS), the most strongly associated variant with CAD has been the locus on chromosome 9p21, which has been identified across a variety of race/ethnic groups. However, the 9p21 locus resides within a gene desert, and the mechanism by which the 9p21 locus is associated with CAD is currently an area of active investigation.
3. Although more than 50 genetic variants have been found to be associated with common forms of CAD, the current utility of genetic testing for predicting CAD risk in the general population does not exceed existing risk models (such as Framingham) based on traditional risk factors.

It has long been recognized that there is a strong familial component to the risk of coronary artery disease (CAD) (1). Indeed, some risk prediction models have demonstrated an improvement when family history of early onset CAD is added to traditional risk factors (2). While some of this familial risk is due to environmental factors that cluster within families, twin and familial aggregation studies clearly highlighted that much of the familial risk is genetic (3–5).

However, the exact genetic markers and mechanisms by which these genetic factors influence the risk of CAD remained largely unknown until recent years. While some of these genetic risk factors were first uncovered through small family-based linkage studies, the recent explosion of genetic data that has occurred since the completion of the first draft of the human genome has allowed large-scale association studies to be performed looking at CAD and associated risk factors such as dyslipidemia, hypertension, and diabetes mellitus (DM). Follow-up studies focusing on the biology underlying the associations of these genetic markers are now starting to produce new pharmacological targets that are poised to revolutionize our treatment of CAD and its risk factors. In addition, genetic studies are shedding light on the genes and pathways that control the response (and potential toxicity) from therapeutic agents, often referred to as "pharmacogenomics." Furthermore, the recent technological advances that have enabled the rapid sequencing of entire human exomes (the portion of DNA that codes for proteins) as well as entire human genomes (the entirety of the DNA sequence) have made the risk prediction of CAD incorporating genetic information a likely future reality.

The Centers for Disease Control and Prevention (CDC) recently established a classification for the use of genetic information based on three tiers, ranked by level of evidence for use (Table 20.1) (6). Tier 1 applications are currently recommended for clinical use based on a systematic review of the evidence and having clear clinical validity and utility. Tier 2 applications were found to have demonstrated clinical validity but at this time still need further evidence for clinical utility on a widespread basis. Tier 3 applications were judged to have sufficient evidence at this time to recommend against widespread clinical use, but could be considered in research applications. In this chapter, we use this framework to describe the current clinical applications of genetic testing available in the setting of CAD, as well as some of the exciting applications that will be available in the foreseeable future.

CASE STUDY #1

Susan is a 35-year-old mother of two. Her father passed away at the age of 45 from a sudden heart attack. She has three brothers and a sister, ages 28 to 40. With the recent birth of her second child, she is worried about her future and comes to you asking if early death due to a sudden heart attack could happen to her or her siblings. A screening lipid panel shows a low-density lipoprotein (LDL) of 200 mg/dL. Physical examination reveals corneal arcus.

TIER I: GENETIC TESTING IN FAMILIAL HYPERCHOLESTEROLEMIA

Susan is a classic example of someone who may suffer from one of the most prevalent inherited cardiovascular conditions, FH. FH is an autosomal codominant condition caused by mutations in genes involved in LDL metabolism and leading to severe elevations in LDL starting in utero. Individuals who inherit FH mutations from both parents (homozygous FH) are rare (prevalence 1 in 1 million) but suffer severe vascular disease by age 20 due to extraordinarily high LDL cholesterol levels (usually above 500 mg/dL) (7).

However, even in the much more common situation in which an FH mutation is inherited from one parent (heterozygous FH), the lifelong elevations in LDL (adult LDL levels average 220 mg/dL) lead to a 20-fold increased lifetime risk of CAD. Approximately 50% of untreated FH males and 30% of untreated FH females will suffer a coronary event before the age of 50 and 60, respectively (8). The prevalence worldwide is now estimated at 1 in 300 to 500 individuals, and is as high as 1 in 100 in some populations (9). The high prevalence and its very notable clinical impact make FH a significant public health problem.

The genetic basis of FH has been extensively studied. Utilizing linkage analysis in families with FH, early studies identified locus coding for the LDL receptor (*LDLR*) as causal for FH, work for which was part of the rationale for Brown and Goldstein's Nobel Prize in 1985 (10,11). Later, apolipoprotein B (*APOB*) was also identified as a susceptibility locus

TABLE 20.1 Genomic Tests and Family History Applications by Level of Evidence Related to CAD

	Clinical Application	Uses
Tier I: Recommended for clinical use	Familial hypercholesterolemia	Cascade cholesterol and/or genetic testing
Good evidence for utility and validity	Family history of premature CAD	Risk stratification and early cholesterol screening
Tier 2: Promise for clinical utility	Pharmacogenomic testing	Personalized drug therapies, avoidance of side effects
Not enough evidence yet for broad use	Example: CYP2C19 variants for clopidogrel response	
Could be useful on case-by-case basis		
Tier 3: No current clinical utility/validity	Common genetic risk factors for CAD	Not currently recommended for use in risk stratification
Broad use not currently recommended	Example: 9p21 locus variants	
May have role in research applications		

for FH (12,13), which accounts for a significant minority of FH in Northern European populations.

More recently, a third locus was identified in the proprotein convertase subtilisin/kexin 9 (*PCSK9*) gene, and using positional cloning, several gain-of-function variants were identified as likely causative of this condition (14). Interestingly, a separate group identified rare loss-of-function mutations in this same gene in African Americans that appeared to be protective for CAD (15). *PCSK9* has been found to bind to the LDL receptor, leading to decreased number of LDL receptors on the hepatocyte cell surface and higher plasma levels of LDL. These genetic studies spurred further studies into the effects of PCSK9 regulation, resulting in the development of inhibitors of PCSK9 that drastically lower LDL levels in mice and primates (16). Studies in patients affected by FH using an antibody to PCSK9 showed great efficacy in lowering LDL levels (17), and this drug is now in late stage clinical trials in patients with FH, a true success story of "base pairs to bedside" medicine.

Genetic testing for FH is widely available. The yield of genetic testing, which now currently focuses on the sequencing (or partial sequencing) of the three genes mentioned (*LDLR*, *APOB*, and *PCSK9*) is thought to be as high as 70% to 80% in patients with a high likelihood of the diagnosis (18). While genetic testing is often not necessary to make the diagnosis of FH in an individual patient, there are several reasons to strongly consider genetic testing, especially for cases where a clinical diagnosis is uncertain (19).

The rationale for the use of genetic testing centers on the "overlap" problem. While FH results in dramatic elevations of LDL cholesterol, there is some overlap in LDL levels with those of the general population (Figure 20.1) (20,21). The discriminatory ability of a lipid panel to distinguish FH is best during childhood but falls over time as environmental exposures cause LDL elevations in the general population. It is important to be able to distinguish true FH from "garden variety" hyperlipidemia for two reasons. First, FH patients have higher risk of CAD due to lifelong exposures to elevated LDL. Second, and perhaps more important, is that the incorporation of genetic testing into "cascade-screening" algorithms has been definitively shown to improve case identification of FH patients in a highly cost-effective manner. Cascade screening is the systematic screening of potentially affected family members of an index case, which is crucial to early identification of FH patients.

In the Netherlands, a national cascade-screening program incorporating genetic testing has resulted in the identification of approximately 75% of all FH cases in the country (over 27,000)—cases with a cost per year of life saved of approximately 8700 € or $14,000 USD. (22). The National Institute for Health and Care Excellence (NICE) in the United Kingdom also recommend the use of genetic testing as part of cascade-screening efforts based on a combination of efficacy and cost effectiveness (18). In the United States, the cost of genetic testing in an index case for FH is approximately $800 to $1300, making it often

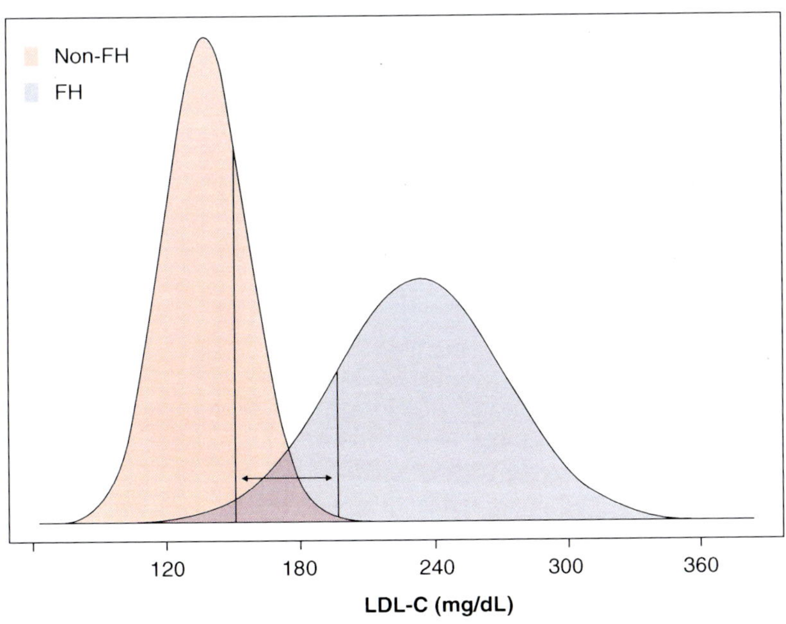

FIGURE 20.1 Distribution of low-density lipoprotein (LDL) levels in patients with and without familial hyperlipidemia (FH). The double arrow refers to a range of significant overlap in LDL levels between patients with and without FH, in which genetic testing may be particularly useful to make a diagnosis.

Source: Adapted from Ref. (20). Starr B, Hadfield SG, Hutten BA, et al. Development of sensitive and specific age- and gender-specific low-density lipoprotein cholesterol cutoffs for diagnosis of first-degree relatives with familial hypercholesterolaemia in cascade testing. *Clin Chem Lab Med*. 2008;46(6):791–803.

prohibitively expensive. However, once the causal mutation is identified the cost drops dramatically (approximately $200 for subsequent mutation testing in relatives). Therefore, the argument for using genetic testing as part of a cascade-screening algorithm is strong. Based on the evidence from the Netherlands and from the United Kingdom, the CDC has classified FH as a Tier 1 condition indicating a high level of evidence for the systematic application of cascade screening (potentially including genetic testing) (6).

In the case of Susan, her lipid levels, physical exam, and family history are suspicious for (but not diagnostic of) FH. Genetic testing could be used to confirm her diagnosis of FH. However, perhaps the more useful aspect of genetic testing for this patient would be for screening potentially affected family members including her children and her siblings (and potentially her aunts, uncles, nieces, and nephews). The benefit would be early case identification and steering affected relatives to aggressive lipid lowering therapy to prevent high morbidity and mortality.

CASE STUDY #2

John is a 65-year-old man with a history of diabetes mellitus and hypertension who comes to your office for the first time. He is concerned because his primary care physician recently told him his cholesterol was a bit high; in addition, he has been having occasional, brief episodes of chest discomfort after exercising, which he thought was just rib pain but now he is not so sure. He undergoes a stress echocardiogram during which he had chest pain again and the results were notable for stress-induced wall motion abnormalities. He agrees to undergo a cardiac catheterization the next day. He also agrees to start on statin for his cholesterol. On his way out of your clinic, he mentions that his wife bought him a direct-to-consumer (DTC) genetic testing panel last year and he remembers something about having a variant that puts him at increased risk for muscle pain with cholesterol medication, and that he was a "nonresponder" to clopidogrel. He asks for your opinion regarding these genetic results and whether he should receive an alternate medication other than clopidogrel if he gets a coronary stent.

TIER 2 APPLICATIONS: PHARMACOGENOMICS

The CDC defines Tier 2 applications of genomic and family history information as those that "hold promise for clinical utility but evidence-based panels have not examined their use or found insufficient evidence for their use...Such applications may provide information for informed decision making by providers and patients." Many of these applications are the direct results of recent genomic studies, such as GWAS, which are poised to be impactful on general clinical care but are in need of further proof of their clinical utility through larger trials. While these applications may not yet be "ready for prime time," they might provide information that, in the correct clinical context, could influence patient care.

One of the most promising of these applications in the cardiovascular realm is pharmacogenomics, the use of genetic and genomic information to guide in the use of particular drugs, including but not limited to predicting optimal dosage, therapeutic efficacy, and side effects. Unlike most of the gene variants for common diseases including CAD, pharmacogenomic variants generally have a larger effect size and therefore impact on predicting response to their associated drugs, making both discovery and clinical application more tractable.

One recent, illustrative example is the value of pharmacogenetic information to guide the use of statins. While the utility of statins in secondary prevention is undoubted (23), there is still a substantial minority of patients for which statins have troublesome side effects of which (by far) the most common are myalgias. Unfortunately, there are no available clinical factors that reliably predict statin-induced side effects in patients, and statin-induced adverse reactions are generally only discovered after a trial of a statin medication. Too often this leads to a patient who is labeled as "statin-intolerant" and this class of drugs and its myriad of benefits are taken off the table as an option for management, despite the availability of other statin formulations that may well be tolerated.

For these reasons, there has been great interest in trying to use pharmacogenetic factors to predict who may be more susceptible to statin side effects. One such effort was the Study of the Effectiveness of Additional Reductions in Cholesterol and Homocysteine (SEARCH), comprising a GWAS of 85 patients of European descent with statin-associated myopathy along with 90 matched controls. This study was able to identify a coding variant (rs4149056, 12% minor allele frequency [MAF] in Caucasians) in *SLCO1B1*, an organic anion transporter found in the liver, that conferred an 18% cumulative risk of myopathy within the first year of therapy with simvastatin, one of the most commonly used statin drugs, in those homozygous for the variant, in comparison to 3% risk in heterozygotes and 0.6% in people without the variant (24). While these findings are not as of yet used routinely in clinical practice, this genetic variant remains the most powerful predictor of an adverse event prior to prescribing a statin. Whether this variant specifically predicts adverse reactions to simvastatin alone or is generalizable to the entire statin class of drugs is still unclear. Some small studies have found that variants in *SLCO1B1* may affect the pharmacokinetics of specific statin formulations differently (25,26). These data could make genotyping of the *SLCO1B1* actionable and in fact, the Clinical Pharmacogenomics Implementation Consortium (CPIC) has released guidelines that suggest a dosing algorithm if the genotype at rs4149056 in *SLCO1B1* is known, recommending a lower dose of simvastatin or an alternate statin in those with the causal genetic variant (27).

In addition, genomics has also been used to find those who also have the potential to respond best to statin therapy. By using the genetic data from the JUPITER trial, Chasman et al used the well-established GWAS method to find single nucleotide polymorphisms (SNPs) associated with larger reductions in LDL cholesterol in those already on a statin (28). They were successful in confirming previously identified associations within the *APOE* gene but also identified promising variants not surprisingly in *ABCG2*, *LPA*, and *PCSK9*, all genes that are intimately involved in the metabolism of cholesterol and/or the statin molecule. One could envision using information regarding individual variants in these genes to determine a risk–benefit ratio prior to initiating statin therapy, which could then be used to guide symptom and lab monitoring after starting statin therapy, as well as possibly to guide the selection of which statin to use.

Another active area of pharmacogenomics investigation in the realm of CAD is clopidogrel, an antiplatelet agent that is (among other things) a mainstay of therapy after percutaneous coronary intervention (PCI). A significant subset of patients manifest some aspect of clopidogrel resistance, estimated in one study to affect up to 25% of all patients presenting with ST-segment elevation myocardial infarction (STEMI)(29). Further mechanistic studies have found that a rate-limiting step in the conversion of clopidogrel, a pro-drug, to its active form is dependent on *CYP2C19*, and variants in this gene can substantially decrease the antiplatelet activity of the drug (30,31). From a clinical standpoint, the Study of Platelet Inhibition and Patient Outcomes (PLATO) trial found those on clopidogrel who had any loss-of-function variant in *CYP2C19* were noted to have an increased rate of cardiovascular death, myocardial infarction (MI), or stroke compared to those who did not have any loss-of-function variants (5.7% vs 3.8%, respectively) (32). In addition, an analysis of genetic data from the Clopidogrel in Unstable Angina to Prevent Recurrent Events (CURE) trial and the Atrial Fibrillation Clopidogrel Trial with Irbesartan

for Prevention of Vascular Events (ACTIVE)-A found that those with gain-of-function mutations in *CYP2C19* actually did better on clopidogrel with fewer cardiovascular events when compared to noncarriers, although those with loss-of-function variants did not appear to do any worse than noncarriers (33).

In an attempt to bring these findings into the realm of clinical care, the RAPID-GENE trial looked at the feasibility of a point-of-care genetic test for common variants in *CYP2C19* loss-of-function alleles, designed to be utilized for patients about to undergo PCI for acute coronary syndrome (ACS) or stable angina, using clopidogrel for those without variation in *CYP2C19* and prasugrel for those found to have loss-of-function variants in one arm as opposed to standard care (no gene testing, clopidogrel for all) in the other arm. In this study of 187 patients with follow-up, the point-of-care test performed admirably with 100% sensitivity and 99.4% specificity, as would be predicted for a genetic test, and 30% of patients in the standard care arm were noted to still have high platelet reactivity on testing as compared to none of those in the gene testing arm. The study was not powered to find a difference in clinical events between the two groups; however, the results show how genetic testing could quite feasibly be incorporated into clinical care in an efficient manner. Further studies are needed at this point to determine if pharmacogenetics in the use of clopidogrel can actually change the clinical outcomes of those receiving PCI.

Regarding John, the patient described in Case Study #2, who is preparing to undergo PCI with a DTC genetic testing panel suggesting variants predicting increased risk of statin adverse effects as well as resistance to clopidogrel, one could certainly consider taking this information into account regarding his management post-PCI. The one caveat is that the information from DTC genetic testing panels at this time for the most part is *not* approved for clinical use by the Food and Drug Administration, so the information may need to be verified by an independent clinically certified laboratory. However, if these variants were verified, one could certainly consider using an alternative antiplatelet agent such as prasugrel or ticagrelor instead of clopidogrel if resistance was a concern and if the patient had no other clinical factors suggesting increased bleeding risk. In addition, one could also consider avoiding the use of simvastatin in John, instead opting to use a statin without a known *SLCO1B1* interaction. Whether these management strategies incorporating pharmacogenetic information will effectively reduce adverse events remains under study, with several centers currently performing randomized controlled trials in this arena.

CASE STUDY #3

Thomas, a 40-year-old researcher with a family history of heart disease comes to your clinic with genome-wide genotyping results of his own DNA that was run by a CLIA-certified laboratory. He has looked at the data himself and cites an article from *Nature Genetics*, according to which he has 10 polymorphisms that are associated with increased risk of coronary disease. He would like to know his future chances of having a heart attack.

TIER 3 APPLICATIONS: CAD GENETIC RISK FACTORS

Through the Herculean efforts of the Human Genome Project, the HapMap Project, and other subsequent genomic projects, millions of common SNPs were identified. Circa 2007, with the development of high-density arrays, it became possible to perform GWAS for complex traits by genotyping hundreds of thousands of SNPs simultaneously. Since 2007, GWAS have identified many SNPs that are associated with CAD and associated risk factors (34–37). More recently, multiple groups have collaborated to perform meta-analyses of GWAS on CAD pooling more than 100,000 subjects of European descent (38) making the total number of associated SNPs greater than 50 (Figure 20.2).

The effect sizes for the SNPs associated with CAD are modest at best, with most risk alleles conferring an odds ratio of 1.05 to 1.20 range. The identification of several well-known lipid genes as CAD susceptibility loci, for example, *LDLR*, *PCSK9*, and *LPA*, supported GWAS as a viable approach to identify mechanisms of pathogenesis in common, complex diseases. Interestingly, most of the loci have no association with known risk factors or pathways previously connected to atherosclerosis, suggesting that there is still much to learn about the disease process.

In these GWAS, the most strongly associated variant with CAD has been the locus on chromosome 9p21. The locus has since been found to be generalizable to different ethnicities, and also showed associations with abdominal and intracranial aneurysms (35,39–42). The effect size of these 9p21 variants, like most other genetic variants implicated with CAD, is however modest. In one meta-analysis of GWAS for CAD, pooled odds ratios were 1.29 for one disease risk allele and 1.67 for two disease risk alleles (43). These odds ratios were independent of other known risk factors, thus further supporting the potential predictive power of genetic factors to add to our current CAD risk prediction algorithms. Nevertheless, the incremental utility for risk prediction using these genetic factors is currently limited.

The mechanism behind how the 9p21 locus affects global risk of CAD is still under investigation. These variants reside in a region referred to as a gene desert

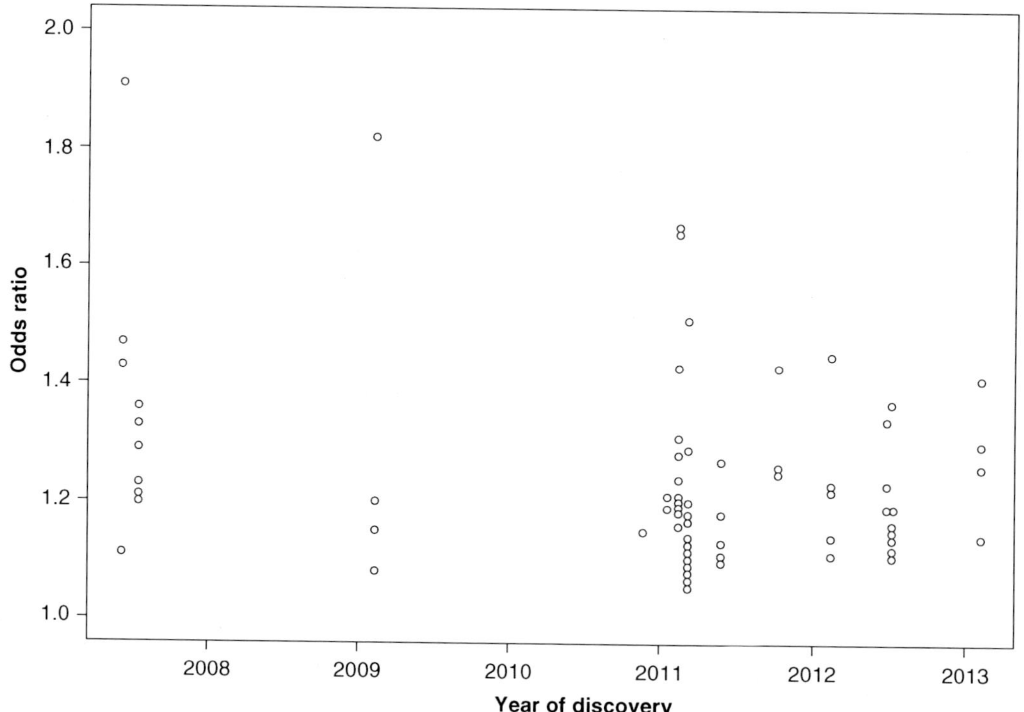

FIGURE 20.2 Data from the National Human Genome Research Institute (NHGRI) GWAS catalog showing number and odds ratios for CAD susceptibility variants identified during the GWAS era.

that is distant from known coding regions, with the nearby genes being the tumor suppression genes *CDKN2A, CDKN2B,* and *MTAP.* Several investigations have used *in vitro* studies to show differences in smooth muscle cell proliferation between carriers and noncarriers of risk alleles, and also with knockdown of *CDKN2A* and *CDKN2B* genes (44,45). Subsequently, studies using hyperlipidemic mice showed that alterations of *CDKN2A, CDKN2B,* and *MTAP* genes modulate atherosclerosis (46–48). It has also been suggested that these variants may act via controlling expression of genes located distally such as *STAT1* (49), or via changes in expression of a noncoding ribonucleic acid (RNA) named *ANRIL,* which also lies within the 9p21 region and may itself regulate expression of *CDKN2A* and *CDKN2B* (44,50). The latter theory has been strengthened by recent efforts to explore the 9p21 region further both through resequencing in select individuals followed by genotyping in a larger sample of patients from the Framingham Heart Study (51). These efforts resulted in the finding that several relatively common genetic variants within this region were linked to CAD risk and also were linked with the expression of *ANRIL* and *CDKN2B,* although interestingly those variants with the strongest association with CAD risk were not those that were most strongly associated with altered expression of *ANRIL* and *CDKN2B* in leukocytes, raising the possibility of tissue-specific effects of these variants.

The genetics approach has also proved to provide elegant answers to important biological questions that had previously been elusive. For instance, significant epidemiological data have suggested that low high-density lipoprotein (HDL) is associated with an increased risk of CAD. However, a causal relationship of low HDL to CAD remained unknown. Importantly, GWAS did not seem to implicate variants associated with low HDL levels, such as the cholesterol ester transfer protein (*CETP*), with CAD risk as was originally expected. Subsequently, the large amount of available genetic data empowered scientists to perform a "Mendelian randomization" study to answer questions related to causality of associations. Mendelian randomization can be thought of as a "natural" randomized control trial, using common genetic polymorphisms with effects that mimic those produced by modifiable exposures (eg, HDL), obtaining unbiased estimates of the effects of a putative causal variable without conducting a traditional randomized trial. Voight et al showed that when comparing a large cohort with variants in genes that increase HDL level, including the endothelial lipase gene (*LIPG*), with those without the variant, the decrease in HDL level in the group with risk variant was not associated with increased risk of MI (52). These results support the findings from recent phase 3 clinical trials involving CETP inhibitors or niacin, which found that raising HDL did not result in improved outcomes (53–55), raising doubt in the role of low HDL as a causal factor for CAD.

In terms of risk prediction, the common variants discovered related to CAD have not currently been shown to be of major clinical benefit. Some of the major limitations are related to the small effect size of the variations and the difficulty in uncovering the "causal" variant responsible for the statistical association (Figure 20.3). However, perhaps the biggest issue

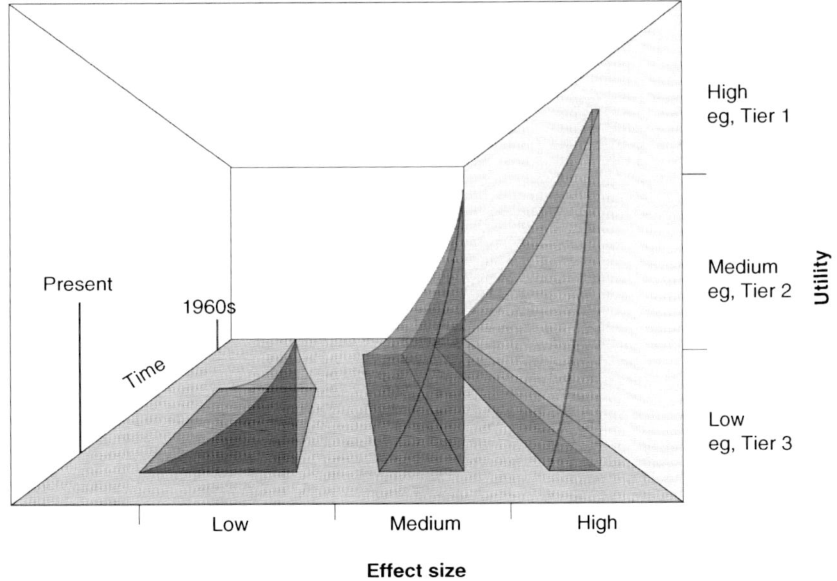

FIGURE 20.3 Schematic representation of susceptibility variant discovery over time as a function of effect size and how these variants translate in terms of clinical utility. In general, larger effect genes discovered many years ago (typically through linkage studies) are more commonly used in clinical settings. Large effect genetic variants are being more frequently discovered now with whole-exome- and whole-genome-sequencing approaches. Many genetic variants of low effect size have been found in the GWAS era but these are seldom used clinically.

relates to the fact that so far, despite massive studies, we have still only identified a fraction of the genetic variation underlying CAD. This quandary has been referred to as the problem of the *missing heritability*. Based on twin studies, the genetic heritability of CAD (the percentage of the phenotypic variation that can be attributed to genetic variation) has been estimated to be somewhere in the range of 40% to 60% (4,56). Altogether, the approximately 50 SNPs identified to be associated with CAD explain only around 10% of CAD heritability (38). Recent findings suggest that the degree of epistasis (ie, nonadditive form of genetic contribution or gene–gene interactions) is more profound than previously thought, and that the current estimate of heritability, predicted based on strictly additive models, may have been overestimated.

A component of the missing heritability may be attributed to the effect of gene–environment (GxE) interaction. GxE interaction, that is, context-dependent changes in the expression of genes and their regulatory networks, has been well documented in yeast studies (eg, in response to varying temperature, drug treatments, and so on) and in different model organisms (57,58). The widespread GxE interactions, and the difficulty of assessing such environmental factors in human studies, likely reduce the power to detect important susceptibility variants in GWAS and partly explain the marginal effect size of these polymorphisms.

A supportive finding is the observation that diet modulates the effect size of the genetic polymorphisms. In the INTERHEART study looking at approximately 8000 patients, Do et al found that the 9p21 polymorphism had lower effect size among patients on diets high on fruits, vegetables, nuts, desserts, and berries (59). The authors subsequently validated this finding with a larger cohort of 19,000 patients from the FINRISK study, demonstrating that the deleterious effect of genetic influence can be increased with an unhealthy lifestyle.

Another component of the missing heritability is thought to reside within the rare (1%–5% MAF) and very rare variants (less than 1% MAF) not included in the GWAS. With the increasingly lower costs of more in-depth sequencing, the search is on for the missing heritability by surveying rare variants. The National Heart, Lung, and Blood Institute (NHLBI)-funded exome-sequencing projects in early MI and hyperlipidemia are both near completion, and the U.S.-based 1000 Genomes Project has been completed. Furthermore, now the U.K. 10,000 Genomes Project and the Million Veterans Program in the United States are also under way, which will increase the power of whole-genome-sequencing (WGS) association studies, as well as build a complete genomic index of all variants, including SNPs, insertions, deletions, and repeats.

No matter what the explanation, it is clear that a large component of CAD heritability remains unexplained, which may explain why the incorporation of CAD risk variants thus far has had minimal effects on risk models based on traditional risk factors (60). It is possible that this situation will improve as we discover more about the genetic basis of CAD.

Clinical Utility of GWAS Results— Incorporating Genetic Data Into Patient Care

What should be done with all the data from these GWAS? While these studies have highlighted many new potential gene targets and pathways for research, applying these results to predict an individual patient's disease risk has not entered mainstream clinics. For instance:

- Will a SNP associated with CAD in individuals of European descent be of utility in an African American patient? While some loci such as 9p21 seem to cross races/ethnicities, other loci have not been reproduced as easily in more diverse cohorts, while some have not been reproduced at all even in cohorts of the same race/ethnicity.
- Can we improve on existing risk prediction or reclassification? As discussed, the small effect size of the individual variants limits their clinical utility in improving risk classification above and beyond traditional risk prediction models (61,62). An early study showed that genetic scores based on a 13-SNP model had predictive ability comparable (but not superior) to traditional risk factor models (60).

Despite the challenges, many genotyping services, including those offered by several DTC services such as 23andme, Navigenics, and deCODEme, have incorporated GWAS results for many cardiovascular phenotypes such as CAD and hypertension into their reports. There continues to be a need, however, to study the impact genotyping tests can have on actual patient care in a prospective fashion. One study by Bloss and colleagues has shown that the use of DTC genotyping did not seem to increase any disease or test-related anxiety to a significant degree, although the results of these tests did not seem to positively affect patient behavior to counteract any additional disease risk inferred from these tests (63). There are now pilot projects at several centers evaluating clinical outcomes and patient satisfaction after receiving genotyping information. Recently, we and others initiated a clinical trial looking at the effect of patients' knowledge of genotype on their health-promoting or harming behaviors (64).

Genotyping tests are quickly becoming widely available, and physicians will need to be well equipped to understand how to use these genetic data in clinical practice. As with all testing, before the findings are incorporated into routine clinical care, widespread use of genetic testing should be subjected to the same scrutiny as we apply to other forms of testing and should be buttressed when possible by robust clinical trial data supporting their use.

As for Thomas, our researcher with his own genetic data including variants associated with CAD, we would predict his overall future risk of CAD using well-defined risk stratification scoring systems such as the Framingham Risk Score or the Reynolds Risk Score. While the hope is that one day genotyping data could be incorporated into these risk scoring systems with improvement in their sensitivity and specificity in predicting CAD events, as mentioned previously much more work has to be done before the use of such genetic data in predicting CAD can be recommended widely.

FUTURE DIRECTIONS

The ongoing massively parallel sequencing of exomes and whole genomes will identify ever more increasing number of rare variants that are related to CAD extending our knowledge of the heritable risk. Whole-genome and exome-sequencing approaches have already proven to be valuable for "medical mysteries," including certain forms of cancer and some Mendelian disorders (65–69). However, currently there are no data to support these approaches for common, complex disease.

Undoubtedly, larger whole-exome and whole-genome-sequencing approaches will allow the identification of the causal CAD disease genes and variants. However, functional characterization of these genes and risk variants are important for these findings to reach their full potential. Similarly, as we increase our knowledge of the genetic signature of heritability, we may be able to develop better risk prediction models, with the hope of moving closer toward the goal of personalized preventive medicine.

System genetics is another approach to understand the mechanism in complex traits, utilizing multiple layers of intermediate molecular phenotypes (eg, RNA, protein, and blood metabolites) in addition to clinical phenotypes. Several large studies are currently ongoing, with goals of creating large system-level biobanks and datasets as well as systematic approaches to investigating the effect of modifying the variants or genes on phenotypes using in vitro systems and mice models.

A point to consider is that the genomic data (especially whole-genome and whole-exome-sequencing data) will produce many incidental findings. These incidental findings may cause confusion and psychosocial harms without improving health outcomes (70,71) (see Chapter 10). Determining which incidental findings should be reported to the patients from the genotyping and/or sequencing studies is likely impossible for most practitioners. For the past 8 years, an independent working group on the Evaluation in Genomic Applications in Practice and Prevention (EGAPP) has been developing evidence-based recommendations on the use of genomic tests in clinical practice and disease prevention (72). The goal would be to systematically curate a list of highly penetrant, actionable gene findings that should be reported back to patients and physicians for further workup.

It is also unclear how much time and effort will be required for routine analysis of whole-genome data. While there are some recent data indicating that this would be feasible (69,70,73,74), the expertise for this process is not available to the general practitioner and data analyses and reports would need to be highly automated, preferably with easy-to-understand decision support tools. And ultimately, there will be a need to do cost-effectiveness analyses to determine whether the widespread use of genetic information will reduce health care expenditures and improve outcomes.

CONCLUSIONS

In summary, we find that the Tier 1, 2, 3 model created by the CDC Office of Public Health Genomics provides a valuable framework for thinking about the current and future applications of genetic testing in CAD. There are clearly areas where genetic testing is useful and cost effective. Other areas are extremely promising but additional research is needed if we are to apply genomic testing to large numbers of patients with, or at risk for, CAD.

REFERENCES

1. Friedlander Y, Siscovick DS, Weinmann S, et al. Family history as a risk factor for primary cardiac arrest. *Circulation*. 1998;97(2):155–160.
2. Ridker PM, Buring JE, Rifai N, Cook NR. Development and validation of improved algorithms for the assessment of global cardiovascular risk in women: the Reynolds Risk Score. *JAMA*. 2007;297(6):611–619.
3. Feinleib M, Garrison R, Fabsitz R, et al. The NHLBI twin study of cardiovascular disease risk factors: methodology and summary of results. *Am J Epidemiol*. 1977;106(4):284–285.

4. Marenberg ME, Risch N, Berkman LF, et al. Genetic susceptibility to death from coronary heart disease in a study of twins. *N Engl J Med.* 1994;330(15):1041–106.

5. Myers RH, Kiely DK, Cupples LA, Kannel WB. Parental history is an independent risk factor for coronary artery disease: the Framingham Study. *Am Heart J.* 1990;120(4):963–969.

6. Khoury MJ, Coates RJ, Evans JP. Evidence-based classification of recommendations on use of genomic tests in clinical practice: dealing with insufficient evidence. *Genet Med.* 2010;12(11):680–683.

7. Nordestgaard BG, Chapman MJ, Humphries SE, et al. Familial hypercholesterolaemia is underdiagnosed and undertreated in the general population: guidance for clinicians to prevent coronary heart disease: Consensus Statement of the European Atherosclerosis Society. *Eur Heart J.* 2013;34(45):3478–3490.

8. Marks D, Thorogood M, Neil HAW, Humphries SE. A review on the diagnosis, natural history, and treatment of familial hypercholesterolaemia. *Atherosclerosis.* 2003;168(1):1–14.

9. Hopkins PN, Toth PP, Ballantyne CM, Rader DJ; National Lipid Association Expert Panel on Familial Hypercholesterolemia. Familial hypercholesterolemias: prevalence, genetics, diagnosis and screening recommendations from the National Lipid Association Expert Panel on Familial Hypercholesterolemia. *J Clin Lipidol.* 2011;5(3)(Suppl):S9–S17.

10. Hobbs HH, Brown MS, Russell DW, et al. Deletion in the gene for the low-density-lipoprotein receptor in a majority of French Canadians with familial hypercholesterolemia. *N Engl J Med.* 1987;317(12):734–737.

11. Hobbs HH, Brown MS, Goldstein JL. Molecular genetics of the LDL receptor gene in familial hypercholesterolemia. *Hum Mutat.* 1992;1(6):445–466.

12. Soria LF, Ludwig EH, Clarke HR, et al. Association between a specific apolipoprotein B mutation and familial defective apolipoprotein B-100. *Proc Natl Acad Sci USA.* 1989;86(2):587–591.

13. Innerarity TL, Mahley RW, Weisgraber KH, et al. Familial defective apolipoprotein B-100: a mutation of apolipoprotein B that causes hypercholesterolemia. *J Lipid Res.* 1990;31(8):1337–1349.

14. Abifadel M, Varret M, Rabès J-P, et al. Mutations in PCSK9 cause autosomal dominant hypercholesterolemia. *Nat Genet.* 2003;34(2):154–156.

15. Cohen J, Pertsemlidis A, Kotowski IK, et al. Low LDL cholesterol in individuals of African descent resulting from frequent nonsense mutations in PCSK9. *Nat Genet.* 2005;37(2):161–165.

16. Chan JC, Piper DE, Cao Q, et al. A proprotein convertase subtilisin/kexin type 9 neutralizing antibody reduces serum cholesterol in mice and nonhuman primates. *Proc Natl Acad Sci USA.* 2009;106(24):9820–9825.

17. Stein EA, Mellis S, Yancopoulos GD, et al. Effect of a monoclonal antibody to PCSK9 on LDL cholesterol. *N Engl J Med.* 2012;366(12):1108–1118.

18. DeMott K, Nherera L, Shaw EJ, Minhas R, Humphries SE et al. Clinical Guidelines and Evidence Review for Familial hypercholesterolaemia: the identification and management of adults and children with familial hypercholesterolaemia. 2008. London: National Collaborating Centre for Primary Care and Royal College of General Practitioners.

19. Goldberg AC, Hopkins PN, Toth PP, et al. Familial hypercholesterolemia: screening, diagnosis and management of pediatric and adult patients: clinical guidance from the National Lipid Association Expert Panel on Familial Hypercholesterolemia. *J Clin Lipidol.* 2011;5(3):133–140.

20. Starr B, Hadfield SG, Hutten BA, et al. Development of sensitive and specific age- and gender-specific low-density lipoprotein cholesterol cutoffs for diagnosis of first-degree relatives with familial hypercholesterolaemia in cascade testing. *Clin Chem Lab Med.* 2008; 46(6):791–803.

21. Hopkins PN. Defining the challenges of familial hypercholesterolemia screening: introduction. *J Clin Lipidol.* 2010;4(5):342–345.

22. Wonderling D, Umans-Eckenhausen MAW, Marks D, et al. Cost-effectiveness analysis of the genetic screening program for familial hypercholesterolemia in The Netherlands. *Semin Vasc Med.* 2004;4(1):97–104.

23. Cholesterol Treatment Trialists' (CTT) Collaborators; Mihaylova B, Emberson J, Blackwell L, et al. The effects of lowering LDL cholesterol with statin therapy in people at low risk of vascular disease: meta-analysis of individual data from 27 randomised trials. *Lancet.* 2012;380(9841):581–590.

24. SEARCH Collaborative Group; Link E, Parish S, Armitage J, et al. SLCO1B1 variants and statin-induced myopathy--a genomewide study. *N Engl J Med.* 2008;359(8):789–799.

25. Brunham LR, Lansberg PJ, Zhang L, et al. Differential effect of the rs4149056 variant in SLCO1B1 on myopathy associated with simvastatin and atorvastatin. *Pharmacogenomics J.* 2012;12(3):233–237.

26. Pasanen MK, Fredrikson H, Neuvonen PJ, Niemi M. Different effects of SLCO1B1 polymorphism on the pharmacokinetics of atorvastatin and rosuvastatin. *Clin Pharmacol Ther.* 2007;82(6):726–733.

27. Wilke RA, Ramsey LB, Johnson SG, et al. The clinical pharmacogenomics implementation consortium: CPIC guideline for SLCO1B1 and simvastatin-induced myopathy. *Clin Pharmacol Ther.* 2012;92(1):112–117.

28. Chasman DI, Giulianini F, MacFadyen J, et al. Genetic determinants of statin-induced low-density lipoprotein cholesterol reduction: the Justification for the Use of Statins in Prevention: An Intervention Trial Evaluating Rosuvastatin (JUPITER) Trial. *Circ Cardiovasc Genet.* 2012;5(2):257–264.

29. Matetzky S, Shenkman B, Guetta V, et al. Clopidogrel resistance is associated with increased risk of recurrent atherothrombotic events in patients with acute myocardial infarction. *Circulation.* 2004;109(25):3171–3175.

30. Hulot J-S, Bura A, Villard E, et al. Cytochrome P450 2C19 loss-of-function polymorphism is a major determinant of clopidogrel responsiveness in healthy subjects. *Blood.* 2006;108(7):2244–2247.

31. Simon T, Verstuyft C, Mary-Krause M, et al. Genetic determinants of response to clopidogrel and cardiovascular events. *N Engl J Med.* 2009;360(4):363–375.

32. Wallentin L, James S, Storey RF, et al. Effect of CYP2C19 and ABCB1 single nucleotide polymorphisms on outcomes of treatment with ticagrelor versus clopidogrel for acute coronary syndromes: a genetic substudy of the PLATO trial. *Lancet.* 2010;376(9749): 1320–1328.

33. Paré G, Mehta SR, Yusuf S, et al. Effects of CYP2C19 genotype on outcomes of clopidogrel treatment. *N Engl J Med.* 2010;363(18):1704–1714.

34. McPherson R, Pertsemlidis A, Kavaslar N, et al. A common allele on chromosome 9 associated with coronary heart disease. *Science.* 2007;316(5830):1488–1491.

35. Helgadottir A, Thorleifsson G, Manolescu A, et al. A common variant on chromosome 9p21 affects the risk of myocardial infarction. *Science*. 2007;316(5830):1491–1493.

36. Samani NJ, Erdmann J, Hall AS, et al. Genomewide association analysis of coronary artery disease. *N Engl J Med*. 2007;357(5):443–453.

37. Teslovich TM, Musunuru K, Smith AV, et al. Biological, clinical and population relevance of 95 loci for blood lipids. *Nature*. 2010;466(7307):707–713.

38. CARDIoGRAMplusC4D Consortium; Deloukas P, Kanoni S, Willenborg C, et al. Large-scale association analysis identifies new risk loci for coronary artery disease. *Nat Genet*. 2013;45(1):25–33.

39. Shen G-Q, Li L, Rao S, et al. Four SNPs on Chromosome 9p21 in a South Korean population implicate a genetic locus that confers high cross-race risk for development of coronary artery disease. *Arterioscler Thromb Vasc Biol*. 2008;28(2):360–365.

40. Assimes TL, Knowles JW, Basu A, et al. Susceptibility locus for clinical and subclinical coronary artery disease at chromosome 9p21 in the multi-ethnic ADVANCE study. *Hum Mol Genet*. 2008;17(15):2320–2328.

41. Zhou L, Zhang X, He M, et al. Associations between single nucleotide polymorphisms on chromosome 9p21 and risk of coronary heart disease in Chinese Han Population. *Arterioscler Thromb Vasc Biol*. 2008;28(11):2085–2089.

42. Lettre G, Palmer CD, Young T, et al. Genome-wide association study of coronary heart disease and its risk factors in 8,090 African Americans: the NHLBI CARe Project. *PLoS Genet*. 2011;7(2):e1001300.

43. Schunkert H, Gotz A, Braund P, et al. Repeated replication and a prospective meta-analysis of the association between chromosome 9p21.3 and coronary artery disease. *Circulation*. 2008;117(13):1675–1684.

44. Holdt LM, Beutner F, Scholz M, et al. ANRIL expression is associated with atherosclerosis risk at chromosome 9p21. *Arterioscler Thromb Vasc Biol*. 2010;30(3):620–627.

45. Motterle A, Pu X, Wood H, et al. Functional analyses of coronary artery disease associated variation on chromosome 9p21 in vascular smooth muscle cells. *Hum Mol Genet*. 2012;21(18):4021–4029.

46. Kuo C-L, Murphy AJ, Sayers S, et al. Cdkn2a is an atherosclerosis modifier locus that regulates monocyte/macrophage proliferation. *Arterioscler Thromb Vasc Biol*. 2011;31(11):2483–2492.

47. Kim JB, Deluna A, Mungrue IN, et al. Effect of 9p21.3 coronary artery disease locus neighboring genes on atherosclerosis in mice. *Circulation*. 2012;126(15):1896–1906.

48. Leeper NJ, Raiesdana A, Kojima Y, et al. Loss of CDKN2B promotes p53-dependent smooth muscle cell apoptosis and aneurysm formation. *Arterioscler Thromb Vasc Biol*. 2013;33(1):e1–e10.

49. Harismendy O, Notani D, Song X, et al. 9p21 DNA variants associated with coronary artery disease impair interferon-γ signalling response. *Nature*. 2011;470(7333):264–268.

50. Cunnington MS, Santibanez Koref M, Mayosi BM, et al. Chromosome 9p21 SNPs associated with multiple disease phenotypes correlate with ANRIL expression. *PLoS Genet*. 2010;6(4):e1000899.

51. Johnson AD, Hwang S-J, Voorman A, et al. Resequencing and clinical associations of the 9p21.3 region: a comprehensive investigation in the Framingham heart study. *Circulation*. 2013;127(7):799–810.

52. Voight BF, Peloso GM, Orho-Melander M, et al. Plasma HDL cholesterol and risk of myocardial infarction: a Mendelian randomisation study. *Lancet*. 2012;380(9841):572–580.

53. Barter PJ, Caulfield M, Eriksson M, et al. Effects of torcetrapib in patients at high risk for coronary events. *N Engl J Med*. 2007;357(21):2109–2122.

54. Schwartz GG, Olsson AG, Abt M, et al. Effects of dalcetrapib in patients with a recent acute coronary syndrome. *N Engl J Med*. 2012;367(22):2089–2099.

55. AIM-HIGH Investigators; Boden WE, Probstfield JL, et al. Niacin in patients with low HDL cholesterol levels receiving intensive statin therapy. *N Engl J Med*. 2011;365(24):2255–2267.

56. Lusis AJ. Atherosclerosis. *Nature*. 2000;407(6801):233–241.

57. Mackay TFC, Stone EA, Ayroles JF. The genetics of quantitative traits: challenges and prospects. *Nat Rev Genet*. 2009;10(8):565–577.

58. Romanoski CE, Lee S, Kim MJ, et al. Systems genetics analysis of gene-by-environment interactions in human cells. *Am J Hum Genet*. 2010;86(3):399–410.

59. Do R, Xie C, Zhang X, et al. The effect of chromosome 9p21 variants on cardiovascular disease may be modified by dietary intake: evidence from a case/control and a prospective study. *Plos Med*. 2011;8(10):e1001106.

60. Ripatti S, Tikkanen E, Orho-Melander M, et al. A multilocus genetic risk score for coronary heart disease: case-control and prospective cohort analyses. *Lancet*. 2010;376(9750):1393–1400.

61. Paynter NP, Chasman DI, Buring JE, et al. Cardiovascular disease risk prediction with and without knowledge of genetic variation at chromosome 9p21.3. *Ann Intern Med*. 2009;150(2):65–72.

62. Brautbar A, Ballantyne CM, Lawson K, et al. Impact of adding a single allele in the 9p21 locus to traditional risk factors on reclassification of coronary heart disease risk and implications for lipid-modifying therapy in the Atherosclerosis Risk in Communities study. *Circ Cardiovasc Genet*. 2009;2(3):279–285.

63. Bloss CS, Schork NJ, Topol EJ. Effect of direct-to-consumer genomewide profiling to assess disease risk. *N Engl J Med*. 2011;364(6):524–534.

64. Knowles JW, Assimes TL, Kiernan M, et al. Randomized trial of personal genomics for preventive cardiology: design and challenges. *Circ Cardiovasc Genet*. 2012;5(3):368–376.

65. Ng SB, Buckingham KJ, Lee C, et al. Exome sequencing identifies the cause of a mendelian disorder. *Nat Genet*. 2009;42(1):30–35.

66. Worthey EA, Mayer AN, Syverson GD, et al. Making a definitive diagnosis: Successful clinical application of whole exome sequencing in a child with intractable inflammatory bowel disease. *Genet Med*. 2011;13(3):255–262.

67. Lupski JR, Reid JG, Gonzaga-Jauregui C, et al. Whole-genome sequencing in a patient with Charcot–Marie–Tooth neuropathy. *N Engl J Med*. 2010;362(13):1181–1191.

68. Musunuru K, Pirruccello JP, Do R, et al. Exome sequencing, ANGPTL3 mutations, and familial combined hypolipidemia. *N Engl J Med*. 2010;363(23):2220–2227.

69. Yang Y, Muzny DM, Reid JG, et al. Clinical whole-exome sequencing for the diagnosis of Mendelian disorders. *N Engl J Med*. 2013;369(16):1502–1511.

70. Dorschner MO, Amendola LM, Turner EH, et al. Actionable, pathogenic incidental findings in 1,000 participants' exomes. *Am J Hum Genet.* 2013;93(4): 631–640.

71. Green RC, Berg JS, Grody WW, et al. ACMG recommendations for reporting of incidental findings in clinical exome and genome sequencing. *Genet Med.* 2013;15(7):565–574.

72. Khoury MJ, McBride CM, Schully SD, et al. The Scientific Foundation for personal genomics: recommendations from a National Institutes of Health-Centers for Disease Control and Prevention multidisciplinary workshop. *Genet Med.* 2009:11(8):559–567.

73. Ashley EA, Butte AJ, Wheeler MT, et al. Clinical assessment incorporating a personal genome. *Lancet.* 2010;375(9725):1525–1535.

74. Dewey FE, Chen R, Cordero SP, et al. Phased whole-genome genetic risk in a family quartet using a major allele reference sequence. *PLoS Genet.* 2011; 7(9):e1002280.

21

Genetics of Valvular Heart Disease

Andreas C. Mauer, David S. Owens, Christopher J. O'Donnell, Wendy S. Post, and George Thanassoulis

TAKE HOME POINTS

1. Although genetic studies of valvular heart disease have only recently identified associated genes, polymorphisms for mitral and aortic valve disease have been identified. However, replication of findings is critical, as several published candidate gene studies have not been replicated.

2. A genome-wide association study (GWAS) for aortic valve calcification (AVC) identified a single nucleotide polymorphism (SNP) within intron 25 of the apolipoprotein(a) gene (*LPA*); this SNP was observed to double the odds of AVC in several cohorts and across ethnicities, achieved genome-wide significance, was validated prospectively in two cohorts demonstrating a marked increased risk of incident aortic stenosis (AS) or aortic valve replacement (AVR), and in a Mendelian randomization analysis, higher "genetically determined" levels of Lp(a) were strongly associated with AS, all findings which support a causal relationship between genetic variation at *LPA* and the development of AS.

3. Candidate gene studies have identified plausible genetic contributions to selected cases of bicuspid aortic valve (BAV), and in particular provide support for a causative role of *NOTCH1* in some individuals, and although some associations such as *TGFB2* and *SMAD6* have not been validated, it is clear that genes involved in valvular embryogenesis play an important role in the development of BAV.

4. There is strong evidence that myxomatous mitral valve disease and mitral valve prolapse (MVP) have a genetic basis, but demonstrate considerable genetic heterogeneity and variable expressivity; to date, the genetic basis of isolated, nonsyndromic MVP has not been elucidated, providing tremendous opportunity for future gene discovery using gene-network, pathway-specific analyses, and/or whole-exome sequencing.

CASE STUDY

A 48-year-old man with no known history of major medical problems presented to the emergency department (ED) with severe shortness of breath. The patient was in his usual state of health until 2 hours prior to presentation when he developed progressive, severe shortness of breath while exercising at home. He subsequently developed a cough productive of frothy pink sputum at which time his wife called emergency medical services. The patient was brought to the ED with the following vital signs: blood pressure 158/92 mmHg, heart rate 122 bpm, respiratory rate 30 breaths/minute, and oxygen saturation of 76% while breathing room air. The patient was quickly intubated and chest radiography and electrocardiogram (ECG) were performed. A more complete physical examination, performed after intubation, revealed an elevated jugular venous pressure of 14 cmH$_2$O, diffuse bilateral lung rales, a tachycardic but regular rhythm, an S3, a loud 3/6 holosystolic murmur at the apex radiating to the axilla, and a soft diastolic blowing murmur at the left upper sternal border. The abdominal and extremity exams were benign. The chest radiograph revealed a normal cardiac size, left atrial enlargement, and bilateral diffuse pulmonary edema with pulmonary vascular congestion. The ECG demonstrated sinus tachycardia and nonspecific ST-T-wave changes.

After intubation, the patient's heart rate normalized but even with the ventilator set at an FiO$_2$ of 1.0, he had trouble maintaining an oxygen saturation greater than 85%. An urgent bedside Doppler echocardiogram was obtained. The mitral valve was not well visualized but eccentric mitral regurgitation was visualized. Its severity was graded as at least moderate but difficult to quantify based on the eccentric nature of the mitral regurgitation jet. The patient also had mild-to-moderate aortic regurgitation.

Because of his clinical deterioration, an intra-aortic balloon pump was inserted at the bedside with improvement in oxygenation and overall clinical status. The patient was stabilized overnight and taken to the operating room the next day. The pre-operative transesophageal (TEE) images (Figure 21.1, Videos 21.1–21.2 [http://www.demos medical.com/media/videos/Shah_Video_21_1.mp4; http://www.demosmedical.com/media/videos/Shah_Video_21_2 .mp4]) demonstrated myxomatous degeneration of the mitral valve with a flail P3 segment of the posterior leaflet and a ruptured chord resulting in severe mitral regurgitation. The aortic valve was trileaflet and also appeared thickened and myxomatous with degenerative changes of the aortic valve leaflets. Mild aortic regurgitation was present. The patient underwent successful mitral valve repair and had an uneventful recovery.

Postoperatively, the patient's cardiologist inquired about his family history. The patient noted that he had a brother with mitral valve prolapse (MVP) and two sisters who were healthy. His mother was also healthy but both of his maternal uncles also had MVP, and one required mitral valve surgery. His father had diabetes and hypertension but was otherwise healthy. Because of an apparent X-linked recessive inheritance of mitral valve disease based on the patient's family pedigree, further genetic testing was obtained. The patient, his brother, and two uncles had an 862 (G→A) mutation in exon 5 of the *FLNA* gene, resulting in a Gly288Arg substitution at a highly conserved residue within the Filamin A protein. The patient's mother also underwent genetic testing and was found to be a carrier of the *FLNA* mutation. The patient and her affected family members were diagnosed with X-linked mitral valve dystrophy.

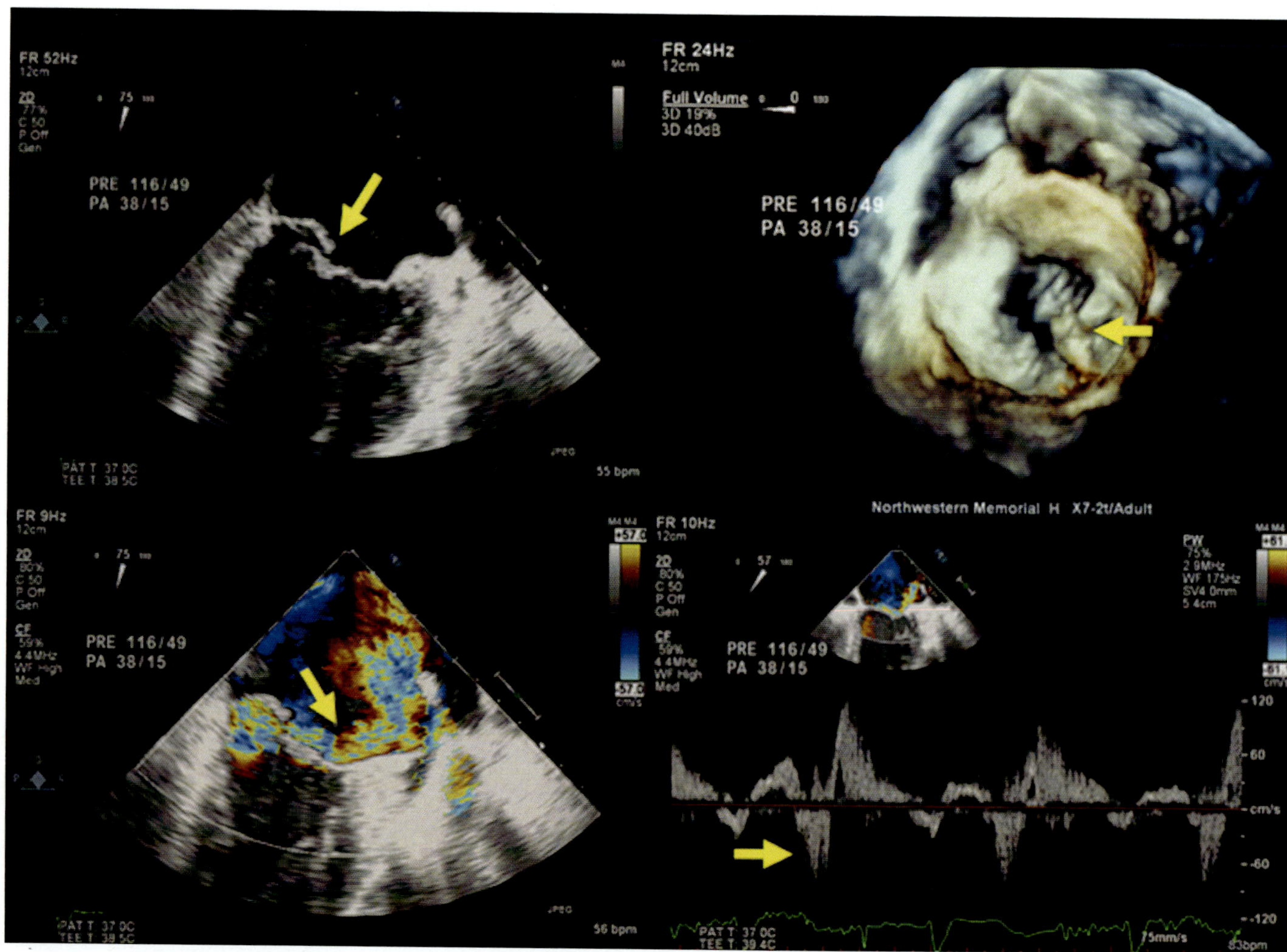

FIGURE 21.1 Transesophageal echocardiography images of the mitral valve of the patient described in the case study. Clockwise from the top left: 2D image demonstrating a flail P3 segment of the posterior mitral valve leaflet (arrow); 3D images demonstrating flail P3 segment with associated ruptured chord (arrow); eccentric mitral regurgitation jet on color Doppler imaging (arrow); and systolic flow reversal in the left upper pulmonary vein (arrow). Findings are consistent with severe mitral regurgitation.

Valve disease represents a common form of cardiovascular disease that continues to increase with the aging population. Over 4% of individuals over the age of 65 have some form of valve disease and its prevalence increases to over 11% in those aged over 75 years. In the United States, there are approximately 85,000 hospital admissions yearly and over 20,000 deaths attributed directly to valve disease (1). Despite the high burden of valve disease, the pathophysiology of many forms of valve disease remains incompletely understood, and insights into the genetic factors involved in its development are only beginning. For example, the largest genetic studies of myocardial infarction (MI) have enrolled over 63,000 MI cases and over 126,000 controls from around the world and uncovered approximately 50 novel genetic associations with major implications for understanding the biology of coronary artery disease (CAD). In contrast, until very recently, limited large-scale genetic studies of valve disease had been performed. However, much work in the genetics of valve disease is ongoing, which will undoubtedly bring new insights into its development, prognosis, and treatment. In this chapter, we discuss in detail the epidemiology, heritability, linkage studies, candidate gene studies, and genome-wide association (and other omic) studies of various types of aortic and mitral valve disease.

AORTIC VALVE DISEASE

Aortic stenosis (AS) is the most common form of valvular heart disease in the Western world, comprising roughly 33% of all cases of valve disease (1). The prevalence of AS (of any severity) is estimated at 0.4% of the general adult population (2), yielding an estimate of 1.2 million affected Americans based on the current U.S. population. The associations between AS, cardiovascular morbidity, and death are well established (3). Because no medical therapies have proven beneficial in preventing or retarding the progression of AS, aortic valve replacement (AVR) is the only treatment. Thus, AS presents a significant public health challenge.

Aortic valve leaflet thickening ("aortic sclerosis") and aortic valve calcification (AVC) precede AS (3) and are highly prevalent among older adults, with 53% of patients over the age of 75 having at least mild AVC (4). Although it is clinically silent and was long considered benign, AVC is progressive (5) and even mild aortic sclerosis confers an approximately 50% increase in risk for cardiovascular death (3).

The clearest risk factor for AVC and AS is age, with an AS prevalence of 1.3% to 1.4% in individuals over the age of 65 and 2.6% to 4.6% in those over the age of 75, depending on the definition of stenosis (2,4,6). Male sex has been associated with AS (6), but this finding has not been consistently observed (2). CAD is more prevalent among patients with AS and AVC than among controls, as are renal insufficiency, hypertension, smoking, and lipid abnormalities (2,3,6,7). However, it is not clear to what extent these comorbidities are the result of a common pathophysiological process, as opposed to being causal in the development of AS. Unicuspid and bicuspid aortic valve (BAV) morphology also substantially increases the risk of AS; the latter is considered separately, see the section Genetics of Bicuspid Aortic Valve Disease.

Several constructs for the pathogenesis of AS have been advanced. Given the strong association of AS with advanced age, one traditional notion is that aortic sclerosis and AS are consequent to passive "wear and tear" over the course of an individual's life (8). However, a more modern view of the genesis of AS is one of a highly coordinated and regulated active process that shares many similarities with atherosclerosis and bone formation.

The coexistence of traditional CAD risk factors has led to intense scrutiny of the role of lipids in AS. For instance, the association between calcific AS and familial hypercholesterolemia (FH) has long been recognized (9,10) (see Rare Mendelian Forms of Aortic Valve Disease). Histologic analysis of sclerotic and stenotic valves has demonstrated the presence of an active inflammatory process with similarities to atherosclerosis, including lipid deposition, macrophage and T-cell infiltration, and basement membrane disruption (8). Low-density lipoprotein (LDL)-receptor deficient apolipoprotein B-100 (LDLR$^{-/-}$ApoB$^{100/100}$) mice develop calcific AS (11), a phenotype that can be suspended with genetic inactivation methods to normalize plasma cholesterol levels (12). These findings prompted investigators to conduct initially promising studies of lipid-lowering medications for the retardation of AS (13,14), though subsequent multicenter trials reported largely negative results (15–17).

The mechanisms that promote valvular calcification remain incompletely understood. As reviewed by Towler (18), cholesterol crystals (19) as well as fragmented elastin fibers and collagen (20) may serve as focal points for mineral deposition. In a process analogous to bone mineralization, matrix proteins such as osteopontin, osteocalcin, and osteonectin that are involved in the regulation of mineralization are expressed in calcified native and bioprosthetic valves (21). Matrix vesicles, which are enriched in alkaline phosphatase, annexin A5, annexin A6, phosphatidyl serine (22) and PHOSPHO1 (23), have also been demonstrated in native and bioprosthetic valves (24). Additionally, circulating cells expressing CD45 and Type I collagen (25), as well as the so-called myeloid calcifying cell (26), have been shown to play a role in ectopic calcification in a variety of tissues including valves. Although a more precise understanding of the mechanisms regulating mineralization will be invaluable in further defining the pathogenesis of AVC and AS, variation in genes related to calcium homeostasis has not been demonstrated except in two nonvalidated candidate gene studies (27,28).

Our current understanding of the pathobiology of AS remains incomplete. Identifying novel genetic associations that underpin AS may provide new insights into the molecular pathways involved in this process and could highlight novel therapeutic targets to treat this condition.

Genetics of Calcific Aortic Valve Disease

Heritability

AS is a complex phenotype, and assessment of heritability is hindered by the late presentation in life and the need for echocardiographic or invasive hemodynamic diagnostic studies. Investigators have therefore encountered difficulty in identifying kindreds with AS, rendering linkage and other traditional approaches difficult. Nonetheless, several studies suggest that there is a heritable basis for the disease.

First, Horne and colleagues searched the Utah Population Database for cases of death attributable to nonrheumatic aortic and mitral valve disease and conducted a genealogic index of familiality analysis (29). They found high relatedness among individuals with nonrheumatic valvular mortality, a finding that extended beyond second-degree relatives of cases with death attributable to aortic valve disease.

Bella and colleagues performed a genome-wide linkage mapping study for valve calcification phenotypes, including AVC and mitral annular calcification (MAC) (30). Evaluating echocardiograms and genotypes in 1871 siblings participating in the Hypertension Genetic Epidemiology Network (HyperGEN) Study, they identified 148 individuals comprising 78 sibships from 68 families that contained at least two affected siblings with valve calcification. The investigators identified a modest degree of familial aggregation, calculating a sibling recurrence risk of 0.25 and recurrence ratio of 2.31 for aortic valve sclerosis. The highest logarithm (base 10) of odds (LOD) score was centered at chromosome 16q22.1-q22.3, within the 1 LOD support interval of the high-density lipoprotein cholesterol 3 (*HDLC3*), musculoaponeurotic fibrosarcoma oncogene (*MAF*), and cadherin 13 (*CDH13*) genes.

In a less quantitative analysis, Le Gal et al calculated the rates of AVR for AS among native residents of the French region of Finistere (31). Because of a "static" population, they were able to identify 756 individuals requiring surgery for AS, who appeared to be concentrated in 40 high-risk areas. Though heritability could not be quantified with this approach, the findings were supportive of the heritability hypothesis.

Finally, Probst and colleagues also analyzed the geographical distribution of patients with AS in France and detected specific regions with high rates of AS (32). They further identified five families with at least three affected siblings. The largest family included 13 cases of severe AS and 20 with less severe aortic valve disease. The investigators then identified a further 35 cases of severe AS within the same geographic region. These 48 cases were all traced back to a single common ancestor born in 1650. Furthermore, none of the families had evidence of confounding disorders such as renal failure or FH (see Rare Mendelian Forms of Aortic Valve Disease). Vitamin D and apolipoprotein E gene analysis did not disclose any association with AS in this family, but the investigators were able to calculate a 33% first-degree recurrence rate, providing the most robust evidence of heritability in calcific AS.

Candidate Gene Studies

Candidate gene studies for AS have focused primarily on genes involved in lipid metabolism and calcium homeostasis and are summarized in Table 21.1. It should be noted that significant concerns have been raised about the methodology of genetic association studies in the era of many of these investigations (33), and few such identified associations have been replicated.

Given the long-standing hypotheses regarding a role for lipids in the pathogenesis of AS, many candidate gene investigations of AS genetics have centered on lipid genes with generally conflicting results. For instance, Avakian and colleagues performed a case–control study of apolipoprotein A1, B, and E polymorphisms in 62 controls and 62 patients with severe AS (34). They found an excess of an apolipoprotein B XbaI restriction site polymorphism and of the apolipoprotein E2 allele among AS patients. However, these results have not been replicated in subsequent studies. Novaro et al also evaluated the apolipoprotein E (*APOE*) gene, querying whether there was an association between *APOE* polymorphisms and valvular calcification in 802 individuals undergoing echocardiography, of whom 43 had AVC (35). They found that age and the E4 allele (but not the E2 allele identified by Avakian) were independent predictors of AS.

Given that inflammatory pathways are also implicated in the pathogenesis of CAD, Ortlepp and colleagues postulated an association between AS and polymorphisms in the chemokine receptor 5 (*CCR5*), connective tissue growth factor (*CTGF*),

TABLE 21.1 Summary of Candidate Gene Association Studies for Calcific Aortic Stenosis

Gene	Polymorphism	Population	Cases	Risk Variant	Odds Ratio	P Value
Apolipoprotein B (*APOB*) (34)	2p24.1 Defined by *XbaI* restriction enzyme	Brazilian	62	XbaI X+/X+	–	0.007
Apolipoprotein B (*APOB*) (37)	2p24.1 E4181K, 3′ untranslated region	French Canadian	457	–	–	1.03×10^{-5}; 1.32×10^{-5}
Apolipoprotein E (*APOE*) (34)	19q13.2 Defined by *HHa1* restriction enzyme	Brazilian	62	E2 allele	–	0.034
Apolipoprotein E (*APOE*) (184)	19q13.2 Defined by *HHa1* restriction enzyme	Predominantly White European	43	E4 allele	1.94 (1.01–3.71)	0.046
Estrogen receptor alpha (*ESR1*) (38)	6q25.1 Defined by *PvuII* and *XbaI* restriction enzymes	Postmenopausal White European Women	41	p allele	3.38 (1.13–10.09)	0.03
Vitamin D receptor (*VDR*) (27)	12q13.11 Defined by *BsmI* restriction enzyme	Not reported	100	B allele	–	0.001
Interleukin-10 (*IL-10*) (36)	1q31-32 G1082A, C819T, and C592A	Not reported	187	A, T, and A alleles	–	0.026; 0.027; and 0.028
Interleukin-10 (*IL-10*) (37)	1q31-32 6 polymorphisms in 5′ region, introns, and 3′ region	French Canadian	457	–	–	$0.016–6.21 \times 10^{-11}$
Connective tissue growth factor (*CTGF*) (36)	6q23.1 G447C	Not reported	457	GC	–	0.099 (0.037 in combination with CCR5 polymorphism)
Chemokine (C-C motif) receptor 5 (*CCR5*) (36)	3p21.31 Δ32 deletion	Not reported	457	Deletion	–	0.213 (0.037 in combination with CTGC polymorphism)
Parathyroid hormone (*PTH*) (28)	11p15.3-p15.1 rs6254	Not reported	538	AA	–	0.007
Myosin VIIA (*MYO7A*) (42)	11q13.5 rs2276288	Not reported	265	AA	–	0.001
Angiotensin II receptor, type 1 (*AGTR1*) (42)	3q24 rs5194	Not reported	265	AA	–	0.004
Elastin (*ELN*) (42)	7q11.23 rs2071307	Not reported	265	AA	–	0.005

and interleukin-10 (*IL-10*) genes (36). Studying 187 patients undergoing surgical AVR, they identified somewhat paradoxically that a haplotype made up of *IL10* polymorphisms known to associate with high Il-10, an anti-inflammatory cytokine, was associated with AS. However, subsequent studies by Gaudreault identified the pro-inflammatory haplotype as associated with AS, highlighting the inconsistencies of these candidate gene studies (37). Ortlepp et al also reported that the combination of the high IL-10 haplotype with rare *CCR5* or *CTGF* alleles was associated with a greater degree of calcification. However, this gene–gene interaction has never been replicated and most likely represents a false positive.

Because of the association of AS with atherosclerosis and advancing age, Nordstrom and colleagues tested the hypothesis that postmenopausal hormone changes, specifically polymorphisms in the estrogen receptor α (*ESR1*), were associated with AS. Because of its roles in fibrosis and atherosclerosis, they also studied transforming growth factor β1 (*TGFB1*) (38). An intronic estrogen receptor mutation that was associated with increased AS risk was identified in 41 women with moderate-to-severe AS. Although a genotype defined by the combination of this polymorphism and two additional *TGFB1* restriction sites was identified as high risk, no high-risk *TGFB1* polymorphisms were identified.

Based on previously established associations between bone mineral mass and a well-characterized polymorphism in the vitamin D receptor (*VDR*) gene known as the BsmI polymorphism, Ortlepp and colleagues studied BsmI status in 100 patients with AS (27). They reported that the frequency of BsmI allele B was 35% higher among AS patients than controls. Schmitz et al also queried a possible role for calcium homeostasis, studying *VDR* as well as the parathyroid hormone (*PTR*) gene (28). In that study, *VDR* variants were marginally significant, but an intronic *PTR* mutation was associated with AS. Although highly plausible due to the importance of calcium homeostasis in aortic valve disease, neither of these associations has been subsequently replicated.

Mutations in the signaling gene *NOTCH1* have been identified in some cases of accelerated valve calcification and BAV disease (39) (see Genetics of Bicuspid Aortic Valve Disease). Therefore, Ducharme and colleagues genotyped *NOTCH1* in 457 individuals with severe tricuspid AS (40). One previously reported *NOTCH1* mutation was found, and 32 other individuals were found to be heterozygous for two other mutations. One polymorphism in intron 2 was found to be associated with AS, but this was not significant after correction for population stratification. Though less evidence associates *NOTCH1* with calcific AS than with BAV, it is interesting to note

that *NOTCH1* represses activation of *Runx2*, a transcription factor that is upregulated in animal models of valvular calcification and that regulates expression of osteoblast-specific genes (39). Moreover, the same group found that *NOTCH1* regulates osteogenic pathways in murine aortic valve cells via bone morphogenic protein 2 (*BMP2*) (41). Thus, *NOTCH1* remains a gene of interest in calcific AS.

As described throughout this chapter, many of these candidate genes have not been validated. Gaudreault et al attempted to replicate several associations with *APOB*, *APOE*, *CTGF*, *IL10*, *PTH*, *TGFB1*, and *VDR* in 457 patients undergoing AVR, the only associations that were reproduced were for two polymorphisms in *APOB* (37). Although the identified polymorphism in *APOB* was different (and not in linkage disequilibrium) with previously identified *APOB* polymorphisms, the concordance identified with two independent *APOB* polymorphisms does support a role for this gene in AS. One possibility is that *APOB* operates via its effects on various lipoproteins (eg, low-density lipoprotein and lipoprotein(a) that have been associated with AS). Finally, Ellis et al also tested previously identified candidate genes in a study of 660 SNPs in 265 AS patients (42). Using stringent criteria, none of the polymorphisms in eight genes previously reported to be associated with AS were replicated. Although they did find novel associations with several genes not previously associated with AS including *MYO7A* (a myosin variant), *AGTR1* (angiotensin receptor II, type 1), and *ELN* (elastin), they cautioned that these findings could not be regarded as definitive and require subsequent replication. In particular, the association with variants in *AGTR1* is interesting, given evidence that angiotensin 2 forming enzymes are upregulated in calcific aortic valves (43), as well as preliminary clinical data suggesting that angiotensin-converting enzyme (ACE) inhibitors improve survival in patients with AS (44).

Emerging Omic Technologies: Genome-Wide Association Studies

With complete characterization of the human genome sequence, GWAS have been conducted since approximately 2005 and can provide strong statistical evidence combined with replication evidence for genetic determinants of valve disease. Only one GWAS to date has been performed for aortic valve disease. Thanassoulis et al genotyped 6,942 individuals in three cohorts with computed tomographic (CT) assessment of aortic valve calcium and replicated the results in several additional large cohorts (45). One SNP within intron 25 of the apolipoprotein(a) gene (*LPA*) reached genome-wide significance and was robustly replicated across several ethnicities. This SNP was observed to double the odds of AVC in all cohorts tested. This

SNP was also validated prospectively in two cohorts demonstrating a marked increased risk of incident AS or AVR. This SNP is in linkage disequilibrium with fewer Kringle IV-type 2 (KIV-2) repeats, an important regulator of LPA expression, leading to higher levels of Lp(a). Based on the known biology of this SNP and using a Mendelian randomization analysis, they demonstrated that "genetically determined" Lp(a) level, as determined by *LPA* genotype, strongly associated with AS, with a hazard ratio of 1.68 per risk allele. These findings support a causal relationship between genetic variation at *LPA* and the development of AS and point to a potential novel therapeutic strategy to reduce aortic valve disease. Forty-four additional SNPs were identified with suggestive evidence for association but did not reach genome-wide significance; of note, the majority of these were not within 60 kB of the candidate genes discussed previously.

Future Directions

Science has only started to tease apart the genetic basis of AVC; unraveling the mechanisms underlying AVC and AS will require investigators to deploy innovative methods to overcome the many difficulties that have hindered the study of this disease.

First, technologies to facilitate accurate phenotyping are vital. The accurate identification of AVC at young ages is a necessary step for assembling reliable pedigrees and quantifying heritability, as is the exclusion of contributing conditions such as BAV morphology. In addition to the obvious utility of echocardiography for these purposes, accurate quantification of AVC by multi-detector computed tomography (MDCT) has already proven useful in the study of aortic valve genetics (45). Advances in positron emission tomography (PET) permit the measurement of valve inflammation (46) and osteoblast activity (47). In addition to generating "deep phenotypes" for scientific purposes, early and accurate detection would also open the path to new treatment avenues such as the testing of therapies (eg, statins or ACE-inhibitors) before the onset of advanced disease.

Second, high-throughput omics technologies have yet to make the impact that has been seen in other fields. With only one published GWAS in the field, the scope for additional study using modern methods is immense. Emerging technologies with potential applications in the study of AS include whole-exome capture and sequencing, exome chip technology, and proteomic and metabolomic assays. To leverage these technologies, investigators will need to develop large, carefully phenotyped cohorts like the extracoronary calcium working group of the Cohorts for Heart and Aging Research in Genomic Epidemiology (CHARGE) Consortium (48). As such cohorts and technologies come online, we can anticipate an accelerating understanding of the genetic basis of AS.

Genetics of Bicuspid Aortic Valve Disease

First described by Leonardo Da Vinci in the 16th century, BAV occurs when the aortic valve develops with only two leaflets rather than its normal trileaflet configuration (Figure 21.2). Though the life expectancy

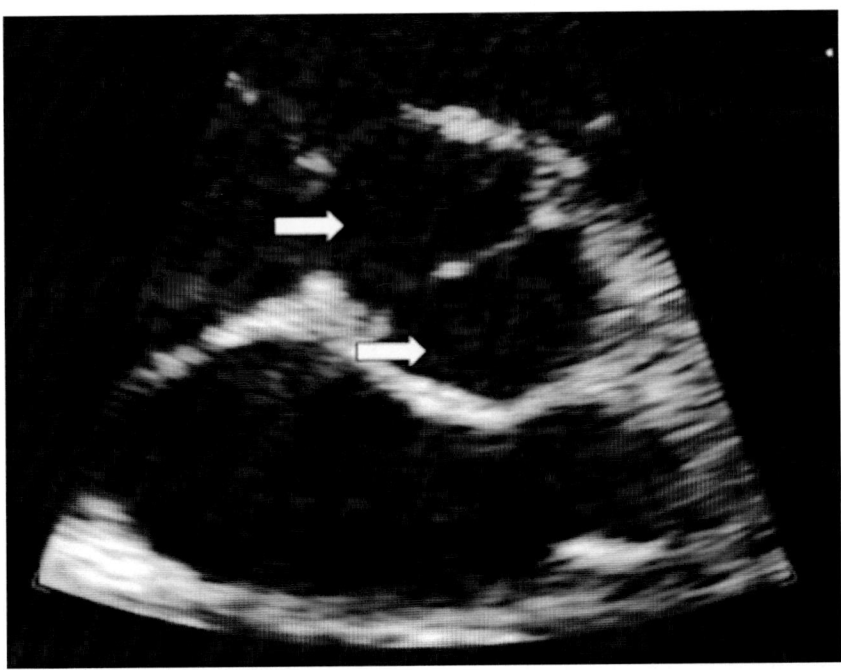

FIGURE 21.2 Bicuspid aortic valve. Only two aortic valve leaflets (arrows) are present on the short-axis parasternal view by transthoracic echocardiography.

of patients with BAV does not differ from that of individuals with normal aortic valves (49,50), up to 75% of patients with BAV experience serious valve-related complications, including AS, aortic insufficiency (AI), endocarditis, aortic coarctation, thoracic aortic aneurysm (TAA), and aortic dissection (50–53). In light of the potential severity of these complications and the relatively high prevalence of BAV, some authors have argued that BAV "may be responsible for more deaths and morbidity than the combined effects of all other congenital heart defects" (54).

Epidemiology of Bicuspid Aortic Valve Disease

In 1886, Osler reported 18 cases of BAV in a series of "over 800" autopsies (55). Over the next century, BAV was found to be the most common congenital heart abnormality, with an estimated prevalence by necropsy or echocardiography of 0.5% to 2% (51,56–58). BAV is approximately three times more common in males than in females (56–58) and may be less common in African Americans than other racial groups (59).

BAV frequently coexists with TAA, and many investigations of the genetic determinants of BAV also include TAA (60–63). It is not certain whether TAAs in individuals with BAV are consequent to similar genetic influences, or whether they result instead from hemodynamic perturbations (64,65).

Though often an isolated abnormality, BAV is also a common finding in patients with more complex congenital heart disease. The spectrum of associated disease includes coarctation of the aorta, ventricular septal defect (66), Turner syndrome (67,68), Williams–Beuren syndrome (69–71), Rubinstein–Taybi syndrome (72), Andersen syndrome (73), and Williams syndrome (74), among others. The association of BAV with other forms of left ventricular outflow obstruction, particularly hypoplastic left heart syndrome (HLHS), has drawn particular interest, with several authors hypothesizing that one or very few genes may underlie diverse left-sided lesions (75–77), or that BAV may represent a forme fruste of more severe malformations (78). Given the broad spectrum of genetic abnormalities and cardiovascular phenotypes associated with these disorders, in such cases BAV is more often considered a component of the underlying syndrome than a primary genetic abnormality.

Heritability

Familial clusters of BAV were first reported in the 1970s (79–81). Emanuel and colleagues provided systematic evidence that BAV was heritable in a study of the families of 41 patients with surgically proven BAV (82). Upon identifying definite or possible BAV in 18 first-degree relatives, they estimated a "minimal family incidence" of 13.6% to 31.7%.

The hypothesis that some forms of BAV are inherited was subsequently confirmed by reports of BAV in pairs of identical twins (83,84), and by descriptions of additional families with multiple affected members (77,85–87).

Huntington and colleagues performed echocardiographic screening in first-degree relatives of 30 patients with BAV and found that 9.1% of family members were affected, with 36.7% of families having at least two affected members (88). In families with familial clustering, they identified a pattern of autosomal dominant inheritance with reduced penetrance, a finding that was substantiated in subsequent linkage and candidate gene studies (39,60,89). More recently, Robledo-Carmona and colleagues found that 4.6% of first-degree relatives of BAV patients also had BAV, with 15% of families having at least two affected members (90).

McBride et al studied the heritability of mixed left-sided obstructive lesions including BAV in 413 members of 124 affected families and found that the relative risk of obstructive disease for first-degree relatives was 36.9, with a heritability of 71% to 90% (75). They proposed a complex but oligogenic inheritance pattern.

Cripe and colleagues studied 309 individuals from the families of 50 probands with BAV (91). The prevalence of BAV in these families was 24%, and additional abnormalities such as coarctation or septal defects were identified in several other family members, with a combined prevalence of 31%. The heritability of BAV was estimated at 89%, and the heritability of BAV in conjunction with another cardiovascular malformation was 75%. They concluded that both BAV and associated malformations were heritable, and that family members of affected individuals should undergo screening.

Calloway and colleagues used echocardiography to study BAV patients from families that had multiple members affected by BAV or HLHS (92). They calculated that BAV heritability was 61% and that the heritability of development of AS and/or AI in BAV patients was 23%. Neither association reached statistical significance, and the authors therefore concluded that, while there is likely a heritable contribution to BAV, the development of AS and/or AI is due to nongenetic factors.

Rare Mendelian syndromes with complex phenotypes that include BAV have also been reported. For example, Gelb and colleagues described a family with limb defects, patent ductus arteriosus, and BAV with a probable autosomal dominant pattern of inheritance (93), and Guo and colleagues reported that BAV was a variable feature of autosomal dominant, *ACTA2*-mediated thoracic aortic aneurysm and aortic dissection (TAAD) syndrome (94,95).

Linkage

Linkage studies for BAV-related loci have produced mixed results. In the most cited study, Garg and colleagues studied a family with nine cases of congenital heart disease, including six cases of isolated BAV, with an autosomal dominant pattern of inheritance (39). A genome-wide scan revealed linkage evidence to a locus on chromosome 9q34-35, a region that included *NOTCH1*, a signaling gene with roles in cell differentiation during organogenesis. Sequencing of *NOTCH1* revealed a mutation that produced a premature stop codon (R1108X) in affected family members. A second *NOTCH1* mutation was found in an unrelated family with BAV, and murine models demonstrated a role for the gene in embryonic valve development.

Goh et al identified a patient with BAV and other anomalies with a deletion at 15q25 (96). Subsequent linkage analysis in 10 families with BAV and TAA confirmed the *NOTCH1* locus as a region of interest.

Following these results, Ellison et al queried a multigenerational cohort with familial BAV and found no evidence of linkage for markers near *NOTCH1* or chromosome 15q25 (89). They concluded that these findings, in conjunction with the phenotypic heterogeneity exhibited by patients with BAV, were evidence for genetic locus heterogeneity.

Andelfinger et al performed linkage analysis in patients with Andersen syndrome, a syndrome that features multiple dysmorphic features as well as arrhythmia and cardiovascular malformations including BAV (73). They identified an association with *KCNJ2*, a gene that encodes for the inward-rectifying potassium channel.

Martin and colleagues genotyped 324 individuals in 38 families with BAV and found evidence in favor of linkage with loci at 18q22, 5q21, and 13q34 (97). Linkage to regions near the previously identified *NOTCH1* and *KCJN2* loci was identified in single families, but mutation analysis did not identify any sequence alterations in either gene that segregated with BAV kindreds.

Hinton and colleagues performed linkage analysis in two family-based cohorts with HLHS or BAV probands. They confirmed Martin's finding of linkage at 18q22 and 5q21, and also found evidence of association between BAV and HLHS loci, suggesting genetic overlap in these two disorders (76).

Finally, McBride et al studied 289 individuals in 43 families with left-sided obstructive lesions and identified three loci (16p12, 2p23, and 10q21) (98). Specific linkage to BAV could not be established, and previously identified loci were not replicated.

In summary, linkage analyses identified several loci that associate consistently with BAV and have provided a firm basis for subsequent candidate gene studies. These findings, particularly those surrounding *NOTCH1*, are consistent with the hypothesis that BAV and more severe left-sided congenital lesions are different points on a spectrum of disease with a common heritable basis.

Candidate Gene Studies

Following the leads established by animal models and the linkage studies described, investigators have studied a number of candidate genes. Among these, the strongest evidence surrounds mutations in *NOTCH1*. All reported mutations associated with BAV are summarized in Table 21.2.

In addition to the seminal studies by Garg, Mohamed et al also sequenced *NOTCH1* in 48 patients with "sporadic" BAV. They identified 57 variants, of which two appeared to be plausible pathogenic mutations following comparison with healthy controls (99). McBride and colleagues also sequenced *NOTCH1* in patients with left ventricular outflow malformations, including BAV (100), confirming the pathogenic role of the two mutations previously identified by Garg and identifying an additional two mutations associated with BAV. They also performed mechanistic studies that suggested that the mechanism of these mutations is related to a loss of ligand-induced *NOTCH1* signaling. McKellar and colleagues sequenced *NOTCH1* in patients with BAV with and without TAA and identified four additional mutations (101). Finally, Foffa et al sequenced *NOTCH1*, *GATA5*, *TGFBR1*, and *TGFBR2* in 11 patients with familial BAV. No plausible pathogenic mutations were found in *GATA5*, *TGFBR1*, or *TGFBR2*, but two novel *NOTCH1* mutations were identified (78).

Two studies did not find *NOTCH1* in BAV cohorts: Martin and colleagues did not identify *NOTCH1* or *KCNJ2* mutations in any of the kindred studies in their linkage analysis (97), and Kent and colleagues sequenced *NOTCH1* in 13 families with BAV and TAA but did not find any pathological mutations (61).

Given the association of mutations in the transforming growth factor-beta (TGF-beta) receptor with aortic aneurysms in patients with Loeys–Dietz syndrome (102), two different groups tested the families of patients with BAV and TAA for mutations in *TGBR1* and *TGFBR2* and found no evidence of association (60,63). Contrariwise, Girdauskas et al sequenced *TGFBR2* in a patient with BAV and aortic dissection and identified a missense mutation (65).

Several investigators have pursued genes identified in murine models, with varying degrees of success. For instance, Lee and colleagues described an increased incidence of BAV in mice with deficiency of endothelial nitric oxide synthase (103); to date, this observation has not been replicated in humans. Majumdar sequenced the transcription factor *NKX2-5* in 19 patients with BAV and TAA but did not find evidence

TABLE 21.2 Summary of Candidate Gene Association Studies for Bicuspid Aortic Valve

Gene	Polymorphism	Inheritance Pattern	Population	Cases
Notch1 (*NOTCH1*) (39)	9q34.3 R1108X	Familial	European–American	9
Notch1 (*NOTCH1*) (39)	9q34.3 H1505del	Familial	Hispanic	3
Notch1 (*NOTCH1*) (99)	9q34.3 T596M	Sporadic	German	48
Notch1 (*NOTCH1*) (99)	9q34.3 P1797H	Sporadic	German	48
Notch1 (*NOTCH1*) (78)	9q34.3 P284L	Familial	Italian	11
Notch1 (*NOTCH1*) (78)	9q34.3 Y1619X	Familial	Italian	11
Notch1 (*NOTCH1*) (100)	9q34.3 G661X	Sporadic	European–American	91
Notch1 (*NOTCH1*) (100)	9q34.3 V2536I	Sporadic	European–American	91
Notch1 (*NOTCH1*) (101)	9q34.3 A1343V	Sporadic	Not reported	60
Notch1 (*NOTCH1*) (101)	9q34.3 P1390T	Sporadic	Not reported	60
Notch1 (*NOTCH1*) (101)	9q34.3 R1350L	Sporadic	Not reported	60
Notch1 (*NOTCH1*) (101)	9q34.3 P1337S	Sporadic	Not reported	60
GATA-binding protein 5 (*GATA5*) (104)	20q13.33 Q3R	Sporadic	Predominantly Caucasian	100
GATA-binding protein 5 (*GATA5*) (104)	20q13.33 S19W	Sporadic	Predominantly Caucasian	100
GATA-binding protein 5 (*GATA5*) (104)	20q13.33 Y142H	Sporadic	Predominantly Caucasian	100
GATA-Binding Protein 5 (*GATA5*) (104)	20q13.33 G166S	Familial	Predominantly Caucasian	100
Transforming growth factor, beta receptor II (*TGFBR2*) (65)	3p22 V387M	Familial	Not reported	2
SMAD family member 6 (*SMAD6*) (106)	15q21-q22 C484F	Sporadic	Caucasian/British	436
SMAD family member 6 (*SMAD6*) (106)	15q21-q22 P415L	Sporadic	Caucasian/British	436

to support any association in humans (62). Padang et al sequenced *GATA5*, a transcription factor with roles in cardiac embryogenesis, in 100 patients with BAV. Among these, four patients were found to have nonsynonymous polymorphisms in *GATA5* (104).

Saravanan and Kadir reported a case study of monozygotic twin brothers with BAV and AS with a shared point mutation in the *APOE* gene (105); this finding has not been replicated as a BAV association in systematic analyses.

Finally, Tan and colleagues sequenced *BMPR2*, *BMPR1*, and *SMAD6*, a second messenger in TGF-beta pathways, in a cohort of patients with heterogeneous congenital heart disease phenotypes. They identified three mutations in *SMAD6*, of which two were identified in patients with BAV (106).

Taken together, candidate gene studies have identified plausible genetic contributions to selected cases of BAV, and in particular provide support for a causative role of *NOTCH1* in some individuals. Though some associations such as *TGFB2* and *SMAD6* have not been validated, it is clear that genes involved in valvular embryogenesis play an important role in the development of BAV. For instance, *NOTCH1* is highly expressed in the embryonic aortic valve and left ventricular outflow tract, and is also known to regulate the expression of *SOX9* and *BMP2*, both of which have important roles in valvulogenesis (107). Intriguingly, both *SOX9* and *BMP2* also have demonstrated roles in calcification (41,108), and *NOTCH1* also represses *RUNX2*, a gene involved in osteoblast gene expression that is upregulated in animal models of valvular calcification (39). Further exploration of genetic variation within the *NOTCH1* pathway may therefore provide insights into common mechanisms underlying BAV disease and calcific AS.

Emerging Omic Technologies (Including GWAS)
To date, no GWAS have been reported for BAV phenotypes. However, LeMaire and colleagues included patients with BAV in stages two and three of a three-stage GWAS for TAA (109). In stage two, 192 of 385 (49.5%) patients with TAA also had BAV, and in stage three 157 of 163 (96.3%) patients with nondissection TAA had BAV. The authors identified multiple SNPs that were strongly associated with TAA at chromosome 15q21.1, which encompassed *FBN1*, the gene that encodes fibrillin-1. These associations held in the presence and absence of BAV, and were not suspected of a primary association with BAV.

To compensate for this difficulty and to limit multiple comparisons, Wooten and colleagues used gene-network analysis techniques to study 68 patients with BAV (110). They identified an association between BAV and a concentration of SNPs at 16p13.3, in the region of the genes *AXIN1* and *PDIA2*, as well as a second association with SNPs localizing to the endoglin (*ENG*) gene. Though *AXIN1* is known to participate in signaling pathways that contribute to cardiac development, the function of *PDIA2* is unknown. Endoglin is involved in neural crest cell differentiation, and, interestingly, is downregulated in the presence of defects in *NOTCH* signaling.

Hitz and colleagues used a variety of bioinformatic techniques to examine the role of rare copy number variants (CNVs) in 59 families with left-sided congenital heart disease, including BAV (111). Though they identified variants in 25 candidate genes and calculated that CNVs contributed to an estimated 10% of cases, specific data regarding BAV were not reported.

Future Directions
Most studies of BAV genetics have utilized linkage or candidate gene approaches. With the increasing sophistication of high-throughput approaches, as well as the growing availability of large, well-characterized cohorts with a variety of cardiovascular phenotypes, genome- or exome-wide studies should become increasingly feasible. Such approaches will not only permit more comprehensive characterization of the genetic determinants of BAV, but will also allow investigators to explore related questions such as the interplay between BAV and TAA, as well as resolving the borders between BAV and the larger spectrum of left-sided obstructive malformations.

Rare Mendelian Forms of Aortic Valve Disease

Though, as noted, some cases of BAV follow an autosomal dominant/reduced penetrance mode of inheritance, as noted in the section Genetics of Bicuspid Aortic Valve Disease, most cases of AV disease do not follow any clear pattern of inheritance. An exception is the occurrence of AV disease in certain rare Mendelian syndromes. These disorders constitute an extremely small fraction of AV disease cases, and may have little in common with most cases that are encountered clinically.

Familial Hypercholesterolemia
The association of FH and CAD is well established. Though the mechanisms are less well defined, AS has long been known to be an additional consequence of this disorder (9,10). Though evidence from animal and human studies are supportive of a lipid hypothesis in AS, phenotypes in FH differ from the typical presentation of AS in several ways. For instance, FH patients frequently manifest supravalvular, rather than valvular, aortic calcification (112).

Alkaptonuria
Alkaptonuria is a rare, autosomal recessive disorder that derives its name from the characteristic darkening of affected individuals' urine with standing and alkalinization. Caused by a mutation in the homogentisate 1,2 dioxygenase (HGD) gene, patients with alkaptonuria suffer from abnormalities of tyrosine catabolism that result in tissue deposition of homogentisic acid (HGA). This causes black pigmentation of collagenous tissues, as well as arthritis, urolithiasis, and vascular and aortic valve mineralization. Premature AS is a common manifestation (113,114).

Aortic Insufficiency
In addition to AS, AI is a feature of connective tissue diseases such as osteogenesis imperfecta (115), as well as Marfan (116,117), Loeys–Dietz (102), and Ehlers–Danlos syndromes (118). In contrast to the mitral valve disease in these populations, AI may be a secondary consequence of primary aortic pathology.

X-linked myxomatous valvular dystrophy (XMVD) is characterized by myxomatous degeneration and insufficiency of the mitral and aortic valves. This syndrome has been traced to defects in the gene encoding the actin-binding protein Filamin A (119) and is described in further depth in Mitral Valve Prolapse.

MITRAL VALVE DISEASE

Mitral Valve Prolapse

MVP is a clinical finding affecting 2.4% of the population in which alterations in mitral leaflet structural integrity or tethering forces allow the leaflets to extend past the valve annulus and into the left atrium under the high pressure load of ventricular systole (120,121). With this, leaflet coaptation may be disrupted resulting in mitral regurgitation that characteristically begins in mid-to-late systole. MVP is the most common cause of isolated mitral valve regurgitation, and when severe, MVP can lead to the need for either mitral valve repair or replacement.

MVP may be either acquired or due to congenital/genetic causes that are seen in isolation or with familial inheritance, or in the context of genetic syndromes. MVP may develop whenever there are alterations in the chordae tendinae or subvalvular support apparatus, such as from endocarditis, trauma, papillary muscle dysfunction (such as from acute ischemia) or, rarely, from rheumatic heart disease (120). However, the most frequent cause of MVP is myxomatous degeneration of the mitral valve leaflets, leading to redundant leaflets and chordal structures that characteristically billow into the left atrium during systole.

Myxomatous mitral valve disease is characterized histologically by disruption of the normal extracellular architecture within the *spongiosa* layer of the leaflet, with disorganized fragmentation of collagen and elastin, and accumulation of proteoglycans and glycosaminoglycans resulting in increased water retention and thickened, gelatinous leaflets (104,120,122). These changes often extend into the *fibrosa* layer, which provides much of the structural integrity of the valve leaflets, and into the chordae, which become elongated and thickened (predisposing to rupture) with repetitive stress. The biological mechanisms underlying these changes have not been fully elucidated, but appear to involve activation of myofibroblast-like cells and increased secretion of collagenases, matrix metalloproteinases, and cysteine endoproteases (cathepsins) leading to progressive leaflet remodeling (123).

Initial reports of MVP likely overestimated the prevalence of the disease, in large part because of the reliance on cardiovascular imaging for noninvasive diagnosis and an incomplete understanding of the three-dimensional structure of the mitral annulus (121,122). The current, best estimate for MVP prevalence comes from 3,491 participants of the Framingham Offspring Study, which found an overall prevalence of 2.4% using 2D rather than M-mode echocardiography (124). Subjects with MVP were slightly more likely to be female (59.5% vs 52.7%, $P = .21$) and had lower body mass index (BMI) (24.6 vs 27.2 kg/m^2, P less than .001). The Study of Health Assessment and Risk in Ethnic Groups (SHARE) study extended these findings to a non-Caucasian population involving Canadians of South Asian and Chinese descent with prevalence estimates (2.7% and 2.2%, respectively) similar to that seen in Canadians of Caucasian descent (3.1%, $P = .79$) (125).

Myxomatous MVP can be sporadic, familial, or seen in the context of cardiovascular syndromes such as Marfan, Loeys–Dietz, Ehlers–Danlos, and Williams syndromes (Table 21.3). The higher prevalence of MVP in these Mendelian disorders implicates the fibrillin and TGF-beta signaling pathways and extracellular collagen homeostasis in the development of

TABLE 21.3   Genetic Syndromes Associated With Mitral Valve Prolapse

Syndrome	Genes Implicated	Protein	MVP Prevalence
Marfan syndrome	*FBN1* *TGFβR2*	Fibrillin TGF-β receptor 2	40%–54% (186,187)
Loeys–Dietz syndrome	*TGFβR1* *TGFβR2*	TGF-β receptor 1 TGF-β receptor 2	21% (102)
Ehlers–Danlos syndrome (types I, II, and III)	*COL5A1, COL5A2* *TNXB*	Collagen type V Tenascin X	~ 6% (188)
Williams syndrome	7q11.23 deletion (including *ELN*)	Elastin deletion	15% (189)

Abbreviation: MVP, mitral valve prolapse.

TABLE 21.4 Summary of Familial Linkage Analyses for Myxomatous Mitral Valve Prolapse (MMVP)

Study	Pedigrees (Generations)	Subjects (Affected/ Unaffected)	Population	LOD Score	Gene Loci
MMVP-1 (138)	2 (3–4)	15/17	Ashkenazi/European	5.6	16p12.1-p11.2
MMVP-2 (140)	1 (5)	12/13	European	3.1	11p15.4
MMVP-3 (139)	1 (5)	9/26	European	3.2	13q31.3-q32.1
XMVD (142)	1 (5)	21/71	European	6.6	Xq28

Abbreviations: MMVP, myxomatous mitral valve prolapse; XMVD, X-linked myxomatous valvular dystrophy.

MVP. It has been shown that fibrillin stabilizes and actives the TGF-β complex, which in turn regulates extracellular matrix content and remodeling seen in MVP (126). However, these pathways have not been definitively linked to nonsyndromic MVP. While musculoskeletal abnormalities (eg, pectus excavatum or hyperlaxity) are not uncommon in isolated MVP (127), less than 1% of patients with MVP have an associated syndrome. This review of the genetic basis of MVP will focus on the familial and sporadic forms of nonsyndromic MVP.

Heritability

The original description of MVP as a potentially heritable disorder predates noninvasive imaging, when Hancock and Cohn described the characteristic auscultatory findings of MVP (mid-systolic clicks and late, holosystolic regurgitation) (128). Since then, both autosomal dominant and X-linked patterns of inheritance have been described (129–131), with variable age- and sex-dependent penetrance (130,132). The overall prevalence of MVP among first-degree relatives has been estimated at 30% to 47%, (130,132,133) but these studies were performed prior to the adoption of more precise phenotyping. Additional evidence for heritability comes from twin studies showing increased prevalence of MVP among twins compared with other siblings (134,135). Put together, these studies provide strong evidence that MVP is a genetic, heritable disorder with age- and sex-dependent penetrance.

Linkage

Several attempts to map the genetic loci for MVP using familial linkage studies have been performed. Early investigations focused on candidate regions surrounding the collagen genes, but no linkage with these regions was observed (136,137). In more recent studies using genome-wide linkage analysis and contemporary MVP phenotyping, three potential loci have been identified. These myxomatous mitral valve prolapse (MMVP) loci are referred to as MMVP-1, MMVP-2, and MMVP-3 in temporal order of their description (Table 21.4).

Disse and colleagues were the first to identify a potential MMVP locus using four pedigrees of autosomal dominant, heritable MVP (138). The most informative of these pedigrees involved 24 individuals over 3 generations, with linkage to the 16p12.1-p11.2 region (MMVP-1). This finding was replicated in a second pedigree, with a combined conservative LOD score of 5.5, but was unable to be replicated in the other two pedigrees, attesting to the genetic heterogeneity of this disease. Additional analyses of single large pedigrees by Freed and colleagues and by Nesta and colleagues identified potential MVP loci at 11p15.4 (MMVP-2) and at 13q31.3-q32.1 (MMVP-3) (139,140). These studies are all limited by small number of pedigrees, reduced ethnic diversity, and lack of independent replication. Moreover, causally plausible genes at these loci have yet to be identified.

A rare form of myxomatous mitral valve disease demonstrating X-linked inheritance was first described in 1969 by Monteleone et al (141). Additional investigation of a large family with MVP that cosegregated with mild hemophilia A permitted mapping of the loci to Xq28 (142). Subsequent investigations have identified Filamin-A *(FLNA)* gene mutations as a cause of this XMVD (142,143). Filamin-A is an intracellular cytoskeletal protein that interacts with the extracellular matrix via cell surface integrins, but also plays an important role in the localization and activation of Smad2 and serves as a positive regulator of TGF-beta signaling (119). Subsequent animal models have shown the importance of Filamin-A in leaflet remodeling during fetal development (144).

Candidate Gene Studies

Several investigators have utilized a candidate gene approach to test whether SNPs are associated with nonsyndromic MVP (Table 21.5). Because of the high prevalence of MVP in Marfan syndrome and collagen vascular disorders, there was particular interest in

TABLE 21.5 Summary of Candidate Gene Association Studies for Mitral Valve Prolapse (MVP)

Gene	Polymorphism	Population	Phenotype	Cases	Risk Variant	Odds Ratio	P Value
Fibrillin-1 (*FBN1*) (145)	15q21.1 Exon 15	Taiwan Chinese Han	MVP	100	TT	2.40 (1.32, 4.39)	.01
Fibrillin-1 (*FBN1*) (145)	15q21.1 Exon 27	Taiwan Chinese Han	MVP	100	GG	3.61 (1.01, 12.89)	.04
Fibrillin-1 (FBN1) (148)	15q21.1 Intron 56	Turkish children	MVP	77	GC	5.04 (2.60, 9.76)	.0001
Collagen IIIα1 (*COL3A1*) (146)	2q31 Exon 31	Taiwan Chinese Han	MVP	100	GG	7.42 (4.40, 12.52)	.0001
Transforming growth factor-β1 (*TGFB1*) (147)	19q13.1 C509T, T869C	Taiwan Chinese Han	MVP	100	–	–	NS
Urokinase plasminogen activator (*PLAU*) (149)	10q22.2 T4065C	Taiwan Chinese Han	MVP	100	TC	6.03 (2.11, 14.83)	.0001
Matrix metalloproteinase (*MMP1*) promoter (150)	11q22.3 1607 1G/2G	Taiwanese	MCTR + MVR	64	1G/1G	3.22	.009
Angiotensin II receptor, type 1 (*AGTR1*) (152)	3q24 A1166C	White European	MVP syndrome	76	C allele	4.0 (1.4, 12.1)	–
Angiotensin II receptor, type 1 (*AGTR1*) (153)	3q24 A1166C	Taiwan Chinese Han	MVP syndrome	100	CC	–	.89
Angiotensinogen (*AGT*) (154)	1q42.2 M235T	Taiwan Chinese Han	MVP syndrome	100	TT	8.55 (4.51, 16.2)	< .001
Angiotensin-I-converting enzyme (*ACE*) (155)	17q23.3 Insertion/Deletion	Taiwan Chinese Han	MVP syndrome	100	I/I	2.14 (1.20, 3.80)	–

Abbreviations: MCTR, mitral chordae tendinae rupture; MVP, mitral valve prolapse; MVR, mitral valve replacement.

these genes and the regulatory pathways implicated in extracellular matrix remodeling.

Chou and colleagues examined a number of candidate SNPs using a case–control study of 100 Taiwanese subjects of Chinese descent who had myxomatous MVP. They found positive associations between MVP and polymorphisms in *Fibrillin-1* (exon 15 and 27) (145) and the collagen type III-alpha 1 gene (*COL3A1*, exon 31) (146) but failed to find an association with TGF-β1 polymorphisms (147). A separate case–control study of a pediatric Turkish cohort (n = 77) with early onset MVP also showed an association with an *FBN1* polymorphism in intron 51 (148). The urokinase plasminogen activator gene (*PLAU*) is involved in elastin and collagen degradation, and positive associations with *PLAU* polymorphisms were also seen in this cohort (149). Finally, investigators examined the association between matrix metalloproteinase-1 (*MMP1*) gene promoter polymorphisms, MMP1 expression, and the risk of mitral chordae tendinae rupture among a population of Taiwanese subjects undergoing mitral valve replacement (150). They found a positive association between MMP1 promoter polymorphisms, and increase in MMP1 gene expression and the risk for chordal rupture. Together, these intriguing findings suggest potential genetic modifiers on the extracellular matrix remodeling seen in MVP, but these studies need to be interpreted with caution given the small sample sizes, unique ethnic groups studied, and the lack of independent replication.

A separate line of inquiry has focused on the role of the renin–angiotensin–aldosterone system (RAAS) on the "MVP syndrome," which includes the MVP valve phenotype along with a constellation of nonspecific signs and symptoms unrelated to mitral valve regurgitation, including atypical chest pain, dyspnea, palpitations, anxiety, and/or recurrent syncope (120,121). Alterations in the autonomic and neuroendocrine systems and in RAAS have been implicated in this clinical syndrome, which has also been linked with ventricular arrhythmias and sudden death (151). Szombathy and colleagues were the first to examine the role of genetic variability in RAAS on the MVP syndrome. In a small cohort (n = 76) of MVP patients of European ancestry, the A1166C polymorphism within the angiotensin II type 1 receptor (*AGTR*) gene was found to be associated with MVP syndrome (152). This finding could not be replicated in Chou's Taiwan-Chinese population with MVP syndrome (153), but both the M235T polymorphism in the angiotensinogen (*c*) gene (154) and the insertion/deletion polymorphism in the angiotensin-1-converting enzyme (*ACE*) gene (155) were associated with MVP syndrome. Because these data have not been replicated in larger, more diverse populations,

it remains unclear whether genetic variability in the RAAS pathway plays an important, causal role in the development of the MVP syndrome.

Emerging Omic Technologies (Including GWAS)

To date, no GWAS have been published examining genetic associations with myxomatous mitral valve disease or MVP in human subjects. The low prevalence of MVP in the general population, the large number of cases needed to achieve sufficient statistical power, and the genetic heterogeneity of the disease process pose challenges to these types of large-scale association studies.

Future Directions

There is strong evidence that myxomatous mitral valve disease and MVP have a genetic basis, but demonstrate considerable genetic heterogeneity and variable expressivity. MVP is highly prevalent in Mendelian disorders involving the fibrillin and TGF-beta receptor pathways, but these account for less than 1% of all MVP cases. To date, the genetic basis of isolated, nonsyndromic MVP has not been elucidated, providing tremendous opportunity for gene discovery. Gene-network and pathway-specific analyses could be considered to leverage standard GWAS statistical power, or whole-exome sequencing of affected trios with sequence analysis for rare variations may provide new means of gene discovery.

Calcific Mitral Valve Disease

Calcific mitral valve disease is defined by calcium deposition along the mitral annulus, valvular, and/or subvalvular structures. MAC is a common finding among older adults that has been associated with increased cardiovascular morbidity and mortality (156–158), including coronary artery and peripheral vascular disease (159–163), atrial fibrillation (164–166), and stroke (167–169). Although extreme calcification occasionally leads to obstruction of blood flow into the left ventricle (170,171), most cases of MAC remain subclinical and detected incidentally on routine noninvasive imaging such as echocardiography or computed tomography. Prevalence estimates for MAC vary greatly based on imaging technology and population studied (ranging between 3% and 42%) (157,164,172), but increase significantly with age (156,164,172).

Several clinical factors have been associated with the presence and severity of MAC, including traditional cardiovascular risk factors (7,173), end-stage renal disease with impaired mineral metabolism (174,175), and chronic inflammatory conditions such as lupus erythematosus (176,177). Previously attributed to simple "wear-and-tear" degeneration, valvular calcification is now recognized as an active metabolic process, and the complex biological pathways regulating progressive calcification have begun to be elucidated in recent years. Inflammatory cytokines, reactive oxygen species, and oxidized lipids promote a pro-osteogenic milieu involving bone morphogen and wnt/beta-catenin signaling (178). In response, local myofibroblasts and vascular smooth muscle cells develop a pro-osteogenic phenotype and begin to respond to additional paracrine and circulating factors, including calcium phosphate (via the PIT1 receptor) (179) and the osteoprotegerin/receptor activator of NF-κβ axis (180,181). This in turn leads to biomineralization of the extracellular matrix, a process that is inhibited by Fetuin A (182,183). While the pathobiology of MAC has been less well studied than vascular or AVC, the biological processes underlying these changes are likely shared.

Heritability

Because MAC has been presumed to be an age-related, degenerative phenomenon, its heritability has not been extensively studied. Perhaps the best evidence for the heritability of MAC comes from the HyperGEN Study, in which a total of 78 sibships from 68 families with 2 or more affected siblings were studied (30). In this study, the sibling recurrence risk and sibling recurrence risk ratio for MAC were 0.25 (±0.035) and 1.78 (95% CI: 1.36, 2.33), respectively. This finding suggests that although environmental factors contribute to MAC, there appear to be strong genetic influences on this process.

Linkage

Because the onset of MAC often begins only during the seventh or eighth decade of life, multigenerational pedigrees of well-phenotyped individuals have not been available for study. The intragenerational, sibpair linkage analysis performed in the HyperGEN Study failed to identify any chromosomal loci associated with MAC (defined as LOD greater than 1.9) (30), though this study may not have been sufficiently powered.

Candidate Gene Studies

Because of the challenges in recruiting large families with MAC, investigators have also utilized a candidate-SNP approach. Novaro and colleagues examined the association between apolipoprotein E (*ApoE*) alleles and MAC among 802 patients, but found no difference in genotype distributions between those with and without MAC (184). An additional study found no association between a SNP locus (rs1333049) that had been linked with MI and MAC (185). Thus to date, no genetic associations of MAC have been identified using a candidate gene approach.

Emerging Omic Technologies (Including GWAS)

The genetic associations of MAC were more thoroughly evaluated by Thanassoulis and colleagues in a CHARGE Consortium collaboration (45). In this study, MAC was identified from computer tomography scans among subjects of White European descent within the Multi-Ethnic Study of Atherosclerosis (n = 2527, 12%) and the Framingham Heart Study (n = 1298, MAC prevalence 20%). A meta-analytic GWAS identified two SNPs that reached the level of genome-wide significance (P less than 5×10^{-8}). These two SNPs (designated rs17659543 and rs13415097) were in high linkage disequilibrium and were located on chromosome 2 near the *IL1F9* gene cluster. After adjustment for age and sex, the per-allele risk of MAC was estimated to be 1.66 (95% CI: 1.39, 1.98). Attempts to replicate these findings were inconsistent, as no association was seen between rs1769543 and MAC among 745 White European participants in the Heinz-Nixdorf Recall (HNR) study or among African American participants in MESA (n = 2495, P = .30), but did replicate among Hispanic Americans in MESA (n = 2025, P = .04). These results suggest a potential role for pro-inflammatory pathways in the development of MAC, but need to be interpreted with caution in view of the inconsistent replication.

Future Directions

MAC is a common finding among older adults, and although historically thought to be an acquired disorder, MAC also appears to have heritable influences. While a single GWAS has suggested that genetic variability in the pro-inflammatory *IL1F9* gene may be associated with MAC, this finding is not well validated. Because of the complex biological pathways that underlie vascular and valvular calcification, it is likely that multiple genetic factors influence the process. While additional GWAS investigations with greater number of participants may increase statistical power to detect associations, a gene-network or pathway-specific approach may also be beneficial.

CONCLUSIONS

Valvular heart disease is a common cardiovascular condition associated with significant morbidity and mortality. Although there are fewer studies on the genetics of valvular disease compared to other cardiovascular traits, several important studies have added considerably to our understanding of the heritability, genetics, and molecular pathogenesis of valvular heart disease, especially left-sided valvular lesions. Here we have presented a comprehensive review of the genetics and genomics of aortic and mitral valve diseases. The robust results of a GWAS showing the role of variation in the *LPA* gene and Lp(a) levels in calcific aortic valve disease demonstrate the power of such studies in the omics era, and provide hope that additional high-quality studies will further illuminate novel molecular pathways and targets in the realm of valvular heart disease.

REFERENCES

1. Roger VL, Go AS, Lloyd-Jones DM, et al. Heart disease and stroke statistics—2012 update: a report from the American Heart Association. *Circulation.* 2012;125(1):e2–e220.
2. Nkomo VT, Gardin JM, Skelton TN, et al. Burden of valvular heart diseases: A population-based study. *Lancet.* 2006;368(9540):1005–1011.
3. Otto CM, Lind BK, Kitzman DW, et al. Association of aortic-valve sclerosis with cardiovascular mortality and morbidity in the elderly. *N Engl J Med.* 1999;341(3):142–147.
4. Lindroos M, Kupari M, Heikkila J, Tilvis R. Prevalence of aortic valve abnormalities in the elderly: an echocardiographic study of a random population sample. *J Am Coll Cardiol.* 1993;21(5):1220–1225.
5. Owens DS, Katz R, Takasu J, et al. Incidence and progression of aortic valve calcium in the multi-ethnic study of atherosclerosis (MESA). *Am J Cardiol.* 2010;105(5):701–708.
6. Stewart BF, Siscovick D, Lind BK, et al. Clinical factors associated with calcific aortic valve disease. cardiovascular health study. *J Am Coll Cardiol.* 1997;29(3):630–634.
7. Thanassoulis G, Massaro JM, Cury R, et al. Associations of long-term and early adult atherosclerosis risk factors with aortic and mitral valve calcium. *J Am Coll Cardiol.* 2010;55(22):2491–2498.
8. Otto CM, Kuusisto J, Reichenbach DD, et al. Characterization of the early lesion of 'degenerative' valvular aortic stenosis. histological and immunohistochemical studies. *Circulation.* 1994;90(2):844–853.
9. Rajamannan NM, Edwards WD, Spelsberg TC. Hypercholesterolemic aortic-valve disease. *N Engl J Med.* 2003;349(7):717–718.
10. Barr DP, Rothbard S, Eder HA. Atherosclerosis and aortic stenosis in hypercholesteremic xanthomatosis. *J Am Med Assoc.* 1954;156(10):943–947.
11. Weiss RM, Ohashi M, Miller JD, et al. Calcific aortic valve stenosis in old hypercholesterolemic mice. *Circulation.* 2006;114(19):2065–2069.
12. Miller JD, Weiss RM, Serrano KM, et al. Lowering plasma cholesterol levels halts progression of aortic valve disease in mice. *Circulation.* 2009;119(20):2693–2701.
13. Novaro GM, Tiong IY, Pearce GL, et al. Effect of hydroxymethylglutaryl coenzyme a reductase inhibitors on the progression of calcific aortic stenosis. *Circulation.* 2001;104(18):2205–2209.
14. Shavelle DM, Takasu J, Budoff MJ, et al. HMG CoA reductase inhibitor (statin) and aortic valve calcium. *Lancet.* 2002;359(9312):1125–1126.
15. Rossebo AB, Pedersen TR, Boman K, et al. Intensive lipid lowering with simvastatin and ezetimibe in aortic stenosis. *N Engl J Med.* 2008;359(13):1343–1356.
16. Moura LM, Ramos SF, Zamorano JL, et al. Rosuvastatin affecting aortic valve endothelium to slow the progression of aortic stenosis. *J Am Coll Cardiol.* 2007;49(5):554–561.
17. Cowell SJ, Newby DE, Prescott RJ, et al. A randomized trial of intensive lipid-lowering therapy in calcific aortic stenosis. *N Engl J Med.* 2005;352(23): 2389–2397.
18. Towler DA. Molecular and cellular aspects of calcific aortic valve disease. *Circ Res.* 2013;113(2):198–208.
19. Laird DF, Mucalo MR, Yokogawa Y. Growth of calcium hydroxyapatite (ca-HAp) on cholesterol and cholestanol crystals from a simulated body fluid: A possible insight into the pathological calcifications associated with atherosclerosis. *J Colloid Interface Sci.* 2006;295(2):348–363.
20. Fartasch M, Haneke E, Hornstein OP. Mineralization of collagen and elastic fibers in superficial dystrophic cutaneous calcification: an ultrastructural study. *Dermatologica.* 1990;181(3):187–192.
21. Srivatsa SS, Harrity PJ, Maercklein PB, et al. Increased cellular expression of matrix proteins that regulate mineralization is associated with calcification of native human and porcine xenograft bioprosthetic heart valves. *J Clin Invest.* 1997;99(5):996–1009.
22. Kapustin AN, Davies JD, Reynolds JL, et al. Calcium regulates key components of vascular smooth muscle cell-derived matrix vesicles to enhance mineralization. *Circ Res.* 2011;109(1):e1–e12.
23. Kiffer-Moreira T, Yadav MC, Zhu D, et al. Pharmacological inhibition of PHOSPHO1 suppresses vascular smooth muscle cell calcification. *J Bone Miner Res.* 2013;28(1):81–91.
24. Kim KM. Calcification of matrix vesicles in human aortic valve and aortic media. *Fed Proc.* 1976;35(2): 156–162.
25. Egan KP, Kim JH, Mohler ER,3rd, Pignolo RJ. Role for circulating osteogenic precursor cells in aortic valvular disease. *Arterioscler Thromb Vasc Biol.* 2011; 31(12):2965–2971.
26. Fadini GP, Albiero M, Menegazzo L, et al. Widespread increase in myeloid calcifying cells contributes to ectopic vascular calcification in type 2 diabetes. *Circ Res.* 2011;108(9):1112–1121.
27. Ortlepp JR, Hoffmann R, Ohme F,et al. The vitamin D receptor genotype predisposes to the development of calcific aortic valve stenosis. *Heart.* 2001; 85(6):635–638.
28. Schmitz F, Ewering S, Zerres K, et al. Parathyroid hormone gene variant and calcific aortic stenosis. *J Heart Valve Dis.* 2009;18(3):262–267.
29. Horne BD, Camp NJ, Muhlestein JB, Cannon-Albright LA. Evidence for a heritable component in death resulting from aortic and mitral valve diseases. *Circulation.* 2004;110(19):3143–3148.
30. Bella JN, Tang W, Kraja A, et al. Genome-wide linkage mapping for valve calcification susceptibility loci in hypertensive sibships: the hypertension genetic epidemiology network study. *Hypertension.* 2007; 49(3):453–460.
31. Le Gal G, Bertault V, Bezon E, et al. Heterogeneous geographic distribution of patients with aortic valve stenosis: arguments for new aetiological hypothesis. *Heart.* 2005;91(2):247–249.

32. Probst V, Le Scouarnec S, Legendre A, et al. Familial aggregation of calcific aortic valve stenosis in the western part of France. *Circulation.* 2006;113(6):856–860.

33. Morgan TM, Krumholz HM, Lifton RP, Spertus JA. Nonvalidation of reported genetic risk factors for acute coronary syndrome in a large-scale replication study. *JAMA.* 2007;297(14):1551–1561.

34. Avakian SD, Annicchino-Bizzacchi JM, Grinberg M, et al. Apolipoproteins AI, B, and E polymorphisms in severe aortic valve stenosis. *Clin Genet.* 2001; 60(5):381–384.

35. Novaro GM, Sachar R, Pearce GL, et al. Association between apolipoprotein E alleles and calcific valvular heart disease. *Circulation.* 2003;108(15):1804–1808.

36. Ortlepp JR, Schmitz F, Mevissen V, et al. The amount of calcium-deficient hexagonal hydroxyapatite in aortic valves is influenced by gender and associated with genetic polymorphisms in patients with severe calcific aortic stenosis. *Eur Heart J.* 2004;25(6):514–522.

37. Gaudreault N, Ducharme V, Lamontagne M, et al. Replication of genetic association studies in aortic stenosis in adults. *Am J Cardiol.* 2011;108(9):1305–1310.

38. Nordstrom P, Glader CA, Dahlen G, et al. Oestrogen receptor alpha gene polymorphism is related to aortic valve sclerosis in postmenopausal women. *J Intern Med.* 2003;254(2):140–146.

39. Garg V, Muth AN, Ransom JF, et al. Mutations in NOTCH1 cause aortic valve disease. *Nature.* 2005; 437(7056):270–274.

40. Ducharme V, Guauque-Olarte S, Gaudreault N, et al. NOTCH1 genetic variants in patients with tricuspid calcific aortic valve stenosis. *J Heart Valve Dis.* 2013; 22(2):142–149.

41. Nigam V, Srivastava D. Notch1 represses osteogenic pathways in aortic valve cells. *J Mol Cell Cardiol.* 2009;47(6):828–834.

42. Ellis SG, Dushman-Ellis S, Luke MM, et al. Pilot candidate gene analysis of patients >/= 60 years old with aortic stenosis involving a tricuspid aortic valve. *Am J Cardiol.* 2012;110(1):88–92.

43. Helske S, Lindstedt KA, Laine M, et al. Induction of local angiotensin II-producing systems in stenotic aortic valves. *J Am Coll Cardiol.* 2004;44(9):1859–1866.

44. Nadir MA, Wei L, Elder DH, et al. Impact of renin-angiotensin system blockade therapy on outcome in aortic stenosis. *J Am Coll Cardiol.* 2011;58(6):570–576.

45. Thanassoulis G, Campbell CY, Owens DS, et al. Genetic associations with valvular calcification and aortic stenosis. *N Engl J Med.* 2013;368(6):503–512.

46. Marincheva-Savcheva G, Subramanian S, Qadir S, et al. Imaging of the aortic valve using fluorodeoxyglucose positron emission tomography increased valvular fluorodeoxyglucose uptake in aortic stenosis. *J Am Coll Cardiol.* 2011;57(25):2507–2515.

47. Dweck MR, Jones C, Joshi NV, et al. Assessment of valvular calcification and inflammation by positron emission tomography in patients with aortic stenosis. *Circulation.* 2012;125(1):76–86.

48. Psaty BM, O'Donnell CJ, Gudnason V, et al. Cohorts for heart and aging research in genomic epidemiology (CHARGE) consortium: design of prospective meta-analyses of genome-wide association studies from 5 cohorts. *Circ Cardiovasc Genet.* 2009;2(1):73–80.

49. Tzemos N, Therrien J, Yip J, et al. Outcomes in adults with bicuspid aortic valves. *JAMA.* 2008;300(11): 1317–1325.

50. Michelena HI, Khanna AD, Mahoney D, et al. Incidence of aortic complications in patients with bicuspid aortic valves. *JAMA.* 2011;306(10):1104–1112.

51. Larson EW, Edwards WD. Risk factors for aortic dissection: A necropsy study of 161 cases. *Am J Cardiol.* 1984;53(6):849–855.

52. Roberts WC, Vowels TJ, Ko JM. Natural history of adults with congenitally malformed aortic valves (unicuspid or bicuspid). *Medicine (Baltimore).* 2012; 91(6):287–308.

53. Sabet HY, Edwards WD, Tazelaar HD, Daly RC. Congenitally bicuspid aortic valves: a surgical pathology study of 542 cases (1991 through 1996) and a literature review of 2,715 additional cases. *Mayo Clin Proc.* 1999;74(1):14–26.

54. Ward C. Clinical significance of the bicuspid aortic valve. *Heart.* 2000;83(1):81–85.

55. Osler W. Bicuspid condition of the semilunar valves and its relation to aortic valve disease. *JAMA.* 1886:49–50.

56. Tutar E, Ekici F, Atalay S, Nacar N. The prevalence of bicuspid aortic valve in newborns by echocardiographic screening. *Am Heart J.* 2005;150(3):513–515.

57. Movahed MR, Hepner AD, Ahmadi-Kashani M. Echocardiographic prevalence of bicuspid aortic valve in the population. *Heart Lung Circ.* 2006;15(5):297–299.

58. Basso C, Boschello M, Perrone C, et al. An echocardiographic survey of primary school children for bicuspid aortic valve. *Am J Cardiol.* 2004;93(5):661–663.

59. Khan W, Milsevic M, Salciccioli L, Lazar J. Low prevalence of bicuspid aortic valve in African Americans. *Am Heart J.* 2008;156(3):e25.

60. Loscalzo ML, Goh DL, Loeys B, et al. Familial thoracic aortic dilation and bicommissural aortic valve: a prospective analysis of natural history and inheritance. *Am J Med Genet A.* 2007;143A(17):1960–1967.

61. Kent KC, Crenshaw ML, Goh DL, Dietz HC. Genotype-phenotype correlation in patients with bicuspid aortic valve and aneurysm. *J Thorac Cardiovasc Surg.* 2013;146(1):158–165.e1.

62. Majumdar R, Yagubyan M, Sarkar G, et al. Bicuspid aortic valve and ascending aortic aneurysm are not associated with germline or somatic homeobox NKX2-5 gene polymorphism in 19 patients. *J Thorac Cardiovasc Surg.* 2006;131(6):1301–1305.

63. Arrington CB, Sower CT, Chuckwuk N, et al. Absence of TGFBR1 and TGFBR2 mutations in patients with bicuspid aortic valve and aortic dilation. *Am J Cardiol.* 2008;102(5):629–631.

64. Foffa I, Murzi M, Mariani M, et al. Angiotensin-converting enzyme insertion/deletion polymorphism is a risk factor for thoracic aortic aneurysm in patients with bicuspid or tricuspid aortic valves. *J Thorac Cardiovasc Surg.* 2012;144(2):390–395.

65. Girdauskas E, Schulz S, Borger MA, et al. Transforming growth factor-beta receptor type II mutation in a patient with bicuspid aortic valve disease and intraoperative aortic dissection. *Ann Thorac Surg.* 2011;91(5):e70–1.

66. Duran AC, Frescura C, Sans-Coma V, et al. Bicuspid aortic valves in hearts with other congenital heart disease. *J Heart Valve Dis.* 1995;4(6):581–590.

67. Hou JW, Hwu WL, Tsai WY, et al. Cardiovascular disorders in turner's syndrome and its correlation to karyotype. *J Formos Med Assoc.* 1993;92(2):188–189.

68. Sybert VP. Cardiovascular malformations and complications in turner syndrome. *Pediatrics.* 1998;101(1):E11.

69. Hallidie-Smith KA, Karas S. Cardiac anomalies in Williams-Beuren syndrome. *Arch Dis Child.* 1988;63(7):809–813.

70. Del Pasqua A, Rinelli G, Toscano A, et al. New findings concerning cardiovascular manifestations emerging from long-term follow-up of 150 patients with the Williams-Beuren-Beuren syndrome. *Cardiol Young.* 2009;19(6):563–567.

71. De Rubens Figueroa J, Rodriguez LM, Hach JL, et al. Cardiovascular spectrum in williams-beuren syndrome: the Mexican experience in 40 patients. *Tex Heart Inst J.* 2008;35(3):279–285.

72. Stevens CA, Bhakta MG. Cardiac abnormalities in the Rubinstein-Taybi syndrome. *Am J Med Genet.* 1995;59(3):346–348.

73. Andelfinger G, Tapper AR, Welch RC, Vanoye CG, et al. KCNJ2 mutation results in Andersen syndrome with sex-specific cardiac and skeletal muscle phenotypes. *Am J Hum Genet.* 2002;71(3):663–668.

74. Lopez-Rangel E, Maurice M, McGillivray B, Friedman JM. Williams syndrome in adults. *Am J Med Genet.* 1992;44(6):720–729.

75. McBride KL, Pignatelli R, Lewin M, et al. Inheritance analysis of congenital left ventricular outflow tract obstruction malformations: Segregation, multiplex relative risk, and heritability. *Am J Med Genet A.* 2005;134A(2):180–186.

76. Hinton RB, Martin LJ, Rame-Gowda S, et al. Hypoplastic left heart syndrome links to chromosomes 10q and 6q and is genetically related to bicuspid aortic valve. *J Am Coll Cardiol.* 2009;53(12):1065–1071.

77. Wessels MW, Berger RM, Frohn-Mulder IM, et al. Autosomal dominant inheritance of left ventricular outflow tract obstruction. *Am J Med Genet A.* 2005;134A(2):171–179.

78. Foffa I, Ait Ali L, Panesi P, et al. Sequencing of NOTCH1, GATA5, TGFBR1 and TGFBR2 genes in familial cases of bicuspid aortic valve. *BMC Med Genet.* 2013;14:44. doi:10.1186/1471-2350-14-44

79. Gale AN, McKusick VA, Hutchins GM, Gott VL. Familial congenital bicuspid aortic valve: secondary calcific aortic stenosis and aortic aneurysm. *Chest.* 1977;72(5):668–670.

80. Feizi O, Farrer Brown G, Emanuel R. Familial study of hypertrophic cardiomyopathy and congenital aortic valve disease. *Am J Cardiol.* 1978;41(5):956–964.

81. McKusick VA. Association of congenital bicuspid aortic valve and Erdheim's cystic medial necrosis. *Lancet.* 1972;1(7758):1026–1027.

82. Emanuel R, Withers R, O'Brien K, et al. Congenitally bicuspid aortic valves. clinicogenetic study of 41 families. *Br Heart J.* 1978;40(12):1402–1407.

83. Godden DJ, Sandhu PS, Kerr F. Stenosed bicuspid aortic valves in twins. *Eur Heart J.* 1987;8(3):316–318.

84. Brown C, Sane DC, Kitzman DW. Bicuspid aortic valves in monozygotic twins. *Echocardiography.* 2003;20(2):183–184.

85. Clementi M, Notari L, Borghi A, Tenconi R. Familial congenital bicuspid aortic valve: a disorder of uncertain inheritance. *Am J Med Genet.* 1996;62(4):336–338.

86. Glick BN, Roberts WC. Congenitally bicuspid aortic valve in multiple family members. *Am J Cardiol.* 1994;73(5):400–404.

87. McDonald K, Maurer BJ. Familial aortic valve disease: evidence for a genetic influence? *Eur Heart J.* 1989;10(7):676–677.

88. Huntington K, Hunter AG, Chan KL. A prospective study to assess the frequency of familial clustering of congenital bicuspid aortic valve. *J Am Coll Cardiol.* 1997;30(7):1809–1812.

89. Ellison JW, Yagubyan M, Majumdar R, et al. Evidence of genetic locus heterogeneity for familial bicuspid aortic valve. *J Surg Res.* 2007;142(1):28–31.

90. Robledo-Carmona J, Rodriguez-Bailon I, Carrasco-Chinchilla F, et al. Hereditary patterns of bicuspid aortic valve in a hundred families. *Int J Cardiol.* 2013.

91. Cripe L, Andelfinger G, Martin LJ, et al. Bicuspid aortic valve is heritable. *J Am Coll Cardiol.* 2004;44(1):138–143.

92. Calloway TJ, Martin LJ, Zhang X, et al. Risk factors for aortic valve disease in bicuspid aortic valve: a family-based study. *Am J Med Genet A.* 2011;155A(5):1015–1020.

93. Gelb BD, Zhang J, Sommer RJ, et al. Familial patent ductus arteriosus and bicuspid aortic valve with hand anomalies: A novel heart-hand syndrome. *Am J Med Genet.* 1999;87(2):175–179.

94. Guo DC, Pannu H, Tran-Fadulu V, et al. Mutations in smooth muscle alpha-actin (ACTA2) lead to thoracic aortic aneurysms and dissections. *Nat Genet.* 2007;39(12):1488–1493.

95. Guo DC, Papke CL, Tran-Fadulu V, et al. Mutations in smooth muscle alpha-actin (ACTA2) cause coronary artery disease, stroke, and moyamoya disease, along with thoracic aortic disease. *Am J Hum Genet.* 2009;84(5):617–627.

96. Goh D, Han L, Judge DP, et al. Linkage of familial bicuspid aortic valve with aortic aneurysm to chromosome 15q. *Am J Hum Genet.* 2002;71:211.

97. Martin LJ, Ramachandran V, Cripe LH, et al. Evidence in favor of linkage to human chromosomal regions 18q, 5q and 13q for bicuspid aortic valve and associated cardiovascular malformations. *Hum Genet.* 2007;121(2):275–284.

98. McBride KL, Zender GA, Fitzgerald-Butt SM, et al. Linkage analysis of left ventricular outflow tract malformations (aortic valve stenosis, coarctation of the aorta, and hypoplastic left heart syndrome). *Eur J Hum Genet.* 2009;17(6):811–819.

99. Mohamed SA, Aherrahrou Z, Liptau H, et al. Novel missense mutations (p.T596M and p.P1797H) in NOTCH1 in patients with bicuspid aortic valve. *Biochem Biophys Res Commun.* 2006;345(4):1460–1465.

100. McBride KL, Riley MF, Zender GA, et al. NOTCH1 mutations in individuals with left ventricular outflow tract malformations reduce ligand-induced signaling. *Hum Mol Genet.* 2008;17(18):2886–2893.

101. McKellar SH, Tester DJ, Yagubyan M, Majumdar R, Ackerman MJ, Sundt TM,3rd. Novel NOTCH1 mutations in patients with bicuspid aortic valve disease and thoracic aortic aneurysms. *J Thorac Cardiovasc Surg.* 2007;134(2):290–296.

102. Attias D, Stheneur C, Roy C, et al. Comparison of clinical presentations and outcomes between patients with TGFBR2 and FBN1 mutations in Marfan syndrome and related disorders. *Circulation.* 2009;120(25):2541–2549.

103. Lee TC, Zhao YD, Courtman DW, Stewart DJ. Abnormal aortic valve development in mice lacking endothelial nitric oxide synthase. *Circulation*. 2000; 101(20):2345–2348.

104. Padang R, Bagnall RD, Richmond DR, et al. Rare non-synonymous variations in the transcriptional activation domains of GATA5 in bicuspid aortic valve disease. *J Mol Cell Cardiol*. 2012;53(2):277–281.

105. Saravanan P, Kadir I. Apolipoprotein E alleles and bicuspid aortic valve stenosis in monozygotic twins. *Interact Cardiovasc Thorac Surg*. 2009;8(6):687–688.

106. Tan HL, Glen E, Topf A, et al. Nonsynonymous variants in the SMAD6 gene predispose to congenital cardiovascular malformation. *Hum Mutat*. 2012; 33(4):720–727.

107. Combs MD, Yutzey KE. Heart valve development: Regulatory networks in development and disease. *Circ Res*. 2009;105(5):408–421.

108. Acharya A, Hans CP, Koenig SN, et al. Inhibitory role of Notch1 in calcific aortic valve disease. *PLoS One*. 2011;6(11):e27743.

109. Lemaire SA, McDonald ML, Guo DC, et al. Genome-wide association study identifies a susceptibility locus for thoracic aortic aneurysms and aortic dissections spanning FBN1 at 15q21.1. *Nat Genet*. 2011; 43(10):996–1000.

110. Wooten EC, Iyer LK, Montefusco MC, et al. Application of gene network analysis techniques identifies AXIN1/PDIA2 and endoglin haplotypes associated with bicuspid aortic valve. *PLoS One*. 2010;5(1):e8830.

111. Hitz MP, Lemieux-Perreault LP, Marshall C, et al. Rare copy number variants contribute to congenital left-sided heart disease. *PLoS Genet*. 2012;8(9): e1002903.

112. Brook GJ, Keidar S, Boulos M, et al. Familial homozygous hypercholesterolemia: clinical and cardiovascular features in 18 patients. *Clin Cardiol*. 1989; 12(6):333–338.

113. Hannoush H, Introne WJ, Chen MY, et al. Aortic stenosis and vascular calcifications in alkaptonuria. *Mol Genet Metab*. 2012;105(2):198–202.

114. Phornphutkul C, Introne WJ, Perry MB, et al. Natural history of alkaptonuria. *N Engl J Med*. 2002; 347(26):2111–2121.

115. Hortop J, Tsipouras P, Hanley JA, et al. Cardiovascular involvement in osteogenesis imperfecta. *Circulation*. 1986;73(1):54–61.

116. Buntinx IM, Willems PJ, Spitaels SE, et al. Neonatal Marfan syndrome with congenital arachnodactyly, flexion contractures, and severe cardiac valve insufficiency. *J Med Genet*. 1991;28(4):267–273.

117. Disabella E, Grasso M, Marziliano N, et al. Two novel and one known mutation of the TGFBR2 gene in marfan syndrome not associated with FBN1 gene defects. *Eur J Hum Genet*. 2006;14(1):34–38.

118. Schwarze U, Hata R, McKusick VA, et al. Rare autosomal recessive cardiac valvular form of Ehlers-Danlos syndrome results from mutations in the COL1A2 gene that activate the nonsense-mediated RNA decay pathway. *Am J Hum Genet*. 2004;74(5):917–930.

119. Kyndt F, Gueffet JP, Probst V, et al. Mutations in the gene encoding filamin A as a cause for familial cardiac valvular dystrophy. *Circulation*. 2007;115(1):40–49.

120. Guy TS, Hill AC. Mitral valve prolapse. *Annu Rev Med*. 2012;63:277–292.

121. Hayek E, Gring CN, Griffin BP. Mitral valve prolapse. *Lancet*. 2005;365(9458):507–518.

122. Levine RA, Slaugenhaupt SA. Molecular genetics of mitral valve prolapse. *Curr Opin Cardiol*. 2007; 22(3):171–175.

123. Rabkin E, Aikawa M, Stone JR, et al. Activated interstitial myofibroblasts express catabolic enzymes and mediate matrix remodeling in myxomatous heart valves. *Circulation*. 2001;104(21):2525–2532.

124. Freed LA, Levy D, Levine RA, et al. Prevalence and clinical outcome of mitral-valve prolapse. *N Engl J Med*. 1999;341(1):1–7.

125. Theal M, Sleik K, Anand S, et al. Prevalence of mitral valve prolapse in ethnic groups. *Can J Cardiol*. 2004; 20(5):511–515.

126. Ng CM, Cheng A, Myers LA, et al. TGF-beta-dependent pathogenesis of mitral valve prolapse in a mouse model of marfan syndrome. *J Clin Invest*. 2004;114(11):1586–1592.

127. Bon Tempo CP, Ronan JA, Jr, de Leon AC, Jr, Twigg HL. Radiographic appearance of the thorax in systolic click-late systolic murmur syndrome. *Am J Cardiol*. 1975;36(1):27–31.

128. Hancock EW, Cohn K. The syndrome associated with midsystolic click and late systolic murmur. *Am J Med*. 1966;41(2):183–196.

129. Cooper MJ, Abinader EG. Family history in assessing the risk for progression of mitral valve prolapse. Report of a kindred. *Am J Dis Child*. 1981;135(7): 647–649.

130. Weiss AN, Mimbs JW, Ludbrook PA, Sobel BE. Echocardiographic detection of mitral valve prolapse. exclusion of false positive diagnosis and determination of inheritance. *Circulation*. 1975;52(6):1091–1096.

131. Shappell SD, Marshall CE, Brown RE, Bruce TA. Sudden death and the familial occurrence of midsystolic click, late systolic murmur syndrome. *Circulation*. 1973;48(5):1128–1134.

132. Devereux RB, Brown WT, Kramer-Fox R, Sachs I. Inheritance of mitral valve prolapse: Effect of age and sex on gene expression. *Ann Intern Med*. 1982; 97(6):826–832.

133. Strahan NV, Murphy EA, Fortuin NJ, et al. Inheritance of the mitral valve prolapse syndrome. discussion of a three-dimensional penetrance model. *Am J Med*. 1983;74(6):967–972.

134. Rizzon P, Biasco G, Brindicci G, Mauro F. Familial syndrome of midsystolic click and late systolic murmur. *Br Heart J*. 1973;35(3):245–259.

135. Girard DE, Girard JB. Mitral valve prolapse-click syndrome in twins. *Am Heart J*. 1977;94(6):813–815.

136. Henney AM, Tsipouras P, Schwartz RC, et al. Genetic evidence that mutations in the COL1A1, COL1A2, COL3A1, or COL5A2 collagen genes are not responsible for mitral valve prolapse. *Br Heart J*. 1989; 61(3):292–299.

137. Wordsworth P, Ogilvie D, Akhras F, et al. Genetic segregation analysis of familial mitral valve prolapse shows no linkage to fibrillar collagen genes. *Br Heart J*. 1989;61(3):300–306.

138. Disse S, Abergel E, Berrebi A, et al. Mapping of a first locus for autosomal dominant myxomatous mitral-valve prolapse to chromosome 16p11.2-p12.1. *Am J Hum Genet*. 1999;65(5):1242–1251.

139. Nesta F, Leyne M, Yosefy C, et al. New locus for autosomal dominant mitral valve prolapse on chromosome 13: Clinical insights from genetic studies. *Circulation*. 2005;112(13):2022–2030.

140. Freed LA, Acierno JS, Jr, Dai D, et al. A locus for autosomal dominant mitral valve prolapse on chromosome 11p15.4. *Am J Hum Genet*. 2003;72(6): 1551–1559.

141. Monteleone PL, Fagan LF. Possible X-linked congenital heart disease. *Circulation*. 1969;39(5):611–614.

142. Kyndt F, Schott JJ, Trochu JN, et al. Mapping of X-linked myxomatous valvular dystrophy to chromosome Xq28. *Am J Hum Genet*. 1998;62(3):627–632.

143. Lardeux A, Kyndt F, Lecointe S, et al. Filamin-a-related myxomatous mitral valve dystrophy: genetic, echocardiographic and functional aspects. *J Cardiovasc Transl Res*. 2011;4(6):748–756.

144. Sauls K, de Vlaming A, Harris BS, et al. Developmental basis for filamin-A-associated myxomatous mitral valve disease. *Cardiovasc Res*. 2012;96(1):109–119.

145. Chou HT, Shi YR, Hsu Y, Tsai FJ. Association between fibrillin-1 gene exon 15 and 27 polymorphisms and risk of mitral valve prolapse. *J Heart Valve Dis*. 2003;12(4):475–481.

146. Chou HT, Hung JS, Chen YT, Wu JY, Tsai FJ. Association between COL3A1 collagen gene exon 31 polymorphism and risk of floppy mitral valve/mitral valve prolapse. *Int J Cardiol*. 2004;95(2–3):299–305.

147. Chou HT, Shi YR, Hsu Y, Tsai FJ. Lack of association between transforming growth factor-beta1 gene polymorphisms and mitral valve prolapse in Taiwan Chinese. *J Heart Valve Dis*. 2002;11(4):478–484.

148. Ozdemir O, Olgunturk R, Karaer K, et al. Fibrillin-1 gene intron 56 polymorphism in Turkish children with mitral valve prolapse. *Cardiol Young*. 2010;20(2): 173–177.

149. Chou HT, Chen YT, Wu JY, Tsai FJ. Association between urokinase-plasminogen activator gene T4065C polymorphism and risk of mitral valve prolapse. *Int J Cardiol*. 2004;96(2):165–170.

150. Lin TH, Yang SF, Chiu CC, et al. Matrix metalloproteinase-1 mitral expression and -1607 1G/2G gene promoter polymorphism in mitral chordae tendinae rupture. *Transl Res*. 2013;161(5):406–413.

151. Sriram CS, Syed FF, Ferguson ME, et al. Malignant bileaflet mitral valve prolapse syndrome in patients with otherwise idiopathic out-of-hospital cardiac arrest. *J Am Coll Cardiol*. 2013;62(3):222–230.

152. Szombathy T, Janoskuti L, Szalai C, et al. Angiotensin II type 1 receptor gene polymorphism and mitral valve prolapse syndrome. *Am Heart J*. 2000;139(1) (Pt 1):101–105.

153. Chou HT, Shi YR, Wu JY, Tsai FJ. Angiotensin II type 1 receptor gene adenine/cytosine1166 polymorphism is not associated with mitral valve prolapse syndrome in Taiwan Chinese. *Circ J*. 2002;66(2):163–166.

154. Chou HT, Hung JS, Chen YT, et al. Association between angiotensinogen gene M235T polymorphism and mitral valve prolapse syndrome in Taiwan Chinese. *J Heart Valve Dis*. 2002;11(6):830–836.

155. Chou HT, Chen YT, Shi YR, Tsai FJ. Association between angiotensin I-converting enzyme gene insertion/deletion polymorphism and mitral valve prolapse syndrome. *Am Heart J*. 2003;145(1):169–173.

156. Fox CS, Vasan RS, Parise H, et al. Mitral annular calcification predicts cardiovascular morbidity and mortality: The Framingham Heart Study. *Circulation*. 2003;107(11):1492–1496.

157. Fox E, Harkins D, Taylor H, et al. Epidemiology of mitral annular calcification and its predictive value for coronary events in African Americans: the Jackson cohort of the atherosclerotic risk in communities study. *Am Heart J*. 2004;148(6):979–984.

158. Barasch E, Gottdiener JS, Marino Larsen EK, et al. Cardiovascular morbidity and mortality in community-dwelling elderly individuals with calcification of the fibrous skeleton of the base of the heart and aortosclerosis (the cardiovascular health study). *Am J Cardiol*. 2006;97(9):1281–1286.

159. Allison MA, Cheung P, Criqui MH, et al. Mitral and aortic annular calcification are highly associated with systemic calcified atherosclerosis. *Circulation*. 2006; 113(6):861–866.

160. Adler Y, Herz I, Vaturi M, et al. Mitral annular calcium detected by transthoracic echocardiography is a marker for high prevalence and severity of coronary artery disease in patients undergoing coronary angiography. *Am J Cardiol*. 1998;82(10):1183–1186.

161. Hamirani YS, Nasir K, Blumenthal RS, et al. Relation of mitral annular calcium and coronary calcium (from the multi-ethnic study of atherosclerosis [MESA]). *Am J Cardiol*. 2011;107(9):1291–1294.

162. Kohsaka S, Jin Z, Rundek T, et al. Impact of mitral annular calcification on cardiovascular events in a multiethnic community: the northern Manhattan study. *JACC Cardiovasc Imaging*. 2008;1(5):617–623.

163. Varma R, Aronow WS, McClung JA, et al. Prevalence of valve calcium and association of valve calcium with coronary artery disease, atherosclerotic vascular disease, and all-cause mortality in 137 patients undergoing hemodialysis for chronic renal failure. *Am J Cardiol*. 2005;95(6):742–743.

164. Savage DD, Garrison RJ, Castelli WP, et al. Prevalence of submitral (anular) calcium and its correlates in a general population-based sample (the Framingham study). *Am J Cardiol*. 1983;51(8): 1375–1378.

165. Fox CS, Parise H, Vasan RS, et al. Mitral annular calcification is a predictor for incident atrial fibrillation. *Atherosclerosis*. 2004;173(2):291–294.

166. Fulkerson PK, Beaver BM, Auseon JC, Graber HL. Calcification of the mitral annulus: Etiology, clinical associations, complications and therapy. *Am J Med*. 1979;66(6):967–977.

167. Kizer JR, Wiebers DO, Whisnant JP, et al. Mitral annular calcification, aortic valve sclerosis, and incident stroke in adults free of clinical cardiovascular disease: the strong heart study. *Stroke*. 2005; 36(12):2533–2537.

168. Rodriguez CJ, Bartz TM, Longstreth WT, Jr, et al. Association of annular calcification and aortic valve sclerosis with brain findings on magnetic resonance imaging in community dwelling older adults: The cardiovascular health study. *J Am Coll Cardiol*. 2011;57(21):2172–2180.

169. Benjamin EJ, Plehn JF, D'Agostino RB, et al. Mitral annular calcification and the risk of stroke in an elderly cohort. *N Engl J Med*. 1992;327(6):374–379.

170. Labovitz AJ, Nelson JG, Windhorst DM, et al. Frequency of mitral valve dysfunction from mitral anular calcium as detected by doppler echocardiography. *Am J Cardiol*. 1985;55(1):133–137.

171. Osterberger LE, Goldstein S, Khaja F, Lakier JB. Functional mitral stenosis in patients with massive mitral annular calcification. *Circulation*. 1981; 64(3):472–476.

172. Barasch E, Gottdiener JS, Larsen EK, et al. Clinical significance of calcification of the fibrous skeleton of the heart and aortosclerosis in community dwelling elderly. the cardiovascular health study (CHS). *Am Heart J.* 2006;151(1):39–47.

173. Kanjanauthai S, Nasir K, Katz R, et al. Relationships of mitral annular calcification to cardiovascular risk factors: The multi-ethnic study of atherosclerosis (MESA). *Atherosclerosis.* 2010;213(2):558–562.

174. Maher ER, Young G, Smyth-Walsh B, et al. Aortic and mitral valve calcification in patients with end-stage renal disease. *Lancet.* 1987;2(8564):875–877.

175. Asselbergs FW, Mozaffarian D, Katz R, et al. Association of renal function with cardiac calcifications in older adults: the cardiovascular health study. *Nephrol Dial Transplant.* 2009;24(3):834–840.

176. Molad Y, Levin-Iaina N, Vaturi M, et al. Heart valve calcification in young patients with systemic lupus erythematosus: a window to premature atherosclerotic vascular morbidity and a risk factor for all-cause mortality. *Atherosclerosis.* 2006;185(2):406–412.

177. Sherif HM. Calcification of left-sided valvular structures: Evidence of a pro-inflammatory milieu. *J Heart Valve Dis.* 2009;18(1):52–60.

178. Bostrom KI, Rajamannan NM, Towler DA. The regulation of valvular and vascular sclerosis by osteogenic morphogens. *Circ Res.* 2011;109(5):564–577.

179. Shanahan CM, Crouthamel MH, Kapustin A, Giachelli CM. Arterial calcification in chronic kidney disease: key roles for calcium and phosphate. *Circ Res.* 2011;109(6):697–711.

180. Weiss RM, Lund DD, Chu Y, et al. Osteoprotegerin inhibits aortic valve calcification and preserves valve function in hypercholesterolemic mice. *PLoS One.* 2013;8(6):e65201.

181. Demer L, Tintut Y. The roles of lipid oxidation products and receptor activator of nuclear factor-kappaB signaling in atherosclerotic calcification. *Circ Res.* 2011;108(12):1482–1493.

182. Chen JH, Simmons CA. Cell-matrix interactions in the pathobiology of calcific aortic valve disease: critical roles for matricellular, matricrine, and matrix mechanics cues. *Circ Res.* 2011;108(12):1510–1524.

183. Jahnen-Dechent W, Heiss A, Schafer C, Ketteler M. Fetuin-A regulation of calcified matrix metabolism. *Circ Res.* 2011;108(12):1494–1509.

184. Novaro GM, Sachar R, Pearce GL, et al. Association between apolipoprotein E alleles and calcific valvular heart disease. *Circulation.* 2003;108(15):1804–1808.

185. Farzaneh-Far R, Na B, Schiller NB, Whooley MA. Lack of association of chromosome 9p21.3 genotype with cardiovascular structure and function in persons with stable coronary artery disease: the heart and soul study. *Atherosclerosis.* 2009;205(2):492–496.

186. Rybczynski M, Mir TS, Sheikhzadeh S, et al. Frequency and age-related course of mitral valve dysfunction in the marfan syndrome. *Am J Cardiol.* 2010;106(7):1048–1053.

187. Faivre L, Collod-Beroud G, Loeys BL, et al. Effect of mutation type and location on clinical outcome in 1,013 probands with marfan syndrome or related phenotypes and FBN1 mutations: An international study. *Am J Hum Genet.* 2007;81(3):454–466.

188. Atzinger CL, Meyer RA, Khoury PR, et al. Cross-sectional and longitudinal assessment of aortic root dilation and valvular anomalies in hypermobile and classic ehlers-danlos syndrome. *J Pediatr.* 2011;158(5):826–830.e1.

189. Collins RT, 2nd, Kaplan P, Somes GW, Rome JJ. Long-term outcomes of patients with cardiovascular abnormalities and Williams syndrome. *Am J Cardiol.* 2010;105(6):874–878.

22

CHAPTER

Genetics of Congenital Heart Disease

Elizabeth M. Bonachea, Courtney E. Vaughn, and Vidu Garg

TAKE HOME POINTS

1. The identification of genetic contributors for congenital heart disease (CHD) is becoming increasingly important in the clinical arena as improved techniques for the medical and surgical management of CHD have resulted in significant increases in survival.
2. Syndromic CHD, defined as the presence of additional noncardiac anomalies, represents the minority of CHD and is thought to have significant genetic etiology. Recent studies are beginning to identify the genetic contributions to nonsyndromic CHD.
3. Whole-exome sequencing is becoming progressively more incorporated into clinical testing of patients with CHD when more traditional approaches have not yielded an identifiable genetic etiology. As the cost of next-generation sequencing (NGS) decreases, such testing may become the standard of care for all individuals with CHD.
4. For patients with CHD, an enhanced understanding of the genetic basis of their disease may have important implications for reproductive planning as well as screening of first-degree relatives; however currently, variable expressivity and penetrance complicate our ability to precisely counsel this population about the genetic implications of these complex phenotypes.

Congenital heart disease (CHD) is the most common type of birth defect, with an estimated incidence of 6 to 19 per 1,000 live births (1). Even with recent improvements in the care of children with CHD, CHD still accounts for nearly one-quarter of birth defects that result in mortality (2). Despite recent advances, the etiology for the majority of cases of CHD remains unknown (3). The role of nongenetic causes such as infectious agents and teratogens has long been recognized, but to date, these causes appear to play a etiologic role in a minority of cases of CHD (4). The role of genetic factors in CHD has become an area of robust investigation (5).

Our increased understanding of the underlying genetics has been the direct result of the molecular dissection of the pathways regulating heart development (6). Investigations utilizing multiple animal model systems have resulted in the elucidation of detailed gene regulatory networks that are important for different aspects of cardiac morphogenesis. The identification of these cardiac development genes has led to a growing cadre of genes that are potential candidate genes for human congenital heart defects. The combination of advances in the understanding of cardiac development and progress in genetic technologies has impacted the clinical genetic evaluation of the child with CHD.

The identification of genetic contributors for CHD is becoming increasingly clinically important in the clinical arena. Improved techniques for the medical and surgical management of CHD have resulted in significant increases in survival and decreased morbidities. Indeed, the number of adults living with CHD has recently surpassed the number of children with CHD (7). For patients with CHD, an enhanced understanding of the genetic basis of their disease may have important implications for reproductive planning as well as screening of first-degree relatives. Variable expressivity and penetrance complicate our ability to precisely counsel this population about the genetic implications of these complex phenotypes.

CHD may be categorized as syndromic or non-syndromic on the basis of coexisting noncardiac anomalies. The majority of CHD is non-syndromic or isolated, with syndromic CHD representing approximately one quarter of all cases (8). While syndromic CHD has been traditionally thought to have genetic etiologies, genetic contributions to nonsyndromic CHD have become increasingly recognized. In this chapter, we discuss the current state of knowledge of the genetic basis of CHD, using illustrative cases. In addition, future areas for investigation and clinical applications of genetic testing in patients with CHD are discussed.

CARDIAC MALFORMATIONS ASSOCIATED WITH CHROMOSOMAL ANEUPLOIDY

Case 1

A male infant was born prematurely at 33 weeks of gestation to a healthy 28-year-old mother. The infant's father and two female siblings were also healthy. The infant was noted to be small for gestational age and to have abnormalities of the hands and feet as well as small, low-set ears. A heart murmur was heard immediately after birth. Over the first days of life, he developed respiratory distress and low blood pressure. He was transferred from the birth hospital to a large children's hospital and underwent testing with echocardiography. He was found have a bicuspid aortic valve, a hypoplastic aortic arch, pulmonary valve stenosis, and an atrial septal defect.

Chromosomal aneuploidy describes a state in which cells have an abnormal chromosome number; patients with aneuploidy typically present with birth malformations involving multiple organ systems. Aneuploidy occurs in approximately 1 in 160 live births, but a large proportion of fetuses with aneuploidy will be miscarried or stillborn (9). Approximately 30% of individuals with aneuploidy will have CHD, but this rate is variable among the aneuploidy syndromes (5) (Table 22.1). Trisomy 21 is the most common aneuploidy syndrome with approximately half of individuals with Trisomy 21 having a cardiovascular malformation, most commonly defects of atrial and/or ventricular septation. Patients with Trisomy 18 have a nearly 100% rate of CHD, usually atrial and ventricular septal defects,

and outflow tract malformations. Trisomy 13, or Patau syndrome, is characterized by midline defects involving the midface, central nervous system, heart, and abdominal wall. Most patients with Trisomy 13 will have CHD, frequently septal defects and heterotaxy. About one-third of females with monosomy X, or Turner's syndrome, have associated left-sided heart defects, most commonly bicuspid aortic valve and coarctation of the aorta. These types of chromosomal aneuploidies are detectable using chromosomal G-banded karyotyping, which has a resolution of 5 to 10 megabases.

In addition to these large chromosome abnormalities, submicroscopic chromosomal deletions or duplications frequently result in clinical syndromes associated with CHD (Table 22.1). When suspicion for a specific deletion syndrome is high, fluorescence in situ hybridization (FISH) may be utilized in the clinical arena. With recent technological advances, comparative genome hybridization (CGH) or chromosomal microarray is also utilized to detect submicroscopic chromosomal deletions or duplications. The best characterized of these microdeletion syndromes associated with CHD is 22q11 deletion syndrome (22q11 DS), often referred to as DiGeorge or velo-cardio-facial syndrome (10). This syndrome involves cardiovascular malformations in addition to characteristic facies, hypocalcemia, and defects of the immune system, all of which result from abnormalities of pharyngeal arch development (11). The typical cardiac malformations result from defects of conotruncal and aortic arch artery development, most commonly tetralogy of Fallot (TOF), truncus arteriosus, and interrupted aortic arch Type B (12). Several studies using mouse models have demonstrated that the cardiac malformations in 22q11 DS are likely due to the loss of *TBX1* (13–15). Another common microdeletion syndrome associated with CHD is Williams–Beuren syndrome (WBS) (16). Affected individuals have supravalvar aortic and pulmonary stenosis as well as peripheral pulmonary stenosis, elfin facies, infantile hypercalcemia, renal anomalies, and cognitive deficits. The syndrome is due to a microdeletion of chromosome 7p11.23 (17). The cardiac phenotype in WBS is due to deletion of *ELN* (encodes for elastin), as mutations in this gene are associated with isolated supravalvar aortic stenosis (18). With the advent of FISH and microarray testing, additional microdeletion syndromes frequently associated with CHD have been identified (a subset of common syndromes are shown in Table 22.1).

Conclusion of Case 1

The clinicians caring for the infant were suspicious for chromosomal aneuploidy, specifically Trisomy 18. In

TABLE 22.1 Common Syndromes Resulting From Aneuploidy and Microdeletions

Syndrome	Cardiac Anomalies	% With CHD	Other Clinical Features
Trisomy 13	ASD, VSD, PDA,HLHS	80%	Microcephaly, holoprosencephaly, scalp defects, severe mental retardation, polydactyly, cleft lip or palata, genitourinary abnormalities,omphalocele, microphthalmia
Trisomy 18	ASD, VSD, PDA, TOF, DORV, CoA, BAV	90%–100%	Polyhydramnios, rocker-bottom feet, hypertonia, biliary atresia, severe mental retardation, diaphragmatic hernia, omphalocele
Trisomy 21 (Down syndrome)	ASD, VSD, AVSD, TOF	40%–50%	Hypotonia, developmental delay, palmar crease, epicanthal folds
Monosomy X (Turner syndrome)	CoA, BAV, AS, HLHS	25%–35%	Short stature, shield chest with widely spaced nipples, webbed neck, lymphedema, primary amenorrhea
47, XXY (Klinefelter syndrome)	PDA, ASD, mitral value prolapse	50%	Tall stature, hypoplastic testes, delayed puberty, variable developmental delay
1p36 deletion	Cardiomyopathy, PDA	50%–70%	Facial abnormalities, mental retardation, microcephaly, hearing loss
5p15.2 deletion (Cri-du-chat syndrome)	ASD, VSD, PDA, TOF	10%–55%	High-pitched cry, mental retardation, microcephaly
7q11.23 deletion (Williams–Beuren syndrome)	Supravalvar AS, PPS	50%–85%	Infantile hypercalcemia, elfin facies, social personality, developmental delay, joint contractures, hearing loss
22q11.2 deletion (DiGeorge syndrome)	IAA Type B, aortic arch anomalies, truncus arteriosus, TOF	75%	Thymic and parathyroid hypoplasia, immunodeficiency, low-set ears, hypocalcemia, speech and learning disorders, renal anomalies

Abbreviations: AS, aortic stenosis; ASD, atrial septal defect; AVSD, atrioventricular septal defect; BAV, bicuspid aortic value; CoA, coarctation of aorta; DORV, double outlet right ventricle; HLHS, hypoplastic left heart syndrome; IAA, interrupted aortic arch; PDA, patent ductus arteriosus; PPS, peripheral pulmonic stenosis; TOF, tetralogy of Fallot; VSD, ventricular septal defect.

addition to the increased prevalence of congenital heart defects previously described for infants with Trisomy 18, clinical features may also include poor prenatal growth, clenched hands with overlapping fingers, and rounded feet with prominent heels. A karyotype was sent, but it resulted as normal 46XY. Subsequent microarray testing identified a deletion of chromosome 8p23-21. One of the many genes in this region is *GATA4*, a zinc-finger transcription factor associated with cardiac septal defects (19). Testing of both of the infant's parents revealed that this deletion was a *de novo* event in the patient.

SYNDROMIC AND NONSYNDROMIC CHD ASSOCIATED WITH SINGLE-GENE MUTATIONS

Case 2

A female infant was born at term to a healthy 25-year-old mother. The pregnancy was uncomplicated, although the prenatal care was limited. The infant was initially well but presented to the local emergency department at 6 days of life when the parents had difficulty waking the baby from sleep. The infant was ashen and mottled with poor perfusion and poor respiratory effort. After aggressive resuscitation, the infant was transferred to the neonatal intensive care unit at the nearest children's hospital. Echocardiography revealed hypoplastic left heart syndrome (HLHS). A detailed family history identified multiple family members on the father's side with CHD. The father of the infant had a bicuspid aortic valve, and his own father had undergone aortic valve replacement. The father's brother had been diagnosed with coarctation of the aorta and had surgical repair in the newborn period. This infant underwent multiple surgical interventions for palliation of her congenital heart malformation.

With the completion of the Human Genome Project and continued advances in sequencing technologies, single-gene defects leading to syndromic and nonsyndromic CHD have been elucidated and they are summarized in Tables 22.2 and 22.3. Some of the earliest work was the discovery that mutation

TABLE 22.2 Common Syndromes Associated With CHD Resulting From Single-Gene Defects

Syndrome	Cardiac Anomalies	Other Clinical Features	Causative Gene(s)
Noonam syndrome	PS with dysplastic pulmonary valve, AVSD, HCM, CoA	Short stature, webbed neck, shield chest, development delay, cryptochidism, abnormal facies	*PTPN11, KRAS, NRAS, HRAS, RAF1, SOS1, NF1, CBL, BRAF, SHOC2*
Costello syndrome	PS, HCM, cardiac conduction abnormalities	Short stature, development delay, coarse facies, nasolabial papillomata, increased risk of solid organ carcinoma	*HRAS*
LEOPARD syndrome	PS and cardiac conduction abnormalities	Lentigines, hypertelorism, abnormal genitalia, growth retardation, sensorineural deafness	*PTPN11, RAF1*
Alagile syndrome	PS, TOF, ASD, peripheral pulmonary stenosis	Bile duct paucity, cholestasis, typical facies, butterfly vertebrae, ocular anomalies, growth delay, hearing loss, horseshoe kidney	*JAG1, NOTCH2*
Marfan syndrome	Aortic root dilation and dissection, mitral value prolapse	Tall stature, arachnodactyly, pectus abnormality, scoliosis, ectopia lentis, spontaneous pneumothorax, striae, dural ectasia	*FBN1, TGFBR1, TGFBR2*
Holt–Oram syndrome	ASD, VSD, AVSD, progressive AV conduction system disease	Preaxial radial ray malformations (thumb abnormalities, radial dysplasia)	*TBX5*
Heterotaxy syndrome	DILV, DORV, d-TGS, AVSD	Intestinal malrotation	*ZIC3, CFC1, ACVR2B, GDF1, LEFTY2, NODAL*
Kabuki syndrome	ASD, VSD, TOF, CoA, PDA, TGA	Growth failure, mental retardation, spinal anomalies	*MLL2*
Char syndrome	PDA	Dysmorphic facies and digit anomalies	*TFAP2*
CHARGE syndrome	ASD, VSD, value defects	Coloboma, choanal atresia, developmental delay, genital and/or urinary anomalies	*CHD7, SEMA3E*

Abbreviations: ASD, atrial septal defect; AV, atrioventricular; AVSD, atrioventricular septal defect; CoA, coarctation of aorta; DILV, double inlet left ventricle; DORV, double outlet right ventricle; HCM, hypertrophic cardiomyopathy; PDA, patent ductus arteriosus; PS, pulmonic stenosis; TGA, transpostion of the great arteries; TOF, tetralogy of Fallot; VSD, ventricular septal defect.

of *Fibrillin 1* (FBN1) was the cause of Marfan syndrome (20–22). Marfan syndrome is a heritable disorder of connective tissue characterized by skeletal, ocular, and cardiovascular malformations (23). Individuals with Marfan syndrome typically exhibit tall stature with long fingers and extremities (known as Marfanoid habitus). They are predisposed to lens dislocation as well as aortic root dilation and subsequent aortic dissection. Genetic testing has led to the early identification of affected individuals, especially among family members, which allows for potentially life-saving early evaluation of the increased risk for aortic root dissection. Subsequently, the genetic etiologies of numerous syndromes have been identified and each is characterized by a unique constellation of birth defects. Holt–Oram syndrome, characterized by atrial and ventricular septal defects, progressive atrioventricular conduction system disease, and radial limb and thumb anomalies, is associated with mutations in the transcription factor gene, *TBX5* (24,25). Alagille syndrome, caused by mutations in *JAG1*, a gene encoding a ligand in the Notch signaling pathway, is characterized by intrahepatic bile duct paucity and cardiovascular malformations, including peripheral pulmonic stenosis, pulmonary valve stenosis, and TOF (26,27). Consistent with this, mutations in a Notch receptor, *NOTCH2*, have also been identified in subjects with Alagille syndrome (28). The phenotype of Noonan syndrome consists of cardiac defects, typically pulmonary valve stenosis and hypertrophic cardiomyopathy, as well as cognitive disability, characteristic facies, and bleeding disorders. Initially, mutations in *PTPN11*, a gene involved in Ras signaling, were identified to be the cause of 50% of cases (29). Subsequent studies have found that mutations of other genes involved in the Ras signaling pathway

TABLE 22.3 Nonsyndromic CHD Resulting From Single-Gene Defects

Gene	Cardiac Abnormalities
BMPR2	ASD, VSD
CRELD1, ALK2	AVSD
ELN	SVAS, AS, PS
GATA4	ASD, VSD, AVSD, TOF
GATA5	BAV, VSD, TOF
GATA6	ASD, VSD, AVSD, TOF
JAG1	TOF, PAS
MYH6	ASD, TA, TGA
MYH7	ASD, Ebstein anomaly
MYH11	PDA
NKX2-5	ASD, VSD, TOF, DORV, conduction abnormalities
NOTCH1	BAV, aortic valve calcification
TBX1	TOF
TBX5	ASD, VSD, AVSD
TBX20	ASD, VSD
PROSIT-240	TGA

Abbreviations: AS, aortic stenosis; ASD, atrial septal defect; AVSD, atrioventricular septal defect; BAV, bicuspid aortic value; DORV, double outlet right ventricle; IAA, interrupted aortic arch; PAS, pulmonary artery stenosis; PS, pulmonic stenosis; SVAS, supravalvar aortic stenosis; TA, truncus arteriosus; TGA, transposition of the great arteries; TOF, tetralogy of Fallot; VSD, ventricular septal defect.

including *RAF1*, *SOS1*, and *KRAS* were also associated with Noonan syndrome (30–32). In addition, LEOPARD and Costello syndromes, which exhibit a similar phenotype as Noonan syndrome, are the result of mutations in Ras signaling pathway members. Another syndrome characterized by dysmorphic facies and digit anomalies along with a cardiac malformation, specifically patent ductus arteriosus, was found to be caused by mutation in the transcription factor gene, *TFAP2* using traditional approaches after the identification of large kindreds (33). Heterotaxy syndrome, which is randomization of cardiac, pulmonary, and gastrointestinal situs, is frequently associated with CHD, specifically atrioventricular septal defects (AVSDs) and transposed great arteries. A subset of these cases have been identified to be caused by mutations in genes such as *ZIC3*, *CFC1*, *ACVR2B*, and *LEFTY2* that regulate left–right asymmetry in the developing embryo (34–36). CHARGE syndrome is a constellation of birth anomalies associated with significant neonatal morbidity. The mnemonic CHARGE stands for coloboma, heart defects, choanal atresia, retarded growth, genital abnormalities, and ear abnormalities (37). Affected neonates face significant challenges in airway management and establishment of feeding. Early investigations indicated an autosomal dominant inheritance pattern (38). While initially identified by analysis of patients with chromosomal microdeletions (see the following), the majority of cases of CHARGE are now known to result from mutations in *CHD7* and clinical testing is readily available (39,40).

Over the past 15 years, single-gene defects associated with isolated or nonsyndromic CHD have been discovered (a subset is shown in Table 22.3). Kindred studies revealed that mutations in *NKX2.5* lead to isolated atrial septal defects with atrioventricular conduction delay (41). Mutations in *GATA4*, a zinc-finger transcription factor known to interact with *NKX2.5*, have been linked to isolated atrial septal defects without conduction system abnormalities (19). Interestingly, a mutation in *Gata4* specifically disrupted an interaction with *Tbx5* suggesting that mutations in any of these interacting transcription factors can lead to CHD (19). Myosin heavy chain 6 (*MYH6*) mutations have been identified as another cause of atrial septal defects (42). MYH6 is known to be activated by TBX5 and GATA4, suggesting that mutations in the common downstream targets of these proteins may be a cause of cardiac septation defects. More recently, investigators have linked mutations in *GATA5* and *GATA6* to CHD. *GATA6* mutations were first identified in patients with truncus arteriosus and subsequently in other types of CHD (43,44).

GATA5 mutations have been identified in individuals with bicuspid aortic valve, ventricular septal defects, and TOF (45–47). TBX20, a member of the T-box family of transcription factors, has also been linked to cardiac septation and valve defects (48). Several genes have been implicated in the genetic etiology of atrioventricular septation defects including *CRELD1*, *ALK2*, and *BMPR2* (22,49). Additionally, mutations in *NOTCH1* have been identified as a cause of aortic valve malformations, including bicuspid aortic valve, via genome-wide linkage analysis of an affected family. Interestingly, family members with trileaflet aortic valves and *NOTCH1* mutations also developed early valve calcification, indicating that *NOTCH1* also plays a role in valvular calcification (50). Numerous additional cardiac development genes have been screened in populations of individuals with CHD and novel sequence variants; many of which exhibit functional deficits in vitro have been uncovered. A recent finding by the National Institutes of Health/National Heart, Lung, and Blood Institute (NIH/NHLBI) sponsored Congenital Heart Disease Genetic Network Study uncovered de novo mutations in histone modifying genes in CHD populations at a higher frequency as compared to a control population (51,52). While these are likely contributing to CHD, the specific contribution in regard to recurrence risk remains unclear (3). Overall, these new developments demonstrate that single-gene defects contribute to both syndromic and nonsyndromic forms of CHD.

Resolution of Case 2

Given the strong family history of left-sided heart lesions, the care team decided to proceed with genetic testing for this infant with HLHS. Karyotype and microarray analysis were normal. A clinical geneticist recommended single-gene mutation testing for *NOTCH1*, which identified a variant in *NOTCH1* likely to affect protein function. Subsequent testing of family members identified the same mutation in the infant's affected father and uncle; the mutation was not present in the infant's mother. The family received extensive genetic counseling as to the risk of CHD in subsequent pregnancies.

COPY NUMBER VARIATIONS IN CONGENITAL HEART DISEASE

Despite these advances and discoveries, the vast majority of individuals with CHD do not have chromosomal aneuploidy or single-gene defects. With continued examination of the human genome using newer technologies, new information about the variability of the genome has been revealed. This new type of genetic variation consists of intermediate-sized chromosomal duplications and deletions, termed CNVs (copy number variants). CNVs lead to changes in gene dosage and affect about 12% of the human genome (53). Comparative genomic hybridization (CGH) or chromosomal microarray (CMA) is a DNA-microarray-based methodology that can detect these submicroscopic CNV genome-wide. Examination of individuals with these methods has led to the identification of a variety of CNVs in different genetic loci associated with diseases such as autism, schizophrenia, and developmental delay (54,55). This work has had clinical implications as it is now recommended that CGH/CMA be performed on individuals with developmental delay/intellectual disability or multiple birth defects (56). Not surprisingly, this type of analysis is increasingly being utilized to examine children with CHD and has led to interesting insights into disease etiology.

Copy Number Variation in Syndromic Congenital Heart Disease

The initial investigations of CNV in CHD have focused on cases of CHD that occur in the presence of additional congenital anomalies (syndromic CHD). One of the first investigations using CGH in CHD was the genetic analysis of patients with CHARGE syndrome.

Analysis of these patients led to the identification of a microdeletion on chromosome 8p21 (40). Whole-genome array was initially used to screen the entire genome of 17 individuals with CHARGE syndrome and detected a 2.3 Mb microdeletion on chromosome 8p21 in one individual. Subsequent sequencing of the critical region revealed heterozygous mutations in the *CHD7* gene in 10 subjects, suggesting that *CHD7* haploinsufficiency leads to CHARGE syndrome. This gene, a chromodomain helicase DNA-binding gene important in early embryonic development, is expressed in all the tissues affected in the syndrome. Later studies identified pathogenic mutations in *CHD7* gene in 58% of cases, confirming the importance of this gene in the pathogenesis of CHARGE syndrome (discussed earlier and listed in Table 22.2) (39).

Using a similar approach, Thienpont and colleagues studied 60 patients with CHD and other birth anomalies and normal chromosomes by standard cytogenetics and detected rare CNVs in 30%. While some of these CNVs were in regions known to contain genes critical for cardiac development such as *NKX2.5* and *NOTCH1*, the majority occurred where no known cardiac genes were located (57). Causation was supported when CNV contained genes important in cardiac development or when they arose de novo, and this analysis suggested that 17% of the patient population harbored causative CNV. The remaining 13% of CNVs were inherited from unaffected parents, so causation could not be established. Expansion of this cohort to 150 patients led to similar results (58). Our group used high-resolution whole-genome array CGH to detect rare CNV in 40 individuals with CHD and normal karyotypes, 20 with CHD and other anomalies or developmental delay and 20 with isolated CHD. We found large causative CNV in five individuals, all with CHD and neurological abnormalities or developmental delay. These results demonstrate that causative CNVs exist in individuals with CHD especially when developmental delay or neurological anomalies are also present (59). These seminal studies demonstrated that whole-genome array CGH could be used to detect disease-associated CNV in individuals with CHD and additional anomalies if conventional genetic evaluation including karyotype is normal.

Copy Number Variation in Nonsyndromic Congenital Heart Disease

CNVs have also been identified in individuals with isolated CHD. The initial publication was by Erdogan and colleagues, who studied 105 individuals with various types of isolated CHD and no evidence of developmental delay using whole-genome array CGH. They detected 18 rare CNVs in this population for a frequency of 17%. The majority were duplications, in contrast to those found in syndromic CHD, which are predominantly deletions. Additionally, 44% were familial, also occurring in parents with no evidence of CHD, perhaps indicating that these CNVs increase susceptibility to CHD but require other modifying factors to manifest the phenotype (60). This work demonstrated that rare CNV could be an important genetic contributor to isolated CHD.

Subsequent studies began to focus on the role of CNV in specific heart defect populations. TOF is one of the most common forms of CHD and has been studied for CNV by multiple investigators. Initial work demonstrated de novo CNV in approximately 10% of individuals with TOF and several of the CNV contained genes implicated in heart development (61). Similarly, adults with TOF had a higher frequency of large rare genic CNV as compared to controls (62). Further support for a role of rare de novo CNV in TOF was found by Soemedi and colleagues as part of a large study examining the contribution of CNV to the risk of CHD (63). A recent study on left–right patterning showed that rare CNVs are overexpressed in individuals with heterotaxy, a CHD that results in abnormal left–right body patterning (64). Within the affected population of 262 subjects, 45 CNVs were identified, of which a minority (7) were large chromosomal abnormalities. Candidate genes within small chromosomal abnormalities were functionally tested using the Xenopus model system and all tested genes demonstrated defects in left–right patterning. These studies supported a role for rare CNV in the etiology of heterotaxy. CNV studies have also been performed in left-sided cardiac malformations, which have been shown to have the highest recurrence risk among CHD (65). Examination of 174 affected individuals for CNV identified 73 unique inherited or de novo CNV, and led the authors to conclude that unique CNV contribute to disease pathogenesis in 10% of cases.

While these studies suggest a role for CNV in disease pathogenesis, it is difficult to assign causality. Different algorithms have been proposed in order to determine causal associations between CNV and CHD. Each algorithm has its own unique strength and weaknesses. Causal CNVs in syndromic CHD are more serious than in nonsyndromic CHD and have been shown to negatively affect human development and reproductive fitness (66). The frequency of causal CNV in the syndromic CHD population is 19%, which is much higher than the 3.6% frequency in the nonsyndromic CHD population. A quantitative study on CHD subjects with phenotyped cardiac malformations analyzed gains and losses over 100 CHD risk genes (67). This study was used to determine the phenotype–gene dosage relationship. Large, rare CNVs that were statistically enriched against the controls were categorized as being causal. In conclusion, the study discovered that 14% of subjects with CHD had causal chromosomal abnormalities and 4.3% had likely causal CNV. We analyzed the clinical utility of using array CGH in a small population of individuals with non-syndromic HLHS. While we identified more CNVs in the HLHS population as compared to controls, the CNVs were likely to be small in size making it difficult to assign causality. The frequency of large CNV was not different in the two populations (68). Further investigation is required to define the role of CNV in nonsyndromic CHD.

THE ROLE OF SINGLE NUCLEOTIDE POLYMORPHISMS IN CONGENITAL HEART DISEASE

Single nucleotide polymorphisms (SNPs) are present throughout the genome and constitute the majority of genetic variation among individuals (69). While SNP genotyping has much of its origin in forensic genetics, it is now being applied to susceptibility studies for a variety of common diseases (70). SNPs are now known to play roles in both human disease and response to therapeutic agents. SNP testing is now emerging in the clinical setting as well. At some large clinical centers, patients with specific oncologic conditions undergo SNP testing to determine a personalized genomic regimen of chemotherapy. The understanding of the role of SNPs in susceptibility to congenital heart defects is still in its early stages.

A large body of work regarding the role of SNPs in CHD centers on folate metabolism. A massive public health initiative was undertaken when folate supplementation in pregnancy was shown to decrease the risk for neural tube defects (71). While less recognized by the lay public, folate supplementation was also shown to decrease the risk of CHD (72). The strongest association was seen in ventricular septal defects and outflow tract malformations such as D-transposition of the great arteries (73). Two common polymorphisms in the methylenetetrahydrofolate reductase (*MTHFR*) gene, C677T and A1298C, have been associated with decreased enzymatic activity. Both have been shown in small studies to be more prevalent in infants with CHD than in those without CHD (74). Two large meta-analyses examining all studies of the *MTHFR* C677T polymorphism demonstrated conflicting results and support the need for further investigation in this field (75,76). Elevated homocysteine is a proposed risk factor for CHD, as homocysteine metabolism is linked to the folate pathway (77,78). Recently a functional SNP in the intron of the methionine synthase reductase (*MTRR*) gene, which encodes for an enzyme critical for homocysteine metabolism, was found to increase the risk of CHD in a large Han Chinese population but further validation is required (79).

An intriguing new field of investigation has emerged regarding the role of folate metabolism pathway SNPs in the modulation of disease susceptibility for CHD in the setting of chromosomal abnormalities. In the case of Trisomy 21, approximately 50% of affected individuals present with CHD. As previously described, the pathognomonic cardiac lesion is an AVSD. However, this lesion is only present in 20% of patients with Trisomy 21 (80). A recent investigation comparing Trisomy 21 patients with AVSD and Trisomy 21 patients without AVSD revealed that the *MTHFR* A1298C polymorphism was overtransmitted to patients with AVSD and undertransmitted to those without AVSD, suggesting a role for folate pathway SNPs in modulating CHD risk in patients with other genetic susceptibility (81).

Similarly, in numerous mouse models vascular endothelial growth factor (VEGF) has been shown to be critical for normal heart development. Three SNPs in the promoter of *VEGF* (C2578A, G1154A, and C634G) had been shown to be a modifier of 22q11

deletion syndrome in a candidate-gene-based study. This haplotype is associated with lower VEGF levels in vivo (82,83). Following this, family-based studies using 148 trios found that the low expression *VEGF* haplotype was transmitted more often than expected to children with TOF suggesting that it may play a role in the development of isolated TOF (84). This has been followed by additional publications demonstrating an association of *VEGF* SNPs with other types of CHD (85–87). As occurred with *MTHFR* polymorphisms, a recent meta-analysis failed to demonstrate an association with *VEGF* genetic variations and CHD (88).

FUTURE DIRECTIONS

The understanding of the genetics of CHD is rapidly expanding, in part owing to new sequencing technologies. The respective roles of CNVs and SNPs in disease susceptibility are areas of growing interest and knowledge. Enhanced availability of NGS and decreasing costs of this technology are allowing for direct sequencing of large cohorts of patients with CHD. These studies have the ability to identify large numbers of candidate genes for further investigation. Another realm of increased emphasis is on the role of epigenetic modulation of genes critical to embryonic cardiac development (89). The NHLBI-funded Pediatric Cardiac Genomics Consortium recently established the Congenital Heart Disease Genetic Network Study to identify specific genetic factors that contribute to CHD on a large scale (52). The field of prenatal genetic testing has recently seen a major development in noninvasive testing through the advent of maternal nucleic acid testing for free fetal DNA. This technology allows for fetal aneuploidy testing using maternal blood samples, potentially obviating the need for invasive testing via amniocentesis or chorionic villus sampling (90). Investigators are now applying NGS technologies to the study of free fetal DNA in maternal blood in order to sequence entire fetal genomes noninvasively (91).

CLINICAL APPLICATIONS

As the understanding of the genetic basis of congenital heart defects has improved, the implementation of genetic testing in the clinical setting has increased. The standard of care in most major clinical centers now includes the use of karyotype and microarray testing in children with significant CHD. A striking example is the 22q11 deletion syndrome. Given the high rates of 22q11 deletion syndrome among patients with conotruncal defects, FISH testing is generally considered the standard of care in patients with lesions such as TOF or interrupted aortic arch type B. Such testing is indicated due to high degree of associated noncardiac defects, which could result in significant morbidity if unrecognized. In cases of apparent syndromic CHD and normal karyotype/microarray, consultation with a clinician geneticist is often utilized to guide further testing for single-gene mutations. Single-gene mutation testing may also be clinically indicated in isolated CHD where there are multiple affected family members with similar lesions. Whole-genome and whole-exome sequencing are becoming increasingly incorporated into clinical testing where more traditional approaches have not yielded an identifiable genetic etiology. At some large pediatric centers, all patients with significant congenital heart defects of unknown genetic etiology undergo whole-exome testing. As the cost of NGS decreases, such testing may become the standard of care for all individuals with CHD. At this time, our ability to identify genetic changes still outstrips our understanding of the relevance of certain variations. Care must be taken in counseling patients and families when genetic testing reveals variations of unknown clinical significance. Ethical consideration must also be given to the use of genetic testing in pediatric patients who have not reached the age of majority to consent for these investigations.

Identification of presumed causative mutations allows for enhanced patient and family counseling. Couples with an increased risk for CHD in subsequent pregnancies may use this information to guide family planning decisions; they might also elect to undergo specialized prenatal testing such as fetal echocardiography and/or amniocentesis with genetic testing. Early identification of fetuses with significant CHD could then improve perinatal outcomes via high-risk obstetric care and early access to tertiary care centers experienced in managing neonates with critical CHD. Additionally, progress in the treatment and palliation of CHD has allowed an increasing proportion of individuals with significant CHD to reach reproductive age. They now face similar challenges in preconception planning and prenatal care.

REFERENCES

1. Hoffman JI, Kaplan S. The incidence of congenital heart disease. *J Am Coll Cardiol.* 2002;39(12):1890–1900.
2. Go AS, Mozaffarian D, Roger VL, et al. Executive summary: heart disease and stroke statistics—2013 update: a report from the American Heart Association. *Circulation.* 2013;127(1):143–152.
3. Fahed AC, Gelb BD, Seidman JG, Seidman CE. Genetics of congenital heart disease: the glass half empty. *Circ Res.* 2013;112(4):707–720.
4. Jenkins KJ, Correa A, Feinstein JA, et al. Noninherited risk factors and congenital cardiovascular defects: current knowledge: a scientific statement from the American Heart Association Council on Cardiovascular Disease in the Young: endorsed by the American Academy of Pediatrics. *Circulation.* 2007;115(23): 2995–3014.
5. Pierpont ME, Basson CT, Benson DW, Jr, et al. Genetic basis for congenital heart defects: current knowledge: a scientific statement from the American Heart Association Congenital Cardiac Defects Committee, Council on Cardiovascular Disease in the Young: endorsed by the American Academy of Pediatrics. *Circulation.* 2007;115(23):3015–3038.
6. Srivastava D. Making or breaking the heart: from lineage determination to morphogenesis. *Cell.* 2006;126(6): 1037–1048.
7. O'Leary JM, Siddiqi OK, de Ferranti S, et al. The changing demographics of congenital heart disease hospitalizations in the United States, 1998 through 2010. *JAMA.* 2013;309(10):984–986.
8. Meberg A, Hals J, Thaulow E. Congenital heart defects chromosomal anomalies, syndromes and extracardiac malformations. *Acta Paediatr,* 2007;96(8):1142–1145.
9. Driscoll DA, Gross S. Clinical practice. Prenatal screening for aneuploidy. *N Engl J Med.* 2009;360(24):2556–2562.
10. Scambler PJ. 22q11 deletion syndrome: a role for TBX1 in pharyngeal and cardiovascular development. *Pediatr Cardiol.* 2010;31(3):378–390.
11. Greenberg F. DiGeorge syndrome: an historical review of clinical and cytogenetic features. *J Med Genet.* 1993; 30(10):803–806.
12. Goldmuntz E, Clark BJ, Mitchell LE, et al. Frequency of 22q11 deletions in patients with conotruncal defects. *J Am Coll Cardiol.* 1998;32(2):492–498.
13. Funke B, Epstein JA, Kochilas LK, et al. Mice overexpressing genes from the 22q11 region deleted in velo-cardio-facial syndrome/DiGeorge syndrome have middle and inner ear defects. *Hum Mol Genet.* 2001; 10(22):2549–2556.
14. Garg V, Yamagishi C, Hu T, et al. Tbx1, a DiGeorge syndrome candidate gene, is regulated by sonic hedgehog during pharyngeal arch development. *Dev Biol.* 2001;235(1):62–73.
15. Lindsay EA, Vitelli F, Su H, et al. Tbx1 haploinsufficieny in the DiGeorge syndrome region causes aortic arch defects in mice. *Nature.* 2001;410(6824):97–101.
16. Pober BR. Williams-Beuren syndrome. *N Engl J Med.* 2010;362(3):239–252.
17. Ewart AK, Morris CA, Atkinson D, et al. Hemizygosity at the elastin locus in a developmental disorder, Williams syndrome. *Nat Genet.* 1993;5(1):11–16.
18. Ewart AK, Jin W, Atkinson D, et al. Supravalvular aortic stenosis associated with a deletion disrupting the elastin gene. *J Clin Invest.* 1994;93(3):1071–1077.
19. Garg V, Kathiriya IS, Barnes R, et al. GATA4 mutations cause human congenital heart defects and reveal an interaction with TBX5. *Nature.* 2003;424(6947): 443–447.
20. Dietz HC, Cutting GR, Pyeritz RE, et al. Marfan syndrome caused by a recurrent de novo missense mutation in the fibrillin gene. *Nature.* 1991;352(6333):337–339.

21. Lee B, Godfrey M, Vitale E, et al. Linkage of Marfan syndrome and a phenotypically related disorder to two different fibrillin genes. *Nature.* 1991;352(6333):330–334.

22. Maslen CL, Corson GM, Maddox BK, et al. Partial sequence of a candidate gene for the Marfan syndrome. *Nature.* 1991;352(6333):334–337.

23. Pyeritz RE. The Marfan syndrome. *Annu Rev Med.* 2000;51:481–510.

24. Basson CT, Bachinsky DR, Lin RC, et al. Mutations in human TBX5 cause limb and cardiac malformation in Holt-Oram syndrome. *Nat Genet.* 1997;15(1):30–35.

25. Basson CT, Cowley GS, Solomon SD, et al. The clinical and genetic spectrum of the Holt-Oram syndrome (heart-hand syndrome). *N Engl J Med.* 1994; 330(13):885–891.

26. Li QY, Newbury-Ecob RA, Terrett JA, et al. Holt-Oram syndrome is caused by mutations in TBX5, a member of the Brachyury (T) gene family. *Nat Genet.* 1997;15(1):21–29.

27. Oda T, Elkahloun AG, Pike BL, et al. Mutations in the human Jagged1 gene are responsible for Alagille syndrome. *Nat Genet.* 1997;16(3):235–242.

28. McDaniell R, Warthen DM, Sanchez-Lara PA, et al. NOTCH2 mutations cause Alagille syndrome, a heterogeneous disorder of the notch signaling pathway. *Am J Hum Genet.* 2006;79(1):169–173.

29. Tartaglia M, Mehler EL, Goldberg R, et al. Mutations in PTPN11, encoding the protein tyrosine phosphatase SHP-2, cause Noonan syndrome. *Nat Genet.* 2001;29(4):465–468.

30. Aoki Y, Niihori T, Narumi Y, et al. The RAS/MAPK syndromes: novel roles of the RAS pathway in human genetic disorders. *Hum Mutat.* 2008;29(8):992–1006.

31. Schubbert S, Bollag G, Shannon K. Deregulated Ras signaling in developmental disorders: new tricks for an old dog. *Curr Opin Genet Dev.* 2007;17(1):15–22.

32. Tartaglia M, Gelb BD, Zenker M. Noonan syndrome and clinically related disorders. *Best Pract Res Clin Endocrinol Metab.* 2011;25(1):161–179.

33. Satoda M, Zhao F, Diaz GA, et al. Mutations in TFAP2B cause Char syndrome, a familial form of patent ductus arteriosus. *Nat Genet.* 2000;25(1):42–46.

34. Gebbia M, Ferrero GB, Pilia G, et al. X-linked situs abnormalities result from mutations in ZIC3. *Nat Genet.* 1997;17(3):305–308.

35. Sutherland MJ, Ware SM. Disorders of left-right asymmetry: heterotaxy and situs inversus. *Am J Med Genet C Semin Med Genet.* 2009;151C(4):307–317.

36. Zhu L, Belmont JW, Ware SM. Genetics of human heterotaxias. *Eur J Hum Genet.* 2006;14(1):17–25.

37. Blake KD, Davenport SL, Hall BD, et al. CHARGE association: an update and review for the primary pediatrician. *Clin Pediatr.* 1998;37(3):159–173.

38. Mitchell JA, Giangiacomo J, Hefner MA, et al. Dominant CHARGE association. *Ophthalmic Paediatr Genet.* 1985;6(1–2):271–276.

39. Lalani SR, Safiullah AM, Fernbach SD, et al. Spectrum of CHD7 mutations in 110 individuals with CHARGE syndrome and genotype-phenotype correlation. *Am J Hum Genet.* 2006;78(2):303–314.

40. Vissers LE, van Ravenswaaij CM, Admiraal R, et al. Mutations in a new member of the chromodomain gene family cause CHARGE syndrome. *Nat Genet.* 2004;36(9):955–957.

41. Schott JJ, Benson DW, Basson CT, et al. Congenital heart disease caused by mutations in the transcription factor NKX2-5. *Science.* 1998;281(5373):108–111.

42. Ching YH, Ghosh TK, Cross SJ, et al. Mutation in myosin heavy chain 6 causes atrial septal defect. *Nat Genet.* 2005;37(4):423–428.

43. Kodo K, Nishizawa T, Furutani M, et al. GATA6 mutations cause human cardiac outflow tract defects by disrupting semaphorin-plexin signaling. *Proc Natl Acad Sci U S A.* 2009;106(33):13933–13938.

44. Maitra M, Koenig SN, Srivastava D, Garg V. Identification of GATA6 sequence variants in patients with congenital heart defects. *Pediatr Res.* 2010;68(4):281–285.

45. Padang R, Bagnall RD, Richmond DR, et al. Rare non-synonymous variations in the transcriptional activation domains of GATA5 in bicuspid aortic valve disease. *J Mol Cell Cardiol.* 2012;53(2):277–281.

46. Wei D, Bao H, Liu XY, et al. GATA5 loss-of-function mutations underlie tetralogy of fallot. *Int J Med Sci.* 2013;10(1):34–42.

47. Wei D, Bao H, Zhou N, et al. GATA5 loss-of-function mutation responsible for the congenital ventriculoseptal defect. *Pediatr Cardiol.* 2013;34(3):504–511.

48. Kirk EP, Sunde M, Costa MW, et al. Mutations in cardiac T-box factor gene TBX20 are associated with diverse cardiac pathologies, including defects of septation and valvulogenesis and cardiomyopathy. *Am J Hum Genet.* 2007;81(2):280–291.

49. Maslen CL. Molecular genetics of atrioventricular septal defects. *Curr Opin Cardiol.* 2004;19(3):205–210.

50. Garg V, Muth AN, Ransom JF, et al. Mutations in NOTCH1 cause aortic valve disease. *Nature.* 2005; 437(7056):270–274.

51. Zaidi S, Choi M, Wakimoto H, et al. De novo mutations in histone-modifying genes in congenital heart disease. *Nature.* 2013;498(7453):220–223.

52. Pediatric Cardiac Genomics Consortium; Gelb B, Brueckner M, Chung W, et al. The Congenital Heart Disease Genetic Network Study: rationale, design, and early results. *Circ Res.* 2013;112(4):698–706.

53. Redon R, Ishikawa S, Fitch KR, et al. Global variation in copy number in the human genome. *Nature.* 2006; 444(7118):444–454.

54. Almal SH, Padh H. Implications of gene copy-number variation in health and diseases. *J Hum Genet.* 2012; 57(1):6–13.

55. Emanuel BS, Saitta SC. From microscopes to microarrays: dissecting recurrent chromosomal rearrangements. *Nat Rev Genet.* 2007;8(11):869–883.

56. Miller DT, Adam MP, Aradhya S, et al. Consensus statement: chromosomal microarray is a first-tier clinical diagnostic test for individuals with developmental disabilities or congenital anomalies. *Am J Hum Genet.* 2010;86(5):749–764.

57. Thienpont B, Mertens L, de Ravel T, et al. Submicroscopic chromosomal imbalances detected by array-CGH are a frequent cause of congenital heart defects in selected patients. *Eur Heart J.* 2007;28(22):2778–2784.

58. Breckpot J, Thienpont B, Peeters H, et al. Array comparative genomic hybridization as a diagnostic tool for syndromic heart defects. *J Pediatr.* 2010;156(5):810–817.

59. Richards AA, Santos LJ, Nichols HA, et al. Cryptic chromosomal abnormalities identified in children with congenital heart disease. *Pediatr Res.* 2008;64(4):358–363.

60. Erdogan F, Larsen LA, Zhang L, et al. High frequency of submicroscopic genomic aberrations detected by tiling path array comparative genome hybridisation in patients with isolated congenital heart disease. *J Med Genet.* 2008;45(11):704–709.

61. Greenway SC, Pereira AC, Lin JC, et al. De novo copy number variants identify new genes and loci in isolated sporadic tetralogy of Fallot. *Nat Genet.* 2009;41(8):931–935.

62. Silversides CK, Lionel AC, Costain G, et al. Rare copy number variations in adults with tetralogy of Fallot implicate novel risk gene pathways. *PLoS Genet.* 2012; 8(8):e1002843.

63. Soemedi R, Wilson IJ, Bentham J, et al. Contribution of global rare copy-number variants to the risk of sporadic congenital heart disease. *Am J Hum Genet.* 2012; 91(3):489–501.

64. Fakhro KA, Choi M, Ware SM, et al. Rare copy number variations in congenital heart disease patients identify unique genes in left-right patterning. *Proc Natl Acad Sci USA.* 2011;108(7):2915–2920.

65. McBride KL, Garg V. Heredity of bicuspid aortic valve: is family screening indicated? *Heart.* 2011; 97(15):1193–1195.

66. Breckpot J, Thienpont B, Arens Y, et al. Challenges of interpreting copy number variation in syndromic and non-syndromic congenital heart defects. *Cytogenet Genome Res.* 2011;135(3–4);251–259.

67. Tomita-Mitchell A, Mahnke DK, Struble CA, et al. Human gene copy number spectra analysis in congenital heart malformations. *Physiol Genomics.* 2012; 44(9):518–541.

68. Payne AR, Chang SW, Koenig SN, et al. Submicroscopic chromosomal copy number variations identified in children with hypoplastic left heart syndrome. *Pediatr Cardiol.* 2012;33(5):757–763.

69. International HapMap 3 Consortium; Altshuler DM, Gibbs RA, Peltonen L, et al. Integrating common and rare genetic variation in diverse human populations. *Nature.* 2010;467(7311):52–58.

70. Sobrino B, Carracedo A. SNP typing in forensic genetics: a review. *Methods Mol Biol.* 2005;297:107–126.

71. Czeizel AE, Dudas I. Prevention of the first occurrence of neural-tube defects by periconceptional vitamin supplementation. *N Engl J Med.* 1992;327(26):1832–1835.

72. Hernandez-Diaz S, Werler MM, Walker AM, Mitchell AA. Folic acid antagonists during pregnancy and the risk of birth defects. *N Engl J Med.* 2000;343(22): 1608–1614.

73. Botto LD, Yang Q. 5,10-Methylenetetrahydrofolate reductase gene variants and congenital anomalies: a HuGE review. *Am J Epidemiol.* 2000;151(9):862–877.

74. Goldmuntz E, Woyciechowski S, Renstrom D, et al. Variants of folate metabolism genes and the risk of conotruncal cardiac defects. *Circ Cardiovasc Genet.* 2008;1(2):126–132.

75. van Beynum IM, den Heijer M, Blom HJ, Kapusta L. The MTHFR 677C->T polymorphism and the risk of congenital heart defects: a literature review and meta-analysis. *QJM.* 2007;100(12):743–753.

76. Wang W, Wang Y, Gong F, et al. MTHFR C677T polymorphism and risk of congenital heart defects: evidence from 29 case-control and TDT studies. *PLoS One.* 2013;8(3):e58041.

77. Beksac MS, Balci S, Guvendag Guven ES, et al. Complex conotruncal cardiac anomalies consecutively in three siblings from a consanguineous family possibly associated with maternal hyperhomocysteinemia. *Arch Gynecol Obstet.* 2007;276(5): 547–549.

78. Han M, Serrano MC, Lastra-Vicente R, et al. Folate rescues lithium-, homocysteine- and Wnt3A-induced vertebrate cardiac anomalies. *Dis Model Mech.* 2009; 2(9–10):467–478.

79. Zhao JY, Yang XY, Gong XH, et al. Functional variant in methionine synthase reductase intron-1 significantly increases the risk of congenital heart disease in the Han Chinese population. *Circulation.* 2012; 125(3):482–490.

80. Freeman SB, Taft LF, Dooley KJ, et al. Population-based study of congenital heart defects in Down syndrome. *Am J Med Genet.* 1998;80(3):213–217.

81. Locke AE, Dooley KJ, Tinker SW, et al. Variation in folate pathway genes contributes to risk of congenital heart defects among individuals with Down syndrome. *Genet Epidemiol.* 2010;34(6):613–623.

82. Lambrechts D, Storkebaum E, Morimoto M, et al. VEGF is a modifier of amyotrophic lateral sclerosis in mice and humans and protects motoneurons against ischemic death. *Nat Genet.* 2003;34(4):383–394.

83. Stalmans I, Lambrechts D, De Smet F, et al. VEGF: a modifier of the del22q11 (DiGeorge) syndrome? *Nat Med.* 2003;9(2):173–182.

84. Lambrechts D, Devriendt K, Driscoll DA, et al. Low expression VEGF haplotype increases the risk for tetralogy of Fallot: a family based association study. *J Med Genet.* 2005;42(6):519–522.

85. Xie J, Yi L, Xu ZF, et al. VEGF C-634G polymorphism is associated with protection from isolated ventricular septal defect: case-control and TDT studies. *Eur J Hum Genet.* 2007;15(12):1246–1251.

86. Smedts HP, Isaacs A, de Costa D, et al. VEGF polymorphisms are associated with endocardial cushion defects: a family-based case-control study. *Pediatr Res.* 2010;67(1):23–28.

87. Vannay A, Vasarhelyi B, Kornyei M, et al. Single-nucleotide polymorphisms of VEGF gene are associated with risk of congenital valvuloseptal heart defects. *Am Heart J.* 2006;151(4):878–881.

88. Griffin HR, Hall DH, Topf A, et al. Genetic variation in VEGF does not contribute significantly to the risk of congenital cardiovascular malformation. *PLoS One.* 2009;4(3):e4978.

89. Vallaster M, Vallaster CD, Wu SM. Epigenetic mechanisms in cardiac development and disease. *Acta Biochim Biophys Sin.* 2012;44(1):92–102.

90. Go AT, van Vugt JM, Oudejans CB. Non-invasive aneuploidy detection using free fetal DNA and RNA in maternal plasma: recent progress and future possibilities. *Hum Reprod Update.* 2011;17(3):372–382.

91. Fan HC, Blumenfeld YJ, Chitkara U, et al. Noninvasive diagnosis of fetal aneuploidy by shotgun sequencing DNA from maternal blood. *Proc Natl Acad Sci USA.* 2008;105(42):16266–16271.

Index

ACE. *See* angiotensin converting enzyme
ACE DD/ID genotyping, 148
ACE gene, 279
AF. *See* atrial fibrillation
Affymetrix GeneChip, 68–69, 74
Affymetrix Genome-Wide Human SNP Array, 26
A/G alleles, 8
Agilent Dual-Mode system, 68
agnostic analysis, 74
AGTR gene, 279
Algorithm for the Reconstruction of Accurate Cellular
    Networks (ARACNE), 74
alkaptonuria, 275
allele sharing, 15
allelic heterogeneity, 188
allelic variants, 196
alpha helix, 37
alpha-tropomyosin (*TPM1*) gene, 197
angiopoietin-like 3 (*ANGPTL3*) gene, 38, 247
angiotensin converting enzyme (ACE), 62, 145
angiotensin II receptor blocker (ARB), 62
AntagomiRs, 147
anticipation, 101
anti-microRNA technology, 64
aortic insufficiency, 275–276
aortic stenosis (AS), 267
aortic valve calcification (AVC), 267
aortic valve disease, 267–276
   accurate identification of, 271
   candidate gene studies for, 268–270
   coexistence of traditional CAD risk factors, 267
   genetics of bicuspid, 271–275
   genome-wide linkage mapping study, 268
   GWAS findings, 270–271
   heritability, 268
   high-throughput omics technologies and, 271
   prevalence, 267
   quantitative analysis, 268
   rare Mendelian forms of, 275–276
   risk factor, 267
   SNPs, 270–271
   valvular calcification, 267
aortic valve replacement (AVR), 267

APOE gene, 274
ARACNE. *See* Algorithm for the Reconstruction of
    Accurate Cellular Networks
ARB. *See* angiotensin II receptor blocker
arrhythmic dilated cardiomyopathy, 207
arrhythmogenic right ventricular cardiomyopathy
    (ARVC), 7, 196
arrhythmogenic right ventricular dysplasia/
    cardiomyopathy, 210–211. *See also* inherited
    cardiomyopathies
  gene mutations associated with, 211
ARVC. *See* arrhythmogenic right ventricular
    cardiomyopathy
AS. *See* aortic stenosis
ascertainment bias, 16–17
association analysis, 9
atherosclerosis, 41
ATP1B1 gene, 136
atrial fibrillation (AF), 14, 197
   application, 147
   case example, 140–141
   genetic basis for, 141–142
   GWAS findings, 144
   heritability, 141–142
   integrating genetic information into clinical
     practice, 147–148
   metabolomics of, 147
   miRNAs, role of, 146–147
   missing heritability of, 143–144
   mRNA, role of, 144–145
   potential opportunities for genotype-guided
     therapies, 148
   proteomics of, 147
   risk of, 142
   risk prediction, 143
   seminal Icelandic GWAS, 142
   systems biology approach to, 145
   transcriptomics of, 144–147
   warfarin therapy for, 148–149
autosomal dominant inheritance, 6
autosomal recessive inheritance, 6
AVC. *See* aortic valve calcification
AVR. *See* aortic valve replacement

Bayesian analysis, 102–103
Bazett's formula, 156
beta-myosin heavy chain gene (*MYH7*), 197
bicuspid aortic valve disease
    candidate genes in, 273–274
    epidemiology of, 272
    GWAS findings, 275
    heritability of, 272
    linkage studies for, 273
    prevalence of, 272
BioDiscovery Nexus Expression, 73
bioinformatic software, 72
biomarkers, 35
    epigenetic, 50
    in HF patients, 46
    microRNAs (miRNAs) as, 60–63
blood pressure genomics
    as a complex trait, 112–121
    future of, 121–122
    genomic and environmental factors on BP,
        111–112
    monogenic BP disorders, 112
    novel BP phenotypes, 121–122
    risk of cardiovascular disease, 111
    severe monogenic hypertension, case study, 111
Bonferroni correction, 28–29, 73
BP. *See* blood pressure genomics
BRCA1 and *BRCA2* genes, 30
BrS. *See* Brugada syndrome
Brugada-like ECG phenotype, 136
Brugada syndrome (BrS), 8, 79, 130, 132
    clinical manifestations, 160
    diagnosis of, 160–161
    genetics of, 159–160
    genetic testing of, 161
    risk stratification for, 161
    treatment strategies, 161
Burden tests, 247

CACNA1C gene, 132
CAD. *See* coronary artery disease
calcific mitral valve prolapse, 279–280
    candidate gene studies, 280
    GWAS findings, 280
    heritability, 280
    linkage studies, 280
calcineurin A, 58–59
calcium channel genes, 132–133
CALM1 and *CALM2* mutations, 81
candidate-based strategy, 81
candidate genes, 36–37
    in aortic valve disease, 268–270
    associated with cardiac ion channelopathies, 131
    in bicuspid aortic valve disease, 273–274
    of BP, 115
    calcific mitral valve prolapse, 280
    characterizing, 37–38
    in mitral valve prolapse, 277–279
    studies, 23
    variation in plasma lipids, 242
cardiac hypertrophy, 58–59
cardiac repolarization/QT interval, 158
cardiac structure and function
    approaches to phenotyping, 181
    case example, 168–169
    challenges of accurate phenotyping of, 170–173
    electrocardiographic LVH, 169
    genetics of, 169–170
    GWAS of, 169
    heritability of indices of, 173
    HyperGEN and Genetic Epidemiology Network of
        Arteriopathy (GENOA) studies, 178
    NHLBI CARe study, 178
    pathway analysis, 180
    published studies, 173–181
    signal intensity plots for echocardiographic parameters,
        179
cardiac troponin T (*TNNT2*) gene, 197
cardiogenetics, 106
    evaluation, 104
cardiomyopathies, genetic basis of. *See also* heart failure
    dilated cardiomyopathy, 189
    hypertrophic cardiomyopathy, 188–189
    research and clinical practice, 211–212
cardiovascular disease (CVD), 39–40, 67
cardiovascular genetic testing, 105–106
CASQ2 gene, 133
catecholaminergic polymorphic ventricular tachycardia
        (CPVT), 7, 130, 161–163
    clinical manifestations, 162
    diagnosis of, 162
    flecainide therapy, 163
    genetics of, 162
    genetic testing, 163
    ICD implantation, 163
    risk stratification, 163
    treatment strategies, 163
CDK2NA/B gene, 31
CHARGE, 116
CHD. *See* congenital heart disease
cholesterol homeostasis and metabolism, 60
CLCNKA gene, 190
clinical utility of genetic information, 9–10
clustering analysis, 70–72
CNV. *See* copy number variant
COGENT. *See* Continental Origins and Genetic
        Epidemiology Network
"common disease, rare variant" hypothesis, 84
concordance, 15, 18
    of MZ vs. DZ twins, 16
congenital heart disease (CHD), 41–42
    cardiac malformations associated with, 288–289
    clinical applications of genetic basis of, 294–295
    copy number variations in, 292–293
    genetic contributors for, 287
    role of single nucleotide polymorphisms in, 293–294
    syndromic and nonsyndromic, 289–292
Continental Origins and Genetic Epidemiology Network
        (COGENT), 121
copy number variant (CNV), 17
    in CHD, 292–293
coronary artery disease (CAD), 9, 43, 46–47, 62
    case example, 254, 256, 258–260
    clinical utility of GWAS results, 260–261
    early onset, 253
    genetic risk factors, 258–261
    genetic testing in familial hypercholesterolemia,
        254–256
    genotyping tests for, 261
    pharmacogenetic information on, 256–257
    statins and, 256
    trials, 257

Cox regression, 27
CPVT. *See* catecholaminergic polymorphic ventricular tachycardia
Crohn-disease-like illness, 77
cross-breeding experiments, 1
CVD. *See* cardiovascular disease
CYP2C9 gene, 92
cytochrome P450 2C9 (*CYP2C9*), 148

dabigatran, 31
Danon disease, 203
Database for Annotation, Visualization and Integrated Discovery (DAVID), 73
data normalization, 69–70
dChip. *See* DNA-Chip Analyzer
DCM. *See* dilated cardiomyopathy
dendrogram, 70–71
de novo strategy, 80–81
differentially expressed gene analysis, 72–73
dilated cardiomyopathy (DCM), 7, 189, 196. *See also* hypertrophic cardiomyopathy
    clinical and pathophysiologic characteristics, 204
    disease genes and putative functional consequences, 205–208
    gene mutations associated with, 206–207
    genetic basis of, 204
    genetic testing for, 209
    genotype–phenotype correlation, 208–209
discordance, 15
    in MZ twins, 17
dizygotic twins (DZ), 15
DMD. *See* Duchenne muscular dystrophy
DMPK gene, 101
DNA-Chip Analyzer (dChip), 69
DNA methylation, 17, 122
    evidence for, 40
    gene-specific, 44
    global, 40–44
    relation to CVD, 40–44
    SAM/SAH ratio, 42
    WBC global, 49
double-hit strategy, 81
Drosha and Di George Syndrome, 57
Duchenne muscular dystrophy (DMD), 101–102
DZ twins. *See* dizygotic twins

early repolarization pattern (ERP), 136–137
ECG traits. *See* electrocardiographic traits
ectrodactyly–ectodermal dysplasia–cleft syndrome, 101
EEC syndrome. *See* ectrodactyly–ectodermal dysplasia–cleft syndrome
Ehlers–Danlos syndrome, 275–276
EIGENSTRAT, 30
electrocardiographic traits (ECG traits)
    case example, 128
    case resolution, 137–138
    early repolarization pattern interval, 136–137
    genetic architecture of complex, 130
    ion-channel candidate genes associated with, 130–133
    phenotypes, 129
    PR interval, 133
    QRS interval, 134
    QT interval, 135–136
    RR interval, 137
electrolyte homeostasis, mutations affecting, 205–207
Encyclopedia of DNA Elements (ENCODE), 232

ENG gene, 275
epigenetic changes in CVD, 92
epigenetics, 17, 20. *See also* DNA methylation; microRNAs
    challenges and opportunities for CVD, 48–50
    of pulmonary hypertension, 232–233
    role of diet and genetic factors in, 49–50
Eppendorf Dualchip system, 68
ERP. *See* early repolarization pattern
estrogen receptor alpha gene, 44
exome chips, 78–79
exome-sequencing studies, 80–81, 84
exosomes, 61
expressivity, 7–8

Fabry disease, 203
false discovery rate (FDR), 73
familial aggregation, measuring of, 13–15
familial combined hypolipidemia (FCH), 34–35, 37
familial hypercholesterolemia, 1, 7, 275
family history, role in genetic counseling, 97–100, 103
family linkage study, 24
FBN1 gene, 9
FBN1 polymorphism, 279
fibrillin 1 (*FBN1*) gene, 8
Freeman–Sheldon syndrome, 77

gap junction–connexin coupling, 134
GATA5 gene, 274
GeneCards, 37
gene–diet interaction, 92
gene–environment interactions, 122
    analysis, 90–91
    in cardiovascular disease, 91–92
    defining, 86–88
    family-based studies, 88–89
    pathway-based, 90
    population-based studies, 89–90
    promise and challenges of, 92–93
    statistical power, 91
gene–environment-wide interaction studies (GEWIS), 90–91
gene expression
    data analysis, 70–73
    data normalization, 69–70
    mutations affecting, 208
    ontology and pathway analysis, 73–74
    quality control of microarray data, 69
    using microarray technologies, 67–68
    validation and meta-analysis, 74–75
Gene Expression Omnibus (GEO), 75
gene–gene interactions, 122
gene ontology (GO), 73
Gene Set Enrichment Analysis (GSEA), 74
gene-specific DNA methylation and CVD, 44
genetic counseling, 9
    cardiovascular genetic testing, 105–106
    definition of, 96–97
    extracardiac findings in genetic syndromes, 105
    and family health history, 97–100
    genetic counselors in cardiovascular practice, 103–105
    genetic risk assessment for, 100–103
    partners in practice, 106
    variant of uncertain significance (VUS or VOUS), 106
genetic linkage analysis, 8–9, 34–35
genetic profiling, 147

genetic risk
  assessment, 100–103
  defined, 100
genocopy, 15
genome-wide association studies (GWAS), 20–21, 35–36, 77, 90–91, 129, 137
  AF, 142, 144
  aortic valve disease, 270–271
  application of results, 31
  bicuspid aortic valve disease, 275
  of BP, 115–116
  calcific mitral valve prolapse, 280
  cardiac structure and function, 169
  in cardiovascular disease, 31
  challenges, 28–30
  coronary artery disease, 260–261
  data analysis, 27–28
  genetic basis of, 24–26
  genomic inflation, 29–30
  limitations, 30
  lipids and lipoproteins, 243–245
  meta-analyses in populations
    of European ancestry, 116
    of non-European ancestry, 116–121
  need for, 23–24
  outcome variable, 26
  population stratification, 30
  power and significance, 29
  predictor variable, 26–27
  reaching significance, 28
  simple, 26–28
  subjects and design, 26
genome-wide significance, 28
genomic control, 29
genomic inflation, 29–30
genomics databases, 72
genotype–phenotype correlation, 208–209
genotype-positive/phenotype-negative individual, 202, 209
genotypes, 26–27
genotyping errors, 28
GEO. *See* Gene Expression Omnibus
germline mosaicism, 100–101
GEWIS. *See* gene–environment-wide interaction studies
GLA mutations, 202–203
Global Blood Pressure Genetics (Global BPgen), 116
global DNA hypomethylation and CVD, 41–43
global GWAS meta-analyses, 121
global hypermethylation, and CVD, 43–44
global hypomethylation, in CVD, 40
glycogen storage disease. *See* Pompe disease
GO. *See* gene ontology
GSEA. *See* Gene Set Enrichment Analysis
GWAS. *See* genome-wide association studies

HapMap project, 30
Hardy–Weinberg equilibrium (HWE), 29
HCM. *See* hypertrophic cardiomyopathy
HD. *See* Huntington's disease
HDL-C. *See* high-density lipoprotein cholesterol
heart failure (HF), 46, 84. *See also* cardiomyopathies, genetic basis of
  case example, 186–187
  common genetic variation with, 191
  disease susceptibility, 187
  genetic basis of idiopathic DCM and, 189–190

GWAS study, 190
  molecular investigations, 190–192
  monogenic traits in, 187–188
  polygenic complex trait in, 188
  prevalence in United States, 187
hereditary cancer syndrome, 99
heritability, 229
  aortic valve disease, 268
  atrial fibrillation (AF), 141–142
  of bicuspid aortic valve disease, 272
  of BP, 112
  calcific mitral valve prolapse, 280
  cardiac structure and function, 173
  common misconceptions regarding, 18
  of complex CVD, 18–19
  complex traits estimation, 20
  of ERP, 136
  estimation of, 18
  implication of high, 17
  missing in CVD, 19–21
  phenotypic measures, 18
  quantifying, 17–18
heterogeneity, 7
heterotaxia, 6–7
HF. *See* heart failure
HFE C282Y mutation, 101
hierarchical clustering algorithms, 70
high-density lipoprotein cholesterol (HDL-C), 34, 36
histone acetylation, 17
histone modifications, 48, 122
homogentisate 1,2 dioxygenase gene, 275
homozygosity-based strategy, 81
HSD11B2 gene, 122
HTT gene, 103
HUGO Nomenclature Committee. *See* Human Genome Organization Gene Nomenclature Committee
Human Ether-a-go-go Related Gene, 132
Human Genome Organization Gene Nomenclature Committee, 12
Human Genome Project, 12, 24, 78
Huntington's disease (HD), 102–103
3-hydroxy-3-methylglutaryl-CoA reductase gene, 246
HYPERGENES Project, 116
hypertension, 9, 43, 47–48, 84, 112, 121
hypertriglyceridemia, 241
hypertrophic cardiomyopathy (HCM), 9, 79, 103, 188–189, 196–197. *See also* dilated cardiomyopathy
  clinical and pathophysiologic manifestations, 197
  disease genes and putative functional consequences, 205–208
  genetic basis of, 197
  genetic heterogeneity, 198–199
  genetic testing of, 201–202
  genotype–phenotype correlation, 208–209
  left ventricular outflow tract (LVOT) obstruction in, 197
  mutations, 198–200
  natural history of, 197
hypotension disorder, 112

identical by descent (IBD), 8, 15
identical by state (IBS), 8
idiopathic atrial fibrillation, 130
Illumina BeadChip, 68
Illumina BeadStudio software, 69
independent assortment, principle of, 6

inflammatory markers, 42–44
inflammatory proteins, 147
Ingenuity Pathway Analysis (IPA), 73
inheritance as a physical unit, 1
inheritance modes
   association analysis, 9
   linkage analysis, 8–9
   segregation analysis, 8
inherited cardiomyopathies. *See also* metabolic
     cardiomyopathies
   case example, 195
   disease genes, 197–198
   functional consequences of gene mutations, 199–201
   genetic heterogeneity, 198–199
   genetic testing of, 201–202
   genotype–phenotype correlation, 201
   hypertrophic cardiomyopathy (HCM), 196–197
   thick and thin filament proteins, mutations in, 200–201
inherited ventricular arrhythmias
   case example, 153–154
International Consortium of BP (ICBP), 116
International Normalized Ratio (INR), 92
intraclass correlation coefficient (ICC), 13–14
ischemic heart disease, 42

KCJN2 gene, 273
KCNA5 gene, 231
KCND3 gene, 132, 137
KCNE1 gene, 132
KCNH2-A116V gene, 155
KCNH2 gene, 132, 154–155
KCNH2-K897T, 155
KCNH2 mutation, 135, 138
KCNQ1 gene, 154–155
KCNQ1 (KvLQT1) gene, 132

LAMP2 mutations, 202–203
LD. *See* linkage disequilibrium
LDL-C. *See* low-density lipoprotein cholesterol
LDLR. *See* low-density lipoprotein receptor
left ventricular noncompaction (LVNC), 6–7, 196,
     209–210
linear regression, 14
LINE-1 DNA methylation, 42
linkage analysis, 8–9
linkage disequilibrium (LD), 6, 24–26, 36, 116
linkage strategy, 80
lipids and lipoproteins, genetics of
   case example, 239–240
   complex diseases and phenotypes, 242–246
   epigenetics, 247–248
   evolution in, 248
   gene variants of, 244
   GWAS findings, 243–245
   microsatellite linkage findings, 242–243
   single-gene diseases, 241
   TG-rich lipoproteins, 241
   whole-exome and whole-genome sequencing studies,
     246–247
LMNA gene, 208
LNA. *See* locked nucleic acid modification
locked nucleic acid (LNA) modification , 64
LOD. *See* logarithm (base 10) of odds score
Loeys–Dietz syndrome, 9, 273, 275–276
logarithm (base 10) of odds score (LOD), 8, 112

long QT syndrome (LQTS), 7
   beta-blocker therapy, 156, 158
   cardiac events in, 156–157
   clinical manifestations, 155
   diagnosis of, 155–156
   genetic modifiers in, 155
   genetics of, 154–155
   genetic testing of, 156–157
   ICD therapy, 159
   nadolol therapy, 158
   propranolol therapy, 158
   risk stratification for, 157
   role of gender and genotype in, 157
   treatment strategies, 158–159
low-density lipoprotein cholesterol (LDL-C), 34–36
low-density lipoprotein receptor (LDLR), 1
LQTS. *See* long QT syndrome
LVNC. *See* left ventricular noncompaction

MAF. *See* minor allele frequency
major depression (MD), 44
Manhattan plot, 27–28
Marfan syndrome, 8–9, 103, 275–276
MD. *See* major depression
Mediterranean diet, 92
meiosis, 8
Mendelian CVD, 1–6
   expressivity in, 7–8
   penetrance of mutations in, 7
Mendelian disease screening, 9
Mendelian disorders, 80
Mendelian phenotype, 7
Mendelian traits, 80
Mendel's Laws, 1, 6
meta-analysis, 74–75
metabolic cardiomyopathies, 202–204. *See also* inherited
     cardiomyopathies
MetaCore, 73
MI. *See* myocardial infarction
microarray gene expression profiling, 68
microarray technologies, 67–68
microRNA–mRNA target pairs, 58
microRNAs (miRNAs), 44, 122
   binding to messenger RNA, 58
   biogenesis, 56–57
   as biomarkers, 60–63, 146–147
   and cardiac function, 58–60
   in cardiac hypertrophy, 58–59
   circulating, 61–62
   and CVD, 44–47
   and CVD-related markers, 47–48
   diagnostic performance, 63
   extravesicular, 62
   in HF, 192
   muscle-specific, 63
   processing pathway, 57
   in pulmonary hypertension, 232
   role in cholesterol homeostasis and metabolism, 60
   structure and function, 56–58
   testing for MI, 63
microsatellites, 24
minor allele frequency (MAF), 35
miRNA–mRNA expression networks, 147
miRNAs. *See* microRNAs
missense mutation, 38

missing heritability
    of AF, 143–144
    problem, 19–21
      of BP, 121
mitochondrial inheritance, 7
mitral valve prolapse, 276–280
    candidate genes in, 277–279
    familial linkage analyses for, 277
    genetic syndromes associated with, 276
    GWAS findings, 279
    heritability, 277
    myxomatous, 276
    prevalence of, 276
MMP1 gene, 279
model-based algorithms, 69
monocarboxylate transport (*MCT*) gene, 44
monogenic BP disorders, 112–114
monozygotic twins (MZ twins), 15
move, zoom in, and zoom buttons, 35
MTHFR gene, 116
multidimensional scaling, 72
multilevel modeling analysis of variance, 14
muscle-specific microRNAs, 63
mutant allele, 6
MutationTaster, 38
MVP syndrome, 279
MyBPC. *See* myosin-binding protein C
MYBPC3 gene, 199
MYH gene. *See* myosin heavy chain genes
MYH7 gene, 199
MYH7 mutations, 200
myocardial infarction (MI), 44–46, 62–63, 267
myosin-binding protein C (MyBPC), 200–201
myosin heavy chain genes (MYH genes), 59
MYOZ1 gene, 145
MZ twins. *See* monozygotic twins

natural selection, 130
NCI-NHGRI Working Group on Replication, 30
next-generation sequencing (NGS) studies, 78, 80, 122
    for clinical diagnostics, 81–84
    DNA sequencing techniques, 211
NGS technique. *See* next-generation sequencing studies
nonparametric linkage analysis (NPL), 8
nonrandom ascertainment, 14
NOS1AP gene, 135–136
NOS1API gene, 155
NOS3 gene, 116
NOTCH1 gene, 273–275
NPL. *See* nonparametric linkage analysis
NPPA and *NPPB* genes, 116
NSGC Code of Ethics, 106

obesity, 43
oligonucleotides, 67
OMIM. *See* Online Mendelian Inheritance in Man catalog
one-channel detection systems, 68
Online Mendelian Inheritance in Man catalog (OMIM), 37
ontology and pathway analysis, 73–74
overlap strategy, 80

pairwise LD, 36
PANTHER. *See* Protein ANalysis THrough Evolutionary
    Relationships
parametric linkage analysis, 8

parent–offspring pair, 15
PCA. *See* principal component analysis
PCR. *See* polymerase chain reaction
PCSK9 gene, 38
pedigree, 97–100
    symbols, 98–99
pedigree analysis, 10
penetrance, 7, 15, 101
PH. *See* pulmonary hypertension
phenocopy, 15
phenotype reclassification, 20
phenotypes, 26
phenylketonuria (PKU), 9
PKU. *See* phenylketonuria
PLATO, 27
pleiotropy, 8
PLINK, 27–28
polymerase chain reaction (PCR), 146
PolyPhen-2, 38
POMC gene, 247–248
Pompe disease, 103, 203–204
population stratification, 30
posterior probability, 102
potassium channel genes, 132, 231
pre-microRNA, 57
principal component analysis (PCA), 72
PRKAG2 mutations, 204
ProbABEL, 27
proopiomelanocortin gene, 247–248
Protein ANalysis THrough Evolutionary Relationships
    (PANTHER), 73
protein coding, 12
psychosocial counseling, 103
pulmonary hypertension (PH)
    BMPR2 mutations, 223–224, 233
    calcium signaling in, 229
    case example, 219–220
    CAV1 mutation, 224–229
    cerebellin-2 (*CBLN2*) mutation, 229
    coding of genes, 232
    endothelin G-protein-coupled receptors, ET_A and
      ET_B, role of, 230
    epigenetics of, 232–233
    genetic testing of, 233
    Kv expression in, 231
    mean pulmonary artery pressure (mPAP), 221
    microRNAs (miRNAs) expression profiling, 232
    nitric oxide and reactive oxygen species, role of, 230
    overview, 221–222
    pathways, 231
    serotonin, role of, 231
    sex hormones, role for, 230
    TGF-beta superfamily, 223–224

Q–q plot, 29–30
QT syndrome, 79
quality control of microarray data, 69
quantitative phenotypes, 13
quantitative traits, 12–13

RAAS. *See* renin–angiotensin–aldosterone system
RCM. *See* restrictive cardiomyopathy
"red flags" in a family history, 99–100
RegulomeDB, 232
relative risk ratio, 14

renal-salt-handling genes, 122
renin–angiotensin–aldosterone system (RAAS), 112, 279
repolarization, 132
restrictive cardiomyopathy (RCM), 204. *See also* inherited cardiomyopathies
reverse transcription polymerase chain reaction (RT-PCR), 74
ribonucleic acid sequencing technologies, 146
RISC. *See* RNA-induced silencing complex
RMA. *See* Robust Multi-array Average
RNA-induced silencing complex (RISC), 44, 57
RNA-seq. *See* RNA transcriptome
RNA sequencing technologies. *See* ribonucleic acid sequencing technologies
RNA transcriptome, 211
Robust Multi-array Average, 69

S-adenosylmethionine (SAM), 41
Sanger sequencing, 78
sarcomere mutations, 205
sarcoplasmic reticulum (SR), 132
SCD. *See* sudden cardiac death
SCN5A gene, 8, 79, 131, 133–134, 154–155
SCN10A gene, 133–134
SCNN1B and *SCNN1G* genes, 112
segregation analysis, 8
sequence kernel association test (SKAT), 93
sequencing technologies, 77
SERCA2A gene, 133
short QT syndrome (SQTS), 130
    clinical manifestations, 163–164
    diagnosis of, 163–164
    genetics of, 163
    ICD implantation, 164
    quinidine therapy, 164
    treatment strategies, 164
short tandem repeats (STRs), 8
SIFT, 38
simple sequence repeat (SSR) markers, 34
single nucleotide polymorphisms (SNPs), 8, 18, 24, 30–31, 35–36, 77, 129, 137, 246
    alleles of, 24
    aortic valve disease, 270–271
    associated with lipid levels, 31
    genotyping of specific, 29
    imputed, 29
    role in congenital heart disease, 293–294
    VEGF, 294
SKAT. *See* sequence kernel association test
SLCO1B1 genotype, 92, 257
SMD. *See* Stanford Microarray Database
"smoking gun" mutation or mutations, 35
SNAP server. *See* SNP Annotation and Proxy Search server
SNP Annotation and Proxy Search server, 36
SNPs. *See* single nucleotide polymorphisms
SNPTEST, 27
sodium channel gene abnormalities, 131–132
SQTS. *See* short QT syndrome
SR. *See* sarcoplasmic reticulum
SSR. *See* simple sequence repeat markers
Stanford Microarray Database (SMD), 75
STEMI. *See* ST-segment elevation myocardial infarction
straightforward genotype–phenotype relationship, 7

stroke, 42, 47
STRs. *See* short tandem repeats
structural proteins, mutations affecting, 207–208
ST-segment elevation myocardial infarction (STEMI), 63
sudden cardiac death (SCD), 154, 197
supervised clustering, 71–72

TAAD. *See* thoracic aortic aneurysm and aortic dissection
TAC. *See* transverse aortic constriction
targeted DNA sequencing, 197
T1D. *See* type 1 diabetes
T2D. *See* type 2 diabetes mellitus
TDT. *See* transmission disequilibrium test
TGBR1 and *TGFBR2* mutations, 273
TGFBR1 and *TGFBR2* genes, 9
TGs. *See* triglycerides
thoracic aortic aneurysm and aortic dissection (TAAD), 7
TNNT1 (the troponin T type 1) gene, 248
total phenotypic variance, 13
transmission disequilibrium test (TDT), 9, 24
transverse aortic constriction (TAC), 59
triglycerides (TGs), 34, 36
    TG-rich lipoproteins, 241
T-test, 73
twin studies, 16
    ascertainment bias in, 16–17
    environmental influences in, 17
    limitations, 16–17
two-channel detection systems, 68
type 1 diabetes (T1D), 47
type 2 diabetes mellitus (T2D), 47

UCSC Genome Informatics, 35
unicuspid and bicuspid aortic valve morphology, 267
unsupervised clustering, 70

validation of gene expression, 74–75
valvular heart disease
    aortic valve disease, 267–276
    case example, 265–266
variable expressivity, 101
VEGF SNPs, 294
virtual karyotyping, 245
vitamin K epoxide reductase complex (*VKORC1*), 148
VKORC1 gene, 92

Watson–Crick base-pairing rules, 64
WBC-derived circulating cytokines, 40
Weighted Gene Co-expression Network Analysis (WGCNA), 74
WES. *See* whole-exome-sequencing studies
WGCNA. *See* Weighted Gene Co-expression Network Analysis
WGS. *See* whole-genome sequencing
whole-exome-sequencing (WES) studies, 77–78, 106
    application of, 81–83
    available strategies for conduct, 80–81
    future prospects, 84
    severe monogenic hypertension, case example, 111
    technologies, 78–79
whole-genome linkage analyses of BP, 112–115
whole-genome sequencing (WGS), 77–79, 84, 106

Wiki-Genes, 37
Wilcoxon–Mann–Whitney test, 73
Williams syndrome, 276
Wnt signaling in CAD, 46–47

X inactivation, 17, 101
X-linked DMD mutation, 101
X-linked dominant inheritance, 6

X-linked myxomatous valvular dystrophy (XMVD),
    275
X-linked recessive inheritance, 6
XMVD. *See* X-linked myxomatous valvular dystrophy

Y-linked dominant inheritance, 6–7

Z-disc genes, 205